ANNALS OF THE NEW YORK ACADEMY OF SCIENCES

Volume 879

EDITORIAL STAFF

Managing Editor
JUSTINE CULLINAN

Associate Editor
LINDA HOTCHKISS MEHTA

The New York Academy of Sciences
2 East 63rd Street
New York, New York 10021

THE NEW YORK ACADEMY OF SCIENCES
(Founded in 1817)

BOARD OF GOVERNORS, October 1998 – September 1999

ELEANOR BAUM, *Chairman of the Board*
BILL GREEN, *Vice Chairman of the Board*
RODNEY W. NICHOLS, *President and CEO* [ex officio]

Honorary Life Governors
WILLIAM T. GOLDEN   JOSHUA LEDERBERG

JOHN T. MORGAN, *Treasurer*

Governors

D. ALLAN BROMLEY	LAWRENCE B. BUTTENWIESER	PRAVEEN CHAUDHARI
JOHN H. GIBBONS	RONALD L. GRAHAM	HENRY M. GREENBERG
ROBERT G. LAHITA	MARTIN L. LEIBOWITZ	JACQUELINE LEO
WILLIAM J. McDONOUGH	KATHLEEN P. MULLINIX	SANDRA PANEM
CHARLES RAMOND	SARA LEE SCHUPF	JAMES H. SIMONS
	TORSTEN WIESEL	

RICHARD A. RIFKIND, *Past Chairman of the Board*
HELENE L. KAPLAN, *Counsel* [ex officio]   PETER KOHN, *Secretary* [ex officio]

TEMPOS IN SCIENCE AND NATURE: STRUCTURES, RELATIONS, AND COMPLEXITY

ANNALS OF THE NEW YORK ACADEMY OF SCIENCES
Volume 879

TEMPOS IN SCIENCE AND NATURE: STRUCTURES, RELATIONS, AND COMPLEXITY

Edited by Claudio Rossi, Simone Bastianoni, Alessandro Donati, and Nadia Marchettini

The New York Academy of Sciences
New York, New York
1999

Copyright © 1999 by the New York Academy of Sciences. All rights reserved. Under the provisions of the United States Copyright Act of 1976, individual readers of the Annals *are permitted to make fair use of the material in them for teaching or research. Permission is granted to quote from the* Annals *provided that the customary acknowledgment is made of the source. Material in the* Annals *may be republished only by permission of the Academy. Address inquiries to the Executive Editor at the New York Academy of Sciences.*

Copying fees: *For each copy of an article made beyond the free copying permitted under Section 107 or 108 of the 1976 Copyright Act, a fee should be paid through the Copyright Clearance Center, Inc., 222 Rosewood Drive, Danvers, MA 01923. The fee for copying an article is $3.00 for nonacademic use; for use in the classroom it is $0.07 per page.*

⊖*The paper used in this publication meets the minimum requirements of the American National Standard for Information Sciences—Permanence of Paper for Printed Library Materials, ANSI Z39.48-1984.*

Cover: *The cover art is a detail from a fifteenth century wedding ceremony fresco that can be found in the Santa Maria della Scala Hospital in Siena, Italy. The two hands symbolize the unification of two persons in marriage and in this context relate to the fundamental theme of the conference: the need for the reunification of the two paths of human perception, the world of science and the world of the arts.*

Library of Congress Cataloging-in-Publication Data

Tempos in science and nature: structures, relations, and complexity / editors, Claudio Rossi ... [et al.].
  p. cm. — (Annals of the New York Academy of Sciences, ISSN 0077-8923 ; v. 879)
  Includes bibliographical references and index.
  ISBN 1-57331-196-0 (cloth : alk. paper). — ISBN 1-57331-197-9 (pbk. : alk. paper)
  1. Science—Philosophy Congresses. 2. Chaotic behavior in systems Congresses. 3. Complexity (Philosophy) Congresses. I. Rossi, C. II. Series.
Q11.N5 vol. 879
[Q141]
501—dc21   99-28222
                                                                        CIP

GYAT / PCP
Printed in the United States of America
ISBN 1-57331-196-0 (cloth)
ISBN 1-57331-197-9 (paper)
ISSN 0077-8923

ANNALS OF THE NEW YORK ACADEMY OF SCIENCES
Volume 879
June 30, 1999

TEMPOS IN SCIENCE AND NATURE: STRUCTURES, RELATIONS, AND COMPLEXITY[a]

Editors
CLAUDIO ROSSI, SIMONE BASTIANONI, ALESSANDRO DONATI,
AND NADIA MARCHETTINI

Advisory Board and Conference Organizers
R. BARBUCCI, S. BASTIANONI, C. BONECHI, C. DEJAK, A. DONATI,
E. FERRONI, A. GOLDBETER, T. HALLAM, S. E. JØRGENSEN,
A. M. LIQUORI, S. LOISELLE, N. MARCHETTINI, M. P. PICCHI, I. PRIGOGINE,
C. ROSSI, M. RUSTICI, AND P. TOSI

CONTENTS

Preface. By CLAUDIO ROSSI AND ENZO TIEZZI ix

Part I. Epistemological Concepts of Complexity

Fundamental Sources of Unpredictability. By MURRAY GELL-MANN 1

Laws of Nature and Time Symmetry Breaking. By I. PRIGOGINE AND
I. ANTONIOU 8

Perspective and Metrics of Time. By GIUSEPPE ARDRIZZO 29

Complexity, Complex Systems, and Adaptation. By F. T. ARECCHI 45

Complexity and the Unfinished Nature of Human Evolution. By GIANLUCA
BOCCHI AND MAURO CERUTI 63

Complexity VII: Composing Natural Science from Natural Numbers, Natural
Kinds, and Natural Affinities. By JERRY L. R. CHANDLER 75

The Paradox of the Visibly Irrelevant. By STEPHEN JAY GOULD 87

The Interplay of Cyclic and Linear Time in the Biological World. By
PIER LUIGI LUISI 98

The Human Sciences and the Natural Sciences: Toward a New Asymmetry.
By ANTONIO MELIS 110

[a]This volume is the result of a conference entitled **Tempos in Science and Nature: Structures, Relations, and Complexity**, sponsored by the Università degli Studi di Siena and the Consorzio Interuniversitario C.S.G.I. and held September 23 through 26, 1998 in Siena, Italy.

Organization and Complexity. *By* EDGAR MORIN 115

Complex Adaptive Theology: From Sinai to Santa Fe. *By*
HAROLD J. MOROWITZ 122

Complexity and Evolutionary Law for Natural Systems: A "New Dialogue" with Nature—In Looking for a Language as a Means of Intercourse with Nature. *By* S. F. TIMASHEV 129

A Dimly Perceived Horizon: The Complex Meeting Ground between Physical and Inner Time. *By* FRANCISCO J. VARELA 143

Poster Papers

Analytical Setting: Time, Relation, and Complexity. *By* GENOVINO FERRI AND GIUSEPPE CIMINI 154

The Consciousness of Time and the Conquest of Space: The Dawn of Civilization, Anthropology, and Archaeoastronomy. *By* LUCIA G. GREGORI AND GIOVANNI P. GREGORI 158

Time Complexity and Learning. *By* ELDA GUALA AND PAOLO BOERO 164

Einstein's Twins. *By* ENZO TIEZZI AND NADIA MARCHETTINI 168

Fractal Time and the Gift of Natural Constraints. *By* SUSIE VROBEL 172

Part II. Complex Molecular and Biological Behavior

Alternating Oscillations and Chaos in a Model of Two Coupled Biochemical Oscillators Driving Successive Phases of the Cell Cycle. *By* PIERRE-CHARLES ROMOND, MAURO RUSTICI, DIDIER GONZE, AND ALBERT GOLDBETER 180

Self Organization in Synthetic Polymeric Systems. *By* JOHN A. POJMAN 194

Poster Papers

Relevance of the Protein–Lipid Interaction on the Functioning of the Bacterial Reaction Center. *By* A. AGOSTIANO AND M. DELLA MONICA 215

Time-Dependent Properties of Isotactic Polypropylene Fibers. *By* G. BARRA, C. D'ANIELLO, R. RUSSO, AND V. VITTORIA 220

Suspended Life in Biological Systems: Fragility and Complexity. *By* C. BRANCA, A. FARAONE, S. MAGAZÙ, G. MAISANO, P. MIGLIARDO, AND V. VILLARI 224

Coupling of Erratic Belousov-Zhabotinsky Oscillators Observed by Spectrophotometric Measure. *By* M. BRANCA, A. BRUNETTI, C. CARAVATI, AND M. RUSTICI 228

Complexity at the Molecular Level: A Dynamic View of Biomolecules Obtained by NMR. *By* ALESSANDRO DONATI, LYDIA BELLIK, MARIA PIA PICCHI, AND CLAUDIA BONECHI 235

Thermodynamics of Extremely Diluted Aqueous Solutions. *By* VITTORIO ELIA AND MARCELLA NICCOLI 241

Linear and Nonlinear Properties of Heart Rate Variability: Influence of Obesity. *By* A. GASTALDELLI, R. MAMMOLITI, E. MUSCELLI, S. CAMASTRA, L. LANDINI, E. FERRANNINI, AND M. EMDIN 249

Fractal Analysis in Human Pathology. *By* P. LUZI, G. BIANCIARDI, C. MIRACCO, M. M. DE SANTI, M. T. DEL VECCHIO, L. ALIA, AND P. TOSI 255

Recurrence Quantification Analysis in Molecular Dynamics. *By* CESARE MANETTI, MARC-ANTOINE CERUSO, ALESSANDRO GIULIANI, CHARLES L. WEBBER, JR., AND JOSEPH P. ZBILUT 258

Photodegradation of Organic Waste Coupling Hydrogenase and Titanium Dioxide. *By* G. M. MURA, M. L. GANADU, G. LUBINU, AND V. MAIDA... 267

Complex Rotational Dynamics of the Cu(II)-Bleomycin System in Mobile and Viscous Media. *By* REBECCA POGNI, ELENA BUSI, GIOVANNI DELLA LUNGA, AND RICCARDO BASOSI 276

On Crystallization Kinetics of Polytetrafluoroethylene. *By* RACHELE PUCCIARIELLO AND CLAUDIA MANCUSI 280

Conformational Chaos of an Elastin-Related Peptide in Aqueous Solution. *By* VINCENZO VILLANI AND A. M. TAMBURRO 284

Vanadium Uptake by Yeasts. *By* MARIA ANTONIETTA ZORODDU 288

Part III. Complexity: Ecological and Environmental Aspects

A Modeling Approach to the Understanding of Very Complex Dynamics in a Planktonic Community. *By* GRACIELLA ANA CANZIANI 292

Complexity and Emergence in Models of Chemically Stressed Populations. *By* THOMAS G. HALLAM AND ERIC T. FUNASAKI 302

The Influence of Classic Sciences on Ecology and Evolution of Ecological Studies. *By* VINCENT HULL AND MARGHERITA FALCUCCI 312

A Tentative Fourth Law of Thermodynamics, Applied to a Description of Ecosystem Development. *By* SVEN E. JØRGENSEN 320

Energy Analyses as a Tool for Sustainability: Lessons from Complex System Theory. *By* MARIO GIAMPIETRO, KOZO MAYUMI, AND GIANNI PASTORE 344

Virtual Biospheres: Complexity versus Simplicity. *By* YURI M. SVIREZHEV. 368

Poster Papers

A Comprehensive Atmospheric Chemistry Model for the Description of Dynamics of Reactive Pollutants. *By* G. BARONE, P. D'AMBRA, D. DI SERAFINO, G. GIUNTA, AND A. RICCIO 383

Modeling the Complex Behavior of Microorganisms. *By* S. BASTIANONI, S. MARTINI, M. PORCELLI, AND C. ROSSI 387

The Effects of Vertical Mixing Parameterization on 3-D Models of a Pelagic Ecosystem. *By* A. BELLUCCI, A. CRISE, G. CRISPI, AND C. SOLIDORO. 392

Identification, Characterization, and Remediation of Contaminated Sites: A Case Study. *By* ELENA COLLINA, MARINI LASAGNI, AND DEMETRIO PITEA 396

Classical Thermodynamic–Statistical Approach of Open Systems Deviating from Maximum Entropy. *By* C. DEJAK, R. PASTRES, G. PECENIK, AND I. POLENGHI 400

Use of Exergy and Structural Exergy as Ecological Indicators for the Development State of Homogeneous Lake Ecosystems. *By* A. LUDOVISI AND A. POLETTI 406

Determination of the Italian Industrial Production Index from a Bayesian Point of View Using Electrical Consumption as Data. *By* L. MAGNONI AND M. MIROLLI 412

The Relational Project: A Goal for the Ecology of City Designing and Planning. *By* VASCO MANCINI, RITA MICARELLI, AND GIORGIO PIZZIOLO 416

Correlation between Greenhouse Effect and Exceptionally High Tides in Venice. *By* N. MARCHETTINI, C. MOCENNI, V. NICCOLUCCI, AND E. TIEZZI 422

Ecodepuration Performances of a Small-Scale Experimental Constructed Wetland System Treating and Recycling Intensive Aquaculture Wastewater. *By* S. PANELLA, I. CIGNINI, M. BATTILOTTI, M. FALCUCCI, V. HULL, N. MILONE, M. MONFRINOTTI, G. A. MULAS, G. PIPORNETTI, L. TANCIONI, AND S. CATAUDELLA 427

Fish Exclusion and PCB Bioaccumulation in Bear Creek, East Tennessee. *By* M. PANZIERI AND T. G. HALLAM 432

Evolution as Computation. *By* G. MICHELE PINNA AND ELISA B. P. TIEZZI . 435

Mathematical Modeling and Numerical Simulation of Space-Dependent Multispecies Interactions. *By* R. C. SOSSAE, J. F. C. A. MEYER, S. LOISELLE, AND C. ROSSI 440

Relations between Energy Consumption and Air Quality in an Urban Environment: Spatial Scenarios of Emission Reduction. *By* LAURA VANNUCCINI, MIRIA BRACALI, AND RICCARDO BASOSI 444

Index of Contributors 449

Financial assistance was received from:
Major Funders:
- MONTE DEI PASCHI DI SIENA BANK
- INTER-UNIVERSITARY CONSORTIUM CSGI
- ITALIAN NATIONAL RESEARCH COUNCIL (CNR)
- UNIVERSITY OF SIENA

Supporters:
- CONSORZIO VENEZIA NUOVA
- WHIRLPOOL
- BRUKER
- OIKOS
- META S.p.A.
- CAMERA DI COMMERCIO DI SIENA
- COMUNE DI SIENA

The New York Academy of Sciences believes it has a responsibility to provide an open forum for discussion of scientific questions. The positions taken by the participants in the reported conferences are their own and not necessarily those of the Academy. The Academy has no intent to influence legislation by providing such forums.

Preface

CLAUDIO ROSSI AND ENZO TIEZZI
*Department of Chemical and Biosystems Sciences, University of Siena,
Pian dei Mantellini, 44, 53100 Siena, Italy*

As Thomas Kuhn remarked, the passing of time often brings anomalies that existing theories are no longer able to explain. The divergence between theory and reality may become enormous and a source of serious problems. This is exactly what is happening today between current scientific theories and the natural situation of the planet.

In Kuhn's terms this means a shift toward a new paradigm. As Palomar in Italo Calvino's novel observed: "If the model does not succeed in transforming reality, reality must succeed in transforming the model." If we want to avoid this paradox, a shift toward a new interrelated paradigm is necessary: this is a challenge for science in the new millennium.

Based on the assumption that the interaction between biochemical-physical constraints and ecological global change is not satisfactorily described by current scientific theories, this book is an attempt to present a different *gestalt* from physics, chemistry, mathematics, ecology, biology, epistemology, and the humanities.

The theme, Tempos in Science and Nature: Structures, Relations, and Complexity, deals with cross-fertilization among different disciplines, keeping in mind the categories of evolution (time), structure, systemic approach, and complexity.

The book presents contributions from some outstanding scientists in physics, physical chemistry, evolutionary biology, and philosophy and some papers from borderline disciplines such as mathematical ecology, evolutionary physics, systems theory, ecological modeling, comparative literature, and cognitive science. The following main topics can be identified: (a) complexity: an opportunity for re-unification of the scientific culture and the humanities; (b) complex molecular behavior and structure; (c) ecological and environmental modeling.

It is important to underline the fact that the complexity of nature has made it necessary to develop concepts and theories that have little in common with the reductionist and mechanical approach that has dominated scientific thought for two centuries. Science has not so far been able to unravel the molecular and macroscopic complexity of natural phenomena, with their nonlinear and irreversible patterns. The objective of this book is to promote cross-fertilization between the humanities and the sciences, to interweave the world of the arts with the world of natural laws and theories. Although the process is still in the early stages and the outcome is uncertain, much of the scientific community is aware of the need to open new paths of exchange between the sciences and the humanities.

Moreover, nonlinear effects, feedback interactions, and irreversible processes now come into the scientific picture and play a very particular role. Natural dynamics are in strict relation with forms, colors, shapes, supramolecular networks, dissipative structures, processes far from equilibrium, and evolutive behavior.

In order to achieve these goals, it is fundamental to harmonize our scientific knowledge with aesthetics, the humanities, genetic epistemology, and everything else dealing with complexity.

Fundamental Sources of Unpredictability[a]

MURRAY GELL-MANN

Santa Fe Institute, 1399 Hyde Park Road, Santa Fe, New Mexico 87501, USA
Theory Division, Los Alamos National Laboratory, Los Alamos, New Mexico 87545, USA
Department of Physics and Astronomy, University of New Mexico, Albuquerque, New Mexico 87131, USA

In discussing the fundamental sources of unpredictability, I shall concentrate on those indeterminacies that are definitely required by theory. Let me therefore begin by eliminating from consideration supposed indeterminacies stemming from doubts that some people may entertain about basic principles or from certain kinds of ignorance that I believe to be temporary and likely to be remedied in the relatively near future. I assume the following:

1. Quantum mechanics is correct. The formulation and interpretation of quantum mechanics are still undergoing some necessary evolution, especially in order to accommodate quantum cosmology in a comfortable way, but the basic character of quantum mechanics has always been the same and we may suppose it will remain unchanged.

2. The elementary particles and their interactions obey a definite dynamical law, discoverable by inquiring complex adaptive systems such as the human scientific enterprise. Although the process of discovery involves a sequence of approximate schemata, there is an endpoint to the process after a finite amount of research. (Of course, it can never be possible to prove that the resulting unified theory is perfect; one can only verify it in the usual way by comparing predictions to available observations.) Humans may already have found this unified quantum field theory in the form of superstring theory, which has, to begin with, no arbitrary parameters. (Of course, spontaneous symmetry breaking may give rise to some parameters and even to a choice of solutions, with probabilities for the various alternatives. I will deal with that possibility further on.) This second assumption is equivalent to stating that there is no necessary fundamental unpredictability stemming from ignorance of the universal dynamical law.

3. The density matrix (in the Schrödinger picture) of the universe near the beginning of its expansion is also knowable. It must in any case be comparatively simple and very far from equilibrium. The second law of thermodynamics and the other associated arrows of time are explained by these properties of the initial density matrix along with the fact that the universe is still very young—the interval of ten billion years is extremely short compared with the relaxation time from the special initial condition.

---

[a]This article reproduces the introductory talk at the meeting on Fundamental Sources of Unpredictability at the Santa Fe Institute, March 19, 1996. It is reprinted by permission from John Wiley & Sons, Inc. (Originally published in *Complexity* 3(1): 9–13, 1997.)

Hartle and Hawking have proposed an ingenious pure state candidate for the density matrix, such that the wave function is in principle calculable from the action functional of the fundamental dynamical theory—the two basic laws of physics would then become one. Another possibility is indicated by the work of Linde and others, namely that the universe emerged and become isolated from a larger system (which I call a multiverse) as a kind of bubble, one among very many. The density matrix would then be impure, with probabilities corresponding to the statistics of such bubbles, but it would still have a simple character. Here "simple" means expressible by means of a concise formula[1–3] in terms of the fundamental fields at the time.

Let us return to the promising candidate for the role of fundamental dynamical law for all matter. Superstring theory was thought for a number of years to exhibit several different mutually exclusive forms, of which the "heterotic" one was the most likely to agree with nature. However, it now appears that all the different forms are related.

Also, the heterotic form, studied in a simple approximation, exhibits many different solutions, corresponding to different numbers of spatial dimensions, different sets of elementary particles, and/or different interactions among particles. Many of those solutions may prove to be spurious when the approximation is improved, but what if multiple solutions remain? It looks as if they will all turn out to be related, just as the different forms of the theory are related, in a way that is reminiscent of the relationship of phases in condensed matter theory.

If there really are multiple solutions, as well as multiple forms of the theory, all manifestations of the same basic law, what determines the "phase" that characterizes our universe? One might imagine a principle that determines the choice, say a principle of least S where S indicates the quantum-corrected, Euclideanized action. Another possibility is a probabilistic situation, in which the likelihood of a particular solution is proportional to $\exp(-2S)$. The one with the lowest value of S would then be the most likely, but not the only possible choice.[1] These likelihoods, or *a priori* probabilities, would be converted into statistical probabilities if our universe is really a multiverse containing an enormous number of largely independent universes. If not, we could still imagine such a multiverse as a mathematical abstraction, an aid to thinking about probability in connection with the universe.

Suppose we know the two basic laws that govern the behavior of all matter. What then? Can we predict, in principle, the history of the universe? Of course not. Because the laws of physics are quantum-mechanical, prediction is limited to probabilities for alternative histories of the universe. The limitations go far beyond the famous, but rather trivial uncertainty principle of Heisenberg. In order to give an account of those limitations, we must sketch the decoherent histories approach to quantum mechanics, as developed by James Hartle and me.[4–6] Our approach can be regarded as part of an ongoing effort by a number of theoretical physicists to construct a modem interpretation of quantum mechanics, one that is compatible with quantum cosmology and that explains in a convincing way how the quasiclassical world of familiar experience emerges from the underlying quantum universe. Some of the other authors who are part of this movement are R. Omnès, R. Griffiths, D. Zeh, W. Zurek, J. P. Paz, C. Isham, and N. Linden.

Viewed in the most general way, from the standpoint of the whole universe (necessary in order to assure compatibility with quantum cosmology), the application of quantum mechanics is always to histories, since predictions of the probabilities of future events are always made subject to the explicit or implicit assumption that certain things have already happened, things that have meaning only if certain other things happened earlier, and so forth.

A very important advantage of the decoherent histories method is that it permits quantum mechanics to be formulated in a straightforward manner while meeting the requirements of general relativity. The correct unified quantum field theory must, of course, include quantized Einsteinian gravitation. Superstring theory does so. Indeed, it predicts, in a suitable approximation, the general-relativistic theory of gravitation within the framework of quantum mechanics. Moreover, the preposterous infinite corrections in perturbation theory that plagued previous attempts to incorporate gravitation into quantum field theory are absent.

Of course, there is still the apparent difficulty that quantum mechanics is usually formulated in terms of a temporal succession of spacelike surfaces, hard to define when the metric of spacetime is quantized. The problem with the usual formulation is even more serious when the topology of Euclideanized spacetime undergoes changes in the course of quantum fluctuations. Fortunately, it has been shown by Hartle[7] that our approach can be used to create a slight generalization of quantum mechanics in which the quantized metric poses no difficulties. In what follows, let us, for the sake of simplicity, ignore the complications that arise from general relativity and deal with the approximation in which we have a conventional Hilbert space. an ordinary time variable, and a Hamiltonian.

Furthermore, in outlining the decoherent histories interpretation of quantum mechanics, let us make the simplifying assumption of a pure density matrix ρ of the universe in the Heisenberg picture (equal to the initial density matrix in the Schrödinger picture), so that

$$\rho = |\Psi\rangle\langle\Psi|, \tag{1}$$

as in the Hartle–Hawking situation.

Probabilities enter only because $|\Psi\rangle$ is being compared to states representing alternative histories α. The absolute square of the scalar product between $|\Psi\rangle$ and each of those (normalized) states is proportional to the probability of the corresponding history.

Without a great loss of generality, we can construct the states representing histories by means of sequences of projection operators at a succession of times t_1, t_2, and so forth. At each time t_k, we have a set of mutually exclusive and exhaustive alternatives α_k, which may depend on the previous alternatives, in accordance with causality. Thus the projection operators at time t_k are labeled $P(\alpha_k; \alpha_{k-1}...\alpha_2\alpha_1)$. A state corresponding to a history $\alpha = \alpha_1\alpha_2\alpha_3\alpha_4...\alpha_{n-1}\alpha_n$ is then proportional to $C_\alpha|\Psi\rangle$, where

$$C_\alpha = P(\alpha_n; \alpha_{n-1}...\alpha_2\alpha_1)P(\alpha_{n-1}; \alpha_{n-2}...\alpha_2\alpha_1)......P(\alpha_2; \alpha_1)P(\alpha_1) \tag{2}$$

Because the alternatives at each time t_k are mutually exclusive and exhaustive, we have

$$P(\alpha_k; \alpha_{k-1}...\alpha_2\alpha_1)P(\alpha_k'; \alpha_{k-1}...\alpha_2\alpha_1) = \delta(\alpha_k, \alpha_k')P(\alpha_k; \alpha_{k-1}...\alpha_2\alpha_1) \tag{3}$$

and
$$\sum_{\alpha_k} P(\alpha_k; \alpha_{k-1}...\alpha_2\alpha_1) = I \qquad (4)$$

where I is the unit operator. As a result, we obtain

$$\sum_{\alpha} C_\alpha = I, \qquad (5)$$

so that

$$|\Psi\rangle = \sum_{\alpha} C_\alpha |\Psi\rangle. \qquad (6)$$

For probabilities to be assignable to the histories α, there must not be any interference terms between them, since the probability of history α or history β has to equal the sum of the probabilities of the two histories α and β. The set of histories without interference terms is called a decoherent set. There are various degrees of decoherence, but for simplicity let us deal mainly with what we call medium decoherence, which means that the various states $C_\alpha|\Psi\rangle$ are all orthogonal to one another. The norm of each of those (unnormalized) states is its probability.

Except for trivial cases, sets of fine-grained histories are not decoherent, and the histories α in a decoherent set must be coarse-grained. By definition, a set of fine-grained histories would specify the values of a complete set of variables at every instant of time. For example, in nonrelativistic quantum mechanics without spin, the positions or momenta of all particles might be given at all times. (The Heisenberg uncertainty principle makes it impossible, of course, to specify exactly both momentum and position for the same particle at the same time.) The coarse-grained histories may be regarded as bundles of fine-grained histories, in which, for example, all times except a discrete set are eliminated by summing over projections onto all values of all variables at all the times that are not in the discrete set. At the discrete times that remain, projections onto all values of many of the variables are summed over. The surviving projection operators at the discrete times would project onto ranges of values of the variables (at those times) that do not have their projections summed over. (In a recent article,[6] Hartle and I have shown that in place of medium decoherence a much stronger form of decoherence is actually needed, implying a much coarser graining of histories.)

We define a realm to be an exhaustive set of mutually exclusive, decoherent, coarse-grained histories. We call it a quasiclassical realm if the projection operators tend, over long stretches of time, to be onto similar ranges of values of similar operators (i.e., roughly related by time displacement) obeying, with high probability, deterministic laws of time development modified by frequent small fluctuations and occasional major branchings. The word "branching" refers to the metaphor of a tree of possible decoherent coarse-grained histories, branching as time goes forward, with probabilities for the different alternatives at each branching. Of course, the events actually experienced in this universe select only one outcome at each branching, unpredictable in advance except for probabilities. Once the particular outcome has occurred, if the result is ascertained, then those probabilities "collapse" to one and zero. When we deal with decoherent histories, there is no mysterious "collapse of the wave function," only the familiar collapse of probabilities that occurs at the race track when we see which horse actually wins a race.

A maximal quasiclassical realm is maximally fine-grained subject to decoherence and quasiclassicality. The quasiclassical realm of familiar experience is a coarse graining of a particular maximal quasiclassical realm, which we can describe in terms of projections onto ranges of values of so-called hydrodynamic variables. Those variables are defined as follows. At each of a set of discrete times, we consider a set of conserved quantities, such as momentum density, energy density, and electric charge density, and nearly conserved quantities, such as nuclear species densities, dislocation densities, and so forth. These quantities are integrated over small volumes, large enough to give sufficient inertia to offer resistance to most fluctuations, but small enough to have rough internal equilibrium. The ranges of values of these quantities and the time intervals are adjusted for decoherence, quasiclassicality, and maximality.

There are obviously many possible variations in detail in the description of this usual maximal quasiclassical realm, but presumably the details do not matter much, so that we are describing essentially a single set of coarse-grained histories. Every complex adaptive system we know has evolved to utilize some coarse-graining of this realm. It is fascinating to speculate about whether other, entirely different quasiclassical realms are exhibited by the theory and, if so, whether complex adaptive systems could evolve to utilize them, but that is a subject to be treated elsewhere. In any case, probabilities and statements about events happening have meaning only within a given realm.

What is crucial in the study of the fundamental sources of unpredictability is that any complex adaptive system,[1,2] (including a composite complex adaptive system such as the human scientific enterprise) makes use of an extreme coarse graining of the usual maximal quasiclassical realm. We use the term IGUS (information gathering and utilizing system) to describe a complex adaptive system as observer. Most of the variables in the universe are inaccessible to an IGUS, referring as they do to remote or hidden places, such as the interiors of distant stars, or to small scales at which measurements are unlikely to be made. Thus, as the maximal set of histories unfolds, in an unimaginable long sequence of accidents with probabilistic outcomes, most of the information about outcomes that have actually occurred (or, in the language of special relativity, occurred in the past light cone) is unknown to the IGUS and unavailable for helping to predict the future. Such information must be summed over, with the result that the actual realm utilized by the IGUS is very coarse-grained indeed as regards the past, and of course similar considerations apply to the future. (The use of statistical mechanics is an example of such extra coarse graining.)

Ignorance as a source of unpredictability has been understood for hundreds of years, if not longer. In this discussion, we are merely putting unavoidable ignorance in the context of quantum mechanics.

The indeterminacies we have discussed are all exacerbated by amplification mechanisms. Such mechanisms are responsible for connecting events at the quantum level with quasiclassical histories. Take a measurement situation such as the Stern–Gerlach experiment, where a sodium atom with valence electron spin up develops a certain photographic grain, whereas it would develop a different photographic grain if its spin were down. Here, a quantum variable is correlated directly with a branching in quasiclassical history. This is true whether a human being (or a chinchilla or a cockroach) actually looks at the result or not. There are, of course, many natural situations, not set up by human beings, in which events at the atomic or subatomic scale cause branchings in the quasiclassical realm. For instance, the decay of a radioactive nucleus in a sheet of mica can produce a long-lasting track, recording the direction of the decay as well as the fact that it took place. That phenomenon is the basis of fission-track dating.

An important source of amplification is chaos, widespread in nonlinear systems. Classically, the term refers to cases in which the outcome of a process is sensitive to the tiniest details of the input, through divergence of classical orbits from one another, a divergence that is exponential in time. Chaos in quantum mechanics is more subtle, but it can lead to amplification of quantum fluctuations so that they materially affect quasiclassical history. The most obvious effect of chaos, however, is to enhance enormously the effect of ignorance of prior outcomes (including the effects of measurement errors).

Now, in addition to everything we have discussed so far, we must deal with the issue of computation. Difficulty of calculation can be a source of unpredictability. Suppose we are given the theory and the maximal realm coarse-grained—for past and future—in accordance with information that can be available to an IGUS (with uncertainties of measurement taken into account). There is then at any time t_k a definite set of probabilities for the fully coarse-grained future histories. But is the calculation of those probabilities possible? Obviously, it is not practical at present unless a gigantic amount of further coarse graining is introduced. It is probable that, as time goes on and calculational techniques improve, the further coarse graining could be progressively reduced. But does that make the whole problem approach tractability in principle? Even if we look just at the cases of degrees of freedom with high inertia, as in orbits of heavy objects, where classical physics applies to an excellent approximation, it is not clear that accurate prediction over very long periods of time is a tractable problem in principle, because of classical chaos. It is possible therefore that questions of calculability in principle will always have to be discussed with reference to requirements for additional coarse graining and to limited accuracy for calculated probabilities.

To recapitulate, we assume that the fundamental theory of matter and the initial condition of the universe are simple and knowable and that quantum mechanics is correct, apart from slight generalization. The fundamental sources of unpredictability are then the following:

1. Possible indeterminacy from the initial condition of the universe if it is impure.

2. Possible indeterminacy from the choice of solutions of the fundamental theory (for this universe) if the choice is probabilistic.

3. Coarse graining required to achieve decoherence of histories in a maximal realm, say the usual maximal quasiclassical realm. (The decoherence should really be strong decoherence, with the extra coarse graining that implies.) The uncertainty principle is automatically included.

4. The probabilistic character of all the branchings in this realm in the future.

5. The huge amount of additional coarse graining resulting from unavoidable ignorance on the part of any given IGUS about the results of many of the branchings in the past.
6. Still more coarse graining to make calculations practical with available computational tools.
7. Limitations on accuracy of calculation with available computational tools.

We have also discussed the enhancement of indeterminacies by amplification mechanisms, including chaos. Now it is possible to regard (1) and (2), if they are present, as representing initial accidents for this universe on a par with all the subsequent accidents. Let us take that point of view. Then we can say that the history of the universe is co-determined by the basic laws (the dynamical theory of all matter and the initial condition of the universe) and the outcomes of an inconceivably long sequence of accidents, and we can describe the fundamental sources of unpredictability as the following: (a) the coarse graining necessary for a maximal realm, say the usual maximal quasiclassical realm, with all its accidents; (b) the probabilistic character of all the accidents (branchings) of that realm in the future; (c) ignorance on the part of a given IGUS of the outcomes of most of the accidents that have already occurred, together with the exacerbation of the resulting unpredictability by amplification mechanisms; and (d) approximations and limitations on accuracy imposed by computational tools available.

REFERENCES

1. GELL-MANN, M. 1994. The Quark and the Jaguar. W. H. Freeman. New York.
2. GELL-MANN, M. 1995. Complexity 1/1: 16–19.
3. GELL-MANN, M. & S. LLOYD. 1996. Complexity 2/1: 44–52.
4. GELL-MANN, M. & J.B. HARTLE. 1990. Quantum mechanics in the light of quantum cosmology. In Complexity, Entropy, and the Physics of Information, SFI Studies in the Sciences of Complexity. Vol. VIII. W. Zurek, Ed. Addison-Wesley. Reading, MA.
5. GELL-MANN, M. & J.B. HARTLE. 1993. Phys. Rev. D 47: 3345.
6. GELL-MANN, M. & J.B. HARTLE. 1997. Strong decoherence. In Proceedings of the 4th Drexel Symposium on Quantum Non-Integrability—The Quantum-Classical Correspondence. D.-H. Feng, Ed. International Press. Cambridge, MA.
7. HARTLE, J.B. 1991. The quantum mechanics of cosmology. In Quantum Cosmology and Baby Universes: Proceedings of the 1989 Jerusalem Winter School for Theoretical Physics. S. Coleman, J.B. Hartle, T. Piran & S. Weinberg, Eds. World Scientific. Singapore.

Laws of Nature and Time Symmetry Breaking

I. PRIGOGINE[a,b] AND I. ANTONIOU[a,c]

[a]*International Solvay Institutes for Physics and Chemistry, Campus Plaine ULB—CP 231, Boulevard du Triomphe, 1050 Brussels, Belgium*
[b]*Center for Statistical Mechanics and Thermodynamics, The University of Texas at Austin, Robert Lee Moore Hall, Austin, Texas 78712, USA*
[c]*Theoretische Natuurkunde (T.E.N.A.), Vrije Universiteit Brussels (V.U.B), 1050 Brussels, Belgium*

INTRODUCTION: IRREVERSIBLE PROCESSES IN NATURE

In recent years, a radical change of perspective has been witnessed in science since the realization that large classes of systems may exhibit abrupt transitions, a multiplicity of states, coherent structures, or a seemingly erratic motion characterized by unpredictability often referred to as deterministic chaos. Classical science emphasized stability and equilibrium; now we see instabilities, fluctuations, and evolutionary trends in a variety of areas ranging from atomic and molecular physics through fluid mechanics, chemistry, and biology to large scale systems of relevance in environmental and economic sciences.[1,2] Concepts such as "dissipative structures" and "self-organization" have become quite popular. The distance from equilibrium, and therefore the arrow of time, plays an essential role in these processes, somewhat like temperature in equilibrium physics. When we lower the temperature, we have various states of matter in succession. In nonequilibrium physics and chemistry, when we change the distance from equilibrium, the observed behavior is even more varied.

How can these findings be interpreted from the point of view of the basic laws of physics? Newtonian dynamics, relativity, and quantum physics, make no distinction between past and future. Time is simply a bookkeeping parameter without any direction. This puzzle has led to an unending series of controversies. This may be called the time paradox. It is interesting that the time paradox was only identified in the second half of the 19th century. It was then that the Viennese physicist Ludwig Boltzmann tried to emulate what Charles Darwin had done in biology and to formulate an evolutionary approach to physics. But at that time, the laws of Newtonian physics had been accepted as expressing the ideal of objective knowledge. As they imply the equivalence between past and future, any attempt to confer on the arrow of time a fundamental significance was resisted as a threat to the ideal of objective knowledge. Newton's laws were considered final in their domain of application, somewhat as quantum mechanics is today considered to be final by many physicists. How then can we introduce unidirectional time without destroying these amazing achievements of the human mind?

How then to make the bridge with dynamics? We shall describe two popular procedures.

It has been well known ever since the pioneering work of Gibbs and Einstein that we can describe dynamics from two points of view. On the one hand, we have the individual description in terms of trajectories in classical dynamics, or of wavefunctions in quantum theory. On the other hand, we have the collective statistical description in terms of ensembles represented by a probability distribution ρ (a probability density in classical mechanics or a density operator in quantum theory).

The probability ρ satisfies a well-defined equation both in classical and quantum theory. This is the Liouville equation, which is the starting point of statistical physics and which we shall study in the subsequent sections of this paper. The Ehrenfests[3] have introduced, in addition to ρ (called the "fine-grained" distribution), a coarse-grained distribution, ρ_{cg}, which results from averaging ρ over the microscopic states. It is straightforward to see that this coarse-grained distribution leads to equilibrium in the future $t > 0$ as well as in the past $t < 0$, because as the laws of dynamics are time reversible, every conclusion obtained from ρ_{cg} for the future $t > 0$ would be valid for the past $t < 0$. Moreover, the introduction of ρ_{cg} introduces arbitrary approximations and a loss of information. Adopting this point of view, we ourselves assume the responsibility for the appearance of the arrow of time. There is also the "cosmological" argument. For example, in his *Lectures on Physics*, Feyman[4] wrote: "For some reason, the universe at one time had a very low entropy ... that is the origin of all irreversibility ... ." This argument is somewhat strange. Whatever the past history of our universe, we observe today both reversible, time-symmetric processes as well as irreversible processes. Our task, therefore, is to understand the origin of the difference between reversible and irreversible processes that exist at present whatever the assumptions on cosmology.

In our opinion, the only possibility is a formulation of the laws of nature which explicitly includes time symmetry breaking and a characterization of the systems that lead to irreversible processes. Therefore, let us first consider in more detail what we may call a "law of nature."

LAWS OF CLASSICAL AND QUANTUM MECHANICS

As is well known, the laws of classical mechanics describe evolution in terms of point transformations. A point ω in phase space is transformed after a time t into the point S_t ω:

$$\omega \mapsto S_t \omega \tag{1}$$

In quantum mechanics the situation is different. As is well known the basic equation is the Schrödinger equation:

$$\frac{\partial \psi}{\partial t} = -iH_{op}\psi \tag{2}$$

The time change is determined by the action of the Hamiltonian operator on the wavefunction ψ. In order to see this action, we need the spectral analysis of the operator H_{op}, that is, the determination of its eigenfunctions and eigenvalues, for example (discrete spectrum),

$$H_{op} = \sum_n |u_n\rangle \omega_n \langle u_n| \tag{3}$$

where the $|u_n\rangle$ (and the conjugates $\langle u_n|$) are the eigenfunctions and ω_n the corresponding energy levels.

Traditionally,[5] quantum theory was associated with Hilbert space, that is, with functions that are square integrable:

$$\int_{-\infty}^{+\infty} dx |f(x)|^2 < \infty \tag{4}$$

The solution $\Psi \mapsto U_t \Psi$ of the Schrödinger Equation (2) preserves the scalar product:

$$\langle f_i, f_j \rangle = \int_{-\infty}^{+\infty} dx f_i^*, f_j,$$

Therefore, U_t is a unitary group.

An essential point, to which we shall come back several times, is that the spectral decomposition, Equation (3), depends on the function space. The Hamiltonian operator is "Hermitian" and has therefore only real eigenvalues in the Hilbert space. That leads to eigenfunctions $e^{\pm i\omega t}$ which are time symmetric as they are also eigenfunctions of the unitary group U_t: $U_t |u_n\rangle = e^{\pm i\omega_k}|u_n\rangle$.

This symmetry breaks, however, when we extend the evolution beyond the Hilbert space and consider a wider class of functions (often called generalized functions[6,7]). A simple example is the δ-function.[8] As is well known, we have

$$\int dx f(x)\delta(x) = f(0) \tag{5}$$

Spectral analysis of operators is the natural tool for the statistical problems associated with the time evolution of the statistical distribution ρ in classical systems as well as for the ones discussed in the Introduction.

In a famous article B.O. Koopman[9] showed that the evolution of probability densities of classical systems is described by a unitary group U_t implemented by the point transformation S_t (Eq. 1):

$$U_t \rho(\omega) = \rho(S_t^{-1}\omega) \tag{6}$$

for all functions ρ in the Hilbert space of square-integrable phase densities. The two descriptions in terms of trajectories and in terms of ensembles involving ρ are equivalent.[10] But this is usually no longer so when we go outside the Hilbert space. Examples will be given in the sections that follow.

In both classical and quantum mechanics, ρ satisfies the Liouville–von Neumann equation (L-N equation)

$$\frac{\partial \rho}{\partial t} = -iL\rho \quad \text{or} \quad \rho(t) = U_t\rho(0) = e^{-iLt}\rho(0) \tag{7}$$

In classsical dynamics L is the differential operator acting on the Hilbert space of square integrable density functions,

$$L\rho = -i\frac{\partial H}{\partial p}\frac{\partial \rho}{\partial q} + i\frac{\partial H}{\partial q}\frac{\partial \rho}{\partial p}$$

whereas in quantum mechanics L is the commutator with the Hamiltonian operator H_{op} acting on the Hilbert-Schmidt space of density operators:

$$L\rho = [H_{op}, \rho]$$

The L-N operator L is Hermitian and e^{-iLt} unitary. As long as we stay in the Hilbert space, the L-N equation gives nothing new with respect to the classical (1) or quantum mechanical (2) descriptions. There is no directed flow of time, but when we go outside the Hilbert space we find spectral decompositions of L that include time symmetry breaking. The L-N equations lead then to new solutions that can no more be implemented by trajectories or wavefunctions. We obtain, then, new dynamical laws that provide the microscopic basis for irreversible processes.

DETERMINISTIC CHAOS: THE BAKER TRANSFORMATION

For the first example we will consider deterministic chaos, in which two trajectories, set as closely as we choose at time $t = 0$, diverge exponentially as time goes on (the Lyapounov exponents). The simplest example that captures the essential characteristics of classical systems is the well-known Baker transformation of the unit square.[1]

Let us begin with a square with sides of length 1; then we cut it in half and build a new square. If we examine the lower part of the square, we see that after one iteration of this process (or mapping), it splits into two bands (see FIG. 1). Moreover, the transformation is reversible. The inverse transformation, which first reshapes the square into a rectangle with length ½ and height 2, returns each point to its initial position. For the Baker map, the equations of motion are very simple. At each step, the coordinates (x, y) become $(2x, y/2)$ for $0 \leq x \leq 1/2$ and $(2x - 1, (y + 1)/2$ for $1/2 < x \leq 1$. To obtain the inverse Baker transformation, we only have to permute x and y.

On the Baker map, the two coordinates play different roles. The horizontal coordinate x is the expanding coordinate, as it is multiplied by 2 (mod 1) at each iteration. The area of the square is preserved, because we have the contracting coordinate y. In the transverse direction, the points draw closer together while the square is being flattened into a rectangle. Because the distance between two points along the horizontal coordinate x doubles with each transformation, it will be multiplied by 2^n after n transformations. If we rewrite 2^n as $e^{n\log 2}$, because the number n of transformations measures time, we see that the Lyapunov exponent is $\log 2$.

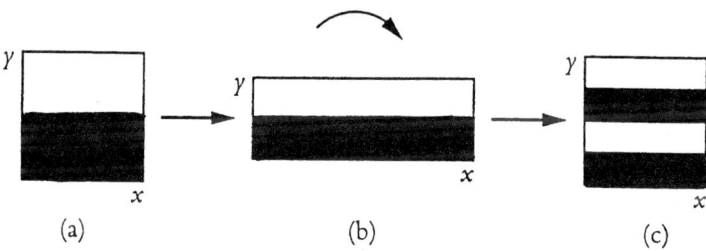

FIGURE 1. The Baker transformation.

There is also a second Lyapunov exponent with the negative value $-\log 2$, which corresponds to the contracting direction y (see FIG. 1). Successive iterations of the Baker transformation lead to fragmentation of the shaded and unshaded areas, producing an increasing number of disconnected regions. Let us consider the successive probability distributions. We have

$$\rho_{n+1} = U\rho_n \tag{8}$$

U is called the "Perron-Frobenius" operator.[11] It has been recently shown that there exist two spectral representations for the evolution U^n, $n \in Z$. The first one is in Hilbert space and can be written

$$U^n = \int_{-\infty}^{+\infty} dk\, e^{ikn} |f_k\rangle\langle f_k| \tag{9}$$

Because the eigenvalues are e^{ikn}, there is no time symmetry breaking. This solution is equivalent to the solution of the equation of motion.

But it has been recently established[12,13] that there is a second spectral decomposition outside the Hilbert space:

$$U = \sum_{n=0}^{\infty} |F_n\rangle \frac{1}{2^n} \langle \tilde{F}_n| \tag{10}$$

(We simplify somewhat the actual situation, which involves Jordan blocks; see Appendix.) The main point is that the eigenfunctions F_n and $\tilde{F}_n$ are distributions. The eigenvalues $1/2^n$ are expressing the rates of approach to equilibrium.[d] Here we have a formulation of dynamics that includes the arrow of time. Note that Equation (10) leads to

$$U\rho = \sum_{n=0}^{\infty} |F_n\rangle \frac{1}{2^n} \langle \tilde{F}_n|\rho\rangle \quad \text{for } t > 0 \tag{11}$$

$\langle \tilde{F}_n|$ being a distribution, this formula is only meaningful when ρ is not a distribution. $\langle \tilde{F}|\rho\rangle$ is the scalar product of $\tilde{F}$ with ρ, such as $\int dx \tilde{F}(x)\rho(x)$, and we have seen in the section on laws of classical and quantum mechanics that products of distributions may lead to divergence. This excludes trajectories that are δ-functions. The initial condition has to be a region of finite dimension (as small as we want). The fundamental quantity is now the probability distribution. Irreversibility is associated with semigroups. The probability distribution ρ tends toward equilibrium for $t \to \infty$. Using Equation (11), $\rho \to \rho_{eq}$, $t \to \infty$.

There is also a spectral decomposition in which $\rho \to \rho_{eq}$ for $t \to -\infty$. Equilibrium would be reached in the past. We have two different semigroups corresponding to different test functions for the distributions. (More details are given in the APPENDIX.) In accordance with our observations, we choose the semigroup (Eq. 11) that leads to equilibrium in the future.

---

[d] $1/2^n = e^{-n \lg 2}$; by comparison with Equation (9) this corresponds to k imaginary.

These conclusions can be extended to an important class of dynamic systems, the so-called K flows (K for Kolmogorov). Both formulations (Eq. 10 and Eq. 11) of dynamics are rigorous. Equation (11) applies to the realistic initial condition which corresponds to a finite area. Equation (11) gives us the first example of a classical dynamical law that is not reducible to trajectories and contains the arrow of time.

POINCARÉ INTEGRABILITY

As is well known, a dynamical system is characterized in terms of the kinetic energy of its particles plus the potential energy due to their interactions. The simplest example would be free, noninteracting particles for which there is no potential energy and the calculation of trajectories is trivial. Such systems are by definition integrable. Poincaré then asked the questions, Are all systems integrable? Can we choose suitable variables to eliminate potential energy? By showing that this was generally impossible, he proved that dynamic systems were generally nonintegrable. Poincaré not only demonstrated nonintegrability, but also identified the reason for it: the existence of resonances between the degrees of freedom. We are all more or less familiar with the concept of resonance from elementary acoustics. Also, all radiation processes are related to resonances. Think about atoms where excited states occur when we introduce radiation with frequency ω_k, which coincides with the frequency $\omega_1 - \omega_0$ separating the excited state $|1\rangle$ from the ground state $|0\rangle$. We shall come back to this problem in the section on particles, fields, and irreversibility. The significant fact is that most dynamic systems are not integrable.

In the following section, we shall concern ourselves primarily with nonintegrable, large Poincaré systems (LPS). Poincaré resonances are associated with frequencies corresponding to various modes of motion. A frequency ω_k depends on the wavelength k. (Using light as an example, ultraviolet has a higher frequency ω and shorter wavelength k than infrared light.) When we consider nonintegrable systems in which the frequency varies continuously with the wavelength, we arrive at the very definition of LPS. This condition is met when the volume in which the system is located is large enough for surface effects to be ignored. This is why we call these systems *large* Poincaré systems.

Thermodynamics and entropy are also usually associated with large systems containing a large number of particles. We shall now consider more closely the dynamics of these systems.

THERMODYNAMIC SYSTEMS

Thermodynamic systems are characterized by a large, practically infinite number of degrees of freedom. Note that it is only in the so-called thermodynamic limit

$$N \to \infty \quad V \to \infty \quad \frac{N}{V} = C: \quad \text{finite} \tag{12}$$

that we obtain a correct formulation of phase transitions.[14] We want to consider the effect of the thermodynamic limit for dynamic systems. We shall show that the interaction between the degrees of freedom leads indeed beyond the Hilbert space structure.

In the thermodynamic limit, we require in addition that there are extensive variables such as the average potential energy U:

$$\lim_{N \to \infty} \frac{\langle U \rangle}{N} = \text{finite} \tag{13}$$

and intensive variables such as the pressure p

$$\lim_{N \to \infty} p = \text{finite} \tag{14}$$

We have recently considered the examples of anharmonic lattices[16] as well as of large systems of interacting particles.[17,18] Let us summarize our results for one-dimensional anharmonic lattices. The equilibrium distance a between two neighboring lattice sites being given, Equation (12) reduces to the limit $N \to \infty$ (as $V = aN$).

Harmonic lattices correspond to integrable systems, as there is no coupling between the normal modes. The potential energy U is given by the quadratic form

$$U - U_0 = \frac{1}{2} \sum_{nn'} A_{nn'} u_n u_{n'} \tag{15}$$

where u_n is the displacement of the nth atom from its equilibrium position (we impose cyclic boundary conditions $u_{n+N} = u_n$). We may solve the dynamical problem either by point transformations using angle–action variables α and J or by the Liouville equation introducing the Hilbert space structure. With obvious notations

$$\rho(J, \alpha) = \sum \rho_{\{n\}}(J) e^{i\Sigma n_k \alpha_k} \tag{16}$$

The Hilbert norm is therefore

$$\|\rho\|^2 = \int dJ \sum_{\{n\}} |\rho_{\{n\}}(J)|^2 \tag{17}$$

To obtain a finite Hilbert norm for $N \to \infty$, well-defined conditions have to be satisfied. Indeed, the norm (Eq. 17) contains terms with $n_k = \ldots, -1_k, 0, 1_k, 2_k, \ldots$

$$\|\rho\|^2 = |\rho_0|^2 + \sum_k |\rho_{1_k}|^2 + \sum_{kk'} |\rho_{1_k 1_{k'}}|^2 + \sum_{kk'k''} |\rho_{1_k 1_{k'} 1_{k''}}|^2 + \ldots \tag{18}$$

which have to converge for $N \to \infty$. This implies

$$\rho_0 \approx O(1), \quad \rho_{1_k} \approx \frac{1}{\sqrt{N}}, \quad \rho_{1_k} \rho_{1_{k'}} \approx \frac{1}{N} \tag{19}$$

as well as

$$\rho_{1_k 1_{k'} 1_{k''}} \approx \frac{1}{N} \tag{20}$$

for the components with $k + k' + k'' = 0$ or a vector on the reciprocal lattice, and

$$\rho_{1_k 1_{k'} 1_{k''}} \approx \frac{1}{N^{3/2}} \tag{21}$$

for the components not on the reciprocal lattice.

However, if $\rho_{1_k 1_{k'} 1_{k''}}$ is too large, such as $N^{-1/2}$ instead of $1/N$, the Hilbert norm diverges.

Consider then anharmonic lattices. The potential energy is, at the lowest order,

$$U - U_0 = \frac{1}{2}\sum_{nn'} A_{nn'} u_n u_{n'} + \frac{1}{6}\sum_{nn'n''} B_{nn'n''} u_n u_{n'} u_{n''} \tag{22}$$

Higher order terms in the displacement would not introduce any change. The Hamiltonian H becomes

$$H = H_0 + \lambda V \tag{23}$$

where we have introduced the parameter λ for the coupling constant.

We may calculate the value of the average potential energy.

$$\langle V \rangle \approx \sum_{kk'k''}{}' \int dJ \, V_{kk'k''} \rho_{1_k 1_{k'} 1_{k''}} \tag{24}$$

Using the value (Eq. 24) of the three-mode correlations (Eq. 20) we obtain

$$\langle V \rangle \approx \sqrt{N} \tag{25}$$

which is incompatible with Equation (13). To obtain Equation (13), we need stronger correlations

$$\rho_{1_k 1_{k'} 1_{k''}} \approx \frac{1}{\sqrt{N}} \tag{26}$$

but then the Hilbert space norm diverges. We refer to the original papers for the description how ρ is "ejected" from the Hilbert space as the result of the interactions.

Our conclusion applies as well to systems of interacting particles in the thermodynamic limit and even to interacting fields (see the section on particles, fields, and irreversibility). However, the extension of the functional space outside the Hilbert space alone does not imply time symmetry breaking. For this, we need in addition nonintegrability in the sense of Poincaré, as applied to large thermodynamic systems (LPS). As is well known, Poincaré's nonintegrability is associated with resonances. This leads to new processes taking place at the statistical level.

We may have processes "destroying" correlations as represented graphically in FIGURE 2. We may also have processes creating correlations (FIG. 3).

As the result of Poincaré resonances, we may have in addition processes relating states corresponding to the same degree of correlation (i.e., ρ_0 to ρ_0) (see FIG. 4). We have called these collision processes.

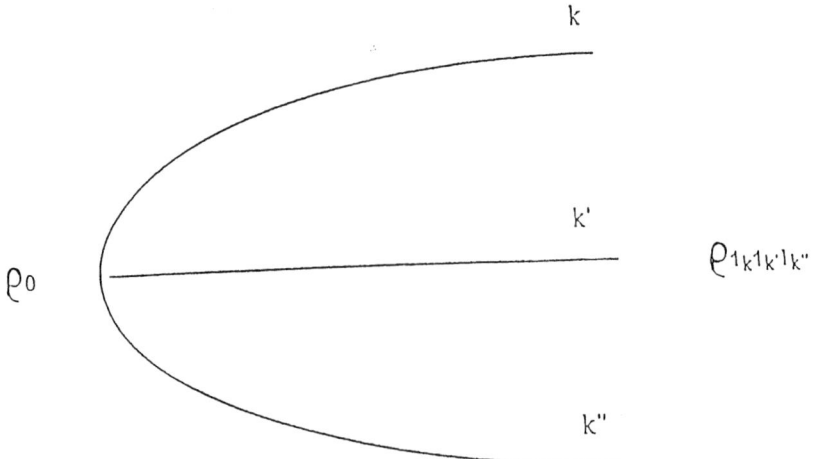

FIGURE 2. Destruction of correlations leading from a three-mode correlation $\rho_{1_k 1_{k'} 1_{k''}}$ to the vacuum of correlation ρ_0.

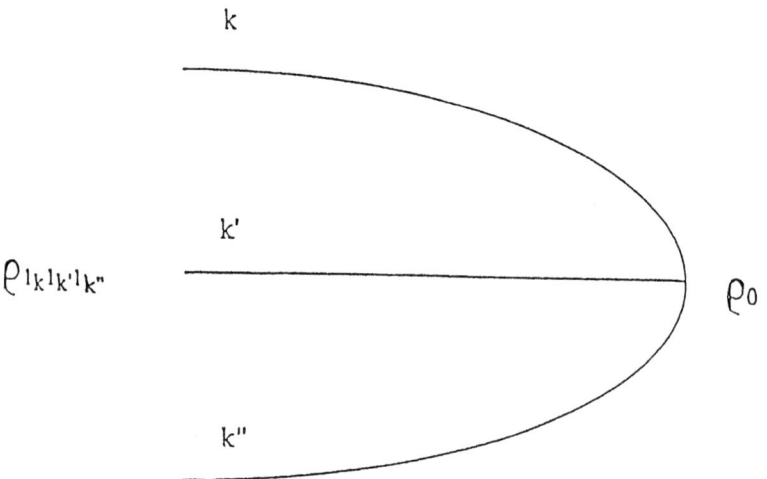

FIGURE 3. Creation of correlations leading from ρ_0 to $\rho_{1_k 1_{k'} 1_{k''}}$.

The collision processes destroy the trajectory description. In FIGURES 2–4, each vertex contains derivative operators $\partial/\partial J$. FIGURE 4 leads therefore to processes containing second-order operators $\partial^2/\partial J^2$ characteristic of diffusive processes.

The main result is that the dynamics of large Poincaré's systems (LPS) are ruled by "Langevin-type" interactions, as they appear, for example, in the Fokker-Planck equation for classical dynamics and in the Pauli's master equation for quantum mechanics. We may qualitatively describe the situation by imagining each observable corresponding to a finite number of degrees of freedom "swimming" in the infinite sea associated with the thermodynamics limit.

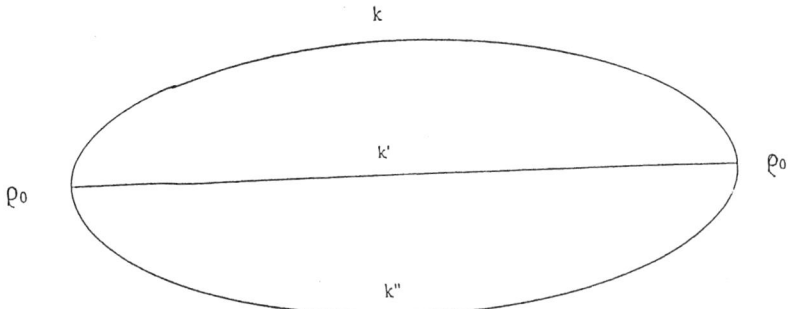

FIGURE 4. Collision processes relating to ρ_0 to ρ_0.

We see that LPS involve new effects not considered in Newtonian or quantum mechanics. As in the case of deterministic chaos, we obtain a formulation of dynamics in terms of probabilities that breaks the time symmetry. LPS can therefore be integrated, not on the level of trajectories or wavefunctions, but at the level of probabilities. This is an important novel aspect we shall briefly describe in the next section.

All these results can easily be extended to quantum mechanics.[18] We then obtain a formulation of quantum mechanics outside the Hilbert space in which the fundamental quantities are density operators and are no longer wave amplitudes.

To go further we have to introduce a class of functions ρ outside the Hilbert space. This class has to include the equilibrium distribution and should not be invariant under time inversion. (This is in contrast with Boltzmann's molecular chaos; indeed, Boltzmann's "molecular chaos" already introduces a privileged direction of time). A class of distributions satisfying these criteria has already been introduced in the monograph "Nonequilibrium Statistical Mechanics"[19] (see also Ref. 20) and is described in detail in recent publications.[16-18]

We shall now describe briefly the spectral representation of the L-N operator when extended outside the Hilbert space.

DYNAMICS OF LARGE POINCARÉ SYSTEMS

We have recently solved the eigenvalue problem for the L-N operator extended outside the Hilbert space. Let's briefly summarize some of the main results. For a system of interacting particles L is given by

$$L = L_0 + \lambda L_V \tag{27}$$

using obvious notations. In classical mechanics L_0 admits eigenfunctions of the form $L_0 \varphi_k = kp/m\, \varphi_k$, $\varphi_k = e^{ikv}$ and a complete set of eigen-projection operators P where

$$L_0 \overset{v}{P} = \overset{v}{P} L_0 \tag{28}$$

v is the number of nonvanishing k vectors. More explicitly, we shall use the notation $\overset{v}{P}_\alpha$ where α denotes the particles associated with nonvanishing wavevectors and v the value of the wavevectors. They satisfy the usual completeness and orthonormality relations.

We also introduce the projection operators $\overset{v}{Q}_\alpha$:

$$\overset{v}{Q}_\alpha = 1 - \overset{v}{P}_\alpha, \quad \overset{v}{P}_\alpha \overset{v}{Q}_\alpha = \overset{v}{Q}_\alpha \overset{v}{P}_\alpha = 0 \tag{29}$$

In particular $\overset{0}{P}_\rho$ is the velocity distribution that is factorized in a product of individual velocity distribution functions. As is well known, the evolution is non-Markovian for thermodynamic systems. We have the so-called master equation for $\overset{0}{P}$, which can be found in many books.[19,21]

$$i\frac{\partial \overset{0}{P}_\rho}{\partial t} = \int_0^t dt' \tilde{\psi}(t') \overset{0}{P} \rho(t-t') + \overset{0}{D}(t, \rho(0)) \tag{30}$$

$\tilde{\psi}(t)$ is the time-dependent collision operator leading from $\overset{0}{P}$ to $\overset{0}{P}$ (see FIG. 4) and the "destruction fragment" leading from $\overset{0}{P}_\mu \to \overset{0}{P}$ ($\mu \neq 0$). This formula includes memory effects ("non-Markovian" behavior). This formula was established by Résibois and Prigogine in 1961.[21] As a result we have three time periods: the Zenon time, the exponential time period, and the long-tail time. A similar classification was introduced earlier for unstable particles.[22]

The experimental discovery of long-tail effects by Alder[23] in the sixties has led to a deep change in the kinetic theory of fluids. The classical kinetic theories (Fokker, Planck, Boltzman) do not contain long tails. This is the subject of the mode–mode coupling theory, which is still in a rather early stage.[21] We shall come back to this question at the end of this section.

In a recent paper[18] we obtained the spectral decomposition of L when applied to the class of functions of the section on Poincaré integrability. We will only quote the result here. The spectral decomposition of the Liouville operator can be written in the form

$$L = \sum_{v\alpha} \left| \overset{v}{F}_\alpha \gg \overset{v}{z}_\alpha \ll \overset{v}{F}_\alpha \right| \tag{31}$$

Therefore, nonintegrable systems (in the sense of Poincaré) can be integrated on the Liouville level. The eigenvalues $\overset{v}{z}_\alpha$ are complex and satisfy the inequality Im $\overset{v}{z}_\alpha \leq 0$. There is, therefore, time symmetry breaking. The diagonalization has been obtained on the level of distribution functions outside the Hilbert space. This representation is *nonreducible* in the sense that it involves probabilities and cannot be reduced to trajectories or wavefunctions.

Of course, one could also obtain a spectral decomposition with Im $\overset{v}{z}_\alpha \geq 0$ which leads to equilibrium in the past. The main point is that the two spectral decompositions are distinct, and they have different domains. The spectral decomposition leads to the basic intertwining relation

$$\Lambda L \Lambda^{-1} = \Theta \tag{32}$$

The operator L is nonunitary ($L^+ \neq L^{-1}$). The collision operator Θ is diagonal in the P_α representation

$$\Theta = \sum_\nu P^\nu \Theta P^\nu = \sum_\nu \overset{\nu}{\Theta} \tag{33}$$

Also, we can obtain a complete set of projectors $\overset{\nu}{\Pi}$ for the Liouville operator L.

$$L\overset{\nu}{\Pi} = \overset{\nu}{\Pi} L \tag{34}$$

$$\overset{\nu}{\Pi} = \Lambda^{-1} \overset{\nu}{P} \Lambda \tag{35}$$

It is quite remarkable that each projection $\overset{\nu}{P} \overset{\nu}{\Pi}$ leads to a Markovian process that we can write in the following form:

$$\frac{\partial \overset{\nu}{P} \overset{\nu}{\Pi} \rho}{\partial t} = -i \overset{\nu}{\Theta} \overset{\nu}{P} \overset{\nu}{\Pi} \rho \tag{36}$$

There is an infinite number of Markovian processes because each distribution function, such as $\overset{0}{P}\rho$ can be written as a superposition of the projection operators $\overset{\nu}{\Pi}$.

$$\overset{0}{P} \rho = \sum_\nu \overset{0}{P} \overset{\nu}{\Pi} \rho \tag{37}$$

We want to emphasize the special role of $\overset{0}{\Pi}$. Indeed, $\overset{0}{\Pi}$ is the only part of the distribution function ρ that contains vanishing eigenvalues $_0 z_\alpha = 0$. Therefore, it is not astonishing that the equilibrium distribution lies in the $\overset{0}{\Pi}$. At equilibrium,

$$\rho_{eq} = \overset{0}{\Pi} \rho_{eq} = (\overset{0}{C} + \overset{0}{P}) \overset{0}{\rho} \tag{38}$$

Also, all classical kinetic equations lie in $\overset{0}{\Pi}$ as well as all invariants of motions such as the Hamiltonian H.

$$H = \overset{0}{\Pi} H = H_0 \tag{39}$$

Causality requires, however, that we take into account $\overset{2}{\Pi}$ (and if necessary $\overset{\nu}{\Pi}$ for $\nu > 2$) for nonequilibrium processes. There are additional points that are important. Instead of the non-Markovian Equation (30), we obtain a superposition of Markovian processes. In this way, the questions left open by the mode–mode coupling approach can be solved (T. Petrosky, private communication). Our approach permits us to introduce causality into kinetic theory. Indeed, $\overset{0}{\Pi}$ leads to correlations over arbitrary distances. Causality is obtained by including the effect of $\overset{2}{\Pi}$ (therefore, $\overset{2}{\Pi}\rho + \overset{0}{\Pi}\rho$). To illustrate this statement consider the time evolution of the binary correlation $g_{12}(r, t)$ with $g_{12}(r, t = 0) = 0$ The results are represented in FIGURES 5 through 7, which show the evolution of the binary distribution functions $g_{12}(r, t)$. Therefore, the classical kinetic theories that retain only $\overset{0}{\Pi}$, always violate causality in the propagation of irreversible processes. To retain $\overset{0}{\Pi}$ is only correct for local properties far from the propagating edge.

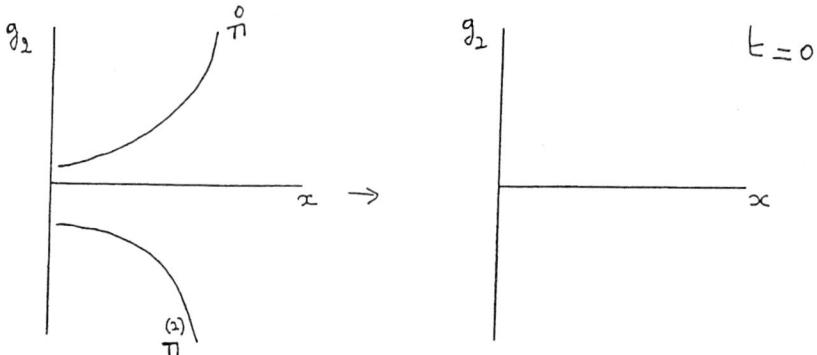

FIGURE 5. $g_2(r, t)$ for $t = 0$. (Complete compensation between $\overset{0}{\Pi}$ and $\overset{2}{\Pi}$.)

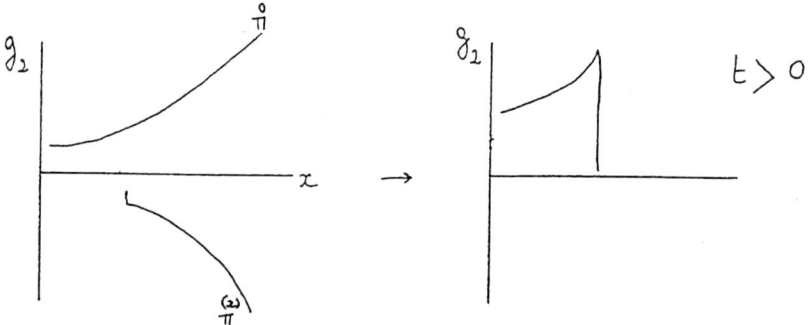

FIGURE 6. $g_2(r, t)$ for $t = 0$. (Partial compensation between $\overset{0}{\Pi}$ and $\overset{2}{\Pi}$.)

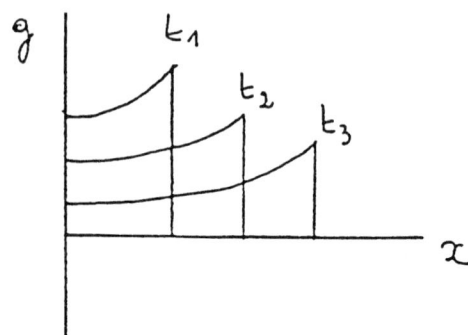

FIGURE 7. Time evolution of $g_2(r, t)$, $t_1 < t_2 < t_3$.

Let us summarize this section. The time evolution of thermodynamic systems is given by the L-N equation applied to a class of generalized functions ρ outside the Hilbert space. In contrast with the Baker transformation studied in the section on deterministic chaos, where we had both a time-reversible and a time-irreversible formulation of dynamics, we have for LPS only a formulation of dynamics in which the central quantity is the probability and which has a broken time symmetry. We have achieved in this way a microscopic formulation of the arrow of time. Once this is done, we can construct this entropy on a microscopic basis, but we cannot present this within the framework of this article.

We want now to demonstrate briefly that our approach has fundamental consequences for basic aspects of modern physics.

PARTICLES, FIELDS, AND IRREVERSIBILITY[e]

Atomic, nuclear, and high energy physics deal with excited states, or unstable particles. These objects cannot be eigenfunctions of the Hamiltonian (in contrast with the ground-state or with stable-state particles). Dirac has recognized this situation very well.[8]

> The fact that we had to use the word "approximately" in stating the conditions required for the phenomena of emission and absorption to be able to occur shows that these conditions are not expressible in exact mathematical language. One can give meaning to these phenomena only with reference to a perturbation method. They occur when the unperturbed system (of scatterer plus particle) has stationary states that are closed. The introduction of the perturbation spoils the stationary property of these states and gives rise to spontaneous emission and its converse absorption.

Let us describe in detail this difficulty using the well-known Friedrichs model, which describes the interaction of a two-level atom with radiation. In this model we have a discrete state $|1\rangle$ representing a bare particle coupled to continuous states $|k\rangle$ corresponding to field modes.[1] The Hamiltonian operator is

$$H = H_0 + \lambda V = |1\rangle \omega_1 \langle 1| + \sum_k |k\rangle \omega_k \langle k| + \lambda \sum_k V_k (|k\rangle\langle 1| + |1\rangle\langle k|) \quad (40)$$

The states $|1\rangle$ and $|k\rangle$ are eigenfunctions of H_0. The effect of the perturbation is to lead to the emission or the absorption of a photon (this model neglects virtual transitions). There are two situations, according to whether ω_1 is less than or greater than zero (FIG. 8).

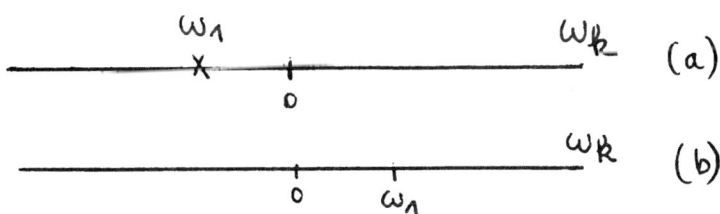

FIGURE 8. (a) Bound state; (b) resonance.

---

[e]This section is a summary of a paper by T. Petrosky, G. Ordonez, and I. Prigogine that will appear soon in *Physica A*.

For $\omega_1 < 0$ we have a bound state; for $\omega_1 > 0$ the state $|1\rangle$ will decay, emitting a photon. The case $\omega_1 < 0$ can be treated easily. The interaction transforms $|1\rangle$ into $|\varphi_1\rangle$ which is an eigenstate of H:

$$H|\varphi_1\rangle = \tilde{\omega}_1 |\varphi_1\rangle \tag{41}$$

The exact expression for $|\varphi_1\rangle$ can be shown to be[24]:

$$|\varphi_1\rangle = N_1^{1/2}\left[|1\rangle - \sum_k \frac{\lambda V_k}{\omega_k - \tilde{\omega}_1}|k\rangle\right] \tag{42}$$

where N_1 is a normalization constant.

Formula (42) clearly shows the photon cloud $|k\rangle$ surrounding $|1\rangle$. Similarly we have the dressed photon. The state $|1\rangle$ evolves time going on to $|\varphi_1\rangle$. FIGURE 9 gives an example of the photon distribution around the stable particle. (For the numerical simulation attributed to G. Ordonez, we use $\lambda = 0.1$ $V_k = L^{-1/2}$, $L = 400$. One clearly sees the cloud surrounding $|1\rangle$.

The two smaller peaks traveling away from the particle are superpositions of dressed photons. These photons are created because the energy of the bare particle is greater than the energy of the dressed particle $\omega_1 > \tilde{\omega}_1$. As the dressed particle emerges, the photons carry away the excess energy.

Now let us consider the case of resonance (see FIG. 8). Friedrichs has shown that then we have the spectral decomposition[24]

$$H = \sum_k \omega_k |\varphi_k^F\rangle\langle\varphi_k^F| \tag{43}$$

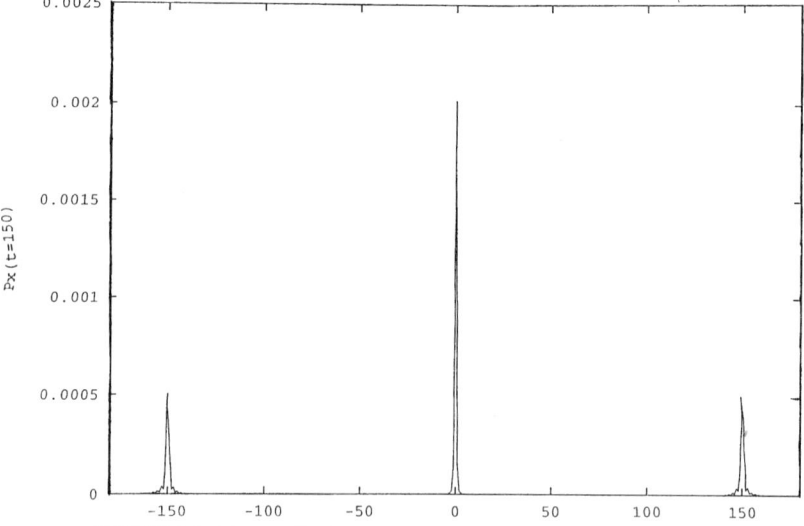

FIGURE 9. Photon distribution around a stable particle ($t = 150$).

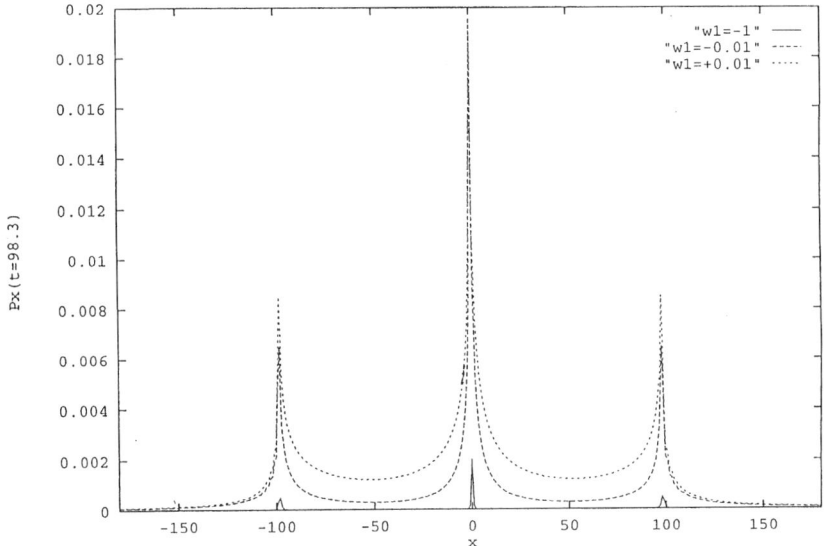

FIGURE 10. Photon distribution near the transition point.

where the $|\varphi_k^F\rangle$ are outgoing waves satisfying completeness and orthonormality. The excited state does not appear explicitly in Equation (43).

As mentioned previously (see FIG. 8), the transition from bound state to resonance takes place at $\omega_1 = 0$. We may visualize the transition through the computer plots shown in FIGURE 10.

In FIGURE 10, we plot the photon space distribution for three cases: (a) $\omega_1 = -1$, (b) $\omega_1 = -0.01$, and (c) $\omega_1 = +0.01$. Case (a) we have already considered. It is given here for reference. Cases (b) and (c) are close to the branching point $\omega_1 = 0$. Case (b) corresponds to the stable particle and (c) to the unstable particle.

It is obvious that the photon distributions are very similar. The resonance also has a photon cloud, but this state (which is the continuation of $|\varphi_1\rangle$) is not an eigenfunction of H. It is interesting how this state is formed. At $t = 0$, we have a wave packet approaching the atom in the ground state. After the contact the atom is excited and then decays exponentially.

Note the asymmetry between the formation period and the decay. This clearly indicates the irreversible character of the process. The question is, What is the nature of the "decaying state" (which appears for $t = 500$ in FIGURE 11)? We can apply the Liouville formulation and show that this state is in Π subdynamics. It is therefore a combination of eigenfunctions of the L-N operator. The main point is that the resonances are described by density operators (and not wavefunctions) that evolve irreversibly. The central quantity is again probability.

This example shows that our theory is an extension of quantum theory, which avoids the difficulty stated by Dirac. It is also interesting to notice that irreversibility appears on the microscopic level. Resonances and excited states are closely related to unstable particles. Let us close this section with some remarks on the relationship between fields and particles.

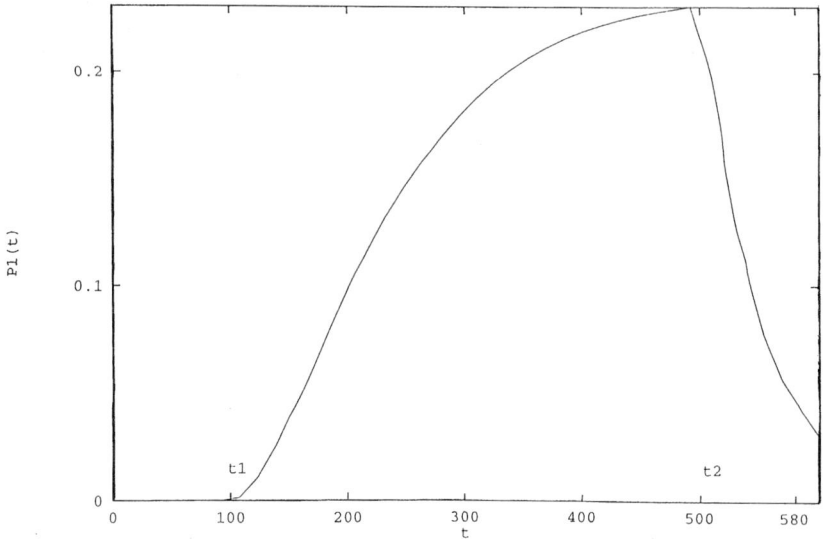

FIGURE 11. Formation decay of a bare particle.

A fundamental problem of modern physics is the relationship between particles and fields. This is the central problem of second quantization. A free quantum field is a *superposition of oscillators*. For the electrodynamic field, the corresponding particles are the photons. We have $H = \Sigma_k \omega_k (n_k + \frac{1}{2})$. Free quantum fields are integrable systems, and the occupation number n_k are invariants. However, as there are no free fields in nature, interactions lead to *nonintegrable systems* where the occupation numbers vary in time.

Nonintegrable fields are systems of an infinite number of degrees of freedom with persistent interactions. There is therefore a strong analogy with thermodynamic systems. Exactly as for thermodynamic systems, we can integrate the equations of motion on the Liouville level of probabilities. (In our presentation we avoid ultraviolet divergencies by means of suitable form factors.) Exactly as for thermodynamic systems, we have to go outside the Hilbert space. (The condition $\langle V \rangle /N$ finite is replaced here by the condition that the zero point energy E_0 of interesting fields E_0 satisfies the condition $V \to \infty$, E_0/V = finite.)

In this way, field theory is deeply changed. According to G. Källen:

> It appears to me that one basic feature of a pure S-matrix theory and also of some of the more extreme versions of the axiomatic approach is just that one completely forgets the development in time. Of course, it is true that many experimental situations, perhaps nearly all of them, can conveniently be described in terms of scattering processes. However, a pure S-matrix theory goes further and assumes that everything can be described as a *scattering during an infinite time interval*.[25]

We can now overcome this difficulty and obtain an evolution equation valid for finite times. Moreover, this introduces time symmetry breaking, which explains the fact that most elementary particles are unstable.

FROM BEING TO BECOMING

The results summarized in this paper show that irreversibility plays a much more important role than had been previously anticipated. The most striking result concerns the dynamics of large systems, thermodynamic systems and fields. Here the laws of classical physics or quantum physics are replaced by the probabilistic Liouville equation associated with the extended Liouville operator beyond the Hilbert space:

$$F = ma$$

$$\frac{\partial \psi}{\partial t} = -iH_{op}\psi$$

$$\xrightarrow[N \to \infty]{} \frac{\partial \rho}{\partial t} = -iL^{ext}\rho$$

Here L^{ext} is the extension of the Liouville operator outside the Hilbert Space. Therefore, the basic laws of physics for large systems are radically different from the classical and quantum mechanics of integrable systems. Our results can be extended to relativity where the Poincaré group splits into two semigroups.[26,27] The main entity is probability, and the time symmetry is broken. Nature evolves as a semigroup. Note also that the conventional approach associated irreversibility and entropy with ensembles of particles, each of which if taken in isolation would satisfy the deterministic laws of classical or quantum mechanics. Now the particles themselves are the outcome of a time symmetry broken formulation except for the ground state.

The inclusion of irreversibility changes our view of nature. The future is no longer given. Our world is a world of continuous "construction" ruled by probabilistic laws and no longer a kind of automaton.

We are led from a world of "being" to a world of "becoming." Nevertheless, why nature has a broken time symmetry is a difficult question. It may be due to the interaction between gravitation and the other fundamental forces. But at this point we are at the frontiers of present science.

ACKNOWLEDGMENTS

We acknowledge Dr. Tornio Petrosky and Dr. Gonzalo Ordonez for their contribution to this work as well as the European Commission ESPRIT Project CTIAC 21042/DGIII, the "Communauté française de Belgique," the Belgian Interuniversity Attraction Poles, and the National Lottery of Belgium for their financial support.

REFERENCES

1. Prigogine, I. 1980. From Being to Becoming. Freeman. New York.
2. Nicolis, G. & Prigogine, I. 1989. Exploring Complexity. Freeman. New York
3. Ehrenfest, P. & T. Ehrenfest. 1959. The Conceptual Foundations of Statistical Mechanics, Enzyklopddie der Matematischen Wissenschaften No. 6, Vol. IV, Leipzig 1912; English translation by M. Moravesik for Cornell University Press, Ithaca, New York.
4. Feynman, R., R. Leighton & M. Sands. 1965. The Feynamn Lectures on Physics. Addison Wesley. Reading, MA.

5. VON NEUMANN, J. 1955. Mathematical Foundations of Quantum Mechanics. Princeton Univ. Press. Princeton, NJ.
6. BOHM, A. 1993. Quantum Mechanics, Foundations and Applications, 3rd edit. Springer. Berlin.
7. ANTONIOU, I. & S. TASAKI. 1993. Generalized spectral decomposition of mixing dynamical systems. Int. J. Quantum Chem. **46:** 425–474.
8. DIRAC, P.A.M. 1958. The Principles of Quantum Mechanics. Oxford University Press. London.
9. KOOPMAN, B. 1931. Hamiltonian systems and transformations in Hilbert spaces. Proc. Natl. Acad. Sci. USA **17:** 315–318.
10. GOODRICH, K., K. GUSTAFSON & B. MISRA. 1980. On a converse to Koopman's lemma. Physica **102A:** 379–388
11. LASOTA, A. & M. MACKEY. 1994. Chaos, Fractals, and Noise. Springer-Verlag. New York.
12. HASEGAWA, H.H. & W. SAPHIR. 1992. Nonequilibrium statistical mechanics of the Baker map: Ruelle resonances and subdynamics. Phys. Lett. A **161:** 477–482.
13. ANTONIOU, I. & S. TASAKI. 1992. Generalized spectral decomposition of the β-adic Baker's transformation and intrinsic irreversibility. Physica A **190:** 303–329.
14. REICHL, L.E. 1980. A Modern Course in Statistical Physics. Arnold. London.
15. PETROSKY, T. & I. PRIGOGINE. 1996. Extension of classical dynamics. The case of anharmonic lattices. *In* Gravity, Particles and Space-Time. P. Pronin & G. Sardanashvily, Eds. World Scientific. Singapore/New Jersey.
16. PETROSKY, T. & I. PRIGOGINE. 1997. The extension of classical mechanics for unstable Hamiltonian systems. Comp. Math. Appl. **39:** 1–44.
17. PETROSKY, T. & I. PRIGOGINE. 1996. Poincaré resonances and the extension of classical dynamics. Chaos Solitons and Fractals **7:** 441–497.
18. PETROSKY, T. & I. PRIGOGINE. 1997. The Liouville space extension of quantum mechanics. Adv. Chem. Phys. **99:** 1–120.
19. PRIGOGINE, I. 1962. Non-Equilibrium Statistical Mechanics. Wiley. New York.
20. BALESCU, R. 1975. Equilibrium and Non-equilibrium Statistical Mechanics. Wiley. New York.
21. RESIBOIS, P. & M. DE LEENER. 1977. Classical Kinetic Theory of Fluids. Wiley. New York.
22. FONDA, L., G.C. GHIRARDI & A. RIMINI. 1978. Decay Theory of unstable quantum systems. Rep. Prog. Phys. **41:** 587–631.
23. ALDER, B. & T. WAINRIGHT. 1969. Phys. Rev. Lett **18:** 988.
24. PETROSKY, T., I PRIGOGINE & S. TASAKI. 1991. Quantum theory of non-integrable systems. Physica A **173:** 175–242.
25. KÄLLEN, G. 1962. *In* La Theorie Quantique des champs. Proceedings of the 12th Solvay Physics Conference. R. Stoops, Ed.: 253. Interscience. New York.
26. BEN-YAACOV, URI. 1995. Lorentz Symmetry of Subdynamics in Relativistic Systems. Physica A **222:** 307–329.
27. ANTONIOU, I., M. GADELLA, I. PRIGOGINE & G. PRONKO. 1998. Relativistic Gamow Vectors. J. Math. Phys. **39:** 2995–3018.

APPENDIX

Time-Asymmetric Spectral Decomposition of the Baker Transformation

The Brussels-Austin groups have shown[12,13] that there is also a new spectral decomposition of the Frobenius-Perron operator U of the Baker transformation involving Jordan blocks:

$$V = |F_{00}\rangle\langle \tilde{F}_{00}| + \sum_{\nu=1}^{\infty}\left\{\sum_{r=0}^{\infty}\frac{1}{2^\nu}|F_{\nu,r}\rangle\langle \tilde{F}_{\nu,r}|\sum_{r=0}^{\nu-1}|F_{\nu,r+1}\rangle\langle \tilde{F}_{\nu,r}|\right\}$$

The vectors $|F_{\nu,r}\rangle$ and $(f_{\nu,r}|$ form a Jordan basis

$$(f_{\nu,r}|V = \begin{cases} \dfrac{1}{2^\nu}\{\tilde{F}_{\nu,r}| + (\tilde{F}_{\nu,r}|\}, & r = 0, \ldots, \nu-1 \\ \dfrac{1}{2^\nu}\{\tilde{F}_{\nu,r}|, & r = \nu \end{cases}$$

(A.1)

$$V|F_{\nu,r}\rangle = \begin{cases} \dfrac{1}{2^\nu}\{F_{\nu,r}\rangle + |F_{\nu,r}\rangle\}, & r = 1, \ldots, \nu \\ \dfrac{1}{2^\nu}|F_{\nu,r}\rangle, & r = 0, \end{cases}$$

$$(\tilde{F}_{\nu,r}|F_{\nu',r'}\rangle = \delta_{\nu\nu'}\delta_{rr'}$$

$$\sum_{\nu=0}^{\infty}\sum_{r=0}^{\nu}|F_{\nu,r}\rangle(\tilde{F}_{\nu,r}| = I$$

The principal vectors $\tilde{F}_{\nu,j}$ and $F_{\nu j}$ are linear functionals over the spaces $L_x^2 \times P_y$ and $P_x \times L_y^2$, respectively. Here, L_x^2 is the space of square integrable functions in the x-variable, and P_y is the space of polynomials in the y-variable.

As a result, the spectral decomposition (Eq. A.1) defines an extension of the Frobenius-Perron operator to the space of functionals over the test functions $L_x^2 \times P_y$. However, only the positive powers U^n, $n = 1, 2, \ldots$, can be extended to the same space. The negative powers U^n, $n = -1, -2, -3, \ldots$, can be extended to the space of linear functionals over the test functions: $P_x \times L_y^2$. Therefore the original unitary group U^n, $n = \pm 1, \pm 2, \ldots$, when extended, splits[13] into two distinct semigroups corresponding to the forward and backward direction of time. The computation of autocorrelation functions using the spectral decomposition (Eq. A.1) gives rise to polynomial contributions to the exponentially decaying components associated with the resonances 2^{-n}, as a result of the Jordan blocks structure. Apart from the intrinsic irreversibility and the natural decomposition of the evolution in terms of the decaying contribution associated with resonances, we can also prove[13] that the trajectories $\delta_{(xy)}(x',y') = \delta((x',y') - (xy))$ are excluded from the domain of the spectral decomposition (Eq. A.1), that is, we cannot compute the deterministic evolution associated with the initial condition localized at the point (x, y) because the expectation value,

$$(U^n \delta | f) = \int_0^1 dx \int_0^1 dy \ \delta((x, y) - B^{-n}(x', y')) f(x'y')$$

diverges. Therefore, the irreversible extension of Baker's dynamics is moreover intrinsically probabilistic. It allows only for predictions associated with probabilities. This is also another generic characteristic of the extended spectral decompositions of unstable or nonintegrable systems.

Perspective and Metrics of Time

GIUSEPPE ARDRIZZO[a]

CIES, c/o University of Calabria, Arcavacata di Rende, Cosenza, Italy

In the biblical passage Daniel 13 (Greek version), better known as the story of "chaste Susannah," the story is told of a virtuous and beautiful woman, who was able to overcome the deception and lies of two elderly judges of the people, up to then considered wise by the entire Jewish community of Babylon. Blinded by passion for her, they "lost their temper, looked away so as not to see the Sky and and not to remember the just judgements" and made an infamous plan hiding themselves in the garden where Susanna used to take her bath, by herself.

The two elderly men put into effect a gesture that could be defined as paradigmatic for the repetitiveness with which it is used, even if unconsciously, in everyday investigation; it is a gesture silently guiding deeds that in common understanding seem not to be directly interested in the facts of ethical–moral nature that pervade the mentioned biblical passage: it is the gesture of "seeing but not being seen." Within the text, that gesture becomes possible because of the presence of three verbs in succession, whose reflective dimension implies getting lost, getting away, and forgetting. Three verbs marking an abandoning of the relational and communicative action between the man who acts and the world, between the man who acts and his way of being in the world. Being in a condition of pure subjectivity, all that can be seen is inevitably seen as an object, therefore inevitably defined and limited. When the wavering and the mechanism of the relational activity fails, one falls into the anomy of "seeing but not being seen," which may be considered the action of a self-referential thought, boundless and originating in the forgetfulness of a primeval value.

The episode of "chaste Susannah" can be used as allegorical bait in order to introduce another episode that is more revealing about the subject we intend to deal with. Between the two stories, the one we referred to and the one we are about to expound, there is no connexion as to the intentionalities of actors or contents; no analogy is possible as to the *fabula* — using a word taken from the theory of fiction — and yet there is something connecting both of them from a structural point of view. This something is "seeing but not being seen," which Filippo Brunelleschi, ingenious inventor and initiator of architecture, puts into effect by drawing two pictures showing the same number of perspectives of the built-up area of Florence: the former represents the baptistery of *S. Giovanni*, seen from the main entrance of the Cathedral; the latter represents the *Palazzo della Signoria*, seen from the Northwest corner of the square of the same name (FIGS. 1 and 2.). Afterwards, he makes a hole next to the vanishing point in each picture, and he puts them, one after the other, in front of a mirror. Then he places his eye next to the hole and sees, "not being seen," that the mirror reflects the represented images but in such a way as to confirm a postulate that guided him in the act of making them and that now gives him the illusion

---

[a]Address for telecomunication: Phone, 39/035 237352; fax, 39/035 231585; e-mail, munizzi@reteitalia.net

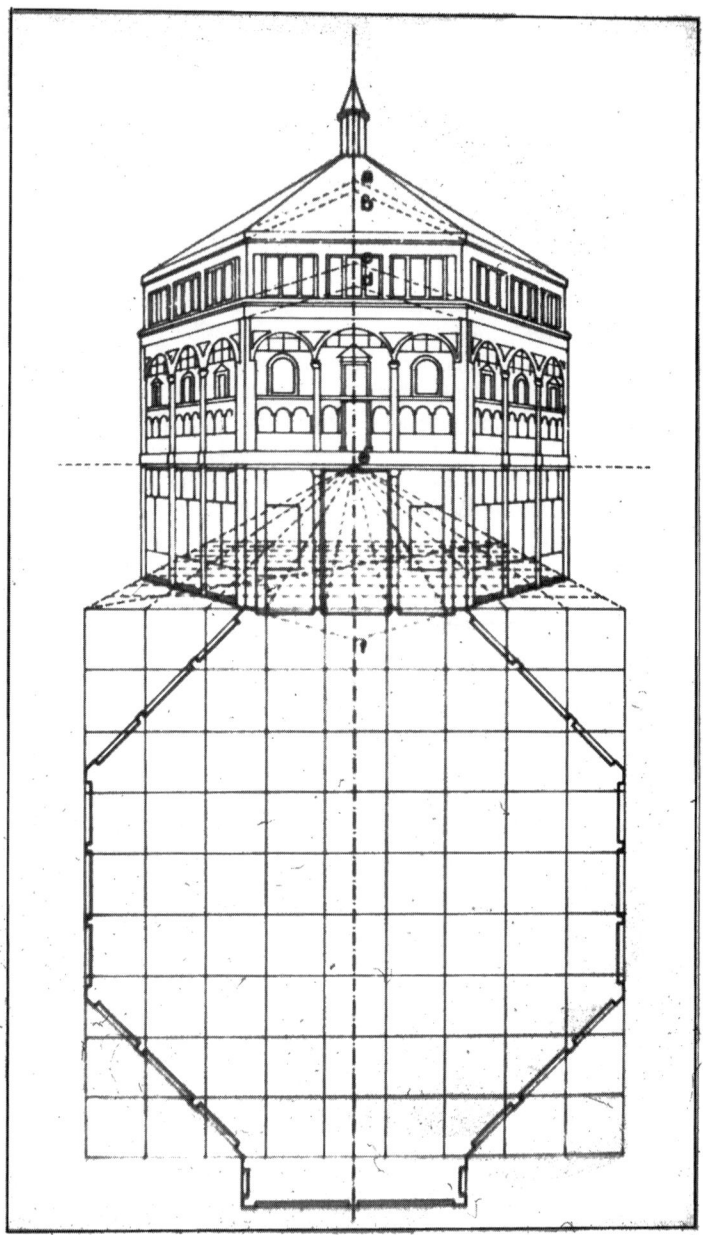

FIGURE 1. Florence. Baptistery of *San Giovanni*, view from the front door of the Cathedral. Reconstructive hypothesis of Filippo Brunelleschi's picture.

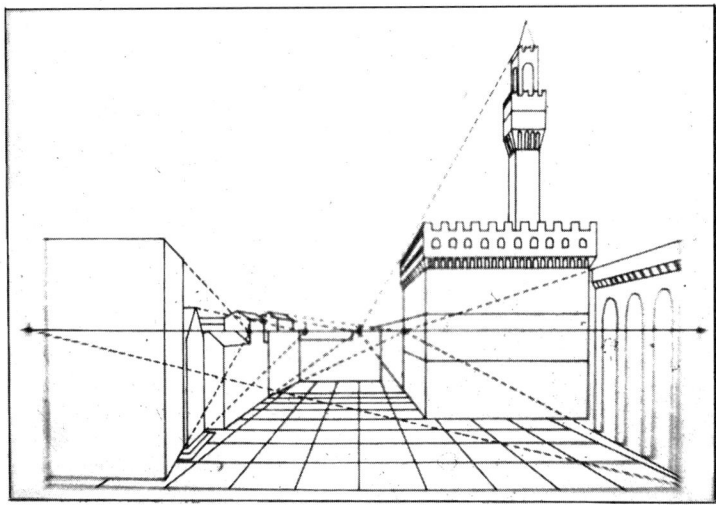

FIGURE 2. Florence. *Palazzo della Signoria,* view from the northwest corner of the square. Reconstructive hypothesis of Filippo Brunelleschi's picture.

of three-dimensionality on a two-dimensional plane, as if he were really looking at the two buildings he had used in order to accomplish his test.

Essentially, he seemed to have reached the truth in representing a systematical space, a "truth" consisting of straight lines producing orthogonal planes, a truth based on mathematics enabling him to define the rules of the perspective representation of an object whose position was spatially determined, with respect to the observer's point of view, according to the basic theorem that will later be part of descriptive geometry.

Therefore, that sense of reality (which before Giotto was sought through an empirical attitude and which expressed itself in the two-dimensionality of its perspective—the closest figures are wider, the furthest are smaller, with a very restricted depth between the twoo groups) now changed giving a theoretical apparatus to the representation (of reality). Briganti points out that "'Seeing' — according to Brunelleschi—is something scientific, corresponding to precise, geometrical, measurable laws. Therefore it is possible to represent reality on a plane, that is to say, to pass from three to two dimensions and to define in a geometrically exact way the ratios of proportion between the various perceived objects, no longer by intuition and approximately but by following mathematical laws."[1] In short, these are the conditions of knowledge from which the central perspective starts, the most doctrinal and involving form even if there are several variations on the theme and innumerable transgressions by the artists themselves. Space reveals itself as a unitary container, inside that objects take form according to a set of syntactical relations which have to find their own logical justification.

"The central perspective superimposes on the vertical and horizontal frontals a system of convergent rays creating a focal point and furnishing a whole range of corners."[2] "The three-dimensional image corresponding to it is an empty space structured as a framework, an endless network consisting of straight lines orthogonal among themselves. Inside it, all the relations between the elements forming it are regulated by Euclidean geometry. The squaring homogeneously extends in all three dimensions—height, width, depth—and can be compared to an immense grid where the position of every division of the framework cannot be individualized in an absolute sense, as on the contrary would be possible in a gigantic chessboard whose limits are known, but it can be individualized only when the position of a box is defined in relation to another."[3] Thus there seemed to be the conclusion of that long path of research towards the realism that the West had started long before Giotto and that now, in the early Renaissance, came to an end with a marked rationalistic tone.

Now, the eye will no longer have to move in order to read the piece of work from the beginning to the end, as happened before; on the contrary, the faculty of the eye will be that of "fixing" on the only point revealing the whole. In other words, it is like reading the organization of the state through the synthetic eye of the Machiavelli's prince, the only motor eye, being appointed and able to link rationally past, present, and future.

The perspective is also referred to as *system of perspective reduction*. Seldom has a concatenation of words appeared to be more effectively revealing: the use of two terms such as *system* and *reduction*—even if read in the technical intentionality defining them—reveals to us a lot about the itineraries undertaken in the context of knowledge. In particular, it is precisely in the act of reducing that such syntactical "virtue" comes out, which is why, about a point of view freely chosen, a representation is organized giving the illusion of coincidence between the represented image and the one given by the direct vision. The whole has to obey the laws of proportionality, otherwise "disharmonies" would be revealed indicating *illegalities*: a mistake in measurement would make the whole untrue. And actually, for a lot of artists of the early Renaissance, the search for perfection in the operation becomes more and more meticulous: a kind of mechanism of purification that binds itself to the abstraction of a law, causing a kind of short-circuit: the more one is faithful to the operation and the more rigorous this operation is, the more one is free in the moment of representing the subject. In the those years "the spatiality of Giotto and Duccio," Panofsky notes, "corresponding to the conceptions of transition of high scholasticism, was overcome through the gradual elaboration of the real central perspective with its space, endlessly extended and organized around a freely chosen point of view."[4] The choice of the central perspective operation agrees precisely with that act which—of course according to the artist—expresses better than others a vision of the world, in conformity with the axiom of a mathematical truth capable of revealing the enigma, finally able to grasp reality in its own objectivity, objectivity that is expressible as the search for a point near equilibrium. The act of establishing the point to infinity, that is to say, the vanishing point coinciding with the point of view, brought about the emergence of a single space-time scale involving in its linearity a directed and homogeneous time, capable, starting from that point, of establishing and seeing that future which ended into *escaton*, the end of times. The initial point of infinity looked directly at the final point of infinity. Therefore a time finally discovered as homogeneous and isochronous came to stand "side by side" with a homogeneous and isotropic Euclidean space.

The charming linearity that symbolically, but not more than this, had found confirmation of its own validity through the inaugural gesture of "seeing but not being seen" seemed to have removed what had been elaborated about vision up to that moment.

> If an epoch, whose vision was conditioned by a representation of space which expressed itself through a severe plane perspective, had to rediscover the curves of our visual world, that is to say spheroid, these curvatures were on the other hand completely obvious for an epoch accustomed to seeing according to perspective, but not according to plane pespective: that is to say classical antiquity. In the writings of ancient opticians and theorists of art ... we continually find of observations of this kind: the straight lines are seen like curves and the curved lines are seen as straight. ... So the basic attitude of ancient optics, which had worked out such notions, was even contrary to the plane perspective."[4]

The gesture disclosing an abstract coherence had therefore led to a kind of blindness, the same one that allowed us to use metaphorically the text of Daniel 13 and that lays down the conditions of "seeing but not being seen" in losing, distracting, not remembering. Therefore, when Brunelleschi looks through the hole made in the painting he gets lost, is distracted, doesn't remember. His relying on what the mirror reflects turns into abandoning himself to geometry, which allows him to determine infinity as interfaces of all the straight lines.

The reflection about *infinity* could boast one of the strongest and most fruitful genealogical trees. Lucretius, in *De rerum natura*, echoing in some way atomists, such as Plato, Democritus, and Leucippus, in their pursuing an initial infinitesimal point that could put an end to the boundless divisibility of continuity, yet Lucrezio took a stand in favour of an idea of a boundless universe (I, 958 e sg.).[5] Leaving aside in this context, a synoptic conception of infinity in the history of thought, we must note that the perspective of the early Renaissance seems to be decisive in inaugurating a new idea of infinity. Panofsky, referring to Cassier, considers the perspective one of those "symbolic forms" through which "a particular spiritual content is linked to a concrete sensitive sign closely identified with it."[4] In some way, infinity represented by the perspective anticipated what later would be formalized in a theoretical field by Cusano, by Giordano Bruno's philosophy of nature, up to Descartes, Liebnitz, and others. As Zellini notes,[6] it is certain that the preparation of a perspective as a symbolic form had remote roots "whereas the Grecian world was essentially discontinuous ..., Neoplatonic metaphysics, Byzantine art and Romanic and Gothic painting and sculpture already revealed a perception of space as continuous '*quantum*,' as homogeneous fluid in growing assonance with the figures emerging from it. The 'infinite' character of this space could not be ignored by the Greeks, but the importance of this infinity for figurative representation was entirely in the negative and material character of the *apeiron*, of infinity."

In coherence with the negative theology of Neoplatonic derivation, Cusano—in *La dotta ignoranza*—comes to the conclusion that, as regards the question about God, it seems to be more convenient to say what is not rather than what is. This is a theological reflection which not only protects against idolatry, but which possesses a term, the only one capable of defining God: "infinity."

Cusano's thought is echoed quite expressly in Giordano Bruno's Neoplatonic reflection about infinity; Bruno's main points of reference are Lucretius and Copernicus, as well as the theology of Cusano himself. Bruno somehow goes beyond Copernicus and shatters his idea of a finiteness of the universe in order to postulate an infinite universe, seat of countless worlds. The infinite universe is emanation of the One-God. The Aristotelian dichotomies between circular and rectilinear motion, between the incorruptibility of eternal places and corruptible heaviness of the earthly places vanish in Bruno. All this is to say that there is no longer a hierarchical scale of places, where one is better than the other, and therefore there are no longer points of reference, such as left and right, up and down. Everything finds its solution in metaphysical equality. And if Copernicus, as he turned everything upside-down had somehow saved a hierarchical vision between the sky and the earth without generating anxious shocks, since man always possesses the means of control in order to investigate the universe, now Bruno could affirm that raising one's eyes to the heaven did not give greater comfort than turning them to the earth.

The conception of infinity, starting from the central perspective of early Renaissance, takes the form of a sort of "boundless" space absolutely autonomous with respect to the objects it contains. And though "still" unattainable by reason, infinity is in some way perceived and expressed in the continuity and homogeneity of surfaces by resorting to mathematical thought. Descartes, later, will theorize this aspect by speaking of a spatial reality as *res extensa*. This will signify that man is inserted in a space environment unitedly and infinitely extended in all directions around him. A space formed like this, in the same way as the conception of infinity, is evidence of its complete extraneousness to man and to his possibility of experiencing. Psychophysiological space and time do not count any more as reference points, but the only reference is represented by the laws of Euclidean geometry. Bruno, by denying any system of absolute orientation, will say over a century and half later what many renaissance painters expressed in their representations. And Cassirer recalls the identities of the two positions by noting that, "In geometrical space, all the oppositions (left–right, up–down) are cancelled because the element, as such does not have a specific content, and all its meaning comes from the relative position that it occupies in the total system. The principle of the absolute homogeneity of spatial points eliminates every difference, such as the difference between up and down, which exclusively concerns the relation of things with our bodies, and thus with a particular object empirically given."[4]

Geometry and perception of reality become inseparable terms for several Renaissance artists, to the point of characterizing in some aspects the peculiarity of the Renaissance itself, at least from our point of observation, which is not that of aesthetics or of art criticism, but only that linked to the way of considering the construction of a textual reality, with respect to the conception of an inner time. From the relation between the geometry and perception of reality, the semantics of representation will derive reasons for organizing a corpus of knowledge, perhaps not totally written or systematized, but anyway such as to permeate the construction of an easily identifiable cultural model. The institution in that historical context of a relation of dependence between the above-mentioned terms will reveal, beneath the phenomenal manifestations, a geometry-of-reality as epiphany of a finally found stability.

The perspective pyramid—the one generating the central point of view—sets itself on the level of the sense as relational operator between its own vertex and the person who looks at the work of art. And in this establishing itself as an element of relation, it may comunicate to the observer a statement of this kind: "The point you are observing, the one given by the convergence of the lines, is the point from which everything begins or in which everything ends." This point, representable only as an implicit ideal, has to be deducible from a compositive structure that calling on the intuition of the observer to collaborate, allows him to see that point as *archè*. The impossibility of representing it in a more explicitly comprehensible reference does not mean its inexistence: its existence is asserted by faith and confirmed by geometry, or perhaps vice versa. On the other hand, if the adopted geometry is that which permits *fidelity*—at the moment of comparison between direct vision and its representation—why ever should it lie when proclaiming unquestionable the existence of an "initial" and a "final" point?

If all that, in the context of "seeing but not being seen," appears to be completely legitimate in the sense of a self-reference establishing an inner legality of its own, exactly as classical physics accepted as an axiom the chance of identifying trajectories once the initial conditions had been fixed, then establishing *a point at will* and making coherent a general argument referring to it in the representation of the universe means creating a real gesture of manipulation. Exactly as in analysis the term manipulation means leveling in a single dimension, to which another person is induced to adhere with a diminution in its subjectivity.

It is a gesture of *unseen* manipulation by the Renaissance treatises, that though speaking of arts of vision it remains some way "blind": it does not see the way as it sees the world.

Essentially the perspective, particularly the central perspective, appears as a device capable of exerting strong control. And this happens not only by respecting compositional syntax, which answers for the inner legality of the work, but also by fixing the user's modalities of vision. He is not free to stand in front of the painting: the painting itself fixes the point from which he *has to* observe. The right perception of the image, Francastel points out, always takes place along the axis and the sequence of the spaces suggested by the picture and irremovably fixed by it. In a painting by Masaccio or Piero della Francesca, the relation between the observer and the image is fixed once and for all by selecting only one among the infinite points of view at the observer's disposal.[7]

For many people, the meaning of control lies in placing the user at the center of the universe: the whole representation turns round his presence, according to a Ptolemaic coherence that becomes a metaphor for the Renaissance: everything is led to a centrality that must be characteristic of man. Things may be like this, but there could be another hypothesis that is perhaps more Keplerian, perhaps more Newtonian. If it is true that the user occupies a position pre-arranged by the work of art and that this work of art is based on a point of view "linking" him to that "infinite" point that is in the structure of the representation, then it is also possible to suppose that the infinite point is a motionless point, just as the representation is motionless in the fixity of the painting. Therefore the only element endowed with mobility becomes the user, who—on the symbolic level—can place himself periodically in front of the work of art and always at that point. In conclusion, the user becomes like a good planet turning round the sun in fixed orbits—speaking of Newtonian skies—giving the sun, in our case, the symbolic dimension of infinity, of the One-God. This recurrence, however, on the historical–liturgical level, was not completely symbolic or arbitrary: particular images of holy art were to be worshipped annually on a fixed day; and, if the representation dictated its own law of vision, they made themselves visible by being exactly at the point from which they had been seen the year before.

"With the Renaissance," Dalai points out, "the character of perspective treatises radically changes: ..., all the theorists of perspective by then mention phenomenal optics, that is to say the laws of 'perspectiva naturalis' [natural perspective], only as an introduction to treatises, whose basic task now consists rather in teaching rules and procedures of perspective reduction, illustrating them by using the greatest possible number of graphic examples."[8][b]

Leonardo, even if not in a systematic way, makes innumerable observations on perspective, nature, anatomy—to quote only the arguments most directly connected to the subject of these pages—observations that in part will be taken up again more organically by Dürer. "Don't read me if you are not a mathematician," Leonardo says. Essentially if painting does not contribute to the discovery of nature, it does not deserve to be approached and, on the other hand, the mathematical principles of perspective allow the latter to be the "bridle and rudder of the painting."

According to the maestro from Vinci, no inquiry held by man can become true science without being screened by mathematical demonstrations. This means, says Geymonat, speaking of Leonardo, that "for the first time in human history the technique–science dialetic has been put into effect with such awareness: both of them give each other aid, suggestions, reasons for serious meditation. ... From the methodological point of view, he can be considered a forerunner of Galileo, because of the basic importance given both to experience and to mathematics; in fact it is not to be excluded that Galileo, in elaborating his mathematical–experimental method, really was even if indirectly, under Leonardo's influence."[9] Despite the clash of opinions expressed by the scholars about Leonardo's procedures a piece of data emerges that we want to underline: for several renaissance artists, the mathematical synthesis of artistic construction appears capable of a very high cognitive power. Looking at the chronology, Brunelleschi and Leonardo seem to follow one another in a temporal continuity thats fully goes beyond the century.

Paolo Giovio, who lived at the same time as Leonardo, reports that the maestro invited the apprentices "to imitate by using very simple lines the power of nature and the contours of the bodies presenting themselves to our eyes with a great variety of motions ... in order to avoid painting in his studio something which was not in accordance with nature." And now we arrive at the moment of the mimetical gesture, a gesture that will constantly follow the artists who see perspective as the objective space of the representation.

From their theoretical reflections and pictorial way of working, it must be inferred that a Euclidean plot of the construction is the one that more than others permits a gesture of absolute fidelity with respect to something that has to be recognized. But Gadamer[10] comes to our aid, preventing us from falling into the same reductionisms; we wanted in some way evidence of other people's way of acting.

> "The cognitive sense of *mimesis,*" he says, "is recognition. .... What recognition is, in its deeper essence, cannot be made out if we simply observe that in it we have known again something which we have already known, that what is known is recognized. The

---

[b]This change, Dalai underlines again, goes "from Alberti, who is the first to codify the Renaissance 'basic construction,' to Piero della Francesca, from Leonardo, to Gaurico, from Jean Pélerin to Dürer, from Daniele Barbaro to Lomazzo, from Vignola to Serlio, up to Guidobaldo dal Monte ... ."

pleasure of recognition lies in the fact that in it we know *more* than we already knew. In recognizing, the known thing emerges if we can say so, like through a new enlightening, from the causality and variability of the conditions in which it is generally overwhelmed, and it is caught in its essence. It is known *as* something. .... Imitation and representation are not only repetition and copy, but knowledge of the essence. Since they are not only repetition but focussing on the thing, it is implicit that they refer to a spectator. They have in themselves a basic deferment to someone for whom the representation is made. ... However in the representation a recognition happens which has the character of a true representation of the essence, and this is founded precisely on the Platonic doctrines according to which every knowledge of the essence is recognition: that is why Aristotele could say that poetry is more philosophical than history ."

Despite the realistic power of the perspective model as a gnosiological instrument, in the sense that it gives the possibility of knowing objects in connection with the choice of logical models, the artistic representation can only exist in relation to an act of *recognition* not reducible to passive "repetition and copy." In any case, it is in this searching for the essence by being "true" to reality that a gesture of breaking with geometrical homogeneity occurs. And it is probably here that *time* can make its space within the piece of work, in the difference between objectified reality and sense of reality, between reality and essence, to refer to Gadamer.

It is strange how the conception of time, in the context of perspective representation, moves into the background, or is even hushed up, in order to give space an almost exclusive right. And yet the whole places itself in a historical period where time was held in great esteem. The humanists—who even went to extremes for many reasons—blame the Middle Ages for having lost the dimension of historical time, calling it "barbarous." From a medieval conception totally turned to the founder Text, which is a unique, exclusive source of knowledge and prompter of behavior—to express it in Lotmanian terms[11]—we passed to a more grammatical conception aimed at instituting bodies of rules that were also rules somehow suggesting ways of understanding time and space. Both of them contributed symptomatically to give the Renaissance a new depth, symbolically recalled in the new perspective dimension. The latter came to show that the distance covered up to this point had impressive, deep, and distant roots and that this course had rebuilt in man's awareness that the last stage worth recognizing had been precisely that of the "classical world."

"The XIV Century," Pomian notices,[12] "is the most important epoch in the history of time from antiquity to the beginning of our century. But not only because it saw the outlining of changes in the way of conceiving time: its importance depends on the fact that from then attitudes towards time, life and death, past and future, started to change." Still in the XIV century, Nicola d'Oresme, White informs us,[13] compares the universe with a big mechanical clock into which God, its Creator, has set such a motion as to make all its mechanisms rotate as harmoniously as possible. In this period, however, the time of the clocks cannot yet be said to be *quantitative time*, because—as Cipolla underlines[14]—their "fidelity" is often regulated by the position of the sun. And yet it seems a course already set towards increasingly sophisticated metrical precision, which will lead to the hand indicating the minutes in the XVI century and finally in the following century, after the discovery of the laws of the pendulum, to the indication of the seconds. Thus, thanks to scientific applications, clock time and quantitative time come to coincide: the clock is now autonomous, no longer exposed to external influences [cf. Ref. 12].

Essentially, the same way in which we designate the temporal marks, first *minute* and *second* minute, warns us about the fact that the reading of time is directed to the semantic field of the trifle. And that with the aim, entirely consistent with that of classical science, of identifying a prime element, finally capable of subduing *Cronos*.

Despite the brevity of this speech, it emerges how these conceptions of time were in course of elaboration—certainly in forms not yet substantial on the theoretical level—during the years most closely affecting the discovery of Euclidean geometry as a perspective device, that is to say, the period of the early Renaissance. Thus it seems difficult to think that faced with such a constant presence of *time*—presence that had interested humanists and merchants, philosophers of nature and craftsmen—it is precisely the artists who have nothing to do with it at the moment of representation. And to use once again the metaphor of "seeing but not being seen," is it not perhaps possible that Brunelleschi in comparing two geometries (the one represented in the painting and the other carried out in the building of the baptistery) forgot that both of them really presented themselves as two abstractions? If the reading given to perspective representation in the artistic field as being void of an inner temporality of its own, as being absolutely not interested in representing time is true, could not all this assume the meaning of a purifying mechanism that denying time, expects to give value of objective truth to what is represented, postulating as self-sufficient an atemporal space mathematically established?

But Gadamer's reflection about *mimesis*, as the reality of an image that is substantiated by the spectator's *re-cognizing* something, tells us something different. In short, we think that what is represented cannot prescind from the mimetic gesture also with regard to time, otherwise every recognition would become impraticable, making every narration impossible. The mimetic gesture itself would become impossible, if not at the risk of depriving precisely what we want to be realistic not only of an attribute but of realistic substantiality. We do not know how much this seeing the work without an inner time of its own depends on the artist's intentionality or on the way of seeing a critical heritage with remote origins. It is certain that, apart from some invitations operated by a few scholars, time inside the piece of work has always been ignored or dealt with using categorial systems confirming its absence, as if time could endanger that purity that brought us to see—and for many the idea is still valid—the work as a "pure" construction and "pure" organization. If the term intends to express that conception of purity constituting XIX century analytical chemistry—and which, incidentally, provided the safety of stability, we believe that nothing could be further from art, also from that which in the Renaissance went in search of geometrical constructions. And with regard to this we are nearer Heidegger in thinking that the work sets itself as "earth," where the "earth" is given not only by the elements that can represent it in plains, trees, animals, and so forth, but also as the same matter with which the work is realized, matter that by its very nature is not "calculable and measurable." Incalculability, not measurability, unforeseeability are also enriched by other elements, for example, the transgression operated by the artist towards his initial intentionalities; in this sense everybody knows that the pen guides the writing just as the brush guides the painting.

In an artistical context the inner time of the perspective, which on level of visions risks being expelled from the door of the geometry, on the level of reality re-enters by that window that opens wide over a radically evolutive world that forces on the artist a mimetic gesture towards reality, so that the work can justify itself and be called credible. Mimetic gesture, which perhaps the artist does not see but which he makes, since it is imposed on him by the ties of a representation that wants to appear as "true," and which can be so only in reaction to a basic "impurity," that same impurity that gives the earth fertility.

Resorting to the conceptions of time elaborated from the XVII century onward, it seems to emerge, as Pomian underlines,[12] that "scientists and philosophers did not wait for the XIX Century to face the problems posed by the introduction of quantitative time in temporal architecture." And perhaps, at least in part, this consideration can be expanded to the context of our speech, in order to verify if a conception of time or of times emerges from the perspective construction of art, and of early Renaissance art in particular, and to which reflections of the philosophers of nature it could be assimilable, always keeping in mind the specific nature of the two ways of knowledge. Ours is an attempt at inquiry, to be interpreted as an invitation to further various readings going back to conceptions of time nearer the contributions of contemporary science. And, in our opinion, art in the very ambiguity of its form can be capable of giving meaningful contributions to the ways of knowledge.

We have said that in the field of artistic representation the perspective construction appears as a pyramid identifying a symbolic "place" of infinity at the point of convergence of its straight lines. The construction of inner spaces consists of a series of wings—the spaces in depth of the so-called "transversals"—which with their following one another to infinity constitute a calculated geometric scanning of space. The user's position marks always and in any case the axis of the present: the work speaks to him whenever he places himself in front of it, according to the space-temporal mark of the *here* and *now*.

This means that the spectator's point of view is linked in a straight line to infinity and the latter seems, although implicitly, always irreversibly directed towards the present-spectator himself. On this line or in strict connection with it, the subject of the representation is placed. In holy subjects, which on the other hand characterize the vast majority of the Renaissance works, particularly the early Renaissance, the foreground character, the one who more directly comes into contact with the present-spectator, comes to show himself, thanks to his privileged location, as a projection of that point to infinity in no way perceptible and identifiable but in the ineffability of the person of God, eternal and creator. Thus, the symbolic point comes to 'cohabit' with the geometric point, generating that "impurity" which implies an unavoidable evolutive condition.[c]

The point to infinity, the foreground subject and the present-spectator—placed in the position indicated and required by the work—are on the same axis, the one given by a straight line which, starting from infinity, does not know discontinuity: to express it in Newtonian language, "it flows uniformly without any connection with anything external." The perpendiculars between the line of the point of view and the intersection of the so-called transversals lead to a rhythmical scanning that could make us think of a metric of time that has been finally found, even though in its still

---

[c]If the symbolic point comes to lie over the geometric point obscuring it, then we would be led in a one-directed way to a reading of a theological–finalistic kind; on the contrary, the reading would come down to the coherence of the geometric constructions. In this sense, the criticism, up to the middle of the XIX century, confines itself to judging the good quality of the work of art with regard to the inner coherences of the perspective virtuosities.

temporary impossibility of expressing itself in the result of a calculation. And this could recall a Newtonian "absolute, true, mathematical" time indirectly achievable, as Pomian says,[12] thanks to calculations, but also achievable in another way, that is by bringing it back to God, whose duration extends from past to future eternity, and whose presence goes from infinity to infinity; but it could also recall a Liebnizian time, as "order of the following Existences."[d]

Generally speaking, the subject of the work is placed in a context appearing as a sort of Aristotelian sublunary world, marked by a qualitative time, to express it like Pomian. The presence of nature, with its own diversities, shows us a temporal complexity of the surroundings clearly contrasting with the absolute time we have pointed out. Here space becomes temporized in evolutive terms; and nature expresses itself in the opening, according to a course moving to the direction of a nondefinable and unstable future, not symmetrical compared to the past. It is the context in which distances from the balances are registered, capable of generating new forms of order and new organizing facts. In this context, no linear vision becomes possible anymore, it is no longer a Newtonian concept, but a place suggesting we should build by building, in the presence of obstacles setting themselves as indispensable conditions for a strategic way of acting from which new solutions can be generated. Thus it is no longer the place of undifferentiated space and time, but the place where both of them directly participate in creating forms of life, according to relations implying a qualitative plurality of times and places. And this does not mean purifying the regularities, but taking note of the fact that the regularities exist and that they exist next to a more widespread disorder that they help to nourish and from which will develop bifurcations generating new, various forms of order.

On this subject, Andrea Mantegna's *San Sebastiano* (FIG. 3) (let's take this work as one of the various possible examples) can help us to test what we have said up to now. The saint martyred at the stake is in the foreground, in the center of the perspective lines, but it is not our intention to dwell on the axis of the absolute time along which he is placed, but on the context. Here we note that not everything rotates around the event (FIG. 4.). On the contrary, the detail of the background informs us about a daily way of life that seems completely unaware of or indifferent to the event: Some people are concentrating on carrying out their activities, others are speaking in pairs. They are all actions presuming qualitatively various times: the conversation time, the working time, and so forth. But all of them also presume various organizing moments, moments of order ready to dissolve in order to generate other moments of order. The noteworthy fact—an unambiguous sign of the direction of time and of the qualitative peculiarity of the temporal diversities—is supplied by the built-up area. Here the latest constructions are spatially postponed in comparison with the classical ones, which are older; but they are anticipated in comparison with the medieval ones, which are temporally intermediate, visible in the ruined castle.[e] This means that time passes making a series of contortions but all the same keeping faith with its very irreversibility. It "moves," going from the foreground construc-

---

[d]"Absolute or relational, Newtonian or Liebnizian, in both cases time is always placed as 'objective': reality in each of its parts or order in conformity with which things follow each other and which has been included in their own development. It is external datum and independent of the consciousness that individuals can have of it ...; the basis of its 'objectivity' is the divine duration or an idea of God."[12]

FIGURE 3. Andrea Mantegna, *San Sebastiano*. Louvre Museum, Paris.

FIGURE 4. Andrea Mantegna, Detail from *San Sebastiano*. Louvre Museum, Paris.

tions to those in the far background, then it passes to those in the intermediary ground, and finally reaches the present-spectator's ordering eye. Moreover, from the scene emerges the awareness that each of these areas shows itself with its qualitative times. This is a clear organizing sign that develops from bifurcations of history—the fall of the ancient world, the fall of the medieval world, and so forth—and that finds in obstacles of various nature the strategic grafting to realize a new order, the one actually shown.

The fact that the spectator takes his place in the point of view fixed by the work and his noting in this position the axis of the present triggers off a geometrical mechanism with regard to the point to infinity. In fact, seeing that point originating everything and finally knowing that he has the possibility to join it in a linear way, through some divine or human figure enjoying this privileged position in the representation,

---

[e]According to Renaissance thought, placing the classical building in the foreground expressed the epoch's esteem towards the Grecian–Roman world. Thus, putting the contemporary buildings in the shelter of the ruins came to mean a spiritual proximity too, a feeling unconnected to everything that could recall a despised Middle Ages.

he can also know what will be waiting for him. So, behind his back, he will have as symmetry the pyramid directed in the opposite sense, and its vertex will coincide with the *escaton*, the last event. The most famous representation of this moment is Michelangelo's *Giudizio Universale*, in the Sistine Chapel. Here, at the back of the judging Christ, there is nothing more because the new paradisical condition has finally been reached, according to the promise. The time of this linearity joining infinity to infinity is an absolute time. It is an irreversible time flowing from a *beginning* to an *end* and which for the believer already from the moment of its birth will assume the form of an oxymoron: its coming into the world will be the beginning of an irreversible return to the *Origin*.

This *time* continuing to flow uniformly in the linearity is ready to be denied by the spectator himself as soon as he moves from the privileged point, because in his, and the world's, constituting complexity he will find the impossibility of keeping faith with a condition of "purity," which moreover has not been required of him. The only form of *purity* resides in the "impurity" of a relating action that has to be expressed in a "seeing having seen."

In spite of this, the temptation to look at the past to predetermine the future in some form will make its way on to flow into the great flowering of philosophies of the history of the last century.

But already the Renaissance, in its maturity, will lose much of its optimism and certainties about the future. Times were changing course, and the past no longer seemed so sure, disclosed, such a holder of great references. On the contrary, after the second decade of the XVI century, turning back will mean seeing a past covered with ruins and mourning. Michelangelo will leave the *Pietà Rondanini* unfinished: on the other hand, which perspective could still style itself as credible to let the sculpture be finished? In which temporal dimension does it place? Caravaggio will be similarly explicit in obscuring the backgrounds of his works leaving a very small perspective depth. No longer a Renaissance man, he will not believe in the chance of finding geometrically a point to infinity as the source of *illumination*. In the *Chiamata di San Matteo*, the *light* will arrive at a slant, hitting a collector in the cumulative gesture of adding money to money, a linear gesture which, as it is, can no longer find reasons.

Once again time imposes different perspectives, a different reading of space. The *freely chosen point* of view, the same that being *"chosen freely"* seems to vehicular a promise of freedom, can really reveal a coercive cage from which it is possible to get out only by that *mimetic* gesture that also imposes fidelity in representing a plurality of times marked by their irreversible sorting out.

REFERENCES

1. BRIGANTI, G. 1988. La Repubblica. Masaccio e Piero. Ed. Roma, Italy
2. ARNHEIM, R. 1962. Arte e percezione visiva. Una nuova grammatica del vedere. Feltrinelli. Milano, Italy.
3. FURNARI, M. 1993. Atlante del Rinascimento. Electa. Napoli, Italy.
4. PANOFSKY, E. 1984. La Prospettiva Come "Forma Simbolica." Feltrinelli. Milano, Italy.
5. TITO, LUCREZIO C. De rerum natura.
6. ZELLINI, P. 1980. Breve storia dell'infinito. Adelphi. Milano, Italy.

7. FRANCASEL, P. 1957. Lo spazio figurativo dal Rinascimento al Cubismo. Einaudi. Torino, Italy.
8. DALAI, M. 1984. *In* La prospettiva come "forma simbolica." E. Panofsky, Ed. Feltrinelli. Milano, Italy.
9. GEYMONAT, L. 1975. Storia del Pensiero Filosofico e Scientifico. Vol. II. Garzanti. Milano, Italy.
10. GADAMER, H.G. 1986. Verità e Metodo. Bompiani. Milano, Italy.
11. LOTMAN, JU. & A. USPENSKIJ. 1975. Tipologia della Cultura. Bompiani. Milano, Italy.
12. POMIAN, K. 1981. Tempo/Temporalità. *In* Enciclopedia Einaudi. Vol. XIV. Einaudi. Torino, Italy.
13. WHITE, L. JR. 1967. Tecnica e Società nel Medioevo. Il Saggiatore. Milano, Italy.
14. CIPOLLA, C.M. 1981. Le macchine del tempo. L'Orologio e la Società, 1300–1700. Il Mulino. Bologna, Italy.

Complexity, Complex Systems, and Adaptation[a]

F.T. ARECCHI

Department of Physics, University of Firenze, and Istituto Nazionale di Ottica, Firenze, Italy

1. INTRODUCTION

A wealth of speculations has recently appeared on complexity and complex systems. The first term has been differently defined in formal languages,[1] computer science,[2] and nonlinear signal analysis,[3] starting in the early 1980s with the intrinsic nonpredictability associated with chaotic time series.[4]

Complexity is associated with epistemic processes. Once a time series of data, coded in a given alphabet, has been assigned, can one retrieve the meaning of the message just by perusal of that sequence? Is "meaning" just (i) discovering the grammatical rules that allow some symbol sequences (words) and forbid some other ones or also (ii) attributing a likelihood of occurrence to each word and hence attempting predictions about the future of the time series? This complexity approach has received different formulations, with different solutions leading to automatic procedures on complexity assignment.[5-9]

On the other hand, natural scientists have rather focussed on the fact that reality uncovers lots of complex structures. Even before any encoding into some alphabet and a consequent mathematical elaboration, any holistic perception of an event implies many possible, not equivalent, ways of describing it. The possibility of many irreducible points of view can be considered as the token of a complex system, independently from any quantitative indicator.

Two different tasks are then faced by investigators of (A) complexity and (B) complex systems problems, namely,

(A) Given an input, what is the optimal use we can make of it? We call "certitude" the subjective confidence that we have done the best in grasping the inner rules of the input.

(B) Facing a piece of world, can we express our knowledge of it in a suitable language, that is, encode phenomena into symbol sequences from some alphabet (which later will be analyzed as in (A))?

As we see, (B) is "prior" to (A). It appears as the problem that any living being has to solve, and it is usually faced by adaptive strategies, which we can later formalize as linguistic procedures, but which arise at a prelinguistic level and even determine the same choice of the most appropriate language.[10] This more fundamental problem is that of "truth" defined as *adaequatio intellectus et rei*, that is, "adjustment of our expectations to the changing world."

---

[a]Elaborated from F.T. Arecchi, "Truth and Certitude in the Scientific Language." *In* F. Schweitzer, Ed., *Self Organization of Complex Structures—From Individual to Collective Dynamics*, Gordon & Breach, London, 1996.

The two approaches are altogether different. In case (A) a learning machine can be foreseen that automatizes the quest for complexity.[7] On the other hand, case (B) hints at the crucial role of a prelinguistic stage where we still have to decide how to encode the stream of perceived phenomena into a linguistic sequence, which is then exposed to the inquiry of the complexity machine (A).

It is beyond our aims to tackle the philosophical problems involved in this distinction. We just limit ourselves to saying that doing science is (B), that is, making it is possible to encode our perceptions into a suitable language, not just building theoretical models to uncover rules and make predictions with regard to given sequences as (A). Our main conclusion is that while there may be a complexity machine for (A), it is in principle impossible to introduce a science machine for (B). Hence (B) remains a human endeavor not reducible to automatic procedures.

This contribution is organized as follows: Section 2 reviews current definitions of complexity showing the virtues and intrinsic limitations of a contextual analysis of a data stream (by "contextual" we mean that we must rely on correlations and symmetries already built in the symbol sequence without the power to modify it). Section 3 is a historical survey of the birth of modem science seen as the emergence of a formalized language out of prescientific observations already expressed in the ordinary language. Section 4 is a dynamic approach to a complex situation. Whereas in Section 2 no *a priori* rules are requested but a data stream is preliminary, in Section 4 we take for granted the general rules of dynamics and prospect a variety of possible data streams, that is, a variety of different physical situations. This variety appears as a natural implementation of the intuitive concept of complexity, as it is representative of what we call complex systems, from economics and sociology to biology and physics.

In Section 5 we present an adaptive strategy recently introduced to recognize[11] and control[12] a chaotic dynamics. Whereas an adaptive procedure was already incorporated in the complexity machine of Section 2, in order to fit the theory to the input sequence, here, in a more radical way, we change the same structure of the measuring apparatus M in order to provide different sets of measurement sequences to be later analyzed. We conclude in Section 6 with some epistemological insights, recalling that a knowledge program based on assigned input data is how to make the best use of our mental representations according to a subjective gnoseology started by Descartes and continued by Hume and Kant. On the contrary, a knowledge based on readjustment of our measuring procedures appears to be the most natural attitude of living beings. It seems more appropriate to the psychology of cognition as described in classical philosophy and as recovered by many naturalistic approaches, under the name of "evolutionary epistemology.[13a,b]

2. COMPLEXITY OF SYMBOLIC SEQUENCES

In computer science, we define as complexity of a word (symbol sequence) some indicator of the cost implied in generating that sequence. There is a "space" cost (length of the instruction stored in the computer memory) and a "time" cost (the CPU time for generating the final result out of some initial instruction).

A space complexity $C_1{}^2$ is defined as the length in bits of the minimal instruction that generates the wanted sequence. This indicator is *maximum* for a random number, since there is no compressed algorithm (that is, shorter than the number itself) to construct a random number. A time complexity C_2, called "logical depth" (Bennett in Ref. 5a) is defined as the CPU time to generate the sequence out of the *minimal* instruction. C_2 is minimal for a random number; indeed, once the instruction has stored all the digits, just command: PRINT IT. Of course, for simple dynamical systems such as a pendulum or the Newtonian two-body problem, both complexities are minimal.

An example of a large C_2 is offered by Wolfram cellular automaton n.86.[14] Whereas C_1 refers to the process of building a single item, C_2 corresponds to finding the properties of all possible outputs from a known source. Following Simon,[10] C_1 refers to a process description and C_2 to a state description. Indeed C_1 corresponds to the effort to arrive at a specific object, and C_2 corresponds to the mental representation of the whole class of objects, that is, for n.86, to all those site states that are necessary for the central site assignment after n time steps. In fact, the exact specification of the whole final outcome is too much for the ambition of the natural scientist, whose goal is more modest. It may be condensed into the two following items:

(i) to transmit some information, coded in a symbol sequence, to a receiver in a compact way, possibly economizing with respect to the actual string length, that is, making good use of the redundancies (this requires a preliminary study of the language style);

(ii) to predict a given span of future, that is, to assign with some likelihood a group of forthcoming symbols.

For this second goal, introduction of a probability measure is crucial[6] in order to design a complexity-machine, able to make the best informational use of a given data set. In view of the difference between (A) and (B) introduced in the beginning, let us sketch the essential elements of knowledge building in natural sciences. First of all, we realize a measuring apparatus M, whose output represents a manageable subset of what is going on in the observed piece of world. We then attribute to knowledge two different meanings:

(i) As we face a phenomenon, M captures (presumably) the relevant aspects of it, so that we can transfer sufficient information, either to another partner or to ourselves if we have to reflect in order to build a possible theory. Knowledge improvement implies trying with different M apparatuses by a suitable program that we will explore in Section 5.

(ii) As observer O_1 is exposed to a symbol stream, it has to transmit a compact explanation to O_2, so that O_2 is able to retrieve the same input sequence. The explanation consists of a tentative theory that we call model m.

The measuring apparatus M is characterized by the following elements[15]:

D: number of probes (dimensionality of the measurement space),
e: resolution in the projected state space (total number of cells is ε^{-D});
τ: time resolution;
$\beta = 1/T$ (T = noise temperature): fuzziness associated with a nonsharp boundary

of each resolution box, yielding some ambiguity in the assignment of an event to a specific cell of state space.

At any time slot of width τ, we extract ε^{-D} different space data that we can encode in a suitable one dimensional string of symbols of an alphabet (e.g., binary). The modeler O_1 is inputted by some sequence s, and it sends an explanation that should enable O_2 to reconstruct an output $s' = s$. Notice that $M = M(D, \varepsilon, \tau, \beta)$ is a whole class of possible instruments, and different individuals will give rise to different data sequences (different words). The explanation[15] consists of a theoretical guess (model in) whose validity is tested by simulating an output and comparing it with the actual input data s. The difference yields an error signal e. Observer O_2 is provided with both m and e and it can reconstruct $s' = s$ upon this information.

The virtue of the explanation X is to have a bit length $\|x\| = \|m\| + \|e\|$ much shorter than the sequence length $\|s\|$ this amounts to extracting a relevant semantics out of the redundant features of s.

The explanatory machine is complex in so far as it spans over a whole class of models in. If one had access to a complete probabilistic description of the modeling universe, then the goal would be to maximize the probability of m conditional to the input s

$$\Pr(m/s)$$

This ideal complete description is not available, but an approximation can be obtained by Bayes' rule

$$\Pr(s/m) = \frac{\Pr(s/m)\mathrm{P}(m)}{\Sigma_m \mathrm{P}(s/m)}$$

All these probabilities are conditioned on the choice of the model class. Furthemore all terms on the right-hand side refer to a *single* data stream s. Here $\Pr(s/m)$ is the probability that a model m produces the given data. With sufficient effort $\Pr(s/m)$ can be estimated. Finally, the normalization in the denominator depends only on the given data and so can be dropped as a constant.

The most likely explanation corresponds to the shortest code of length

$$\|x\| = -\log_2 \Pr(m/s)$$

There are two criteria for a good explanation:

(i) x must explain s, that is, O_2, should resynthesize the original data $s' = s$

(ii) the bit length $\|x\| = \|m\| + \|e\|$ must be minimized.

The efficiency of an explanation is given by the compression ratio

$$C(m, s) = \frac{\|x\|}{\|s\|} = \frac{\|m\| + \|e\|}{\|s\|}$$

C is a cost function. The optimal model minimizes this cost.

There are two limit cases. When the model is trivial ($\|m\| = 0$) the entire data are on the error channel: $\|e\| = \|s\|$. On the contrary a tautological model $m = s$ has no error: $\|e\| = 0$.

The reconstruction of a state space out of an assigned time series and the assignment of transition probabilities among states is a task that can be faced in many ways.[5,6] A suitable topological machine can be foreseen[15] that amounts to a labeled graph $G = \{V, E\}$ made of vertices v connected by labeled edges that assign a probability of going from vertex v to v' on observing a symbol s. Out of this labeled graph, suitable complexity indicators can be extracted.

According to the title of this section, here we have explored the complexity of a given symbolic sequence either from the computational point of view, aimed at reconstructing the single item, or from the probabilistic point of view, aimed at selecting a model within a class.

Preliminary to that, however, the problem arises of how we have obtained a given sequence, and this implies a critical analysis of the measurement apparatus. In the forthcoming section, we show how measuring apparatuses are suitable formalizations of the everyday knowledge expressed in the ordinary language, and we will provide an adaptive strategy to optimize the measurement performance in view of some specific goals.

3. FROM THE ORDINARY LANGUAGE TO THE SCIENTIFIC LANGUAGE

The word of the ordinary language is polysemic. In general it denotes a large variety of different situations, that we call "events," distributed on a "semantic space" (FIG. 1a). If the word is the name of an object, we usually do not mean the isolated object (which would be a mental abstraction) but the object embedded in different environments.

In a given language, the same word can be attributed with different degrees of appropriateness to different events (object plus environment). The different attributions have a different probability, as it emerges from a perusal of a historical dictionary providing for each word the frequencies of occurrence of different connotations in the literary texts of that language. In fact, the histogram is a finitistic approximation, a kind of coarse graining, due to the limited number of available texts.

If, however, we consider the everyday use, the continuous probability curve is more appropriate, since the environment includes the observer with his (her) own mood of the moment, hardly can be reduced to a countable number of states. Thus, whereas an artificial cognitive agent (a collection of detectors with fixed resolution feeding the input of a universal computer) would extract a histogram, thus justifying a finitistic approach, instead finitism seems excluded from the human everyday experience, as we reflect on the variety of nuances that qualifies a poem, or even a private conversation. Whence comes the problem of interpretation, that is, of what is the right meaning to be attributed to a word, within the wide support subtended by its probability distribution? In Indo-European languages, a quasi-univocal, or narrow range space of meaning g is obtained by a "filtering" operation, which consists of supplying the word with a sufficient number of attributions or specifications, as sketched in FIGURE 1a. A discourse, seen as a flow of different words connected by grammar rules, appears as a wide riverbed within which everybody can cut out a different interpretation (FIG. 1b). As well known, there is no unique sense of a given discourse, but one must refer to other sources of information, besides the text itself, in order to narrow the semantic range of each term.

Such ambiguities of the ordinary language were discussed at length by many Renaissance philologists and were well known to Galileo. He provided[16] a way out of the ambiguities through his suggestion of "naming" an event via the number extracted by a suitable measuring apparatus M applied to the event itself (FIG. 2a), that is, limiting himself to quantitative "affections" rather than attempting to grasp the "essence." This procedure apparently provided univocal meanings, since it filtered out

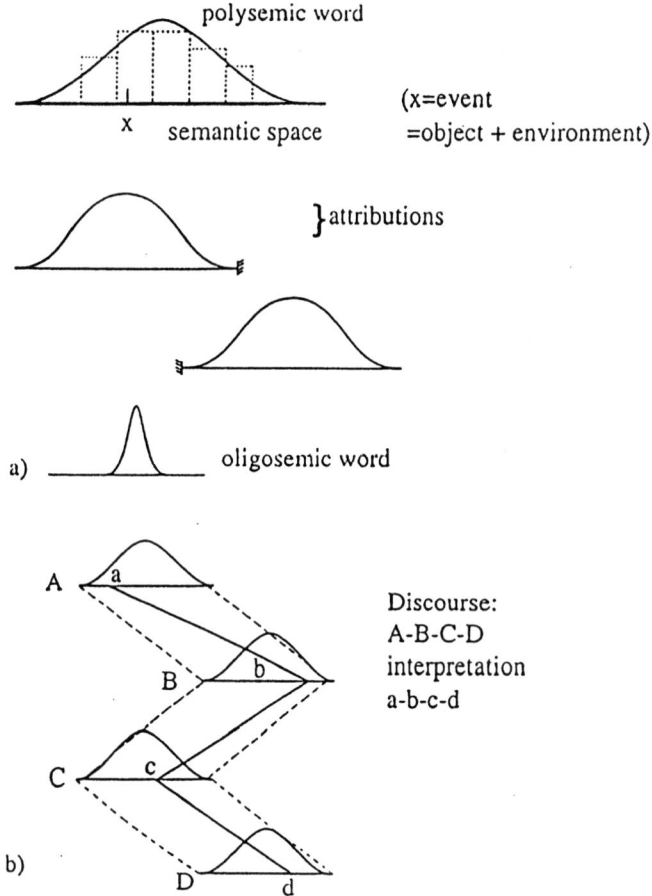

FIGURE 1. (a) Polysemic word represented as a probability curve over a semantic space. Constraining the word by further attributions narrows its semantic range, that is, reduces the probability spread on the semantic space ("oligo-semic" word). (b) A discourse, as the connection of different semantic spaces, yields a wider riverbed, allowing for different interpretations. (From the point of view of a linguist, this sketch is a rough caricature, since it reduces the semantic narrowing to a precontextual session).

a single denotation, clearing away all the context. As an example, a physicist does not speak of a "table" but of the "weight" or the "length" of the table. In this new language, the syntactical rule connecting two words becomes a mathematical relation connecting the output numbers from two measuring apparatuses related to two "objects."

This provides a solid framework for any scientific description, in terms of well-established existence uniqueness theorems. Thus, the flow of scientific discourse consists of sharp, necessary connections among pointlike objects of different semantic spaces, corresponding to different measurements as shown by the solid line in FIGURE 2b. That solid line seems to be a great progress compared to the wide riverbed of FIGURE 1b. It means that the scientific language is free from interpretational ambiguities. The most crucial aspect of Galileo's self limitation is the apparent arbitrariness in placing M over the semantic space, that is, the large number of dif-

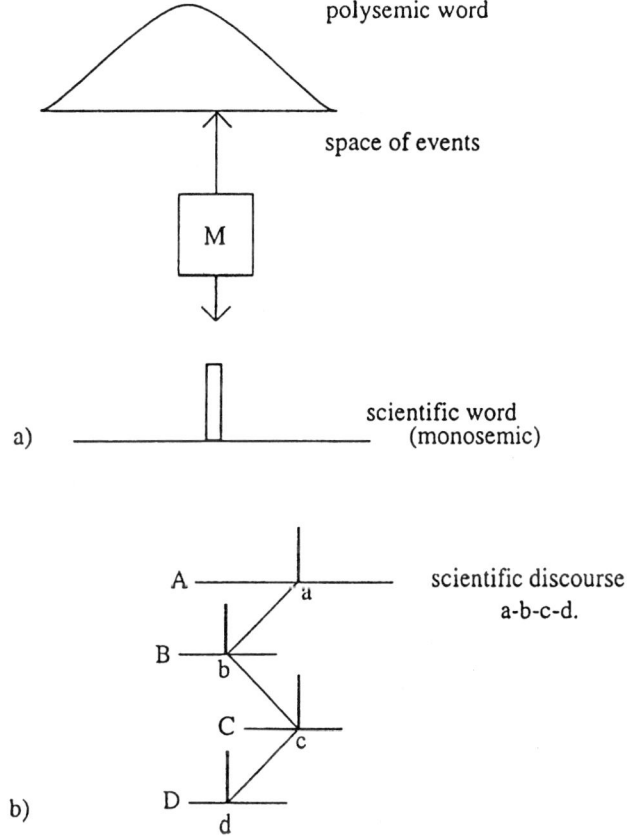

FIGURE 2. (a) A fixed measuring apparatus selects a univocal. denotation, attributing to the event a membership property to a suitable set. (b) A scientific discourse is a chain of necessary connections between univocal terms, thus leading to a unique pattern.

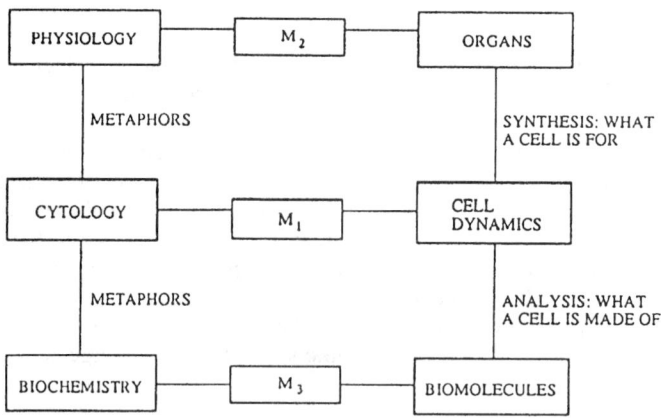

FIGURE 3. A living organ (e.g., a heart) is scientifically approached at different levels, depending on different apparatuses M_i, giving rise to different sciences. Within each science the univocal words are defined by the corresponding M_i. To communicate among different sciences, one must rely on metaphors. [M_i: measuring apparatus that characteristizes a specific science ($i = 1, 2, 3$).]

ferent univocal words extractable for a class of events usually denoted by a single word in an ordinary language. This proliferation of Ms has avoided for centuries the question of complexity. Suppose, for instance, (FIG. 3) that by M_1 we look at the cell dynamics. Thereby, we build the specific science of "cytology" with its own words We may now realize that many cells form an organ, but the organ is observed by a different M_2, which provides different words and hence a different science: "physiology." Similarly, if we look at the biomolecules we build a "biochemistry."

We have thus obeyed Occam's razor (*Entia non sunt multiplicanda praeter necessitatem*) in a economic way, that is, changing language whenever it is no longer appropriate. A more strict obedience would consist in a reductionistic approach. As shown in FIGURE 4, we can build a hierarchy from large to small and say that the behavior of smaller objects should determine that of larger ones. But here a perverse thing, already hinted by Anderson,[17] occurs. If our words were a global description of the object in *any* situation (as the philosophical "essences" in Galileo's letter) then, of course, knowledge of elementary particles would be sufficient to make predictions on animals and society. In fact Galileo's self-limitation. to some "affections" is sufficient for a limited description of the event, but only from a narrow point of view. Even though we believe that humans are made of atoms, the affections that we measure in atomic physics are insufficient to make predictions on human behavior.

The fact that higher levels in the ladder of FIGURE 4 display features not predictable from the lower ones is what colloquially we call complexity. This way complexity is not a property of things (like being red or hot), but it is a relation with our status of knowledge, and for modem science it emerges from Galileo's self-limitation.

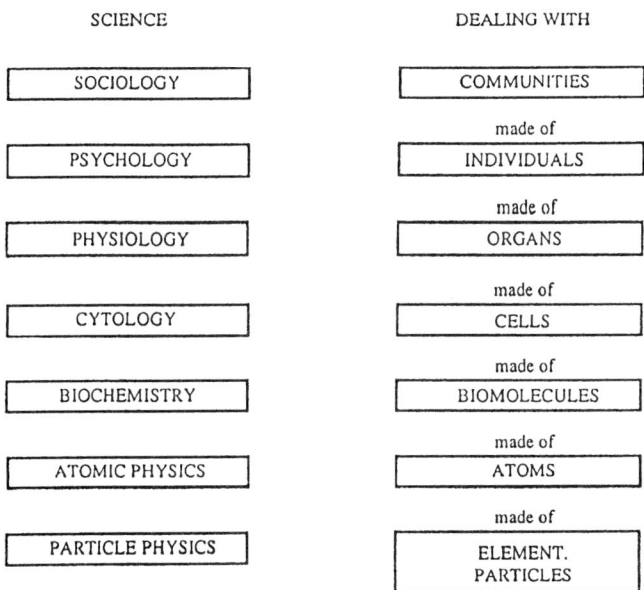

FIGURE 4. Hierarchy of sizes, showing how large objects are made of smaller ones. If the science of the small object explains the larger one, we have a logical reductionism. This occurs only in the absence of complexity (see next figure).

4. A DYNAMICAL APPROACH TO COMPLEXITY

A theory is successful if it is a "compressed" description of the world, that is, if the length of its initial assumption is much shorter than a detailed description of the events themselves. At the start, a physical theory is just mathematics. It becomes a model, that is, it acquires semantic values, whenever we interpret the objects of the theory as elements of reality.[18] Therefore, a scientific theory must be considered as a set of primitive concepts (defined by suitable measuring apparatuses such as M of FIGURE 2a) related by axioms. The deduction of all possible consequences (theorems) provides predictions that have to be compared with the observations. If the observations falsify the expectations, then one tries with different axioms.

The deductive process is affected by a Gödel undecidability like any formal theory, in the sense that it should be possible to build a well-formed statement, but the rules of deduction are unable to decide whether that statement is true or false. Besides that, a second drawback is represented by *intractability*, that is, by the exponential increase of possible outcomes among which we have to select the final state of a dynamic evolution. FIGURE 5a sketches a bifurcation tree well familiar to computer scientists, because one has to perform a complex calculation with branch points implying multiple choices of the type "if–then." I rather consider it to be the bifurcation tree of a complex nonlinear dynamics, as one changes a suitable control parameter α. Going back to the reductionistic tentative of explaining reality out of

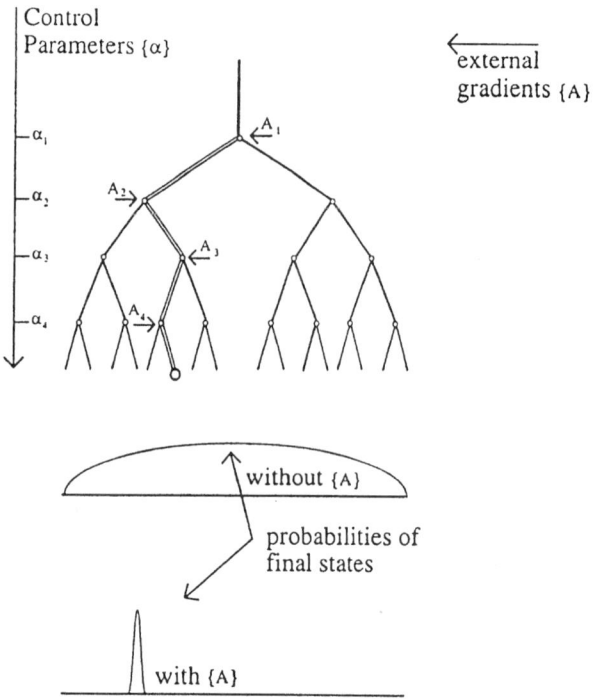

FIGURE 5. Bifurcation tree in nonlinear dynamics. As a control parameter α is tuned through different values, novel steady states appear. By tuning α from α_0 to α_N, the system goes from 1 to 2^N different states. We call dynamical complexity such an ambiguity. It is responsible for the failure of reductionism. In the absence of external gradients, all final outcomes are equally probable. We call "organization" the occurrence of just one event out of 2^N. This implies that at each α_i an external agent A_i has broken the bifurcation symmetry (see FIG. 6).

its constituents, then we find an exponentially high number of possible outcomes, when only one is in fact that observed. This means that, while the theory, that is the syntax, would give equal probability to all branches of the tree, in reality we observe an *organization process*, whereby only one final state has a high probability of occurrence. It is here necessary to recall some descriptive elements on the bifurcation of the stable branches of a dynamics for different settings of a control parameter. These are the necessary ingredients of any complex dynamics. Notice that dynamical bifurcations in a system of interactive identical particles display specific symmetries (FIG. 6a). Only external gradients break this symmetry (FIG. 6b). Thus, during the course of a dynamical evolution, either because some control parameters $\{\alpha\}$ are tuned from the outside to assume different values or because internal feedbacks change the $\{\alpha\}$ set in course of time, starting from some initial conditions we expect an exponential increase of final states. Whenever there has been organization, this means that at each bifurcation vertex of FIGURE 5 the symmetry was broken by an

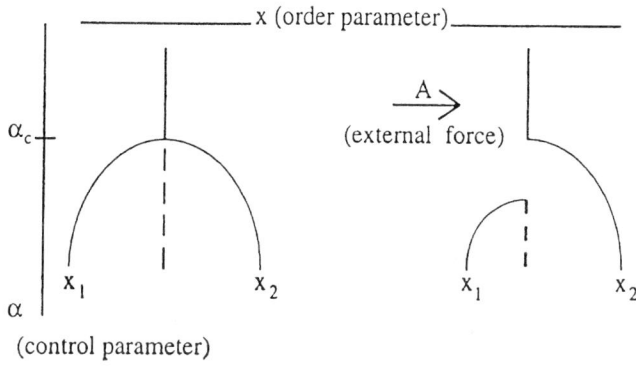

Probabilities $\alpha > \alpha_c$

$P(x_1)=P(x_2)=0.5$ $P(x_1|A)=0$ $P(x_2|A)=1$

FIGURE 6. Examples of bifurcation diagrams. The dynamical variables x (order parameter) varies horizontally, the control parameter o varies vertically. Solid (dashed) lines represent stable (unstable) steady states as the control parameter is changed. *Left*: symmetric bifurcation with equal probabilities for the two stable branches. *Right*: asymmetric bifurcation in presence of an external field A. If the gap introduced by A between right and left branch is wider than the range of thermal fluctuations at the transition point α_c then the right (left) branch has probability 1(0).

agent external to the system under investigation. We can thus stipulate the following things:

(i) A set of control parameters

$$\alpha_1, \alpha_2, ..., \alpha_N = \{\alpha\}$$

is responsible for successive bifurcations leading to an exponentially high number (in the example, of the order of 2^N) of final outcomes. If the system has no boundary effects (considered of infinite size), then all outcomes have comparable probabilities, and we call complexity the impossibility of predicting which one is the state we will observe at the end of the chain of bifurcations.

(ii) A set of external forces

$$A_1, A_2, ... A_N = \{A\}$$

applied at each bifurcation point breaks the symmetries, biasing toward a specific choice and eventually leading to a unique final state.

We are in the presence of a conflict between (i) *syntax* represented by the set of rules (axioms) $\{\alpha\}$ and (ii) semantics represented by the intervening g external agents $\{\alpha\}$. The syntax provides 2^N legal outcomes. But if the system is open to the external world, the presence of which is expressed by $\{A\}$, then it organizes to a unique final outcome. Once the syntax $\{\alpha\}$ is known, evidence of an unique final result implies that the set of external events $\{A\}$ must have occurred. Therefore we can take $\{A\}$ as the element of reality in which our system is embedded. We define "certitude" the correct application of the rules $\{\alpha\}$ and "truth" as the combination $\{\alpha, A\}$ of those *a priori* rules with external influences $\{A\}$ that perform the choices. However, the same final outcome would be reached by a different set of rules $\{\beta\}$. In such a case, retracing back the new tree of bifurcations, we would reconstruct a set $\{\beta\}$ of external agents. Thus, it seems that truth, $\{\alpha, A\}$ or $\{\beta, B\}$, is language dependent. Furthermore, the "emergence" of organization means that we can even build a set of axioms $\{\varepsilon\}$ that succeeds in predicting the correct final state without external perturbations, that is, $\{E\} = \emptyset$ (FIG. 7). This is indeed the pretension of the so called "autopoiesis," or "self-organization,"[19] to which I have opposed the term "heteroorganization."[20]

From a cognitive point of view, the theory $\{\varepsilon\}$ can be reputed to be a "petitio principi," a tricky formulation tailored for a specific purpose and not applicable to slightly different situations. Rather than explicitly listing the elements of reality, as for example $\{A\}$ for $\{\alpha\}$, the user of language $\{\varepsilon\}$ has already exploited at a preformalized level the elements of reality, and has made good use of them in planing the axioms $\{\varepsilon\}$. From a cognitive point of view, an *ad hoc* model may be appropriate for a specific situation, but in general it lacks sufficient breadth to be considered as a general theory. However, in describing the adaptive strategy of a living species, or a community, or so forth, a "self-organization" may be the most successful action. In other words, once the environmental influences have been known, better to incorporate this knowledge in the model from the beginning, this assuming the fast convergence to a given goal. These pieces of knowledge which precede axiomatization have received different names, such as "abduction"[21] or "tacit dimension."[22] Some of them have been memorized as universal tools either in our genetic heritage or dur-

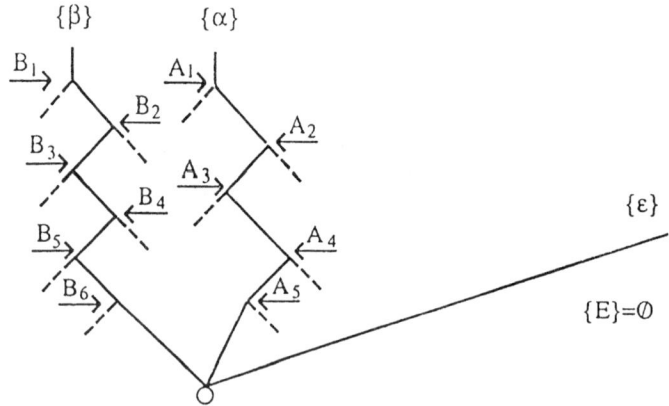

FIGURE 7. Different theoretical models may explain the same final state. The backward reconstruction of the dynamical history will then retrieve different classes of external agents.

ing infancy in our learning age. This seems to be the cognitive value of Jung's "archetypes."[23]

Each one of the separate bifurcation trees, with its collection of external perturbations pruning most of the tree branches, is a point of view on the word or, using terminology introduced in the psychology of cognition,[24,25] a "schema."

5. AN ADAPTIVE MEASUREMENT PROCEDURE

In Section 3 we introduced a measuring apparatus $M = M(D, \varepsilon, \tau, \beta)$ that can be modified by changing the number of probes D, space (ε) or time (τ) resolution, or fuzziness (β). Any change in M leads to a different set of output numbers, and hence to different sequences s (words). We can tailor the M characteristics in order to emphasize a given set of dynamical properties and project out some other ones. Since M acts as a projector into the subspace of the variables measured by M itself, chang-

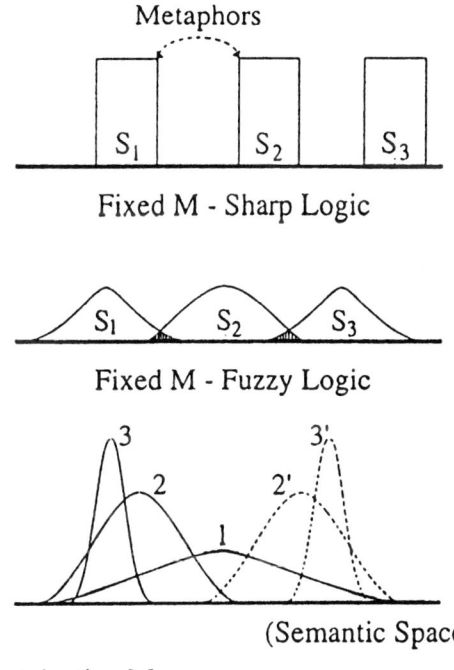

FIGURE 8. Fixed measuring apparatuses giving rise to different sciences. Inter communication is based on metaphors (*top*) unless by fuzzy logic one accepts ill-defined terms with overlaps (*center*). When the measurement is no longer locked to a fixed position of semantic space, nor to a fixed resolution, then the scientific knowledge can cover, with different degrees of detail, different areas of the semantic space, allowing information exchange within a unique language.

ing M means changing the "point of view" under which we observe the world and hence making a different theoretical model.

In the previous section, we show an indeterminacy in the reconstruction of the elements of reality $\{A, B, C, \ldots\}$ that modify the dynamical $\{\alpha, \beta, \gamma, \ldots\}$ of a mentally isolated system. As a result, the truth is represented by a combination of a model developed for the subsystem, plus the external gradients, or boundary conditions, once the subsystem is embedded in a suitable environment.

Which one is the most appropriate among the pairs

$$\{\alpha, A\}, \{\beta, B\}, \{\gamma, C\}, \text{etc.}?$$

The question is equivalent to asking: Among all possible measuring apparatuses M applicable to an event, which one is the most appropriate? If we prefer not to decide, we independently make use of different Ms and correspondingly define different sciences (FIG. 3).

In real life, we have to face problems overlapping different separate sciences. For instance, a cardiac disease may be due to a global offset of the pacemaker or to some local cytopatology or even to a drug effect acting at a biomolecular level.

FIGURE 8 shows the difference between fixed and adaptive M. In the first case we sharply define three separate sciences. What can be exchanged among different specialists is not technical words, which are specific to each science, but just the residual metaphorical part, not filtered into the technical word. Should we say that two scientists of different areas always communicate by metaphors? A tentative way out (fuzzy logic[27]) is to avoid sharp definitions, so that different disciplinary terms have regions of overlap.

The most natural approach, however, seems to start with M at very low resolution, covering all the disciplinary areas, and then—depending on a specific problem—to zoom toward one of the other narrow points of view. A successful line of adaptive measurement has been worked out in my group.[11,12] It consists in developing a measuring procedure M whereby we observe a system only at "almost equal" geometric separations in state space. As a consequence, M is activated only for short time intervals at irregularly distributed times, separated from each other by unequal intervals. The stroboscopic sequence of these intervals has an information content that provides fast, reliable answers to the following questions:

1. recognizing chaotic dynamics (Lyapunov exponents, different unstable periodic orbits (UPOs);
2. discriminating determinism from stochastic noise;
3. controlling chaos, that is, stabilizing one of the UPOs contained within a chaotic attractor.

When applied to the control of chaos,[12] this adaptive algorithm is effective for values of the stroboscopic times much larger than the Runge-Kutta integration steps and smaller that the periods of all UPOs. In other words, the method introduces a natural adaptation time scale that is intermediate between the minimum resolution time of the dynamics and the time scale of the periodic orbits.

The effectiveness of this adaptive method is rooted upon a drastic simplification. Rather than looking at all degrees of freedom of a D-dimensional object, that is, to all the information contained in the invariant set where the motion is confined, this approach extracts only the information contained in a one-dimensional string of data, namely, the stroboscopic series of observation time intervals $\{\tau\}$. A lesson seems to emerge from this example, that is, any successful way of approaching the world relies on reducing the complexity by limiting oneself to a simplified description. This seems peculiar of any scientific description that is by no means holistic, but, sharply confined to a particular point of view.

6. CONCLUSION

In this conclusion we discuss the truth value of scientific statements on the basis of the considerations of the previous sections. In the scientific investigation, we select a quantitative feature by application of an apparatus M at a particular point of the semantic space. The emerging description of reality represents an observation "from one point of view." Due, however, to the variety of possible Ms, we must justify at a metascientific level why we have selected that M rather than another one. This is a general question dealing with the role of those elements of reality which are preliminary to one particular program. In Section 5 we have seen that adaptation is a slalom among different sets or rules $\{\alpha\}$, $\{\beta\}$, ..., under the guidance of a preferential set of external elements that has nonzero intersections with $\{A\}$, $\{B\}$, ..., but does not coincide with either of them.

In the face of the truth problem we can take three attitudes, namely,

(i) Assume the adaptive strategy and its associated reality set as a kind of privileged reference frame. Indeed, being the result of an optimization process, it appears more appropriate than any particular theory, $\{\alpha\}$ or $\{\beta\}$.

(ii) Consider the truth problem as a metatheoretical problem. At this metalevel, the set of all sets of truth values $\{A\}$, $\{B\}$, ..., has to be considered as the truth, but, with the stipulation that any individual set makes sense only if associated with the corresponding theory.

(iii) A more fundamental approach recovers the polysemicity of the ordinary language as a virtue, not a drawback. More than questioning the power of any specific theory we put into question the same set-theoretical approach to the fundamental concepts of the physical description. Going back to FIGURE 2, we have seen that M provides a sharp connotation that allows to classify any observed entity within all appropriate sets. Hence the set-theoretical character of all modern sciences, with the consequent antinomies of modern logic after Cantor, Russel, Gödel, and so forth transferred into the heart of the scientific language.

An adaptive M means that the localization in semantic space is no longer as sharp as whenever it is defined by a precise stipulation as for the *sets*. This degree of smoothness seems to me to be going back to the polysernicity of ordinary language. Hence Epimenides Cretan says, "all Cretans are liars" is no longer an antinomy, since Epimenides is not bound to be always a liar, but a liar in general, even though sometimes even he can tell the truth!

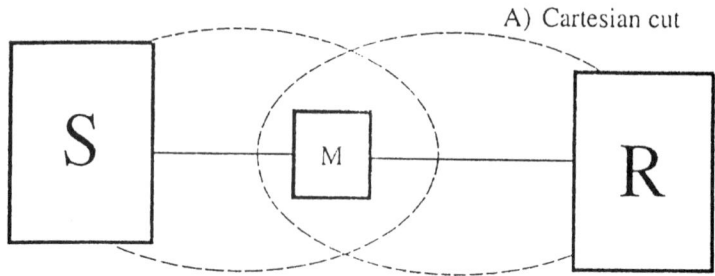

FIGURE 9. Knowledge interpretation: R = reality M = symbol generator (measuring apparatus), S = symbol interpreter (model builder). Dashed line A (cartesian cut) $M + R$ provide representations as symbol sequences, which are interpreted by S. S can be replaced by Turing machine. Dashed line B (realistic approach): $S + M$ globally face R. Before producing outputs, M is readjusted among a class of possible measuring apparatuses by a prelinguistic procedure not expressible within the formal language which later M provides to S.

Going back to the title, the truth versus certitude issue can be summarized by the following scheme (FIG. 9). R stands for reality (whatever this means), S for a symbol interpreter (an intelligent being or even a Turing machine!), and M is the measuring apparatus.

In modern science, M is usually not questioned, and the elaboration takes place on the output of M. This was called as the (A) *complexity* approach, leading eventually to certitude, when the S machine has optimized the explanation necessary to retrieve the M output sequence, according to the discussion of Section 2. From a gnoseological point of view, if the M box corresponds to our senses, as suggested by Hume, then S (called by Descartes *res cogitans*) has to face *not the world, but the representation* already coded by M, and this is a grammatical problem solvable by a machine. This means that Descartes' mind is equivalent to a Turing machine, as already suspected by many experts of artificial intelligence. This strong association of M on the side of R is equivalent to what Atmanspacher called "the Cartesian cut."[28]

On the contrary, the (B) *complex system* approach regards world's knowledge as an endeavor globally faced by the observer $(S + M)$ through an adaptive procedure M, for which, however, a linguistic foundation does not exist, because any linguistic formulation is subsequent to the operation of M. During the scientific operation, $(S + M)$ acts in an *entangled way*. A meta-level of investigation (psychology of cognition) is required to disentangle S from M. In summary, there is a nonlinguistic residue in the scientific operation which then precludes a Turing machine from acting as a creative scientist.

ADDENDUM

The connection from M to S in FIGURE 9, rather than being considered as unidirectional, must be taken as a feedback loop that provides sensory messages $M S$

renormalized by the mental expectations $S\,M$, and not just bare sensations, the way they would be tested in a laboratory session of a behaviorist psychology department. A hot issue debated by early sociologists (Marx, Durkheim) is whether science means knowing the world or trying to modify it. In fact, if the link from R to M is also considered a feedback loop, that would mean that reality R is continuously modified during the interaction with the observer $(S + M)$, and hence only the global $(R + M + S)$ entangled system should be considered. This would happen if, for instance, a smart theoretician had discovered the dynamical rules of the stock market and, rather than writing a paper on it, he/she tries to take advantage of that knowledge by playing in the market. As a consequence, the R situation just observed gets modified, and one must give up on an "objective"(i.e., observer-independent) description. In fact, this seems to be the general strategy of living species or communities, seen as adaptive systems.

Finally, the above-mentioned entanglement problems play a role in microscopic quantum mechanics, but these are still open problems, which represent an intellectual challenge, even though decoherence due to the environment destroys the problems in most practical cases.[29]

REFERENCES

1. HOPCROFT, J.E. & J.D. ULLMAN. 1979. Introduction to Automata Theory, Languages and Computation. Addison-Wesley. Reading, MA.
2. (a) KOLMOGOROV, A.N. 1965. Three approaches to the quantitative definition of information. Probl. Inform. Trans. **1**: 4. (b) CHAITIN G.J. 1966. On the length of programs for computing binary sequences. J. Assoc. Comp. Math. **13**: 547.
3. (a) GRASSBERGER, P., T. SCHREIBER & C. SCHAFFRATH, 1991. Nonlinear time sequences analysis. Int. J. Bif. Chaos **1**: 521. (b) ABARBANEL. H.D.I., R. BROWN, J.J. SIDOROWICH & L.S. TSIMRING. 1993. The analysis of observed chaotic data in physical systems. Rev. Mod. Phys. **65**: 1331.
4. SHAW, R. 1981. Strange attractors, chaotic behavior, and information flow. Z. Naturforsch. **36a**: 80.
5. See the following contributions in the Series of Santa Fe Institute Publications Proceedings Volumes: (a) DAVID PINES, Ed. 1987. Emerging Syntheses in Science. Vol. 1. (b) W.H. ZUREK, Ed. 1990. Complexity, Entropy and the Physics of Information. Vol. 8. (c) M. CASDAGLI & S. EUBANK, Eds. 1992. Nonlinear Modeling and Forecasting. Vol. 12.
6. GRASSBERGER, P. 1986. Toward a quantitative theory of self-generated complexity. Int. J. Theor. Phys. **25**: 907.
7. CRUTCHFIELD, J.P. & K. YOUNG. 1989. Inferring statistical complexity. Phys. Reu. Lett. **63**: 105.
8. D'ALESSANDRO, G. & A. POLITI. 1990. Hierarchical approach to complexity with applications to dynamic systems. Phys. Rev. Lett. **64**: 1609.
9. For an updated critical review, see: BADII, R. & A. POLITI. 1996. Complexity (especially Chapters 8 and 9). Cambridge University Press. Cambridge, England.
10. SIMON, H. 1982. The architecture of complexity. Proc. Am. Philos. Soc. **106**: 467.
11. ARECCHI, F.T., G.F. BASTI, S. BOCCALETTI & A. PERRONE. 1994. Adaptive recognition of a chaotic dynamics. Europhys. Lett. **26**: 327.
12. BOCCALETTI, S. & F.T. ARECCHI. 1995. Adaptive control of chaos. Europhys. Lett. **31**: 127.
13. (a) RIEDL, R. 1979. Order in Living Organisms. Wiley. New York. (b) WUKETITS, F.M. 1990. Evolutionary Epistemology and its Implications for Humankind. State University of New York Press. Albany, NY.

14. WOLFRAM, S. 1986. Theory and Application of Cellular Automata. World Scientific. Singapore.
15. CRUTCHFIELD, J.P. 1992. Semantics and thermodynamics. *In* Nonlinear Modeling and Forecasting. M. CASDAGLI & S. EUBANK, Eds. Santa Fe Institute. Santa Fe, NM.
16. GALILEI, G. Letter to M. Velser on the sun spots. 1 December 1612. Italian original in: 1932. Opere di G. Galilei. Firenze **5:** 186 ff.
17. ANDERSON, P.W. 1972. More is different. Science **177:** 393.
18. TARSKI, A. 1956. The concept of truth in formalized languages. *In* Logic, Semantics, Mathematics: Papers 1923–1938 by A. Tarski. Clarendon Press. Oxford, England.
19. KROHN, W., G. KUPPERS & H. NOWOTNY. 1990. Selforganization, Portrait of a Scientific Evolution. Kluwer Academic Publishers. Dordrecht, the Netherlands.
20. ARECCHI, F.T. 1992. A Critical Approach to Complexity and Self-Organization. La nuova Critica III, Quaderno 19–20, 7–39.
21. PEIRCE, C.S. 1931–1935. Collected papers. Vols I–VI. C. Harshorne & P. Weiss, Eds. Harvard University Press. Cambridge, MA. See also: 1958. Vols. VII and VIII. A.W. Burks, Ed. Harvard University Press. Cambridge, MA.; W.R. Hanson. Patterns of Discovery: An Inquiry into the Conceptual Foundations of Science. Cambridge University Press. Cambridge, England.
22. POLANYII, M. 1958. Personal Knowledge: Towards a Post-Critical Philosophy. Routledge & Kegan Paul. London.
23. VON FRANZ, M.L. 1988. Psyche und Materie. Eranos. Einsiedeln (C.H.), Switzerland.
24. RUMELHART, D.E. & J.L. MCCLELLAND. 1987. Parallel Distributed Processing. MIT Press. Cambridge, MA.
25. GELL-MANN, M. 1994. The Quark and the Jaguar. Little Brown. London.
26. MITCHELL WALDROP, M. 1992. Complexity. Simon & Schuster. New York.
27. ZADEH, L.A. 1987. Fuzzy Sets and Applications. Selected papers. J. Wiley. New York.
28. ATMANSPACHER, H. 1994. Complexity and meaning as a bridge across the cartesian cut. J. Consciousness Stud. **1:** 168.
29. OMNES, R. 1994. The Interpretation of Quantum Mechanics. Princeton University Press. Princeton, NJ.
30. SLOTINE, J.J.E. & LI WEIPING. 1991. Applied System Control. Prentice Hall. Englewood Cliffs, NJ.
31. GROSSBERG, S. 1988. Nonlinear neural networks: Principles, mechanics and architecture. Neural Netw. **1:** 17.
32. LORENZ, E.N. 1963. Deterministic nonperiodic flow. J. Atmos. Sci. **20:** 130.
33. ROESSLER, O.E. 1976. An equation for continuous chaos. Phys. Lett. **57A:** 397.
34. ROESSLER, O.E. 1979. An Equation for Hyperchaos. Phys. Lett. **71A:** 155.
35. MACKEY, M.C. & L. GLASS. 1977. Oscillation and chaos in physiological control systems. Science **197:** 287.
36. GRASSBERGER, P. & L. PROCACCIA. 1983. Characterization of strange attractors. Phys. Rev. Lett. **50:** 346.
37. BADII, R. & A. POLITI. 1984. Hausdorff dimension and uniformity factor of strange attractors. Phys. Rev. Lett. **52:** 1661.
38. KAPLAN, D.T. & L. GLASS. 1992. Direct test for determinism in a time series. Phys. Rev. Lett. **68:** 427.
39. WAYLAND, R., D. BROMLEY, D. PICKETT & A. PASSAMANTE. 1993. Recognizing determinism in a time series. Phys. Rev. Lett. **70:** 580.

Complexity and the Unfinished Nature of Human Evolution

GIANLUCA BOCCHI[a] AND MAURO CERUTI[b]

[a]*CERCO (Centro di Ricerca sulla Complessità), Università di Bergamo, Bergamo, Italy*
[b]*Dipartimento di Epistemologia ed Ermeneutica della Formazione, Università di Milano, Milano, Italy*

THE CHALLENGE OF DIVERSITY: THE IRON AGE OF THE PLANETARY ERA

In Europe and the Western world, the year 1492 is remembered for the discovery of America and the dawn of the modern age. However, the meeting of the histories of the European and American continents represented by that date is really the discovery of the Earth. It marks the threshold of the modern age of Europe, but above all the start of the planetary age of humanity. Before this threshold, the human species spread over the surface of the Earth in the manner of a diaspora. *Homo sapiens*, the most recent hominid species, arose in a particular ecosystem of a particular continent (the East African savanna), and spread from there to the rest of the planet: Asia, Australia, Europe, Alaska, Patagonia, Hawaii, Madagascar. It established in ecosystems where conditions vastly differed from those of the savanna: the Arctic tundra, mountain environments, desert, tropical forests. In this planetary diaspora, *Homo sapiens* became fragmented, generating populations that were diversified both in genetic and in cultural-linguistic terms. The ways of life of each population and their forms of religion and spirituality were marked by the local traits of the ecosystems in which they lived.

The history of the planetary diaspora of *Homo sapiens* is certainly also moulded by the meeting and hybridization of different populations, which gave rise to great civilizations: Mesopotamia, Egypt, the Mediterranean, India, China, Meso-America, and so forth. Empires have risen and fallen, unifying different ethnicities for briefer or longer spans of time. Domestic animals and cultivated crops spread from their zones of origin to much wider areas. The horse spread from the Caspian and Pontic steppe to the Middle East, transforming the art of warfare and the histories of kingdoms and dynasties. Wheat and barley, first cultivated in the Fertile Crescent, reached the British Isles and India in a few millennia.

But modern research into the history of human settlement of the planet, based on genealogical trees constructed both from genetic traits and languages, shows that scissions and separations prevailed for a long time over convergences and hybridizations. It is possible to glimpse certain stages of the migrations of the original nucleus of the human species and of the ways in which they gave rise to distinct populations. For many millennia, the world was very vast for our species. Small groups could freely choose to settle lands uninhabited by other humans: the Austronesian stock that left Indonesia and New Guinea found such situations in New Zealand, Hawaii, Polynesia, and even Madagascar (which was not so far from the place of origin of our species). Except in the seats of great civilizations, population density was low: Different groups and cultures could coexist side by side, unaware of each other. Until relatively recent times, this was still the pattern even in much of Europe and India, where Indo-European tribes of many stocks alternated with even more diversified non-Indo-European tribes.

At the threshold of 1492, the world was still divided into many basically separate subsystems of different sizes. Some covered significant proportions of certain continents, but most occupied smaller spaces: a mountain valley, a series of reliefs, an area of steppe, a strip of forest, a Pacific island, an Arctic coast. The linguistic and cultural fragmentation, which still characterizes impervious areas such as the Caucasus and the New Guinea highlands, reflects the models of settlement prevailing in much of the history of the human species. Among all these subsystems, there were connections and relations, but they were generally sporadic and involved few people for brief periods of their existences. For most people, and in daily life on the whole, the existence of other cultures and peoples passed almost completely unnoticed, or was tranfigured by mystic auras. No one had any notion of the planet as a whole, nor of the patterns of human settlement.

In the very first decades after the landing of Columbus, this separation of cultures and compartmentalization of the world was shattered. The expansionist policies of a few European monarchies destroyed the barriers between Europe and America, Europe and Africa, Asia and America, introducing the guidelines of the first global economy, which had salient features in the silver mines of South America, the sugar plantations of West Africa, the silk production of China, the spice trade of the Moluccas, the trans-Atlantic and trans-Pacific routes of the galleons.

Until 1492, the planet was divided into different, substantially separate farming systems. Different plants were cultivated in their various places of origin (principally the Middle East and China). Their cultivation spread sporadically to other areas but did not cause a homogeneity of farming systems. The areas in which agriculture had developed in America (Central America, the Andes and the present SW United States) remained isolated from each other. Soon after 1492, these barriers were upset. It can be truly said that Columbus caused the agricultural unification of the world. The European farm landscape and the European way of life were quickly transformed by the introduction of maize, potatoes, beans, tomatoes, pumpkins and peppers. Brazilian cassava became a basic food in a wide tropical belt from the Congo to Sumatra. In the first decades of the sixteenth century, crops destined to be of great economic importance began to be cultivated in the New World: coffee, sugarcane, and bananas. Coffee had been cultivated for millennia in Ethiopia, sugarcane and bananas in New Guinea. The flow of animals between the two worlds, however, was almost entrely one-way in the direction of America, deeply transforming ecosystems and ways of life. The horse became a protagonist in the North American prairie, and cattle proliferated in the Argentine pampas.

The decades following 1492 also saw another macroscopic process of convergence between different parts of the world, and this was the microbial unification of the world. It was a painful and destructive process. With the arrival of Columbus, most inhabitants of the Americas were exterminated not by the *conquistadores* but by contact with viruses and germs to which they had no immunity. Until that time, the evolution of diseases in the New World had followed paths that diverged from those of the Old World, not only by virtue of geographic isolation, but also because the domestication of animals played a very minor part in the lives of native Americans. Many Old World diseases were in fact products of farming practices and sedentary life style, transmitted to humans by domestic animals or otherwise dispersed to the new habitats. The land mass consisting of Europe, Asia, and Africa has less rigid geographic barriers and the circulation of farm products and animals had long broken down the confines between microbial subsystems. Europe paid progressively with the epidemics of antiquity and the Middle Ages (often of Oriental origin); the American peoples paid all at once.

The linguistic panorama of the world was also profoundly modified by the breakdown of geographic barriers in the modern age. The Indo-European languages, already diffused in a vast area of Eurasia, continued their expansion, this time in a sort of "new Europe." Four languages originally confined to the extreme western edge of the European continent, English, French, Spanish, and Portuguese, became the dominant languages of most of North America, South America, and Oceania (as well as some areas of Africa and Asia). The Russian language expanded to a vast area of northern Eurasia. This spectacular process of expansion of a few languages (together with their respective cultures and ways of life) caused the extinction of many others. In the last few centuries, the linguistic variety of the world has decreased progressively. Whole families of languages have disappeared. Many others are threatened, being spoken only by a few elderly persons. This is true of the native languages of America, Siberia and the Russian Far East, New Guinea, and Australia.

The centuries of the modern age, those five centuries that Edgar Morin called the "iron age of the planetary era" have been essentially ambivalent. On the one hand, there was a progressive emergence of new links between individuals, groups, and populations (including biological hybridization with the birth of new lineages); a first principle of tolerance; the acknowledgment of the importance and depth of other cultures, an interest in their spirituality, music, and literature. On the other hand, there was also the recurring presumption of certain fragments of the planetary mosaic to speak in the name of the whole. The five centuries of the iron age had been the stage of mass extinctions for most of the cultures inhabiting the planet. They have seen continual attempts to level the aspirations, values, and spirituality of individuals and groups, promoted by a few strong cultures (European, Chinese, Japanese, Arab, Bantu, ...). As strong as ever, this ambivalence continues to mark the hours and days of our lives. With the help of linguistics, genetics, anthropology, and archaeology, we have embarked on the fascinating adventure of reconstructing the history of human colonization of the planet. However, in the meantime, most of the human cultures have become extinct or nearly so.

The meaning of the struggle between opposite trends, of the precarious equilibrium between Eros and Thanatos, has been radically transformed by the events of our century. One of these, the atomic threat, shook many consciences throughout the world, because it showed the perverse attraction of the prospect of annihilation, the last abyss that swallows the many levelings imposed with sword and fire. The sharpened conscience generated by the atomic threat gradually sensitized us to other characteristics and risks of the planetary age.

In 1492, most human populations were truly part of one or a few local ecosystems: they were codependent on other animal species, plant species, and the respective climatic rhythms and variations. Even in Europe and other civilizations where the cultural and technological dimensions were more autonomous, the daily life of most people was dedicated to farm work in a single place, according to the rhythms of local ecosystems. After 1492, with the breakdown of the barriers between human populations, a breakdown of barriers between ecosystems also occurred, particularly a transformation of the nature of relations between human populations and local ecosystems. Individuals depended increasingly less for their survival and welfare on the natural characteristics of particular places. Today, and not only in the West, almost every meal is composed of foods produced in different parts of the world.

The spread of technologies for goods of all kinds has produced an illusion that the human species is no longer dependent on nature. This is not true. The increasing mixing of populations and their growing independence from local ecosystems reveals that the survival of the human species depends on the good functioning of a single, immense, global ecosystem, in which many species of animals, plants, and bacteria cooperate to maintain environmental conditions suitable for life in general.

In the course of this planetary age, and especially in the last hundred years, the economic, technological, and social development of the human species has led to an assault on what remains of local ecosystems as well as on the global ecosystem. Humanity has not only damaged the variety of its own species, but also that of life as a whole. Biologists estimate that the genetic heritage of thousands of animal species is lost through extinction every year along with the delicate equilibria of many local ecosystems. The dramatic question is whether these continual assaults on parts of the global ecosystem will sooner or later affect its overall functioning. This question can be expressed in a way that is even more disquieting. The global ecosystem has enormous resilience, resistance, and powers of self-repair; and its existence is unlikely to be irreparably damaged by imprudent actions of the human species. These acts are much more likely to modify the conditions of the global ecosystem that permitted the human species to flourish, opening the way for new equilibria favorable for other forms of life, but unfavorable or even lethal for ours. Now the threat of the unleashing of planetary forces, such as climate, atmospheric circulation and the composition of the atmospheric, marine, and other compartments is perhaps stronger than the atomic threat that impends over human species since the days of Hiroshima and Nagasaki.

In the course of the first centuries of the planetary era, the human species has therefore been trapped in a paradox that does not yet seem to have a viable solution. On the one hand, scientific advances have shown the evolutionary potential of variety, diversity, interaction, hybridization, flexibility, redundance, and individuality. The history of life on the earth has created long-term mechanisms embodying these values. Specifically, it has developed mechanisms for producing novelties based on wide variations in the characters and behavior of single organisms. The evolutionary value of diversity is also becoming more and more evident for cultural evolution: In individual and collective processes of creation, novelty emerges whenever there is tension, interaction, hybridization, and permanent conflict between different ways of thinking and attitudes.

On the other hand, however, modern mankind seems victim to a compulsion to copy and repeat, which leads him to destroy many sources of cultural and biological variety. This may be anti-evolutionary in the short period (reducing the creativity of the human species) as well as in the long period (upsetting the global ecosystem).

In the face of the dramatic problems of today, it becomes important to take a long-term view, questioning ourselves about the source, among other things, of this compulsion. Certain deterministic explanations underline aspects of earlier stages of the process of hominization: belonging to small, rather homogeneous groups might have had advantages over tackling the world single-handed. Thus, human biological and cultural evolution would have favored mechanisms of internal coherence in groups at the expense of interactions with other groups. In this case, planetary age humanity would be the victim of a side effect of an evolutionary process triggered by environmental conditions that no longer exist. It would be victim of a kind of past heritage and would need to evolve again and to genuinely rehominize: to rethink itself, not through the interminable quarrels of small groups but through the multiplication of connections between the individual and the planetary whole.

These controversial questions call for a global anthropological perspective, a multidisciplinary context for the many dimensions of human existence and evolution.

THE COMPLEXITY OF HUMAN HISTORY: THE MANY BIRTHS OF HUMANITY

At the end of *Le Paradigme Perdu,* Edgar Morin introduces the notion of *peninsular man,* as a powerful tool for uniting nature and culture and filling the gap produced by disciplinary specialization. In his view, basic anthropology runs counter to the "insular notion of man, isolated from nature and from his own nature" and man's self-idolatry. A closed, fragmentary, and simplistic theory of humanity must give way to an open, complex, multidimensional theory.

The notion of peninsular man, in fact, lies at the heart of much innovative research performed in the last 20 years on the process of hominization. A clear image emerges of the many births of humanity. Many "hominization thresholds," that is, origins of the macroscopic traits (e.g., erect posture, technology, language) conventionally chosen to characterize human uniqueness with respect to the animal world, are considered. These thresholds actually arose at quite different times. Erect posture, which enabled new modes of locomotion and feeding in the Primate order and favored colonization of the difficult savanna environment, dates back about four and a half million years. It was first seen in *Australopithecus ramidus,* a species still belonging to the ancestral genus (*Australopithecus*), from which *Homo* eventually emerged. The first signs of technology, of a species that could systematically make and use stone tools, were associated with *Homo habilis,* an ancestor of ours, already belonging to the genus *Homo*. This development emerged about two and a half million years ago. Language, on the other hand, is a trait of our species, *Homo sapiens,* and is largely associated with the spread of the modern variety (sometimes called *Homo sapiens sapiens*). Recent research on the origins of language show that all the languages spoken today may be derived from a single ancestral language, and that the diversification between them is not too great. This suggests that the ancestral language is relatively recent, dating back, say, 100,000 years. Certain anthropologists consider this threshold to be confirmed by archaeological documentation: The origin and development of language is thought to be related to the spread of human settlements, regional differentiation in tools, increased long-distance contacts demonstrated by the presence of exotic objects, and so forth.

The time differences are important, but also the differences between the various sets of species sharing a given "human" trait. We cannot really consider our species, *Homo sapiens*, to be the only carrier of the "hominization" values regarded as essential and definitive. To the contrary, an outside observer might consider that our ancestor *Homo erectus*, who enjoyed great stability and evolutionary success for at least a million years, deserved the title of *Homo* par excellence, whereas his descendent *Homo sapiens* has only been in the world for a short time and shows an inclination for self-destruction.

Moreover, we are discovering that the path from *Australopithecus ramidus* to *Homo sapiens* is more diversified and complex than was previously thought. The number of species attributed to the genus *Homo* has multiplied, certainly due to many new findings, but also to a different methodological attitude. Until very recently, the evolution of hominids was regulated by the paradigm of the shoehorn and the ladder, terms introduced and discussed at length by Stephen J. Gould in *Wonderful Life*. Fossils with quite divergent anatomical traits were nevertheless grouped in a single species, which was regarded as having great internal variability. Ian Tattersall, a major authority on the evolution of the hominids, notes that this procedure violates consolidated taxonomic practice: in taxa far from ours, anatomical differences just as pronounced as those between various hominid specimens are generally regarded as justifying the attribution of fossils to different species. The truth is that the arbitrary hypothesis that hominid species have enormous internal variability made it possible to summarize the genealogy of our species as a linear progression: in the simplest version, *Australopithecus* → *Homo habilis* → *Homo erectus* → *Homo sapiens*. The immediate temptation was to interpret the ladder as a sequence from the less to the more "human," like a progression of degrees of hominization, aimed at a final product.

Application of the normal criteria of taxonomy of animal species to hominid fossils, however, brings out quite a different picture of the process of hominization. Each macrospecies can be divided into a number of species having greater internal homogeneity, narrower distribution, and often a shorter lifetime. According to Ian Tattersall, the traditional progression of three species, *habilis* → *erectus* → *sapiens*, thus gives way to a model with at least seven ascertained species. In fact, there were certainly more than seven, since for all animal groups, only a small percentage of the species that lived have left fossil remains. The temporal coexistence, usually in different areas and habitats, of different species belonging to the same genus *Homo*, may be a quite normal feature of human evolution. Ironically, the present situation may be more anomalous, since we are currently witnessing the survival of only one species of hominid, *Homo sapiens*. Moreover, the three great species of the traditional image of human evolution are not necessarily linked by a linear ancestor–descendent relationship; in other words, *Homo habilis* did not necessarily generate *H. erectus*, nor *H. erectus*, *H. sapiens*. The species of future promise were probably unobtrusive local varieties. Like many other evolutionary processes, the linear progression is currently reconsidered as a branching bush, a history of nonlinear and indirect paths. The bush of the process of hominization is marked as much by pruned and dead branches as by successes and innovations.

Another feature that makes the image of human evolution problematic is biological and cultural gaps and discontinuities. The biological turning points most evident from our fossils concern the mean brain size of the various hominids: *Homo sapiens* has a much greater brain volume than *Homo erectus* (and other contemporary species), which in turn have a much greater brain volume than *Homo habilis*. However, in the evolution of culture, and especially technology, such findings are not so immediate. In fact, nearly two million years of human evolution seem to have been characterized by very stereotyped technology, to the point that different and successive species all seem to be at almost the same stage of "cultural" evolution. Furthermore, the great cultural turning point that characterizes the history of the modern variety of *Homo sapiens*, the emergence of refined and expressive art, linked somehow to religion and spiritual needs, in what was called the Cro-Magnon phase, does not seem to have a biological counterpart. *Homo sapiens* was already anatomically modern and had the linguistic, cognitive, and communicative powers that he has today.

The complexity and unfinished nature of the process of hominization, with all its feedbacks, is the great discovery that recent research in basic anthropology has made. In this new perspective, many important results of different branches of recent anthropological research acquire new meanings and significances, suggesting that humanity should be approached as an evolution rather than as a being, as a process rather than a state. The unfinished nature of the hominization process does not mean incompleteness. In a stronger sense it means that the future outcomes of hominization are not necessities of human nature. They depend on historical factors and the current choices of our species. Humanity is not a destiny: Humanity is a continuous process of invention.

THE COMPLEXITY OF HUMAN HISTORY: CONTINUITY AND DISCONTINUITY IN THE PLANETARY AGE

Together with the recent burgeoning of research into the long process of hominization (which, as we have seen concerns the roots and relations of our species in the broader context of primate and hominid evolution), a new type of literature that aims to reconstruct the "middle periods" of the process of hominization has emerged. This literature deals with the history and planetary spread of the species *Homo sapiens* (especially its "modern" variant) or with the more particular history and spread of what we call civilization, from the cradles of the agricultural revolution (Middle East, northern China, etc.). These "middle periods" are the 100,000–150,000 years of the history of *Homo sapiens*, or the 13,000 years of the history of "civilization," very brief intervals compared to the five-million-year history of the hominids but very long compared to the 500 years of the planetary age or even the 2500 years of Western civilization.

Three recent books on related questions and research make important contributions to the planetary history of *Homo sapiens*: a rethinking of the overall experience of our species in a comparative perspective that brings out the specificities, convergences, and divergences of the many local histories of humanity. *The History and Geography of Human Genes* by Luigi-Luca Cavalli Sforza, Paolo Menozzi, and Alberto Piazza examines the genetic heritages of various human populations, obtaining an articulated view of the paths and timing of the great migrations that led *Homo sapiens* out of its African cradle and into the other continents and islands of the planet. *The Origin of Language* by Merritt Ruhlen uses traditional and controversial questions of relationship and linguistic classification to show that they provide strong indications that all the languages spoken today had a common ancestor. If so, it becomes possible to make plausible hypotheses about the manner in which the present linguistic families diversified. *Guns, Germs and Steel* by Jared Diamond concentrates on the specific characters and relations among the local histories of great areas of the planet, in relation to the cultivation of plants and the domestication of animals. The narrative centers on certain key questions: Why did the planetary age begin the way it did? Why did the Old World (Eurasia) collide with the New World on American soil and not vice versa? Why Pizarro at Cajamarca and not Atahaualpa at Cadiz?

Of the many changes in perspective of planetary history, one is of fundamental importance: The planetary age of humanity appears as a "peninsular" age in the sense described above. Although as we said, the singularity and novelty of the problems and horizons generated by the developments of the planetary age are emerging strongly, they do not mean a total absence of roots and antecedents. To the contrary, they mean a plurality of roots and a nonlinearity in their interactions and mutual codeterminations. They mean amplification, remixing, resignification of ancient trends that become transfigured and new. They indicate a genuine emergence that supersedes traditional oppositions between continuity and discontinuity, proposing an ecology of historical change that is created by dynamic tension between these poles.

First of all, the cultural leveling (with the consequent extinction of languages and ways of life) produced by the spill of Europeans into the world began before 1492 has many illustrious historical antecedents (even on a less vast scale) and is also somehow a process that accompanied the spread of farming. The distribution of the main linguistic families of the world is a clear sign. In most cases these families originated in restricted areas, and their spread may be related to the migration of peoples or plants and animals.

The endemic and global reduction of linguistic variety in the last few centuries was preceded by the spectacular expansion of various languages. The Indo-European languages expanded from their place of origin in the Ukraine, Caucasus, or Anatolia; and all the most feasible hypotheses about their origin presuppose a connection with the spread of agriculture and the domestication of animals. Likewise, the Sino-Tibetan languages spread from a small area in Northern China, and the Austronesian languages (today spoken from Madagascar to Polynesia, via Malaysia, Indonesia, and the Philippines) perhaps originated in southern China or in Taiwan. The Bantu languages originated between Nigeria and the Cameroon and are now spoken in almost all of central and southern Africa. Sometimes the languages that have succumbed can be clearly identified, because they still exist. Speakers of Khoi-san (once known as Bushmen or Hottentotts) were hunters and gatherers that lived in a vast area, stretching from the Cape of Good Hope to Ethiopia, through much of southern and eastern Africa. In the last 3000 years, they have progressively lost ground to Bantu farmers and are now confined to small areas at the southwestern tip of the continent, where European settlers rather than Bantus (their plants were not suited to the temperate zone) reduced them to a threatened minority. Sometimes the ancient languages have disappeared, leaving recognizable ethnic gaps. Such is the case for certain groups of African pygmies, which have continued to live as hunters and gatherers, and have genetic features that distinguish them from the surrounding peoples. Today, however, they speak the languages of their neighbors, which is a sign of recession and fragmentation as a result of the expansion of others.

Sometimes we observe spectacular inversions of roles, the dynamics of which can be deciphered only partially. For example, the Austronesian languages seem to have originated in continental China (whence they have long disappeared, erased by Chinese expansion) and spread to Taiwan (where they had an absolute majority a few centuries ago, but today are on the point of extinction). Although evicted from their land of origin, they have spread to vast areas in the southern hemisphere, in their turn limiting and replacing other ancient linguistic groups, such as the Indo-Pacific group, which includes the languages of central New Guinea and neighboring islands.

Things that at first seem mere curiosities, sometimes turn out to be important links for understanding the routes of *Homo sapiens*. The various groups of languages spoken today in the Indo-Chinese peninsula probably originated in continental China, spreading only later to Southeast Asia. Today they have replaced all the languages once spoken in that area: Even genetically distinct groups, a sign of older autochthony, like the *negritos* of Malaysia and the Philippines, have been assimilated linguistically (as in the case of the pygmies; today their original language is unknown). There is only one exception: Certain indigenous peoples of the Andaman islands, an archipelago in the Gulf of Bengal belonging politically to India, still speak a language that is unrelated to the surrounding Asiatic families. Other related languages became extinct in the recent past. This language turns out to be related to the family of Indo-Pacific languages that still prevail in vast areas of New Guinea. The indigenous people of the Andaman islands are therefore a trace of that remote migration from the African cradle of *Homo sapiens* into Indonesia, New Guinea, and Australia. According to a hypothesis of Cavalli Sforza, Menozzi and Piazza, this route included the Red Sea, Arabian peninsula, Indian Ocean, Indian peninsula, Gulf of Bengal, Indo-Chinese peninsula, and so forth. The Andamanese therefore gain the title of one of the most ancient peoples of the earth. The linguistic discrepancy allows us to open a window on the abyss of time and to look back perhaps 50,000 or 60,000 years.

Linguistic events do not always provide sufficient or reliable evidence for reconstructing the complex fabric of cultural relations. One should beware of over-simplifications of the type: Linguistic and cultural diversity is always a positive value, whereas linguistic or cultural homogeneity is negative or something to avoid. In the history of the human species, the spread and replacement of linguistic families have occurred in many different ways and are not necessarily a sign of bloody invasions and conquests. However, the ethno-linguistic events briefly sketched here are a kind of compendium of the process by which farming populations substituted hunter–gatherer populations, even before 1492.

The history of our species has not been a continuous progression, but a succession of rather ambivalent metamorphoses and revolutions. Crises of ethnic, linguistic, and cultural diversity are not the only products of the planetary era. As we have seen, the crisis of biodiversity is even more serious. Every year, thousands of animal species disappear forever from the face of the earth. Sometimes they have been explicitly preyed upon by our species for food or other purposes. In most cases, however, their extinction depends on the collapse of the ecosystems of which they are part. Relatively intact ecosystems are progressively giving way to partially and totally anthropized areas. Below a certain threshold of size and complexity, they can no longer function as a whole, hosting large numbers of species. The most spectacular and dramatic aspects of the global crisis of biodiversity can be seen in the threat to tropical rainforests, the ecosystem with the highest number and density of animal species. Many species living in the tropical rainforest have not been identified, let alone numerically estimated. Some may never be, falling victim to extinction before science can catalogue them.

The overall result of so many local crises of single habitats is devastating. If considered through parameters such as the number of extinct species, the present crisis of biodiversity has already caused the third mass extinction in terms of severity of the many in the history of the biosphere. The worst occurred at the end of the Permian era (about 245 million years ago), when about 95% of the existing animal species perished. The second occurred at the end of the Mesozoic era (about 65 million years ago) and is the most famous because the victims included all the dinosaurs. The first event is thought to have been due to climatic cooling and other effects of continental drift. The second seems to have been due to the impact of a celestial body. The third, however, is largely due to the effects of a single species on the rest of the biosphere.

The destructive effects of the human species on other plant and animal species have been amplified by the fall of barriers between populations and ecosystems, typical of the planetary age, especially in the last few decades. These effects, however, did not arise with the planetary age nor with the industrial revolution. Many severe ecological crises occurred even before 1492. Events in Madagascar, Hawaii, and New Zealand, the last lands of reasonable size reached by the human species, are particularly significant. In the few decades and centuries after the landing of humans, the diversity of autochthonous fauna declined significantly; and, in particular, larger animals, such as the giant lemurs of Madagascar and the moa of New Zealand, became extinct. When Europeans landed many centuries later, they merely gave the *coup de grace* to disrupted equilibria.

On a larger scale, the impact of *Homo sapiens* seems to have been lethal for the larger mammals of America and Australia. In the Americas, a rapid and severe wave of extinctions of autochthonous mammals seems to have occurred about 11,000–12,000 years ago, tthat is, just after the first arrival in numbers of the human species, in this case from the Asian continent via a strip of land now cut by the Bering straits. This chronology has caused much controversy. Upholders of the "human cause" are opposed by those who attribute the great continental-scale extinctions to post-glacial climatic perturbations. The question is still unanswered, but the correlation between the two events (migration of *Homo sapiens* and large-scale extinctions) looks suspicious, and much evidence points to human responsibility.

A similar problem, relating to remoter times but on a smaller scale, is posed by the fate of the large marsupials that populated the Australian continent when our species first landed there. Why did these animals, which had populated the continent for hundreds of thousands or even millions of years, become extinct within a few millennia of the arrival of *Homo sapiens* (about 40,000 B.C.)? Could it really have been a severe drought? Incidentally, the traditional societies of the American and Australian continents, to whom these extinctions are ascribed, were not farmers bent on the large-scale leveling of ecosystems, but hunters and gatherers who, in the modern world, show particular respect for the environment, given the scarcity and the price in terms of energy of the renewable resources (plants and animals) on which they depend. Could this respect have been sharpened by the experience of an early catastrophe, in which inexpert populations witnessed the disappearance of animal resources that seemed almost inexhaustible when they landed in the new worlds? In any case, the lack of large animals (not only as sources of food, but also as beasts of burden) seems to have hindered some societal developments in these continents and was certainly a source of weakness when Columbus landed in the Americas (just think of the impact of the sudden arrival of the horses of the *conquistadores*).

Two simplified versions of the relations between the planetary age and human history have prevailed, not only in popular literature but also in scientific approaches and the philosophies of history of the last two centuries. The first is centered around the idea of progress, in which history and prehistory are counterposed, and humanity is depicted as gaining in wisdom from technology and the planetary-scale economy. The second proposes various versions of the myth of the golden age and conveys the idea that the technological, economic, and cognitive conquests of the last few centuries have been at the expense of human resources, culture, ecosystems, plant and animal species, knowledge, and wisdom.

From our point of view, both these versions are partial and inadequate. Together, they suggest the need for a complex and holistic perspective that would bring out the uniqueness of the events and processes of the last five centuries. After 1492, everything changed: The range of human potential for creation and destruction increased. Such a perspective would also enable us to understand the conflicting elements of human experience, to glimpse the first equilibrium upset by *Homo sapiens* and his relentless urge to create new equilibria. These dramas are not peculiar to the planetary age; they are ingrained in every human society and every human experience, marking all human adventures on the planetary scene. This is why human adventures and experiences distant in space and time cannot be dismissed as archaeological curiosities. They are often pertinent for formulating, understanding and tackling our present problems.

Homo sapiens was not born human, but in the course of history, he has learned to be human. During his evolution, new forms of humanity have overlapped and replaced each other. As the planetary age proceeds, the emerging web of knowledge and experience may enable our species to learn to be global, to become one with the ecosystems through sustainable relationships, to understand how to exploit the creative side of cultural diversity, and perhaps to walk the narrow ridge of a new stage of hominization.

BIBLIOGRAPHY

1. BOCCHI, G. & M. CERUTI. 1985. La Sfida della Complessità. Feltrinelli, Milano.
2. BOCCHI, G. & M. CERUTI. 1997. Solidarity or Barbarism. Peter Lang. San Francisco.
3. BOCCHI, G. & M. CERUTI. 1999. The Narrative Universe. Hampton Press. New Jersey.
4. CAVALLI-SFORZA, L.L. 1996. Gènes, Peuples et Langues. Odile Jacob. Paris.

5. CAVALLI-SFORZA, L., F. CAVALLI-SFORZA & S. THORNE. 1996. The Great Human Diasporas: The History of Diversity and Evolution. Perseus Press.
6. CAVALLI-SFORZA, L.L., P. MENOZZI & A. PIAZZA. 1994. The History and Geography of Human Genes. Princeton University Press. Princeton, NJ.
7. CERUTI, M. 1994. Constraints and Possibilities. The Evolution of Knowledge and the Knowledge of Evolution. Gordon and Breach. New York.
8. CERUTI, M. 1995. Evoluzione senza Fondamenti. Laterza. Roma-Bari.
9. CROSBY, A. 1972. The Columbian Exchange: Biological and Cultural Consequences of 1492. Greenwood. Westport, CT.
10. CROSBY, A. 1986. Ecological Imperialism. The Biological Expansion of Europe: 900–1900. Cambridge University Press. Cambridge, England.
11. DIAMOND, J. 1997. Guns, Germs and Steel. A Short History of Everybody for the Last 13,000 years. Jonathan Cape. London.
12. EISLER, R. 1987. The Chalice and the Blade. Our History, Our Future. Harper and Row. San Francisco.
13. ELDREDGE, N. 1995. Dominion. Can nature and culture co-exist? Henry Holt. New York.
14. ELDREDGE, N. 1998. Life in Balance: Humanity and the Biodiversity Crisis. Princeton University Press. Princeton, NJ.
15. GOULD, STEPHEN JAY. 1989. Wonderful Life: The Burgess Shale and the Nature of History. Replica Books.
16. LEAKEY, R. & R. LEWIN. 1995. The Sixth Extinction. Doubleday. New York.
17. LEWIN, R. 1993. The Origin of Modern Humans. Freeman. San Francisco.
18. MONTUORI, A. & I. CONTI. 1993. From Power to Partnership. Creating the Future of Love, Work and Community. HarperCollins. San Francisco.
19. MORIN, E. 1973. Le Paradigme Perdu: La Nature Humaine. Seuil. Paris.
20. MORIN, E. & A.B. KERN. 1993. Terre-Patrie. Seuil. Paris.
21. RUHLEN, M. 1996. The Origin of Language: Tracing the Evolution of the Mother Tongue. Wiley. New York.
22. STANLEY, S. 1996. Children of the Ice Age: How a global catastrophe allowed humans to evolve. Harmony Books. New York.
23. TATTERSALL, I. 1995. The Fossil Trail. Oxford University Press. New York.
24. TIEZZI, E. 1984. Tempi Storici Tempi Biologici. Garzanti. Milano, Italy.
25. TIEZZI, E. 1996. Fermare il Tempo. Un'interpretazione Estetico-Scientifica della Natura. Raffaello Cortina. Milano, Italy.
26. TUDGE, C. 1996. The Time before History. Five Million Years of Human Impact. Simon and Schuster. New York.
27. WARD, P. 1997. The Call of Distant Mammoths. Why the Ice Age Mammals Disappeared. Springer-Verlag. New York.

Complexity VII: Composing Natural Science from Natural Numbers, Natural Kinds, and Natural Affinities

JERRY L. R. CHANDLER[a]

Krasnow Institute for Advanced Study, George Mason University, and Washington Evolutionary Systems Society, 837 Canal Drive, McLean, Virginia 22102-1407, USA

INTRODUCTION

To construct, to synthesize, to generate, to compose, to make whole, to create—these activities lie at the root of humanness. The last 50 centuries of natural history provide a record of increasing human capacity for analysis and synthesis. The role of natural science in human affairs has widened as we have improved our capacities to conduct these operations over ever greater domains. During this long span of time, man has learned to ground behaviors in two distinctive patterns. In one pattern, one-to-one correspondence rules predominate over a substantial range of natural dynamics. In the other pattern, man has grounded socio-cultural behaviors in collaborative, cooperative, and communicative structures that tend to generate coherent behaviors within our human communities. Collaborative, cooperative, and communicative structures lead to coherent dynamics that generate creative actions among interdependent members of these communities. Are these two forms of dynamics—one based on one-to-one correspondence rules and the other based on cooperative, collaborative and communicative rules—of the same form? If not, how do they differ? Is this question of form a question of the nature of autonomy (self-law)? Is this question of form a question of the nature of wholeness? Wholeness implies closure, completeness. Natural kinds of various sorts are distinguishable within a closure or within a complete object. Both autonomy and wholeness imply enumerability. Thus, the question of form will be approached within the natural sciences from the perspective counting. I seek to show that simple counting, 1, 2, 3, ..., among well-formed concepts, is a powerful tool for constructing unified view of natural science and to begin to dismantle the artificial academic boundaries.

Natural numbers are at the heart and soul of mathematical constructions. Both structural science and dynamical science are intimately related to counting within a well-formed category. Natural kinds are the objects of study within natural science. A fundamental objective of structural and dynamical science seeks to relate natural kinds to natural numbers.

Co-enumeration of natural kinds and natural numbers can generate one-to-one correspondence relationships between natural kinds and natural numbers. Despite the fact that such enumerations ignore the intertwining of time scales, this co-enumeration of natural numbers with natural kinds can be semantically structured into

---

[a]Address for correspondence: 703-790-1651 (voice); JLRChand@Erols.com (e-mail).

a hierarchy of relationships, thereby contributing to a grounding for a philosophy of science. In this short essay, I seek to construct a compact form of natural science from natural numbers, natural kinds, and natural time spans within the framework of categories. These categories will be formal categories, categories somewhat different from the philosophical categories of Aristotle or Kant. The reader is referred to mathematical works of Ehresmann and Vanbremeersch,[1] Hatcher,[2] Lawvere and Schanuel,[3] and Mac Lane,[4] for formal descriptions of categories.

NATURAL KINDS

The world of our experience can be partitioned into objects that have similar attributes. Here, we are concerned with natural objects, not products of human construction. Natural kinds are objects within classes of similar objects. Examples of natural kinds include atoms, molecules, bacteria, insects, mammals, humans, and societies. Natural objects have structure—that is, potentially distinguishable parts and pieces are fit together to generate a natural form. In other words, objects are composed from parts by organizing the parts into wholes. Construction of new "wholes" from parts can be distinguished into two forms: The object is of the same category as the parts or the object is in a different category as the parts. These two methods of forming new categories correspond to "horizontal" and "vertical" compositions. Horizontal compositions synthesize new members of the same category of a natural kind. (For example, allowing acetic acid and ethyl alcohol to interact generates two new structures, ethyl acetate and water.) Vertical compositions construct objects of a new natural kind, with emergent properties that distinguish the classes of attributes of the wholes from the classes of attributes of the parts. (For example, allowing a cell to interact with nutrients can generate a nascent cell.) The philosophical distinction between horizontal and vertical compositions is profound. Whereas both the horizontal and vertical example are consistent with the conservation of mass and the conservation of charge, a vertical composition substantially increases the functional attributes and the complexity of the dynamics. (For example, cells have many attributes that are not possessed by molecules.) These examples of categories of natural kinds illustrate differences in organizational structure, albeit while ignoring the profound dynamical distinctions between vertical and horizontal compositions.

NATURAL NUMBERS

The natural numbers, as is well known, are defined as 1, 2, 3, ... Historically, the natural numbers appear to have co-evolved with the emergence of language and human socio-cultural/economic systems. Other categories of numbers, such as the rationals, the irrationals, the complex, the odds, and the evens, are constructed from the natural numbers by either expanding the conceptual base (the integers) or by the processes of mathematical operations (rationals, the irrationals, the complex) or both (the odds, the evens). It is possible to construct one-to-one correspondences between natural numbers and objects of categories (such as natural kinds). In a very simple example, the construction of a one-to-one correspondence between the odd numbers and the even numbers associates with each odd number an even number. Another example is the one to one correspondence between positive electrical charges in the nucleus and negative electrical charges in the orbitals of a neutral atom. Still another, but less rigorous, example, is the one-to-one correspondences between adenine and thymine in a DNA duplex. Within the political structure of the United States, a one-to-one correspondence relation exists between congressional districts and congressional representatives in the House of Representatives. Each nonconceptual example operates on a different time scale.

CRITERIA FOR GROUNDING NATURAL SCIENCE

In recent decades the nature of traditional scientific values has come under intense scrutiny. More precisely, some assert that scientific truths have become a primary source of public values, while others assert the opposite, that science is "value free." Such conflicting assertions reveal the conflicts in language, methods and symbols of different academic disciplines and obscure the underlying unity and integrity of the natural sciences. Can natural science be grounded on a logical foundations that provides a coherent structure for the traditional scientific disciplines? The traditional Newtonian perspective does not encompass molecular behavior, let alone biological or human behaviors such as medical practice and should be abandoned as a basis of a philosophy of natural science. Many writers have sought to place physical cosmology as the logical source of natural science. Such efforts are now the source of public ridicule because cosmology itself is in early stages of development as a scientific theory. Relationships between cosmology and biology and medicine are equally primitive.[6] What are the alternative sources of groundings for natural sciences? Logic? Language? Matter? Space? Time? Can any of these concepts be ignored in grounding a natural science?

LOGICAL GROUNDING

Logical relationships within natural science are grounded within a universe of discourse. At present, natural science is composed from many universes of discourse, the relationships among them being rather poorly structured. Here, I propose to bring these discourses into a common logic framework. Just as natural kinds can be formed into horizontal or vertical compositions, logical operations can also be viewed as either horizontal or vertical compositions. Consider the following from *The Primary Logic*,[6]

> We call the totality of beings, individuals, objects, events, etc., to which we refer, a universe, while the totality of signs with which we refer to the beings, individuals, objects, events which are included in the universe, we call a language. We then state that a theory is constituted by the universe and its related language ... .
>
> Let T0 = theory, L0 = language, U0 = universe. In general terms, we have:
>
> $$T0 = \langle L0, U0 \rangle$$
>
> If we take the language L0 of a theory T0 as a universe we need a metalanguage L1. We shall then have a metatheory which we shall call T1.

$$T1 = \langle L1, L0 \rangle$$

But we can take as a universe the metalanguage $L1$, and in that case we need a meta-metalanguage $L2$. The totality of $L2$ and $L1$ will thus constitute the metametatheory $T2$ as follows

$$T2 = \langle L2, L1 \rangle$$

In general, a metatheory will be characterized as follows

$$T(n+1) = \langle L(n+1), Ln \rangle.$$

Within the natural sciences, what are the potential relationships among T0, T1, T2, ..., $T(n+1)$? To approach this question, we need a notation, scientific concepts, and logical grounding within the material structures of nature.

DEGREES OF ORGANIZATION: A NOTATIONAL GROUNDING

Natural kinds range from very simply organized objects, say an electron, to very highly organized objects, say the human body. In order to construct a unified view of natural science, a common notation is necessary in which all natural objects can be represented. I have proposed such a notation, based on the historical emergence of increasing organization over geologic time spans.[6-8] Briefly, a degree of organization is a class of natural kinds, such as an atom, a molecule, a cell, a human being, and so forth. Intrinsic to a degree of organization ($O°(n)$, where n is a natural number) is semantic and syntactic closure in the sense of a theory composed of a universe of discourse and a language. Structural science is generated by operations on categories of natural kinds over degrees of organization. Operations on degrees of organization form a hierarchical structure composed of objects at the same or lesser degrees of organization. At a minimum, the operational rules must be consistent with the conservation rules for mass and charge. A resulting hierarchically organized object constructs a natural hierarchy of scale (mass/volume) as a consequence of the many mappings between the parts and the whole. Formally, an operation from one degree of organization to a another degree of organization can be viewed as a structural operation. It is characterized by the parts, the sum, and the patterns of relationships among the parts.

METHODOLOGICAL GROUNDING

A unified view of scientific methodology is needed to complement the logical and notational groundings. This aspect of a philosophy of science can now be addressed in terms of observations, descriptions, and symbolizations. These three facets of scientific activities must be intimately intertwined in order to generate scientific conclusions from the interdependent works of experimentalists, theoreticians, and practitioners.

1. Observations are classified into structural or dynamic science depending on the ratio between time scales of observation to stability of the object. A complete denotation of the object (mass, form, history, state) under observation is crucial to reproducibility in other circumstances. Within the organizational notation, the observations can be classified as elements of $V°1$, $V°2$,

$V°3$, ... Simultaneous observations from multiple degrees of organization are often necessary for complex systems.
2. Descriptions of observed attributes of objects may use one of several languages. Technical or metalanguages are necessary for logical consistency. Technical language can be recomposed into a series of metalanguages, $M°1$, $M°2, M°3$, ... In order to describe observations in natural language, syntactical language is replaced by analogies and metaphors of the natural language.
3. Symbolizations of observations may also be in one or more forms of logical representations. One-to-one correspondence relationships for structural objects are frequently used. Continuous functions are often used for representations of space and/or time. Pragmatic principles, in terms of hypothetical statistical distributions, are often used to symbolize complex systems.

Potentially, "scientific validity" can be asserted by finding reproducibility, consistency and predictability within the observations, descriptions, and symbolizations where the symbolizations may represent one-to-one correspondence, coherent, or pragmatic truths. The validity of one to one correspondence principles is often given precedence within the scientific community over less rigorous enumerative methods that also create associations between cause and effects.

MATERIAL GROUNDING

I propose that the atomic table is the primary source of unity in natural science and should be used as a universal grounding for structural science. A primary reason for this proposition is that naturally occurring atoms can be placed in one to one correspondence with the natural numbers. For example, the usual symbolic order for the first ten elements of matter is 1, H; 2, He; 3, Li; 4, Be; 5, B; 6, C; 7, N; 8, O; 9, F; and 10, Ne. Patterns of relationships among atomic affinity organize the natural numbers into a table with iterative attributes. For example, element 2, helium, and element 10, neon, are both inert gases, consistent with the electrodynamic predictions. Strong evidence supports the belief that all natural materials, including the human body, are composed from atoms. Natural affinities of atomic structures for one another generate patterns of molecular structures, cellular structures, neural structures, ... From a general systems perspective, atoms have four intrinsic attributes which are common to higher degrees of organization: (1) semantic and syntactical closures; (2) specific, compositionally dependent, geometric forms; (3) affinities among the parts; and (4) form-generating cyclic dynamics.

1. *Closure.* Semantic and syntactical closure specifies the number of protons (+), electrons (−), and neutrons (uncharged) (p, e, n); the masses of the p, e, n sum to the approximate mass of the atom and provide the structural foundation for the conservation rules. Syntactical closure over the same set of particles is used to generate electrodynamic behaviors of the components via the Schrödinger equation. All higher order compositions of material objects inherit homologous attributes.

2. *Conformation.* Atoms possess deterministic geometric forms. Solutions to the Schrödinger equation (and approximations to solutions) provide a (determinis-

tic) three dimensional geometry, for each atom, energy requirements for changes in this geometry and a pseudo-volume in terms of the distribution functions for orbitals. The term pseudo-volume is used because the Schrödinger equation does not generate a distinctive physical boundary, the *sine qua non* of many larger scale systems. Schrödinger functions generate specific geometric angles between individual orbitals. Thus, the sp3 orbitals of methane are exactly 109°28' and point toward the corners of a tetrahedron. In other words, the specific geometric forms of small molecules are deterministic functions of the atom orbitals. The fact that the position and momentum of individual electrons are probability distribution functions has little bearing on the form of atoms or small molecules—despite great impact on physical philosophy. For example, the Woodward-Hoffmann rules for chemical reactions are based on symmetry of the entire molecule, not the dynamics (positions or momentums) of individual particles. All higher degrees of organization inherit the geometric potential of the constituent elements.

3. *Concatenation.* Each part (each p, each e, each n) is linked to the whole in a specific manner—this linkage is described precisely by the quantum numbers for the electrons and less precisely for the quantum states of a nucleus. The parts are concatenated to give a specific three-dimensional geometry to atoms and molecules. A stable three-dimensional conformation is associated with a local energy minimum. Although these linkages specify observables—spectra, relative distance from nucleus and energy level of each electron—they also specify the geometry of the whole. Higher degrees of organization inherit these affinities and potential affinities.

4. *Cyclicity.* Cyclic behavior implies the time dependent genesis of form (i.e., form in time). The stability of an atom emerges from continuous, cyclic motion of the subatomic particles organized into quantum numbers and states. Each electron is assigned a specific quantum number and is observable, in principle, by specific energy-dependent transitions. For example, the 1s2 orbitals of electronic motion form a sphere. Thus, despite the probabilistic dynamics of location and momentum, the geometry of orbitals is determined by the complete set of mass/charge interactions and is a deterministic function of the discrete set of particles. The uniqueness of each electron in an orbital is a critical aspect of structural science. For example, living systems generate energy by organizing the flow of electrons over specific pathways of molecular orbitals. All higher degrees of organization inherit these orbital structures and the potential for more complex orbitals structures.

These four organizational attributes of atoms (closure, conformation, concatenation, and cyclicity) are often attributes of natural kinds that are composed from atoms. Categorical objects (molecules, macromolecules, cells, individuals) composed from atoms organize the atomic attributes into specific patterns that generate analogous semantic concepts. In complex, highly organized systems, the generic concepts of closure, conformation, concatenation, and cyclicity may be fundamental to the genesis of dynamic stability.[6-8] Metaphorically, these four concepts are related to the formal, material, efficient, and final causes of Aristotle.

Structural science is grounded in the notion that natural ordering relationships exist among objects of increasing complexity. Behaviors within structural science are related to causal structures which dynamically interrelate the degrees of organization in accord with affinities. An ordering structure may be imposed on local processes by higher order organizations—leading to the intertwined concepts of bottom-up, top-down, inside-outward, and outside-inward causality.[8,9]

Three examples illustrate applications of structural science to natural numbers and natural systems.

1. The first is composing a chemical structure in terms of one-to-one correspondence relations among atoms of the atomic table ($O°2$) by additive operations. A chemical structure is represented as an $O°3$ object (graph) with molecular orbitals (rather than atomic orbitals) co-ordinating different nuclei in space and time.
2. The second example is the dose–response relationships of the drug ($O°3$) as it dynamically induces health related changes in man, $O°7$.[9] This involves both one-to-one correspondence relationships among chemical structures ($O°3$, $O°4$) and effects on the coherent behavior of the body. Operations of addition and multiplication are active generators of responsiveness.
3. A third example lies in the coherent behavior of life and concerns the origin of mathematical operations. Mathematical operations of addition, multiplication, and exponentiation are conjectured to be a coherent, reproducible composition from subobjects of the pre-responsiveness objects ($O°1$, $O°2$, $O°3$, $O°4$) and from evolutionary, organized dynamics of responsiveness of the mind itself ($O°5$, $O°6$, $O°7$).[12]

NATURAL TIME SCALES

Natural science, as noted at the beginning, must be viewed from two perspectives—structural science and dynamic science. Of course, in the natural world, these two aspects of nature are inseparable. A natural structure implies a natural dynamic. A natural dynamic implies an object undergoing change in material composition or a change in space or both. Nonetheless, some separation of structural science from dynamic science can be achieved in terms of time scales.[9] Time scales can be used to calculate stability or instability with respect to a referential framework, such as Newtonian (absolute) time. From a practical perspective, each natural kind fits into an observable time scale where its organizational features may be studied. Within structural science, one can seek to compose natural kinds of increasing complexity from natural kinds of lesser complexity. It is a synthesis based on the emergence and classification of natural affinities, rather than merely analysis of decomposition products as is common in reductionistic philosophies. In simple, practical terms, syntheses within structural science must relate the parts, the whole, and the surroundings. In order to stabilize the whole, one seeks cooperative and collaborative relationships among the parts. This synthetic grounding of structural science assumes that the traditional philosophical dictum, "the whole is more than the sum of its parts" is inadequate to describe a structural science of living systems for two reasons: it ignores the role of the structural surroundings in generating the whole and it ignores the role of time scales which bind the behaviors of the whole to the surroundings. (In earlier writings,[10] I have termed the structural organizations "third-order cybernetics" because of the necessity of living organisms to sustain relationships between the parts, the whole, and the surroundings—the ecoment.)

ORGANIZATION OF STRUCTURAL SCIENCE

It is well known that structures of natural kinds have been determined by various scientific methods. Examples of structures are atoms, molecules, individual organisms, cells, organs, anatomy, and ecosystems. Structures are characterized by many attributes, but three fundamental attributes of natural structures are geometric form, electrical charge, and physical mass. Mass, as is well known, is a universal character of matter. Conservation implies additivity. Thus, additivity of mass applies to both horizontal and vertical compositions of structures. Electrical charge, another well known attribute of matter, is also conserved and is also an additive attribute of matter. But, charge is not a universal character of matter; the subatomic particle, the neutron, being neutral. Nevertheless, additivity of charge applies to both horizontal and vertical compositions of structures. The conservation of mass and charge constrain the compositions of higher structural objects from lesser objects by selecting objects that are stable over the time scale of operation. The well-known rule of composition of molecules from atoms as ratios of small whole numbers is a familiar example of the constraints relating natural kinds and natural numbers.

But, Form Is Not Conserved

The absence of conservation rules for geometric form is a critical source of biological diversity. The empirical observations of geometrical forms remain one of the cornerstones of science, but it is often mathematically convenient to ignore geometry and represent each object of a natural kind as a point. This misrepresents variability of geometric forms as basic attributes of the structure of natural kinds. To represent an object as a mathematical point is to ignore structural science; such representations lack wholeness or completeness. Both the quantity and the ratio of charge and mass contribute to geometric form. But, charges and masses of structures are not related by any single rule, rather the ratios vary from structure to structure. Changes in the charge associated with mass can change the geometric form of the parts and of the whole. In sharp contrast to the simple rule for the additive character of mass and charge, no simple rule relates charge or changes in charge to geometric form. Rather the contrary, small changes in electrical charge can be accompanied by massive changes in the geometry of the object. For example, the biological functioning of hemoglobin depends on the changes *in form* induced by binding oxygen. Equally importantly, the spatial variability of charge with increasing mass can generate virtually limitless number of geometric patterns of relationships between mass, charge and *form*. The conservation rules imply the one-to-one correspondence rules for both charge and mass for both vertical and horizontal compositions. But specific structural *forms*, which lie at the heart of organizational hierarchies of natural kinds, require more complex representations as mathematical objects known as graphs. Graphs, which are composed of vertices and edges, not merely mathematical points, conserve aspects of geometric form and molecular organization.

COMPOSING VERTICAL RELATIONS AMONG CATEGORIES OF NATURAL KINDS

In order to construct vertical relations from atoms to more complex objects, categories are formed by aggregating particles and atoms into categories of natural kinds that conserve mass and charge but create new geometric forms by combining natural affinities.[6-8,10] The following selection of degrees of organization is guided by relevance to human health:

$O°1$ subatomic particles,
$O°2$ atoms,
$O°3$ molecules,
$O°4$ biomacromolecules,
$O°5$ cell (composed from objects of $O°1$, $O°2$, $O°3$, $O°4$),
$O°6$ organs (composed from cells and cellular interactions),
$O°7$ individuals,
$O°8$ ecoment (defined as the surroundings of the individual organism),
$O°9$ ecosystems,
$O°10$ solar system,
$O°11$ universe.

The universe in its totality is included for logical completeness. The vertical categories of organization parallel the domains of discourse of the metalogics introduced at the beginning of this paper.

DISCUSSION

What is science? It was proposed that natural science could be composed from structural science and dynamic science. Within structural science, I propose that natural numbers and natural kinds can be placed in one-to-one correspondence with one another. Further, four descriptive concepts (closure, conformation, concatenation, and cyclicity) can be used to compose one enumerable category into another enumerable category. Finally, the relations between scientific activity (observation, description, and symbolization) and one to one correspondences are preserved over a hierarchy of natural kinds and natural affinities.

What is the relationship between this construction of natural objects and other views? J. Bronowski[14] wrote: "... science takes its coherence, its intellectual and imaginative strength together, from the concepts at which its laws cross, like knots in a mesh." Structural science seeks to construct one-to-one and many-to-one correspondences within the metaphorical "knots in a mesh" of Bronowski. Dynamic science seeks to construct temporal relationships over sequential "knots of the mesh" (time scales of organizations), which bind the mesh into a logical whole, a category of objects and morphisms that generate complex behaviors.

Western analytical and scientific philosophy is partially grounded in Hume's qualitative interpretations of Newtonian dynamics, often strongly influenced by probabilistic aspects of quantum theory and by Darwinism. Associations between these philosophies and human values remain highly controversial. Such philosophical views often focus on the attributes of space, time, and energy rather than structure. An alternative grounding of science and its relationship to human values, especially for medicine and law, is needed. Structural science is shown here to be directly related to one specific philosophy of truth, the one-to-one correspondence theory. The mathematical grounding of structural science becomes the patterns of relationships among structures rather than Newtonian forces. For structural scientific observations, the power of the correspondence theory is closely associated with its relationship to logical truth tables and hence the logical association between descriptions of observations and symbolizations of structure. In 1936, Birkhoff and von Neumann[15] published a critical synthesis of relations between mathematics and the electrodynamics of particles. Truth as coherence, in addition to truth as a one-to-one correspondence between natural numbers and natural kinds, plays a prominent role in the Birkhoff–von Neumann argument. Can natural science as a composition of structural science and dynamical science provide an alternative grounding for scientific philosophy? The material grounding is based on the atomic table. Higher order material structures are composed by means of categorical operations on simple objects under conservation rules. Dynamic behaviors serve to bind terms to a philosophy of truth relating natural numbers to natural kinds.

Category theory was born in 1945 as the mathematics of natural equivalencies.[16] Technically, category theory is grounded in objects, morphisms, and compositions of morphisms, within a framework of algebraic topology. In more user-friendly terms, category theory can be viewed as a conceptual "shell" for the generation of "paths." The logical potential of Mac Lane's "natural equivalencies" for relating natural kinds to natural numbers is demonstrated by the wide-ranging application of category theory to object-oriented algorithms.[16] Category theory is not just a recent advance in one subfield of mathematics, such as catastrophic or chaotic dynamics. Rather, category theory creates a different foundation for integrating logic, structures, and functions.[2–4] Mac Lane asserts all of mathematics is developable from an axiomization of categories.[4] Whereas set theory is often cited as the modern basis of mathematics and hence the central building material for the logic of the natural sciences, category theory is asserted to be equally powerful.[4] The categorical theory of natural equivalencies can be used to compose material structural "equivalencies" and hence to create relational systems of logics for more highly organized biological and social systems. This paper has contributed to a logical foundation for structural science, a syntactical and semantic grounding, based on the organization of the atoms of the atomic table, concepts describing atomic attributes, and the "inheritance" of these concepts during the vertical composition of higher degrees of organization. This is a necessary foundation for preparing the substrate for theories of more highly organized systems, such as cells, individuals, social systems, and so forth. Importantly, Ehresmann and Vanbremeersch[1,10,17,18] developed a theory of consciousness around hierarchical categorical operations. The logic of the Ehresmann and Vanbremeersch theory and the theory pesented here may be comparable.

A consistent philosophy of natural science as a hierarchical "mesh" can be organized from the concepts expressed here. An intimate intertwining of structural science and dynamic science generates a congruent picture of observations, descriptions, and symbolizations of natural science. A hierarchy of natural objects ranging from the very, very small to the very, very large can be organized within this mesh. Logically, classical, symbolic, and semantic descriptions can all be integrated within the framework of mathematical categories of natural kinds. Concrete examples of the intermingling and intertwining of structural objects and dynamic morphisms are widely known in medical diagnosis and therapy (for example, sickle cell anemia and chronic myelocytic leukemia).[13] Natural science will benefit from carefully analyzing such medical examples and using the knowledge to construct complex systems within a categorical framework.[12,19] In summary, while simple structural systems are often amenable to one-to-one correspondence rules, the coherence of life emerges from the hierarchical organization of cooperative, congenitive, collaborative, and communicative systems.[6-8,10,19]

ACKNOWLEDGMENTS

I am grateful to Professor Andree Ehresmann for numerous exchanges on the relations between mathematics, science, and philosophy. Discussions with Richard Khuri, Paul Kainen, and Andrew Vogt and other WESS members have enlivened and enlightened my explorations.

REFERENCES

1. EHRESMANN, A.C. & J.P. VANBREMEERSCH. 1987. Hierarchical evolutive systems. Bull. Math. Biol. **49**: 13–50.
2. HATCHER, W.S. 1968. Foundations of Mathematics. W.B. Saunders. Philadelphia, PA.
3. LAWVERE, F.W. & S.H. SCHANUEL. 1997. Conceptual Mathematics. Cambridge University Press. Cambridge, England.
4. MACLANE, S. 1986. Mathematics: Form and Function. Springer Verlag. Berlin.
5. MALATESTA, M. 1997. The Primary Logic. Gracewing. Leominster, Herefordshire, England.
6. CHANDLER, JERRY L.R. 1991. Complexity: A phenomenologic and semantic analysis of dynamical classes of natural systems. WESScomm **1**: 34–40.
7. CHANDLER, JERRY L.R. 1992. Complexity II. Logical constraints on the structure of scientific languages. WESScomm **2**: 34–37.
8. CHANDLER, JERRY L.R. 1996. Complexity III. Emergence, symbolization and notation. WESScomm **5**: 1–14.
9. CHANDLER, JERRY L.R. 1985. New mechanistic models for risk assessment. Fundam. Appl. Tox. **5**: 634–652.
10. VANBREMEERSCH, J.P. et al. 1996. Are Interactions between Different Time Scales Characteristic of Complexity? Actes Du Symposium ECHO. Amien, France.
11. CHANDLER, JERRY L.R. 1995. Third order cybernetics. In Proceedings, 14th International Symposium of Cybernetics. Jean Ramaekers, Ed. International Association for Cybernetics. Namur, Belgium.
12. CHANDLER, JERRY L.R. 1998. Semiotics of emergence in a simple cell. In Informational Processing in Cells and Tissues. M. Holcombe & Ray Paton, Eds.: 185–195. Plenum Press. Sheffield, England.
13. ALAVANJA, M.A. et al. 1987. Biochemical epidemiology: potential contributions to cancer risk assessment. J. Natl. Cancer Inst. **78**: 633–643.
14. BRONOWSKI, J. 1956. Science and Human Values. Harper Tourchbook. New York. p. 53.
15. BIRKOFF, G. & J. VON NEUMANN. 1936. Logic of quantum mechanics. Ann. Math. **37**: 823–843.

16. EILENBERG, S. & S. MACLANE. 1945. General theory of natural equivalences. Trans. Am. Mat. Soc. **58:** 231–294.
17. EHRESMANN, A.C. & J.-P. VANBREMEERSCH. 1996. Multiplicity principle and emergence in MES. J. Syst. Anal. Model Sim. SAMS **26:** 81–117.
18. EHRESMANN, A.C. & J.-P. VANBREMEERSCH. 1997. Information processing and symmetry-breaking in memory evolutive systems. Biosystems **43:** 25–40.
19. CHANDLER, JERRY L.R. 1998. Applications of category theory to theoretical biochemistry. Abstracts of the DIMACs Conference on Applications of Graph Theory in Chemistry. March. Princeton University. Princeton, NJ.

The Paradox of the Visibly Irrelevant

STEPHEN JAY GOULD
Museum of Comparative Zoology, Harvard University, Cambridge, Massachusetts 02138, USA

An odd principle of human psychology, well known and exploited by the full panoply of prevaricators, from charming barkers like Barnum to evil demagogues like Goebbels, holds that even the silliest of lies can win credibility by constant repetition. In current American parlance, these proclamations of "truth" by xeroxing fall into the fascinating domain of "urban legends."

My favorite bit of nonsense in this category intrudes upon me daily, and in very large type, thanks to a current billboard ad campaign by a company that will remain nameless. The latest version proclaims: "Scientists say we use 10% of our brains. That's way too much." Just about everyone regards the "truth" of this proclamation as obvious and incontrovertible—though you might still start a barroom fight over whether the correct figure should be 10, 15, or 20% (I have heard all three asserted with utter confidence). But this particular legend can only be judged as even worse than false: for the statement is truly meaningless and nonsensical. What do we mean by "90% unused"? What is all this superfluous tissue doing? The claim, in any case, can have no meaning until we develop an adequate theory about how the brain works. For now, we don't even have a satisfactory account for the neurological basis of memory and its storage—surely the *sine qua non* for formulating any sensible notion about unused percentages of brain matter! (I think that the legend developed because we rightly sense that we ought to be behaving with far more intelligence than we seem willing to muster—and the pseudoquantification of the urban legend acts as a falsely rigorous version of this legitimate, but vague, feeling.)

In my field of evolutionary biology, the most prominent urban legend—another "truth" known by "everyone"—holds that evolution may well be the way of the world, but one has to accept the idea with a dose of faith because the process occurs far too slowly to yield any observable result in a human lifetime. Thus, we can document evolution from the fossil record and infer the process from the taxonomic relationship of living species, but we cannot see evolution on human timescales "in the wild."

In fairness, we professionals must shoulder some blame for this utterly false impression about evolution's invisibility in the here and now of everyday human life. Darwin himself—though he knew and emphasized many cases of substantial change in human time (including the development of breeds in his beloved pigeons)—tended to wax eloquent about the inexorable and stately slowness of natural evolution. In a famous passage from the *Origin of Species*, he even devised a striking metaphor about clocks to underscore the usual invisibility:

> It may be said that natural selection is daily and hourly scrutinizing, throughout the world, every variation, even the slightest; rejecting that which is bad, preserving and adding up all that is good; silently and invisibly working .... We see nothing of these slow changes in progress until the hand of time has marked the long lapse of ages.

Nonetheless, the claim that evolution must be too slow to see can only rank as an urban legend—though not a completely harmless tale in this case, for our creationist incubi can then use the fallacy as an argument against evolution at any scale, and many folks take them seriously because they just "know" that evolution can never be seen in the immediate here and now. In fact, a precisely opposite situation actually prevails: biologists have documented a veritable glut of cases for rapid and eminently measurable evolution on timescales of years and decades.

However, this plethora of documents—while important for itself, and surely valid as a general confirmation for the proposition that organisms evolve—teaches us rather little about rates and patterns of evolution at the geological scales that build the history and taxonomic structure of life. The situation is wonderfully ironic, a point that I have tired to capture in the title of this article. The urban legend holds that evolution is too slow to document in palpable human lifetimes. The opposite truth has affirmed innumerable cases of measurable evolution at this minimal scale—but, to be visible at all over so short a span, evolution must be far too rapid (and transient) to serve as the basis for major transformations in geological time. Hence, the "paradox of the visibly irrelevant"—or, "if you can see it at all, it's too fast to matter in the long run!"

Our best and most numerous cases have been documented for the dominant and most evolutionarily active organisms on our planet—bacteria. In the most impressive of recent examples, Richard E. Lenski and Michael Travisano[1] monitored evolutionary change for 10,000 generations in 12 laboratory populations of the common human gut bacterium, *Escherichia coli*. By placing all 12 populations in identical environments, they could study evolution under ideal experimental conditions of replication, a rarity for the complex and unique events of evolutionary transformation in nature. In a fascinating set of results, they found that each population reacted and changed differently, even within an environment made as identical as human observers know how to do. Yet, Lenski and Travisano did observe some important and repeated patterns within the diversity. For example, each population increased rapidly in average cell size for the first 2000 generations or so, but then remained nearly stable for the last 5000 generations.

But a cynic might still reply: fine, I'll grant you substantial observable evolution in the frenzied little world of bacteria, where enormous populations and new generations every hour allow you to monitor 10,000 episodes of natural selection in a manageable time. But a similar "experiment" would consume thousands of years for multicellular organisms that measure generations in years or decades rather than minutes or hours. So we may still maintain that evolution cannot be observed in the big, fat, furry, sexually reproducing organisms that serve as the prototype for "life" in ordinary human consciousness. (A reverse cynic would then rereply that bacteria truly dominate life, and that vertebrates only represent a late-coming side-issue in the full story of evolution, however falsely promoted to centrality by our own parochial focus. But we must leave this deep issue to another time.)

I dedicate this essay to illustrating our cynic's error. Bacteria may provide our best and most consistent cases for obvious reasons, but measurable (and substantial) evolution has also, and often, been documented in vertebrates and other complex multicellular organisms. The classic cases have not exactly been hiding their light under a bushel, so I do wonder why the urban legend of evolution's invisibility persists with such strength. Perhaps the firmest and most elegant examples involve a group of organisms named to commemorate our standard bearer himself—Darwin's finches of the Galapagos Islands, where my colleagues Peter and Rosemary Grant have spent many years documenting fine-scale evolution in such adaptively important features as size and strength of the bill (a key to the mechanics of feeding), as rapid climatic changes force an alteration of food preferences. this work formed the basis for Jonathan Weiner's excellent book *The Beak of the Finch*, so the story has certainly been well and prominently reported in both the technical and popular press.

Nonetheless, new cases of such short-term evolution still maintain enormous and surprising power to attract public attention—for interesting and instructive, but utterly invalid, reasons as I shall show. I devote this essay to the three most prominent examples of recent publications that received widespread attention in the popular press as well. (One derives from my own research, so at least I can't be accused of sour grapes in the debunking that will follow, though I trust that readers will also grasp the highly positive twist that I will ultimately impose upon my criticisms.) I shall briefly describe each case, then present my two general critiques of their prominent reporting by the popular press, and finally explain why such cases teach us so little about evolution in the large, yet remain so important for themselves, and at their own equally legitimate scale.

GUPPIES FROM TRINIDAD

In many drainage systems on the island of Trinidad, populations of guppies live in downstream pools, where several species of fish can feed upon them. "Some of these species prey preferentially on large, mature-size classes of guppies." (I take all quotes from the primary technical article by Reznick et al.[2] that inspired later press accounts. Other populations of the same species live in "upstream portions of each drainage" where most "predators are excluded ... by rapids or waterfalls, yielding low-predation communities."

In studying both kinds of populations, Reznick and colleagues found that "guppies from high-predation sites experience significantly higher mortality rates than those from low-predation sites." They then reared both kinds of guppies under uniform conditions in the laboratory and found that fishes from high-predation sites in lower drainages matured earlier and at a smaller size. "They also devote more resources to each litter, produce more, smaller offspring per litter and produce litters more frequently than guppies from low-predation localities."

This combination of observation from nature and the laboratory yields two important inferences. First, the differences make adaptive sense, for guppies subjected to greater predation would fare better if they could grow up fast and reproduce both copiously and quickly before the potential boom falls—a piscine equivalent of the old motto for electoral politics in Boston: vote early and vote often. On the other hand, guppies in little danger of being eaten might do better to bide their time and grow big and strong before engaging their fellows in any reproductive competition. Second, because these differences persist when both kinds of guppies are reared in identical laboratory environments, the distinction must record genetically based and inherited results of divergent evolution between the populations.

In 1981, Reznick had transferred some guppies from high-predation downstream pools into low-predation upstream waters then devoid of guppies. These transplanted populations evolved rapidly to adopt the reproductive strategy favored by indigenous populations in neighboring upstream environments: delayed sexual maturity at larger size and longer life. Moreover, Reznick and colleagues made the interesting observation that males evolved considerably more rapidly in this favored direction. In one experiment, males reached their full extent of change within 4 years, while females continued to alter after 11 years. Because the laboratory populations had shown higher heritability for these traits in males than in females, these results make good sense. (Heritability may be roughly defined as the correlation between traits in parents and offspring due to genetic differences. The greater the heritable basis of a trait, the faster the feature can evolve by natural selection.)

This favorable set of circumstances—rapid evolution in a predictable and presumably adaptive direction based on traits known to be highly heritable—provides a "tight" case for well-documented (and sensible) evolution at scales well within the purview of human observation, a mere decade in this case. The headline for the news report on this paper in *Science* magazine (March 28, 1997) read: "Predator-free guppies take evolutionary leap forward."

LIZARDS FROM THE EXUMA CAYS, BAHAMA ISLANDS

During most of my career, my field work has centered on the biology and paleontology of the land snail *Cerion* in the Bahama islands. During these trips, I have often encountered fellow biologists devoted to other creatures. In one major program of research, Tom Schoener (a biology professor at the University of California, Davis) has, with numerous students and colleagues, been studying the biogeography and evolution of the ubiquitous little lizard, *Anolis*—for me just a fleeting shadow running across a snail-studded ground, but for them a focus of utmost fascination (while my beloved snails, I assume, just blend into their immobile background).

In 1977 and 1981, Schoener and colleagues transplanted groups of 5 or 10 lizards from Staniel Cay in the Exuma chain to 14 small and neighboring islands that housed no lizards. In 1991, they found that the lizards had thrived (or at least survived and bred) on most of these islands, and they collected samples of adult males from each experimental island with an adequate population. In addition, they gathered a larger sample of males from areas on Staniel Cay that had served as the source for original transplantation in 1977 and 1981.

This study then benefits from general principles learned by extensive research on numerous *Anolis* species throughout the Bahama islands. In particular, relatively longer limbs permit greater speed, a substantial advantage provided that preferred perching places can accommodate long-legged lizards. Trees, and other "thick" perching places therefore favor the evolution of long legs. Staniel Cay itself includes a predominant forest, and the local *Anolis* tend to be long legged. But when lizards must live on thin twigs in bushy vegetation, the agility provided by shorter legs (on such precarious perches) may outweigh the advantages in speed that longer legs would provide. Thus, lizards living on narrow twigs tend to be shorter-legged. The small Cays that received the 14 transported populations have little or no forest growth and tend instead to be covered with bushy vegetation (and narrow twigs).

J.B. Losos, the principal author of the new study, therefore based an obvious prediction on these generalities. The populations had been transferred from forests with wide perches to bushy islands covered with narrow twigs. "From the kind of vegetation on the new islands," Losos stated, "we predicted that the lizards would develop shorter hindlimbs." Their published study validates this expected result: a clearly measurable change, in the predicted and adaptive direction, in less than 20 years.[3] A news report appeared in *Science* magazine (May 2, 1997) under the title: "Catching lizards in the act of adapting."

This study lacks a crucial piece of documentation that the Trinidadian guppies provided, an absence immediately noted by friendly critics and fully acknowledged by the authors. Losos and colleagues have not studied the heritability of leg length in *Anolis sagrei*, and therefore cannot be certain that their results record a genetic process of evolutionary change. The growth of these lizards may feature extensive flexibility in leg length, so that the same genes yield longer legs if lizards grow up on trees and shorter legs if they always cavort in the bushes (just as the same genes can lead to a thin or fat human being depending upon a personal history of nutrition and exercise). In any case, however, a sensible and apparently adaptive change in average leg length has occurred within 20 years on several islands, whatever the cause of modification.

SNAILS FROM GREAT INAGUA, BAHAMA ISLANDS

Most of Great Inagua, the second largest Bahamian Island (Andros wins first prize), houses a large and ribby *Cerion* species named *C. rubicundum*. But fossil deposits of no great age lack this species entirely and feature instead an extinct form named *Cerion excelsior*, the largest of all *Cerion* species. Several years ago, on a mudflat in the southeastern corner of Great Inagua, David Woodruff (of the University of California, San Diego) and I collected a remarkable series of shells that seemed to span (and quite smoothly) the entire range of form from extinct *C. excelsior* to modern *C. rubicundum*. Moreover, and in general, the more eroded and "older looking" the shell, the closer it seemed to lie to the anatomy of extinct *C. excelsior*.

This situation suggested a local evolutionary transition by hybridization, as *C. rubicundum*, arriving on the island from an outside source, interbred with indigenous *C. excelsior*. Then, as *C. excelsior* declined towards extinction while *C. rubicundum* thrived and increased, the average anatomy of the population transformed slowly and steadily in the direction of the modern form; this hypothesis sounded good and sensible, but we could devise no way to test our idea, because all the shells had been collected from a single mud flat (analogous to a single bedding plane of a geological stratum), and we could not determine their relative ages. The pure *C. excelsior* shells "looked" older, but such personal impressions count for less than nothing (subject as they are to a researcher's bias) in science. So we got stymied and put the specimens in a drawer.

Several years later, I teamed up with paleontologist and geochemist Glenn A. Goodfriend from the Carnegie Institution of Washington. He had refined a dating technique based on changes in the composition of amino acids in the shell over time. By keying these amino acids changes to radiocarbon dates for some of the shells, we could estimate the age of each shell. A plot of shell age versus position on an anatomical spectrum from extinct *C. excelsior* to modern *C. rubicundum* produced a beautiful correlation between age and anatomy: the younger the specimen, the closer to the modern anatomy.

This ten to twenty thousand year transition by hybridization exceeds the time period of the Trinidad and Exuma studies by three orders of magnitude (that is, by a factor of 1000), but even 10,000 years represents a geological eye-blink in the fullness of evolutionary time; whereas this transformation in our snails marks a full change from one species to another, not just a small decrement of leg length or a change in the timing of breeding within a single species. (For details, see G.A. Goodfriend and S.J. Gould.[4]) Harvard University's release (with no input from me) carried the headline: "snails caught in act of evolving."

A scanning of any year's technical literature in evolutionary biology would yield numerous and well-documented cases of such measurable, small-scale evolutionary change, thus disproving the urban legend that evolution must always be too slow to observe in the geological microsecond of a human lifetime. These three studies, all unusually complete in their documentation and in their resolution of details, do not really rank as "news" in the journalist's prime sense of novelty or deep surprise. Nonetheless, each of these three studies became subjects for front page stories in either the *New York Times* or *Boston Globe*.

Now please don't get me wrong. I do not belong to the cadre of rarefied academics who cringe at every journalistic story about science for fear that the work reported might become tainted with popularity thereby. And, in a purely "political" sense, I certainly won't object if major newspapers choose to feature any result of my profession as a lead story—especially, if I may be self-serving for a moment, when one of the tales reports my own work! Nonetheless, this degree of public attention for workaday results in my field (however elegantly done), does fill me with wry amusement, if only for the general reason that most of us feel a tickle in the funny bone when we note a gross imbalance between public notoriety and the true novelty or importance of an event, as when Hollywood spinmeisters manage to depict their client's ninth marriage as the earth's example of true love triumphant and permanent.

Of course I'm delighted that some ordinary, albeit particularly well done, studies of small scale evolution struck journalists as front page news. But I still feel impelled to ask why these studies, rather than 100 others of equal care and merit that appear in our literature every month, caught this journalistic fancy and inspired such prime attention. When I muse over this issue, I can only devise two reasons, both based on deep and interesting fallacies well worth identifying and discussing. In this sense, the miselevation of everyday good work to surprising novelty may teach us something important about public attitude towards evolution, and towards science in general. We may, I think, resolve each of the two fallacies by contrasting the supposed meaning of these studies, as reported in public accounts, with the significance of such work as viewed by professionals in the field.

THE FALLACY OF THE CRUCIAL EXPERIMENT

In high school physics classes, we all learned a heroically simplified version of scientific progress based on a model that does work sometimes, but by no means always: the *experimentum crucis*, or crucial experiment. Newton or Einstein? Ptolemy or Copernicus? Special Creation or Darwin? To find out, perform a single, decisive experiment with a clearly measurable result replete with decisive power to decree yea or nay.

The decision to treat a limited and particular case as front page news must be rooted in this fallacy. Reporters must imagine that evolution can be proved by a single crucial case, so that any of these stories may provide decisive confirmation of Darwin's truth—a matter of some importance given the urban legend that evolution, even if valid, must be invisible on human timescales.

But two counterarguments vitiate this premise. First, as a scientific or intellectual issue, we hardly need to "prove" evolution by discovering new and elegant cases. We do not, after all, expect to encounter a page-one story with the headline "new experiment proves earth goes around sun, not vice versa. Galileo vindicated." The fact of evolution has been equally well documented for more than a century.

Second, and more generally, single "crucial" experiments rarely decide major issues in science, especially in natural history where nearly all theories require data about "relative frequencies" (or percentage of occurrences), not pristine single cases. Of course, for a person who believes that evolution never occurs at all, one good case can pack enormous punch, but science resolved this basic issue more then one hundred years ago. Nearly every interesting question in evolutionary theory asks "how often" or "how dominant in setting the pattern of life"—not "does this phenomenon occur at all?" For example, on the most important issue of all—the role of Darwin's own favored mechanism of natural selection—single examples of selection's efficacy advance the argument very little. We already know, by abundant documentation and rigorous theorizing, that natural selection can and does operate in nature. We need to determine the *relative strength* of Darwin's mechanism among a set of alternative modes for evolutionary change; and single cases, however elegant, cannot establish a relative frequency.

Professionals also commit this common error of confusing well-documented single instances with statements about relative strength among plausible alternatives. For example, we would like to know how often small and isolated populations evolve differences as adaptive responses to local environments (presumably by Darwin's mechanism of natural selection), and how often such changes occur by the random process known as "genetic drift," a potentially potent phenomenon in small populations (just as a small number of coin flips can depart radically from 50–50 for heads and tails, while a million flips with an honest coin cannot stray too far from this ideal). Losos's study on lizard legs provides one vote for selection (if the change turns out to have a genetic basis), because leg length altered in a predicted direction towards better adaptation to local environments on new islands. But even such an elegant case cannot prove the domination of natural selection in general. Losos has only shown the power of Darwin's process in a particular example. Yet the reporter for *Science* magazine made this distressingly common error in concluding: "If it [change in leg length] is rooted in the genes, then the study is strong evidence that isolated populations diverge by natural selection, not genetic drift as some theorists have argued." Yes, strong evidence for these lizards on that island during those years—but not proof for the general domination of selection over drift. Single cases don't establish generalities, so long as alternative mechanisms retain their theoretical plausibility.

THE PARADOX OF THE VISIBLY IRRELEVANT

As a second reason for overstating the centrality of such cases in our general understanding of evolution, many commentators (and research scientists as well) ally themselves too strongly with one of the oldest (and often fallacious) traditions of Western thought: reductionism, or the assumption that laws and mechanics of the smallest constituents must explain objects and events at all scales and times. Thus, if we can render the behavior of a large body (an organism or a plant, for example) as a consequence of atoms and molecules in motion, we feel that we have developed a "deeper," or "more basic" understanding than if our explanatory principles engage only large objects themselves and not their constituent parts.

Reductionists assume that documenting evolution at the smallest scale of a few years and generations should provide a general model of explanation for events at all scales and times—so these cases should become a gold standard for the entire field, hence their status as front-page news. The authors of our two studies on decadal evolution certainly nurture such a hope. Reznick and colleagues end their publication on Trinidadian guppies by writing: "It is part of a growing body of evidence that the rate and patterns of change attainable through natural selection are sufficient to account for the patterns observed in the fossil record." Losos and colleagues say much the same for their lizards "Macroevolution may just be microevolution writ large—and, consequently, insight into the former may result from study of the latter."

We tend to become beguiled by such warm and integrative feelings (for science rightly seeks unity and generality of explanation). But does integration by reduction of all scales to the rates and mechanisms of the smallest really work for evolution, and do we crave this style of unification as the goal of all science? I think not, and I also regard our best general reason for skepticism as conclusive for this particular subject, however rarely appreciated though staring us in the face.

These shortest term studies are elegant and important, but they cannot represent the general mode for building patterns in the history of life. The reason for their large-scale impotence strikes most people as deeply paradoxical, even quite funny, but the argument truly cannot be gainsaid. Evolutionary rates as measured for guppies and lizards, are *vastly too rapid* to represent the general modes of change that build life's history through geological ages. But how can I say such a thing? Isn't this statement ridiculous *a priori*? How could these tiny, minuscule changes—a little less leg, a minimally larger size—represent too much of anything? Doesn't the very beauty of these studies lie in their minimalism? We have always been taught that evolution is wondrously slow and cumulative, a grain by grain process, a penny a day towards the domain of Bill Gates. Doesn't each of these studies document a grain? Haven't my colleagues and I found the "atom" of evolutionary incrementation?

I believe that these studies have discerned something important, but they have discovered no general atom. These measured changes over years and decades are too fast by several orders of magnitude to build the history of life by simple cumulation. Reznick's guppy rates range from 3700 to 45,000 darwins (a standard metric for evolution, expressed as change in units of standard deviation—a measure of variation around the mean value of a trait in a population—per million years). By contrast, rates for major trends in the fossil record generally range from 0.1 to 1.0 darwins. Reznick himself states that "the estimated rates [for guppies] are ... four to seven orders of magnitude greater than those observed in the fossil record" (that is, ten thousand to ten million times faster?).

Moreover and with complete generality, thus constituting the "paradox of the visibly irrelevant" in my title, we may say that any change measurable *at all* over the few years of an ordinary scientific study must be occurring far too rapidly to represent ordinary rates of evolution in the fossil record. The culprit of this paradox, as so often, can be identified as the vastness of time (a concept that we can appreciate "in our heads" but seem quite unable to place into the guts of our intuition). The key principle, however ironic, requires such a visceral understanding of earthly time: if a case of evolution proceeds with sufficient speed to be discerned by our instruments in just a few years—that is, if the change becomes substantial enough to stand out as a genuine and directional effect above the random fluctuations of nature's stable variation and our inevitable errors of measurement—then we have witnessed something far too substantial to serve as an atom of steady incrementation in a paleontological trend. Thus, to restate the paradox: if we can measure it at all (in a few years), it is too powerful to be the stuff of life's history.

If large scale evolution proceeded by stacking Trinidad guppy rates end to end, then any evolutionary trend would be completed in a geological moment, not over the many million years actually observed. "Our face from fish to man," to cite the title of a famous old account of evolution for popular audiences, would run its course within a single geological formation, not over more than 400 million years, as our fossil record demonstrates.

Evolutionary theory must figure out how to slow down these measured rates of the moment, not how to stack them up! In fact, most lineages are stable (*non*-changing) nearly all the time in the fossil record. When lineages do change, their alteration usually occurs "momentarily" in a geological sense (that is, confined to a single bedding plane) and usually leads to the origin of a new species by branching. Evolutionary rates during these moments may match the observed speed of Trinidadian guppies and Bahamian lizards, because most bedding planes represent several thousand years. But, during most of a typical species' lifetime, no change accumulates, and we need to understand why. The sources of stasis have become as important for evolutionary theory as the causes of change.

(To illustrate how poorly we grasp this central point about time's immensity, the reporter for *Science* magazine called me when my *Cerion* article, co-authored with Glenn Goodfriend, appeared. He wanted to write an accompanying news story about the exception I had found to own theory of punctuated equilibrium, an insensibly gradual change over 10 to 20 thousand years. I told him that, although exceptions abound, this case does not lie among them, but actually represents a strong confirmation of punctuated equilibrium! We found all 20,000 years worth of snails on a single mud flat, that is, on what would become a single bedding plane in the geological record. Our *entire* transition occurred in a geological moment and represented a punctuation, not a gradual sequence of fossils. We were able to "dissect" the punctuation in this unusual case, hence the value of our publication, because we could determine ages for the individual shells. The reporter, to his credit, completely revised his originally intended theme, and published an excellent account.)

In conclusion, I suspect that most cases like the Trinidadian guppies and Bahamian lizards represent transient and momentary blips and fillips that "flesh out" the rich history of lineages in stasis, not the atoms of substantial and steadily accumulated evolutionary trends. Stasis is a dynamic phenomenon. Small local populations and parts of lineages make short and temporary forays of transient adaptation, but these tiny units almost always die out or get reintegrated into the general pool of the species. (Losos himself regards the new island populations of lizards as evolutionarily transient in exactly this sense, because for such tiny and temporary colonies are almost always extirpated by hurricanes in the long run. How, then, can such populations represent atoms of a major evolutionary trend? The news report in *Science* magazine ends by stating: "But whether the lizards continue to evolve depends largely on the winds of fate, says Losos. These islet are periodically swept by hurricanes that could whisk away every trace of anolian evolution.")

But transient blips and fillips are no less important than major trends in the total "scheme of things." Both represent evolution operating at a standard and appropriate measure for a particular scale and time—Trinidadian blips for the smallest and the most local moment, faces from fish to human for the largest and the most global frame. One scale doesn't translate into another. No single scale can be deemed more important than any other, and none operates as a basic model for all the others. Each scale embodies something precious and unique to teach us; none can be labeled superior or primary. (Guppies and lizards, in their exposition of momentary detail, give us insight, unobtainable at broader scales, into the actual mechanics of adaptation, natural selection, and genetic change.)

The common metaphor of the science of fractal models—Mandelbrot's familiar argument that the coast of Maine has no absolute length, but depends upon the scale of measurement—epitomizes this principle well. When we study guppies in a pond in Trinidad, we are operating at a scale equivalent to measuring the coastline by wrapping our string around every boulder on every headland of Acadia National Park. When we trace the increase in size of the human brain from Lucy (about 4 million years ago) to Lincoln, we are measuring the coastline as depicted on my page of Maine in Hammond's Atlas. Both scales are exactly right for their appropriate problems. You would be a fool to spend all summer measuring the details in one cove in Acadia, if you just wanted to know the distance from Portland to Machiasport for your weekend auto trip.

I find a particular intellectual beauty in such fractal models—for they invoke hierarchies of inclusion (the single cove embedded within Acadia, embedded within Maine) to deny hierarchies of worth, importance, merit, or meaning. You may ignore Maine while studying the sand grain and be properly oblivious of the grain while perusing the single-page map of Maine on the single pages of your atlas. But you can love and learn from both scales at the same time. Evolution does not lie patent in a clear pond on Trinidad any more than the universe (*pace* Mr. Blake) lies revealed in a grain of sand. But how poor would be our understanding, how bland and restricted our sight, if we could not learn to appreciate the rococo details that fill our immediate field of vision, while forming, at another scale, only some irrelevant and invisible jigglings in the majesty of geological time.

REFERENCES

1. LENSKI, RICHARD E. & MICHAEL TRAVISANO. 1994. Proc. Natl. Acad. Sci. USA **91**: 6808–6814.

2. REZNICK, D.N., F.H. SHAW, F.H. RODD & R.G. SHAW. 1977. Evaluation of the rate of evolution in natural populations of guppies (*Poecilia reticulata*). Science **275:** 1934–1937.
3. LOSOS, J.B., K.I. WARHEIT & T.W. SCHOENER. 1997. Adaptive differentiation following experimental island colonization in Anolis lizards. Nature **387:** 70–73.
4. GOODFRIEND, G.A. & S.J. GOULD. 1996. Paleontology and chronology of two evolutionary transitions by hybridization in the Bahamian land snail *Cerion*. Science **274:** 1894–1897.

The Interplay of Cyclic and Linear Time in the Biological World

PIER LUIGI LUISI[a]

Institute of Polymers, ETH Zentrum CAB, CH-8092 Zürich, Switzerland

INTRODUCTION

The desire to grasp time has been known since antiquity. It has usually been attempted by interpreting time as a god, which conveys the notion that time cannot be explained rationally, and that it is something powerful and mysterious. For the old Greeks, the river Oceanos, flowing around Earth, was the generation of all, including time itself. In fact, Oceanos is Chronos, Time itself, and also Aion, the power of controlling the changes in the world, and the mystical round element Omega that in late antiquity symbolized the "lifetime" and also eternity (FIG. 1). This identity between time and a river that flows around Earth and encompasses all is very suggestive.

All ancient civilizations had various gods of time, generally all represented in a male form. Chronos was called Saturn by the Romans and was often depicted as an old man with a sickle, the symbol of death. A few Roman illustrations portray him with a sickle and, at the same time, with an Orobourous, the snake that bites its own tail, to symbolize circularity and eternity and also perfection, something closed on itself. The theme of time and mythology is fascinating, and I would like to refer interested readers to the excellent volume on the subject by Marie Louise von Franz.[1]

CYCLIC AND LINEAR TIME

The combination of the sickle and the circle is also very suggestive, as the sickle symbolizes death, and the Oroborous symbolizes eternity. In fact, ever since ancient times, time has appeared in two aspects. One aspect is cyclic, or periodic, time—the time that repeats itself over and over again. The other is linear and irreversible time, which never reverses.

We are all familiar with cyclic time. Astronomic observations have implemented this notion since prehistoric time: The sun rises every morning, goes up to its zenith at midday, sets in the west, and in the morning comes up again, and so on and so on. The same is true with the phases of the moon and with the movements of the stars. On Earth, we are all familiar with the periodicity of the seasonal changes: Summer, fall, winter, and spring follow each other; and the tree loses its leaves and makes them again half a year later; the snake sheds its skin and renews it, the birds return in spring and leave again when the weather begins to be cold.

These periodic events give us a hint of eternity: Everything is constant and repeats itself regularly. One does not need to be concerned; everything repeats itself forever.

---

[a]Address for correspondence: luisi@ifp.mat.ethz.ch (e-mail).

FIGURE 1. "Oceanos flows as a great river round the rim of the earth, the psyche of the universe and the generation of all. He is Time (Chronos) itself, and also Aion .... As a boundary of the world, Oceaonos is its compelling destiny (Roman relief, the so-called Bocca della Verità, Santa Maria in Cosmedin, Roma)." (Quoted from Marie Louise von Franz.[1])

But time, as we mentioned already, also has an ominous side. Time runs in one direction only and never comes back. We know all too well that no time-moment can come twice, and we all know that death is part of this implacable arrow of time. The arrow, launched by the bow, goes in one direction only. Birds may come again in spring, but each individual bird is going to die within a relatively short time interval; each living thing has an inner clock that beats and beats towards its own end. *Tempus erit*, as was said in the old times.

The notion of an "arrow of time" is used in this relation, and several books have been published with this title. The physicists keep asking why it is so, Why is there this symmetry in our perception of time? In fact, the laws of physics per se do not demand that time flows irreversibly. Most of the physics laws are beautifully symmetrical with respect to time: Time could very well be negative and run in two directions, and the laws of physics would respect that.

In physics, this is related to to the second principle of thermodynamics, to the continuous increase of entropy in our closed system, linked with the long phase of expansion of the universe, which is a unidirectional process. I said already that the literature in the field of the arrow of time is very large, and much of it is often specialized physics literature. Among the books that are accessible to nonphysicists is one by Coveney and Highfield,[2] as well as a book of quite a different nature written by Stephen Hawking, *A Brief History of Time*.[3]

I would like to make a couple of statements about Hawking's book. The author distinguishes three arrows of time: a thermodynamic one, a cosmological one, and a psychological one. It is interesting to read about the connection between these three concepts. For Hawking, the most important is the cosmological arrow of time, which begins with the Big Bang. The Big Bang has a particular meaning for this symposium, because it marks the beginning of time. The Big Bang is the creation of our universe, and recently the Roman Catholic Church has accepted the idea of the Big Bang, stating, however, that one should inquire what happened after the Big Bang, but not what happened during it, as this is the work of God.

According to one model of the origin of the universe, the expansion of the Big Bang was followed by a period of compression, which eventually will culminate in a big crank, an implosion, returning to a point with infinite density. From this, the entire process should start over again—here, we have an example of cyclic time in cosmic dimensions.

Hawking, in his book, tries to defend a model in which there is no circularity. He states the basic argument in the following excerpt his book: "If the Universe had a beginning, we could assume that there was a Creator who did it. But if the universe is completely closed in itself, without boundary, then there would be no beginning and no end: it would simply be. What place, then, for a Creator?"

THE ORIGIN OF LIFE

Let us leave these cosmic dimensions and go back to Earth. Instead of the Big Bang, the origin of the universe, let us consider the question of the origin of life—a universal question that features the arrow of time. This is illustrated in the following cartoon (FIG. 2), which represents what all our biology textbooks say on the matter of origin of life: that life on our planet is derived from inanimate matter.

The story goes as follows: Once upon a time, when it was formed about 4.5 billion years ago, Earth was a fireball, which could not possibly host life. Then, at a certain point, there was life. Therefore, if we eliminate the possibility of divine creation and of panspermia (the idea that life came from some other planet), the consequence is that life came by itself from the rocks and simple organic matter that were forming the Earth's crust.

How was this implemented? The mechanism goes back to Alexander Oparin, the Russian chemist who, inspired by the theory of Darwinian evolution, came up with

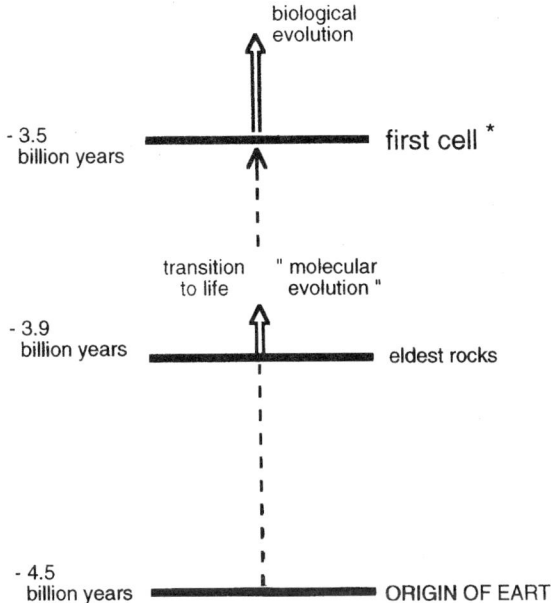

FIGURE 2. Prebiotic molecular evolution, going from the origin of Earth (4.5 billion years ago) until the age of the first cellular fossils, 3.5 billion years ago. Transition to life probably occurred within a few hundred million years. See Ref. 7 for a more detailed discussion.

the idea that inanimate matter might also undergo a process of prebiotic molecular evolution, by means of which there would have been a spontaneous increase of molecular complexity. For example, first small organic molecules were formed, then macromolecules, then their complexes, then their interaction to form a membrane, then the first primitive form of compartmentalized metabolism, and so on. At a certain point, this spontaneous increase of molecular complexity, after a long series of trial and error, gave rise to a complex system possessing the emergent property that we call life. You arrive at life, then, by passing through a series of higher and higher hierarchic levels of molecular architecture.

The statement that life on Earth came from inanimate matter is actually considered so trivial that its implications are barely discussed. I believe this to be wrong, because the implications of this arrow of time are enormous. It represents the foundations of our biological sciences. Let us pursue a couple of these implications.

The first implication is inherent in the observation that life consists of only molecules and their dynamic patterns. Oparin actually simply said that life consists only of molecules, forgetting to add the concept of the dynamic patterns, which we believe today to be of central importance. But still, life can be created, in theory, merely through an interplay of molecules.

It is then a reductionistic scenario: all can be reduced to molecules, including life itself. The term *reductionism,* used in this context, should, however, be qualified. True, the implication is that only molecules and their patterns constitute the structure of all intermediates leading from inanimate matter to life. Nevertheless, here the term reductionism does not imply that the properties of life and all the intermediates—the functioning of the muscle cells, the movement of the cilia in bacteria—can be understood in terms of the constituting molecules. In fact, we have to consider here the notion of *emergence*: At each hierarchic level of complexity, new qualities arise that cannot be explained on the basis of the properties of the constituents. This principle already works at the level of simple things like water: Structurally, water is made out of hydrogen and oxygen. But the properties of water cannot be determined on the basis of the properties of hydrogen and oxygen. If we consider an enzyme, its structure can be understood in terms of atoms and molecules, but its properties cannot be foreseen or understood once the composition or even the structure of the macromolecule is known. Many examples at higher levels of complexity can be given.

All this is tantamount to saying that there is structural reductionism here, which does not imply that the properties can be understood in terms of the molecular components. Whether this lack of understanding is ontological or simply epistemological is a question that for the moment I would like to leave out of the discussion.

Connected to the question of reductionism, there is another implication of this first arrow of time that has to do with a more philosophical question: If life consists only of molecules and their dynamic patterns, has science, then, anything to say about the realm of spirit—about human consciousness and the world of the psyche and soul? Is spirituality consistent with this scientific view that there are only molecules and their dynamic patterns? Or is this realm something that science, as a consequence, has to deny?

This is in fact a central question in our civilization: How can we bring together these two basic realities of our human experience, the scientific belief that there are only molecules and the obvious reality of the existence of the spiritual world? I believe that these two views are not inconsistent with each other.

One possibility for consistency is offered by the previously mentioned notion of emergence. One can say that at a certain level of structural complexity a property emerges that we call consciousness. This is the view to which Francis Crick subscribes, among others, although expressed differently. In his book *The Astonishing Hypothesis,"* he claims[4] that consciousness is the result of a particular complexity of the tridimensional neuronal network.

There are, however, more subtle ways to look at the relation between molecular structure and consciousness; for example, one can refer to the Santiago theory of cognition and to the inherent principle of dual causality between structure and cognition. Here it is not held simply that structure creates consciousness, but rather that living structure and cognition are one, creating each other as one entity out of a process of recursive coupling interactions. There are different levels of cognition, from an amoeba up to the primates, and consciousness arises at the level of the human mind. It would be interesting to dwell on this subject, but I have neither the time nor the competence, two good reasons to go ahead with my main topic, which is the relation between linear and cyclic time.

AUTOPOIESIS AND THE LIVING CELL

Another important implication of this arrow of time going from the inanimate to life lies in the scientific project that life in its minimal form, so-called minimal life, can be reconstructed in the laboratory and that this can take place in a short time, days or hours, once you have the right prebiotic ingredients in the right conditions. The experimental *chemistry of the origin of life* started with the experiments of Stanley Miller, who was able to produce amino acids and other relatively complex biological molecules by sparking electricity into a mixture of simple components: water, ammonia, methane, and hydrogen.

This was 1953, the same year in which Watson and Crick published the structure of the DNA double helix. It is well established now that simple biological monomers can be made under prebiotic conditions. However, the real question is how to go further, so as to arrive at a primitive cell. Many chemists all over the world are engaged in this Faustian enterprise, which until now has seen no ultimate success.

Notice from FIGURE 2 that from the time of the origin of the earth until the first cell fossil, there is no trace of the intermediate steps. We do not know really what happened; we have no clue about the pathway. If we do not know, how can we proceed in the reconstruction strategy?

The pragmatic answer you find in the literature is the following: Each chemist is free to choose his or her own pathway. Do it as you wish, just make life starting from non-life. As I said, this has not succeeded yet, despite the large number of extremely clever chemists who work in the field, and it would be interesting to discuss the reasons for this failure. Of course the problem of the origin of life has been solved several times at the blackboard by biology theoreticians—one says there are more than 30 hypotheses on how life started. All these theoretical schemes are interesting conceptually; some are truly clever. But determining the real answer lies, as always, only with experiments.

Let us now assume that the arrow of life has finished its job and has created a living cell. A living cell is a very complex entity, hosting several thousand reactions, several hundreds of proteins, and several hundreds of various nucleic acid families and thousands of small metabolites. Here, however, we are interested in the time progress of the cellular activity and in its general behavior. The logical pattern of a living cell is relatively easy to capture. A cell is defined by a physical boundary, which is usually of its own making. The main function of a cell in homeostasis is in fact *self-maintenance*: There are very many reactions that bring about the consumption of cell components, but those components are synthesized again by the cell machinery. Thus, a cell remains itself in the face of all these thousands of reactions. The cell is busy maintaining its own identity, its own self. The self-maintenance consists of a process of self-generation from within.

These considerations are the basis of the so-called theory of autopoiesis, developed in the 1960s and 1970s by Maturana and Varela. This theory also provides an early definition of minimal life: a living system is a system provided with self-maintenance by means of self-generation within a boundary of its own making.[5-7] An autopoietic unit *is* minimal life, and it is a system defined by a physical boundary of its own making that is self-maintaining because of its internal processes of component production. This is illustrated in FIGURE 3. Each living cell, and actually each living

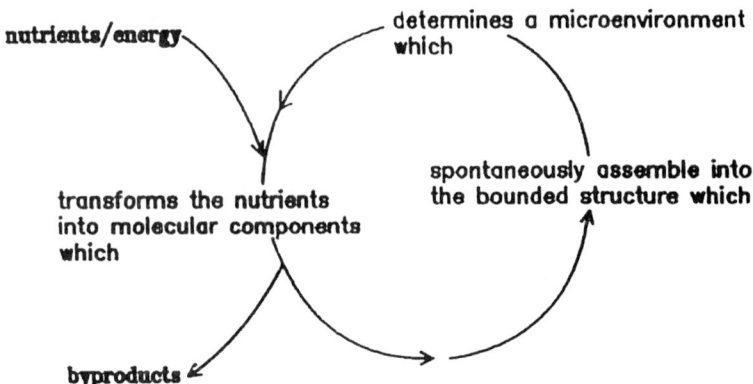

FIGURE 3. The autopoietic logic of life, with the so-called circularity of the self (see Ref. 5).

organism, can be seen in these terms, namely as a factory that constantly remakes itself from the inside. So it is with a tree and with every animal we know of, including, of course, a human being. A discussion about this ensuing definition of life, *vis à vis* alternative definitions of life, is discussed elsewhere.[8,9]

Here, as we explore the concept of cyclic time, it is important to look at the logic of this autopoietic cycle. Its logic is circular, as can be seen in FIGURE 3: The bounded organization produces a network of reactions that produce the molecular structures, which in turn produce the bounded organization, which then produce the structure, and so forth. There is no beginning and no end; it is a time event whose logic is circular.

The origin of life is actually the initiation of this circularity. Our arrow of time is then transformed by the origin of life into a cyclic process, the beginning of the whole life circularity process.

It goes without saying that energy and food coming from outside and permeating inside thanks to the semipermeable membrane augment this circularity. In this sense, a living organism can also be seen as a dissipative structure: The structural pattern remains the same, but the flow of materials is always renewing itself, just like a whirl in the sink or other classic dissipative structures.

What is the simplest possible autopoietic system? This is illustrated in FIGURE 4, which represents a system defined by a boundary constituted by one component S, the boundary being permeable to the nutrient A, which produces S, and one reaction, which destroys S. There are only two reactions, and depending on the balance between these two velocities, you have various time progresses of autopoiesis. The reaction that is more properly consistent with the original definition of autopoiesis is homeostasis, the cyclic maintenance of the self; in addition, we have the process of growth and self-reproduction, a kind of cyclic time projected into the future; alternatively we have death, the arrow of time.

I mention this also to show that the theory of autopoiesis permits one to build simple models, and not only on the black board. These systems have been implemented

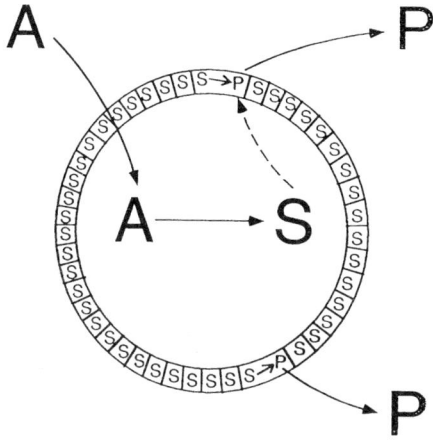

$\upsilon_{gen} = \frac{d[S]}{dt}$; $\upsilon_{dec} = -\frac{d[S]}{dt}$

if $\upsilon_{gen} = \upsilon_{dec}$ **homoestasis**

if $\upsilon_{gen} > \upsilon_{dec}$ **self-reproduction**

FIGURE 4. The minimal autopoietic system. It represents a bounded system formed by only one constituent S, with a chemical reaction inside the boundary that forms S itself. A competitive reaction destroys S into a decay product P. Depending upon the relative velocities of these two competitive processes (generation velocity or decomposition velocity), three different time progresses of the autopoietic system canbe envisaged. For more details, see P.L. Luisi in Ref. 7.

experimentally in our group in Zürich, where based on this kind of reasoning, we have constructed self-reproducing micelles and liposomes.[10–12]

If we look at the human body, we notice that it is continuously reconstituting itself—the skin and every other tissue of the body loses billions of cells every minute, but they are made anew by the body itself. Actually, the body itself is every day a new entity, although we keep saying that we are always the same human being. Thus, projected in time, life is a repetition of the same identity, a kind of spiral—the spiral is the composition of the arrow of time and the circularity of the self in autopoiesis (FIG. 5, top and center). Of course, as we all know, this process of repetition does not go forward indefinitely. There is death, which brings an end to it. So, the best representation is a helix, which dwindles down and dies, as shown in the cartoon at the bottom of FIGURE 5.

Death is the end of each living process, whatever rises to life has to die. Life and death are, in fact, one. An anecdote illustrates this nicely: two Buddhist monks, one elderly and one young, witness the birth of a new baby. "Oh, a new human being was

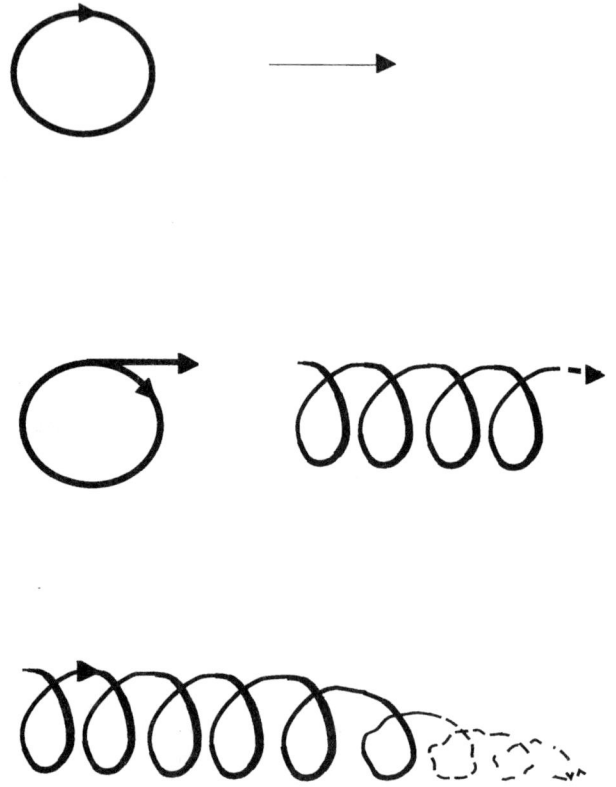

FIGURE 5. A schematic illustration of the formation of a time spiral by the composition of the linear time and the cyclic time. For living organisms, the life spiral is, however, not a constant repetition of the same elements. In fact, decay and death compete with the self-maintenance and self-reproduction cycle, and eventually win.

born!" says the young monk with a beaming smile. "Oh, a new human being has just begun to die!" answers the older monk.

Remaining in the realm of mythology, we can refer to a snake from 13th century Mexican culture (FIG. 6), with two heads that look in opposite directions, symbolizing life and death—or joy, luck, and life, but also misfortune and catastrophe.

REPRODUCTION AND EVOLUTION

I will not discuss at this point the fear of death and its consequences. Let me just mention that perhaps the idea of *reincarnation*, or the idea of eternal life of the soul, which is always present in the major religions, has to do with this fear of death—one wants to believe that there is something beyond death.

FIGURE 6. Double-headed serpent in turquoise mosaic, Mixtec workmanship, 13th–14th century. While in China the play of the opposite is viewed as being harmonious, the idea of the Mexican double serpent is more tragic (the heads look in opposite directions). (From Marie Louise von Franz.[1])

This leads me to make an additional comment concerning the relation between science and religion: Whereas science and religion can find a broad area agreement about the question, What is life? there is a much deeper, perhaps fundamental contrast about the question, What is death? For science, death is the end of individuality: there is nothing after death other than molecules, and nothing can be transmitted to a next life or elsewhere, be it hell or heaven. For religion, it is not so.

It is a fact that each individual dies, but it also true that new individuals are constantly being born. So it is for trees and snakes and animals and fish and human beings. This brings a rather interesting question: Where does all this material for these new beings come from?

It comes from material found on the earth, which in turn comes from dead structures—possibly with the addition of some contribution from meteorites. When a tree or an animal has reached the end of its life cycle, all its molecules will be dispersed and used for building new structures and organisms. Our planet is a huge reshuffling machine, whereby new organisms continuously arise from dead ones. All living organisms—animals and fish and plants—are composed of molecules that have been used many times before. Thus, there are probably molecules in my body that once belonged to Giuseppe Garibaldi, and perhaps the lady in front of me has some molecules that belonged to Marylin Monroe as well as some molecules that belonged to a dinosaur which lived in Africa 70 millions years ago. One can make the point that the very atoms that have been originated in the Big Bang are still with us.

In this way, a species goes from one generation to the next, reproducing itself. It is not, however, a static cycle, as it takes time to go from one generation to the next. Again, the composition of the cyclic time of reproduction with the forward movement of time gives rise to a helix. Each element of a helix turn is composed of a large population of individuals, who all die, feeding other wheels of reproductive cycles. It is a big spiral (FIG. 5) that is generated by the ashes of previous existence. In this sense, science also speaks of rebirth—in this case, however, the individual is completely and definitely destroyed with each cycle.

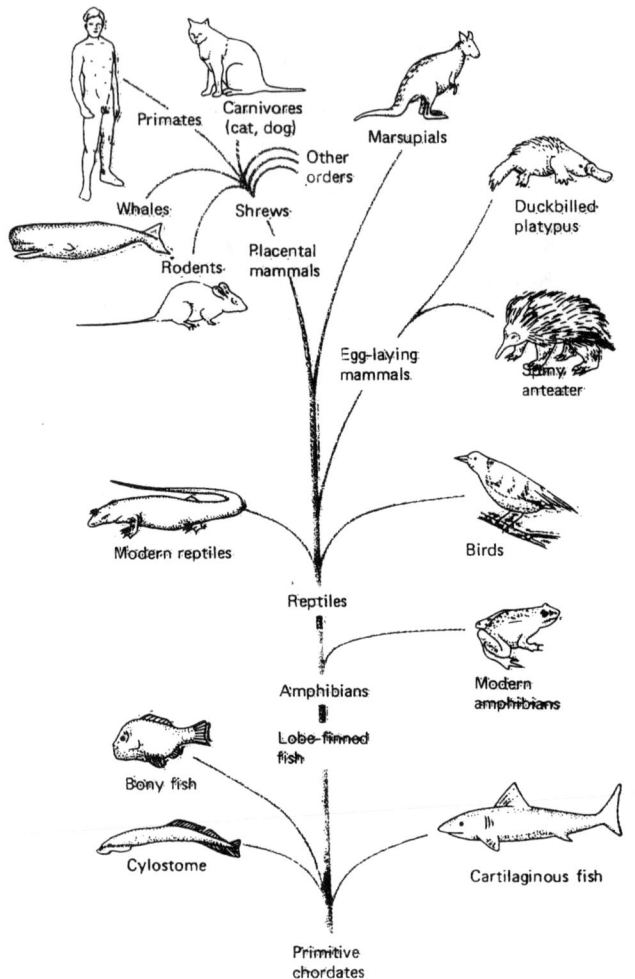

FIGURE 7. One view of the tree of evolution. The primitive chordates derive in turn from more primitive organisms, and eventually all derive from the most ancient prokariotic cells.

One cannot forget that the arrow of time that moves the reproduction cycles has *biological evolution* associated with it. As is well known, not all individuals are genetically identical, and, under certain evolutionary pressures, this phenomenon may give rise to the onset of genetically new species. The global spiral, along the axis of the arrow of time, thus becomes degenerated, and each branch may divide into two or more branches.

This gives rise, as we all know, to evolutionary trees, one example of which is illustrated in FIGURE 7.

This is another powerful image of modern science that we all take for granted: evolution connects all living organisms historically with each other. A tiny, invisible thread binds us to the monkeys and to the amphibia and to the fish, to the insects and to the microorganisms from which all started. We are all relatives to each other—Darwinism is really the most universal theory of brotherhood that ever existed.

It is also, however, a hard narrative of fights for survival, with a long history of biological hecatombs, which a few times in the history of our planet destroyed up to 95% of all living species. In this sense, with this mixture of brotherhood and genocides, Darwinism is really a close representation of our human world.

Let me touch, in concluding, on my last point. This question, which stems directly from the inspection of this evolutionary arrow of time, depicts the evolutionary tree. The question is: Has this arrow of time a meaning, does it go toward some clear end? Or not? Is there a purpose in the plans of nature, in particular in biological evolution?

If we want to remain on scientific ground, one answer, the one that is scientifically more founded, is the one that affirms that mankind is the product of contingency. Let me quote Stephen Jay Gould, who says: "… run the tape again, and the first step from procaryotic to eucaryotic cell may take twelve billion years instead of two …"[13] And, in this re-running of the tape, perhaps mankind would not arise again. There are those, however, who hold the belief that the making of mankind is the *aim* of evolution, and the next step is the perfecting of mankind. For example, see the recent book by Polkinghorne,[14] who had the guts to challenge modern scientists with the idea of God again. Others, remaining within the realm of science, also refute the idea of contingency expressed by Gould and hold rather to the hypothesis of convergence, according to which certain features in evolution, mankind included, are almost an obligatory pathway.

We have several alternatives, then, and I guess that each one of us should find here his/her own answer.

REFERENCES

1. Von Franz, M.L. 1978. Time. Thames and Hudson. London.
2. Coveney, P. & R. Highfield. 1990. The Arrow of Time. W.H. Allen, Ed. London.
3. Hawking, S.W. 1988. A Brief History of Time. Bantam Press. London.
4. Crick, F. 1994. The Astonishing Hypothesis. Simon and Schuster. New York.
5. Varela, F., H. Maturana & H. Uribe. 1974. Biosystems **5:** 187.
6. Varela, F. 1979. Principles of Biological Autonomy. North Holland. Amsterdam.
7. Luisi, P.L. 1993. *In* Thinking about Biology, W. Stein & F. Varela, Eds. Addison-Wesley. Reading, MA.
8. Luisi, P.L. 1998. Origin of Life **28:** 613.
9. Luisi, P.L. 1997. *In* Astronomical and Biochemical Origins and the Search for Life in the Universe. B. Cosmovici, S. Bowyer & D. Werthimer, Eds. New York.
10. Bachmann, P.A., P.L. Luisi & J. Lang, 1992. Nature. **357:** 57.
11. Walde, P., R. Wick, M. Fresta, A. Mangone & P.L. Luisi. 1994. J. Am. Chem. Soc. **116:** 11649.
12. Luisi, P.L. 1996. Adv. Chem. Phys. **XCII:** 425.
13. Gould, S.J. 1991. Wonderful Life: The Burgess Shale and the Nature of History. Pinguine Science. New York.
14. Polkinghorne, J. 1998. Belief in God and Age of Science. Yale University Press. New Haven, CT.

The Human Sciences and the Natural Sciences: Toward a New Asymmetry

ANTONIO MELIS

Dipartimento di Filologia e Critica della Letteratura, University of Siena, Via Roma, 47, 53100 Siena, Italy

At a distance of about 40 years, it may be useful to reconsider a debate that had great impact on relations between the human sciences and the sciences of nature. *The Two Cultures* by C. P. Snow[1] mainly denounced a break between science and the humanities, manifesting as mutual nonrecognition, which made any dialogue impossible. Despite the controversy that it aroused, Snow's book became an international reference point. One reason for this was that Snow had the authority of personal experience of both spheres, from physics to literature. A discussion, in certain aspects still in course, sprang in the wake of this warning cry. It has recently taken a historical perspective, with attempts to pinpoint the decisive moment of the break. For example, Immanuel Wallerstein[2] sees this moment as the revolt against theology which subsequently expanded to include philosophy. Scientists began to emerge as a group with an identity of its own. This process culminated in the divorce between philosophy and science at the end of the 18th century. The science faculties of universities divided from the arts, replicating the division of knowledge at an organizational level. A lacerating division between a presumed world of subjectivity and a just as presumed world of objectivity was postulated.

According to Wallerstein, science had the better in this dispute. It became essential to wear the robes of science in order to gain prestige. Nevertheless, this did not prevent a new type of dualism from arising, with positivism on one side and hermeneutics on the other.

Here, I would like to consider the evolution of the debate from the disciplinary and practical viewpoints. I intentionally use this neutral expression to avoid the equivocal solemnity, especially when used by humanists, of the term "science." Indeed, in the very field of the so-called humanities, there has been a trend towards creating asymmetry with respect to science, however, with inverted roles.

Literary criticism has been particularly affected. In the past, scientific method has certainly been referred to in literary studies, but it was a historical school, in the positivist tradition, based on the cult of data and fact. However, in this approach, there was a rigor that in subsequent forms of scientism has gradually been lost. These forms arose, mainly in the 1970s, under the impetus of linguistics, or rather some of its currents. Besides the direct effect of linguistics on literary studies, we should also consider the effects of an anthropology heavy with linguistic suggestions. Here it is worth looking at the exchange of arguments between Claude Lévi-Strauss and Vladimir Propp.[3] The French anthropologist hailed the *Morphology of the Fable* as a precursor of structuralism; the Russian scholar of folklore was sceptical.

Of course, there were those who disagreed, but because of an excess of ideology, they were not as incisive as they could have been, at least in the long period. This criticism does not imply a negation of the sociopolitical implications of cultural hypotheses, but merely warns against too simple an application of them.

With regard to literary analysis, critics may have had a type of inferiority complex towards the natural sciences. Literary scholars have never been so vocal about "scientificness" as in that period. The word often had a somewhat menacing tone when used in connection with any other type of approach to the literary text. Indeed there was a trend towards excluding or drastically reducing the aesthetic dimension of the literary work. In the more extreme formulations, the text became a pretext. It was not recognized in its nature of complex world, meeting point, and precarious synthesis of different impulses. Its *raison d'être* seemed to be to provide a testing ground for different theories. The basic principle of adapting the critical approach to the nature of the text thus became less important.

Hypertrophic and continually changing terminology was coined and became a new canon, often sustained aggressively, especially in the academic field. Fortunately, the fad now seems to have passed, and its exasperated nominalism, devoid of any true advance in knowledge, appears for what it was.

What is even more interesting to underline here, however, is the new asymmetry that this line produced with respect to the natural sciences. The structuralist and semiotic currents shared the aspiration of giving literary exegesis a scientific statute, but their idea of what was scientific was one that science was just then beginning to radically question. This new failure to meet, now inverted, can be described in terms of oppositions. From the 1970s, the accent in hegemonic criticism was placed on quantity at the expense of quality. The spirit of system superimposed on the complexity of the text, with a clear reductive trend. On this path, some quite clamorous interpretative aporias occur. Among possible examples, I cite the mechanical application of schemes deduced from popular tales, along the lines of Vladimir Propp, to the study of "cultured" narrative, with the inevitable result of transforming the richness of an author's artistic universe into a generic repertoire of motives and situations. Again we observe a systematic reduction in complexity.

Another opposition can be established from the role played by aesthetics. In literary studies, as we have said, this is the trait that tended to disappear, drowned by infinite semiosis, obscuring the destination of the work, which except in pathological cases was not conceived for professors of literature.

While this was happening, contrary processes were occurring in science. Doubt was shed on the dominant paradigm on the same grounds that afflicted the neoscientism of the humanists, especially literary scholars. The qualitative dimension of natural processes was claimed against any merely quantitative consideration. At the same time, a just as radical reappraisal of the aesthetic dimension occurred. In the new perspective, knowledge does not consist only of mathematical relations, but also has odors, tastes, and colors as irreducible components.

This led to a type of "follow the leader" between two approaches that seemed destined to diverge. The question of rebuilding some kind of convergence on a new basis then arose. From the above it is evident that the basic mistake of "humanist scientism" was that of abdicating the specificity of its approach. The two cultures cannot meet and unite if they compromise on their heritages. For a higher synthesis, both must offer their best. The situation is similar to that of interethnic relations, where again, in fact, we often find that the problems are rendered banal. In the name of "political correctness," a kind of generic mutual tolerance, based on an aseptic vision of cultural specificity, is passed off as an ecumenical attitude. In such an approach there is no genuine transcultural opening towards the other that could promote osmosis and contamination, instead of respectful indifference.

In the same way, there must be no renouncing of identity or even polite exchange of courtesies in the meeting of the two cultures. Often, in fact, acclaimed interdisciplinary practice becomes mere juxtaposition of noncommunicating worlds.

There may be a real possibility of change in the acquisition of a common ecological perspective. Needless to say, I do not mean the caricature that often usurps the name of ecology. The kernel of ecological thought, which can make an incisive contribution to a new cultural synthesis, is the idea of solidarity between all the manifestations of nature, in spatial and temporal terms. The prospect of a common destiny applies to the present of all beings, but at the same time invests the future and past. The relation with the future is described by the happy formula of "solidarity between generations." The relation with the past could be expressed as responsibility towards what previous generations have passed down to us. This is again a uniting of generations.

This sense of belonging could mean the start of a direct process to recover an overall rather than mutilated view of culture. Nature and man cannot be regarded as two separate worlds. It is necessary, however, to realize that this is not a painless process. Fundamentally, it casts doubt on anthropocentrism. In all school textbooks, the collocation of man at the center of the universe is regarded as the culminating moment for the humanist perspective. Today, we need to humbly step back so that we can place man in his true place in the global life cycle.

What consequences can this new vision have for literary studies? I think we can call this perspective "an ecology of literature." I have already used this expression in relation to the world of Andean literature.[4] It is quite different from "ecological literature." The accent is not on content and description, but the view of the relation between man and nature is holistic. Radical changes in style are involved. Not by chance is this dimension strongly present in Latin American literature. In the clash of cultures that occurred in these lands, the indigenous substrate has bequeathed, in a very vital way, a different manner of considering man in the world. The very qualities of sound, color and taste discovered today in the ideas of scientists are matched by the cultural experience of American peoples. Spanish-American literature owes its worldwide success, apart from commercial operations, to this re-emergence of repressed cultural layers. The organic relationship with nature is accompanied by a closer relationship between writing and speech. In the highest forms of this literature, the resistance of oral culture emerges, despite the assaults it has received since conquest, when writing was used as an instrument of domination. In this context, writing does not supplant the oral tradition; it cannot even be considered a higher stage of the latter, as in the hierarchical view typical of the linear view of history.

All these factors of cultural originality are linked. They are the particular contribution of Latin American culture to the unity we are interested in. For this message to be received, a common view of intercultural relations needs to be overcome. In the dominant framework, Latin American culture is recognized as an inexhaustible reservoir of fantastic creations. Again we see a version, albeit partial, of the good savage.

A new concept of rationality is needed if we are to resolve this impasse. In fact, among Latin American scholars, especially in the social sciences, the concept of "alternative rationality" has recently emerged with force. This notion is not entirely foreign to European culture; I can cite, for example, Ludwig Wittgenstein. Particularly in his notes on Frazer's *The Golden Bough*,[5] there is quite a different vision from that born of positivism, from which we have notions such as "primitive culture," which relegate the experiences of different peoples to the realm of prelogical thought.

Acceptance of complexity also means recognition of different models of articulation between functions regarded, in the hegemonic culture, as separate and opposite. The claim of subjectivity, not only of individuals but also of peoples, manifests in all forms of knowledge. It tends to break down barriers created by traditions that have often believed themselves the *only* tradition. Also in this case, relationships with "marginal" cultures lead to new perspectives. The fall of barriers between the various literary manifestations has been interpreted as "mixing genres," in violation of the separatist code of official culture. In the case of the Spanish-American literary experience, this mixing feels like a recovery of one's own genetic heritage. In this process, which has sometimes been catalogued under the vague but all-inclusive label of *postmodern*, the force of the indigenous substrate and the African transplant to the American continent again emerges. There is no continuity between the indigenous cultures and the new culture that arises with colonization. The trauma of conquest meant a violent break, though there was underground resistance. In the case of the African slaves, the situation is aggravated by uprooting from their native land, but this did not prevent them from maintaining a link with their culture, albeit in new forms.

In both cases, however, though in different forms, we find a global concept of artistic expression. What we, in Eurocentric terms, read as "texts," are really only part of an indissoluble whole, which is thus inevitably shattered. The destiny of words is not writing. Words assume speech, often set to music. Words and music unite with dance, which in turn means costumes and symbolism. Beyond this weave of different components, however, there is a decisive relation with the collective dimension. This is an essentially social form of art, which tends to break down the divisions between protagonists and spectators. Frequenters of contemporary Latin American theater will have noticed how this ritual substrate emerges strongly, at times combined with elements derived from the classical theater.

This indigenous and partly African background comes to life in country feasts linked to the seasons. Of course, modernization has also affected these ancient rituals. However, the survival of tradition in the long period is determined by its capacity to absorb the stimuli of modernity without denying its ancient roots. Paradoxically, cessation of these internal changes in feasts and rites marks their death; they may continue as fossils for scholars and tourists, as has occurred for many traditions.

After this necessary digression, I shall try to draw some inevitably provisional conclusions. The fact that literary scientism is coming to a natural end does not mean a return to an autarchic view of literary criticism. An alternative path for the recovery of the multidisciplinary dimension is indicated by the encounter with so-called marginal or subordinate literature. This path will make it possible to reopen an authentic dialogue between science and the arts, both enriched by the heritage of their respective identities and complexity.

REFERENCES

1. SNOW, C.P. 1959. The Two Cultures. Cambridge University Press. Cambridge, England.
2. WALLERSTEIN, I. et al. 1986. Open the Social Sciences. Stanford University Press. Stanford, CA.
3. PROPP, V. 1972. Morfologia della Fiaba. Einaudi. Torino, Italy.
4. MELIS, A. 1998. Hacia una ecología de la literatura. III Jornadas Andinas de Literatura Latinoamericana. Quito, agosto 1997.
5. WITTGENSTEIN, L. 1967. Bemerkungen über Frazers "The Golden Bough." Wittgenstein Nachlass Verwalter.

Organization and Complexity

EDGAR MORIN[a]

Centre National de la Recherche Scientifique, Siège: 3, rue Michel-Ange, 75794 Paris cedex 16, France

The universe, in its diversity, has atoms; molecules; stars; galaxies; and—at least on one small planet—living organisms, humans, and societies. These different entities, which are different from each other and not reducible to each other, have in common the fact that each is made up of organized elements that form a whole. What we perceive as the world around us are these wholes.

The significance of the general notion of organization is not immediately evident. It is not in a search for a common minimum that we need to concentrate our efforts. It is in the manner of perceiving, conceiving, and thinking about the things of our world in an organized way. To obtain a generic, rather than general, concept, it is the idea of organization that we have to delve into.

What is organization? Organization binds elements (particles, atoms, molecules, cells, individuals, etc.) in relationships that thus become components of a whole. In the first definition, organization is a structure of relations between components to produce a whole with qualities unknown to these components outside the structure.

Hence, organization connects parts to each other and parts to the whole. This gives rise to the complex character of the relation between the parts and the whole. Dilthey[1] had already stated, "A whole cannot be understood except by understanding its constituent parts, which cannot be understood except by understanding the whole." Two centuries earlier, Pascal[2] referred to this circular relation, "I consider it impossible to know the parts without knowing the whole, or to know the whole without knowing the parts."

There is a close relation among the concepts of organization, interrelation, and system.[b] These three terms, although inseparable, can be distinguished from each other. The concept of interrelation refers to the types and forms of links between elements and between the elements and the whole. The concept of *system* refers to the complex unit of an interrelated whole, to its characters and properties. The concept of *organization* refers to the structuring of the parts within, with and through a whole. The two notions, organization and system, are linked by that of interrelation: the whole interrelation, if it has stability or regularity, acquires an organized character and produces a system. There is a circular reciprocity between these three terms. When the notion of system disperses the quality of being and existing (to say "living systems" tends to take the emphasis away from living beings and their existential dimension), the notion of organization refers to to something concrete.

---

[a]Address for correspondence: 331-42789099 (voice); 331-48048635 (fax).

[b]In some cases, the idea of network seems more pertinent than that of system, in the sense that a system tends to have well-defined borders, whereas a network has limits that vary. Both, however, are interconnections of elements of an organized nature. As well, a network can become a system by closing on itself (Internet) or by acquiring an organizing center (the railways). For me, the primordial and common character is organization.

In any case, we need a concept in three, three concepts in one, a macro-concept, each constituting a definable aspect of the same world, but that we subject to the hegemony of the idea of organization.

Organization is a notion that is dependent and at the same time independent of its constituents. The relative autonomy of the ideal of organization is illustrated by isomers, compounds with the same chemical formula and molecular weight but with different properties only because the atoms are arranged differently in the molecule. We know that differences between atoms are the result of differences in the number and structure of three types of particles and that the diversity of living species depends on differences in the number and structure of four basic elements that form the letters of a "code." We immediately see the importance of organization if it changes the qualities and character of systems or entities consisting of similar, but differently structured elements.

Organization gives the entity autonomy and stability, at least within certain limits. Autonomy is derived from stability, and stability proceeds from autonomy. It is therefore not a mere exercise of the mind to distinguish an organized entity in relation to its environment. Organizational stability is ensured in a fixed way by the chemical bonds in molecules, but in the stars, stability/autonomy is guaranteed by an antagonistic complementarity between implosive and explosive processes. As we shall see, the maintenance of autonomy/stability is a primordial problem for living beings, which preserve it by continuous regulation and reorganization.

UNITAS MULTIPLEX

The relatively autonomous and relatively stable entity that organization produces, maintains, and preserves, is a complex unit, simultaneously one and multiple: unitas multiplex. This holds for the atom, the star, the living being, and the social unit.

The idea of multiple unit embodies two mutually exclusive notions. The concept of unit renders homogeneous and breaks up multiplicity; the concept of multiplicity divides unity into compartments and breaks it up. Hence the organized entity is one and homogeneous from the point of view of the whole, and different and heterogeneous from the point of view of the constituents.

What we have to understand is the complex characteristics of the unitas multiplex: it is a global, nonelementary entity, because it consists of different parts. It is a nonhomogeneous but hegemonic unit because the organized whole dominates the distinct elements and holds them in its power. It is a nonprimitive but original unit: it has its own irreducible properties. It is an individual unit, quite indivisible: it can be decomposed into separate elements, but this changes its existence.

Organization contains the seeds of its own disorganization. Maintenance of the complex unit presupposes the existence of dissociative forces. Binding forces contain or presuppose forces of repulsion. The parts that undergo organizational constraints carry the virtuality of their acquisition of autonomy in relation to the whole; this happens when the constraints relax or break down, destroying the organization, as occurs in cells that evade the constraints of the organism and proliferate in a disordered manner as a malignant cancer. Finally, the whole organized entity tends to a condition of disorder, according to the second principle of thermodynamics. Everything that is composed tends to decompose. However, in the more complex organizations, self-organization maintains homeostasis with negative feedback (elimination of deviations that threaten the stability of internal complexity) and uses the mortal effect of the forces of disintegration to regenerate itself. A living organism ceaselessly degrades its energy; this degradation would be irreversible if its autonomy did not permit it to draw on energy from outside to produce new molecules and young cells to replace those that decompose. Hence anti-organization is not just antagonistic, but necessary to organization.

Moreover, the organizational constraints of the whole cause division in the great living and social polyorganizations, between the universe of the parts and the universe of the whole. None of the cells of Anthony (50–100 billion in number), none of his organs, know that Anthony declared love to Cleopatra, and Anthony knows nothing about the life and functioning of his cells: there is mutual ignorance at the heart of indissoluble unity.

Thus the idea of unitas multiplex acquires density of meaning when we understand that we cannot reduce the whole to its parts or the parts to the whole or the one to the multiple or the multiple to the one, but that we have to try to conceive the notions of whole and parts, one and different, organization and antiorganization, together, in a way that is simultaneously complementary and antagonistic.

EMERGENCES

All organizations produce something beyond their components, considered in an isolated or juxtaposed way: (a) the organization itself; (b) the global unit constituting the whole; (c) the new qualities and properties emerging from the organization and global unit. These can be called "emergences."

As indicated by von Foerster, the rule of composition of elements that interact in organization is superadditive (superadditive composition rule, see von Foerster,[3] pp. 866–867).

It is now important to extrapolate the qualities or new properties that emerge with organization and globality. They are qualities or properties of an innovative character with respect to the qualities or properties of the components taken separately or structured differently in another type of system. Thus the atom has original properties, such as stability with respect to the particles that compose it, and it imparts this quality of stability by feedback to the labile particles that it unites. With regard to molecules, "the new species bears no relation to the primitive constituents: its properties are not the sum of theirs, and it behaves differently in all circumstances. Though the mass (the total quantity of matter) remains the same, its quality or essence is completely new" (see Auger,[4] pp. 130–131).

The mixture of two gases, ammonia and hydrochloric acid, produces solid ammonium chloride. The apparently banal but in fact complex example of water shows that its liquid character (at normal temperatures) is not due to the properties of the atoms but to the molecules of water, bound together in a flexible way.

The association of an atom of carbon in a molecular chain brings about stability, an essential quality for life. As far as life is concerned, "clearly the properties of an organism are more than the sum of the properties of its constituents"[5] and clearly the living cell has emerging properties unknown to macromolecules outside biological orgnization: it feeds, metabolizes, and reproduces. These emerging properties, the group of which is what we call life, affect the whole because it is a whole and affects the parts by feedback because they are parts. From the cell to the organism, from the genome to the gene pool, complex organizing units with emerging qualities constitute themselves.

Finally, the implicit or explicit postulate of human sociology is that society cannot be regarded as the sum of all the individuals that compose it, but is an entity with specific qualities.

It is quite extraordinary that apparently elementary notions, such as matter, life, sense, and humanity, are really emerging qualities of complex organizations. Matter only has consistency at the level of the atomic system. Physical materiality is not the first quality but emerges in and through organization. Life emerges from living organization: living organization does not emerge from a vital principle. The sense that linguists look for in the depths of language is the emergence of discourse, which appears in the unfolding of global units and has feedback on the basic units that made them emerge. The human being is an emergence of a hypercomplex brain system in an evolved primate. To define man in opposition to nature means defining him exclusively for his emerging qualities.

The surprise is that the emerging qualities of a basic system, the atom, become the basic elements of the molecules, the emerging qualities of which become the primary elements for cell organization, which become the basic elements of multicellular organisms, and so forth.

Emerging qualities have feedback on the parts and give them qualities that they could not have if they were isolated from the organizing whole. Thus, the neutron acquires the qualities of duration in the nucleus, electrons acquire the quality of individuality in the atom under the organizational effect of Pauli's exclusion principle. The cell creates the conditions for the full development of molecular qualities not seen in the isolated state. In human society, culture enables individuals to develop their aptitudes for language, crafts, and art: their richest individual qualities emerge within the social system. Thus we see systems in which macro-emergences have feedback on their parts, creating micro-emergences. The whole is not only more than the sum of the parts, but the part of the whole is more than the part by virtue of the whole.

The idea of emergence contains the closely linked ideas of quality, product (the emergence is produced by the organization of the system), globality (because it is indissoluble from the global unit), and innovation (because it is a new quality with respect to previous qualities of the elements). Quality, product, globality, and innovation are therefore the notions that need to be connected to understand emergences.

An emergence has something relative (with respect to the organization that produced it and on which it depends) and something absolute (in its innovation); it must be considered from these two apparently antagonistic points of view.

The emergence is a new quality that arises once the system is constituted and therefore has the property of event. The emergence presents as irrefutable phenomenon. It is empirically irreducible because it cannot be reduced to the qualities of the organized elements. It is not logically deducible because it cannot be deduced from the sum of the qualities of the organized elements. The new properties that arise at cell level are not deducible from their molecules. Hydrogen is irreducible and nondeducible to its constituents, crystals to their constituents, living organisms to their constituents, intelligence to its constituents, awareness to the constituents of the brain. Even when it can be predicted from the conditions in which it arises, the emergence consititutes a logical jump and opens the gap in our minds through which the irreducible can penetrate. The emergence forces us to revise the notion of qualitative leap.

How do we classify emergence? Sometimes it seems to be an epiphenomenon, product or resultant, at other times the main phenomenon constituting the originality of the organized entity. For example, if we consider our awareness, it is the global product of brain interactions and interferences, inseparable from the interactions and interferences of a culture on an individual. It is possible to conceive it as an epiphenomenon, a flash like a will-o'-the-wisp, incapable of modifying programmed behavior (genes, urges, society, etc.). Awareness can also rightly be viewed as a superstructure, resulting from deep organization that manifests in a superficial and fragile way, like all that is secondary and dependent. Such a description, however, does not consider that this fragile epiphenomenon is at the same time the most extraordinary global quality arising from the brain, self-reflection through which "the I emerges from the brain." This description also ignores the feedback of awareness on ideas, behavior, and being, and the revolution it causes (awareness of death). Finally, this description ignores the completely new and sometimes decisive dimension that the self-critical attitude of awareness can bring to personality. The feedback of awareness may be variably uncertain and cause modifications of variable degree. Awareness manifests as a pure epiphenomenon, a superstructure, a global quality, capable or incapable of feedback, depending on the moment, conditions, individuals, the problems faced, and the urges aroused. However, more than anything it is the supreme and richest product of the human intellect, and its value is related to its fragility, like all that is best and most precious to us: love, understanding, the primary virtues, the soul, and the spirit are complex virtues, phenomena of wholeness, emergences; and this is why they cannot survive death, which is the disintegration of the whole and dispersal of the parts.

Thus the concept of emergence is not reduced by those of superstructure, epiphenomenon, or globality, but entertains necessary relations, oscillating and uncertain, with them. Its very irreducibility and this undefined and dialectic relationship make it a complex notion.

THE COMPLEXITY OF THE NOTION OF ORGANIZATION

With the idea of system and organization, things are no longer things, and objects, enriched by complexity, are no longer merely objects.

Organized objects not only obey an external universal order, they produce, in their structure and specific configurations in space and time, their own organizational order. They often arise with the collaboration of disorder and have to struggle not only against external, but also against internal disorder.

Organization binds, forms, transforms, produces, maintains, orders, and renders autonomous. It cannot be reduced to structure. Structure only means rules of invariance and transformation in a system. Organization means structure, relation to wholeness, specific characters, relations between the whole and the parts, unity–multiplicity, and emergences. The idea of structure mutilates the idea of organization, strips the idea of system, enucleates the idea of complexity. The more the organization is complicated, the more the idea of structure becomes inadequate. Hence, it is in anthropo-social hypercomplexity that this small region of organizational truth presumes to erect itself a throne. In biology, the current dominant concept makes genes govern organization, whereas genes are an institution in self-organization.

There is a primordial epistemological interest in the notion of organization. Organization opposes separability (that breaks up the complex unit) and reducibility (which suffocates the microlevels); hence organization itself cannot be reduced to "holism," which suffocates the microlevels of the constituent parts. In its complex nature, it is a key linking concept: it institutes multiple unity, establishing inseparable complementarity between the idea of unity and the idea of diversity or multiplicity, which originally repelled and excluded each other; it establishes a circular relation between the parts and the whole, the whole and the parts, whence the need for circular understanding from the whole to the parts and vice versa.

The whole is more and less than the sum of the parts: this pseudoarithmetic formula suggests that the whole produces qualities unknown to the isolated parts, namely emergences, and at the same time establishes constraints that suffocate qualities and render virtual certain possibilities of the parts. Hence the whole is not necessarily superior to its parts, if, for example, like a totalitarian empire and the nations it dominates, it inhibits the qualities of the parts that were richer than those of the whole, or if the richer emergences belonged to the parts, as for example awareness, which emerges in individuals but not in society. Extending this idea to the cosmos, it really seems that "some small parts of the universe have a greater reflective power than the whole" (see Gunther,[6] p. 383). We have also noted the importance of the idea of emergence from the logical point of view.

The idea of organization invokes the concrete quality, not only of an object, but in the case of organizations perennially self-producing and self-organizing, of a being. As we have said, only things that are organized can be known as beings, and the idea of organization is therefore of ontological importance. The organized being, and especially the self-organizer, is a *"dasein,"* "to be there," *hic et nunc*, depending on an aleatory environment and subject to time the transformer; thus we come to the idea of existence, which is the condition of living beings in a universe where there is risk, danger, and probability.

Hence the organization is rooted in *physis* (the physical world), but at the same time it draws from the observer-inventor who isolates it relatively in a tangle of organizing–disorganizing feedback mechanisms and a web of systems one within the other (see Méthode 1,[7] pp. 139–141). The idea of organization, like that of system, is physical for the feet and mental for the head.

It is understandable that science, based on the reducible, the simple and the elementary, reacted against complexity of organization. It is understandable why the concept of organization was ignored and that of system avoided and neglected. Very few systemists have introduced complexity into the definition of system.

The main thing is that the notion of organization induces us to use a number of "connection keys," which will be increasingly necessary as we load organization with complexity. On the one hand we can predict a break with linear thought and a need to use feedback and self-production cycles, like the need for circular understanding to establish the relation between the whole and the parts. On the other hand, we are induced to tackle logical complexities in the identity of the multiple unit, the product–producer, and the nondeducibility of the emergence. All this leads us towards dialogic, a principle of knowledge that conceives the complementarity of antagonisms, such as in the relation organization–disorganization.

The complex notion of organization allows a great advance in understanding, but this advance opens onto a great cosmic mystery; why does organization, and not just disorder or order, appear in our universe?

BASIC COMPLEXITY

Organization is a complex basic concept of universal importance. The increase in organizational complexity manifests through an increase in the number and internal variety of the constituents and through a process of complication of internal structures. Beyond a certain threshold, when the physico-chemical organization of a complex of macromolecules, for example, can no longer take on more variety, then a more complex organization that becomes self-organizing emerges and makes new qualities emerge: the qualities of life.

We will be able to consider the specific characters of the different types of self-organization when we have clarified the connection keys, especially recurring cycles and dialogic.[c]

REFERENCES

1. DILTHEY, W. 1947. Le Monde de l'Esprit. Aubier. Paris, France.
2. PASCAL, B. 1897. Les Pensées. L. Brunschvicg. Paris, France.
3. VON FOERSTER, H. 1962. Communication Amongst Automata. Am. J. Psychiatry **118**: 866–867.
4. AUGER, P. 1966. L'Homme Microscopique. Flammarion. Paris, France.
5. JACOB, F. 1965. Leçon inaugurale faite le vendredi 7 mai. College de France. Paris, France.
6. GUNTHER, G. 1962. Cybernetical ontology and transjunctional operations. *In* Self-Organizing Systems. Yovits, Jacobi & Goldstein, Eds. Spartan Books. Washington, DC.
7. MORIN, E. 1977. La Méthode, Tome 1: La Nature de la Nature. Le Seuil (coll. Points). Paris, France.

---

[c]Some passages have been modified from the section entitled "Organisation" of Méthode 1 (pp. 94–151: pp. 104, 105 multiple unity; pp. 106–108 emergence). To complete the examination of the notion of organization, the reader is referred to the following passages, not used here: particularly pp. 112–114: "the whole is less than the sum of the parts," pp. 115–123; "organization of the difference, complementarity and antagonism," pp. 123–144. The concept of system, particularly pp. 126–129: "All is not all," pp. 129–136; organization in organization, pp. 138–144; beyond formalism and realism, pp. 150–151.

Complex Adaptive Theology: From Sinai to Santa Fe

HAROLD J. MOROWITZ

Krasnow Institute for Advanced Study, George Mason University, 4400 University Drive, Fairfax, Virginia 22030-4444, USA

In the intellectually sophisticated but prescientific world of the redactors of the Old Testament, the universe was formed and set in place solely to provide a home for mankind. The drama was played out in the geographic domain that is the present-day Middle East, and the heavens and heavenly bodies were a scenic backdrop. Ptolemaic astronomy and biblical cosmology were in essential agreement that Earth was at the center of the universe. The *de novo* beginnings were the creative act of the one God. And so it was that the cultures of the Judeo-Christian–Islamic traditions came to a view that mankind was overwhelmingly important, the very reason for God's acts of creation. These religious beliefs were closely tied to the earth-centered view of the world.

This geocentric universe was the predominant cosmological position until the Renaissance, with the emergence of the science of physics and the growth of a large body of astronomical observations. This led at first to the heliocentric universe of Kopernik (originally formulated in 1514, but not published for 30 years) and the empirical support of this view by Galileo Galilei (1564–1642).

The change in the status of humans or in human self-evaluation over this period must have been enormous. To move from being the center of the universe to the role of being the inhabitant of one planet among many circling the central star is clearly a downgrading of position. To demote mankind is also to lessen an anthropomorphic God, whose importance must rise and fall with the status of those who are created in the divine image.

The strong negative position of the Roman Catholic church toward heliocentrism as seen in Galilei's heresy, which was not in radical contradiction to scripture, had many causes. One of these must have been opposition to the great decrease in the cosmic importance of man and thus of his God, as noted above.

In the century that followed Galilei, the total orderliness of celestial mechanics was discovered by Brahe and Kepler and was explained as being derivable from the first principles of physics in the work of Newton. In addition, significant developments in the optics of telescopes and the concomitant charting of the heavens led to yet another blow to the human ego. The size of the universe as judged by the developing techniques of astronomy steadily grew. We were not only travelers on a small planet orbiting a star; our star itself was one among a vast number that spread out on all sides to unfathomable distances.

The bad news continued into the 19th century. Fossil finds indicated a much older earth than had previously been envisioned. In both time and space, the position of *Homo sapiens* was being diminished. The mid-19th century brought two more crushing blows to mankind's inner sense. The *Origin of Species* by Charles Darwin,[1] followed by *The Descent of Man*[2] a few years later, robbed man of a unique act of creation by the creator and left us among our simian relatives. The second law of thermodynamics, as formulated by Rudolf Clausius and William Thompson, indicated that the universe that God had given us in perpetuity was going one day to suffer a heat death, a decay into molecular disorder.

In 2400 years or less, mankind's self image had moved from the central purpose of creation, a little lower than the angels and living in the center of the universe, to a hairless ape on a small planet circling a rather ordinary star in a huge universe of countless stars that had existed for a very long time and would someday end in a whimper.

The unhappiness with the theory of evolution on the part of many Western religious leaders was in part due to the difficulties in shifting from our special creation to visualizing man as an ordinary animal arising like other animals in an unfolding evolutionary sequence. If they were unhappy with Darwin in discussing our origins, they seemed less aware of Clausius' robbing us of our eternity in his dictum that the energy of the universe is constant and the entropy goes to a maximum.

In the 19th century, the full impact of being demoted was felt deeply in Western culture. This angst and its effects are illustrated in the life of Philip Henry Gosse, a biologist of distinction who was elected to the Royal Society for his work on marine organisms. Gosse was also a lay preacher of the Plymouth Brethren, a fundamentalist Christian group. In the 1840s his two ideological foci were being driven into conflict by the developing geological evidence of the great age of the earth and amassing paleontological finds of extinct species of organisms. He responded with the book *Omphalos*,[3] published in 1847 just one year before *The Origin of the Species* appeared.

Omphalos: On Untying the Geological Knot solves the age problem by asserting that God had created the world five millennia or so years ago with evidence of previous times that had not passed in order to provide continuity and solve problems of annual rings on trees in the Garden of Eden for years that had not existed. This somewhat tragic author was rejected by both scientists and his co-religionists and spent his remaining days in pastoral duties and in growing orchids. In Gosse we find in one man the dilemma that had an impact on the entire world.

As the 19th century drew to a close, another discovery furthered the human loss of self-image. Natural radioactivity, at first a curiosity, led to the eventual possibility of dating the earth and paleontological samples with some accuracy. Within the understanding of geophysics, the question of time that had been so important to Gosse was now resolved. The earth is about 4.6 billion years old, and the oldest fossils, the Warawoona Stromatolites of Australia, are 3.56 billion years old.

Developments in astrophysics and particle physics have led to an understanding of the life and death of stars. Evidence from a number of sources led to the postulate of the universe forming in a Big Bang somewhere between 10 and 15 billion years ago. We clearly live in a very old and very large universe.

The gulf between science and theology was in part a historical one. Western religion, which we can trace back about 4000 years, predates the understanding of science, and hence saw the world in different categories than we currently use. It deals with knowledge that it regards as absolute, and hence must respond to any change with great caution.

Two of the final domains where special creation has been invoked in the 20th century have been the origin of life and the origin of mind. A series of experiments and reconceptualizations of the problem have led to a widely held view that the origin of life is a solvable problem in terms of organic chemistry operating under plausible conditions on Earth. As such, it can be reduced to an experimental problem accessible in the laboratory. Life, as we understand it, violates no laws of physics or chemistry. It is now thought of as an emergent property of complex adaptive systems.[4]

The construct of emergence has been an important idea in complexity theory.[5] It is a formalization and extension of the idea that the whole is more than the sum of the parts. So from a set of interacting agents and nonlinear rules of interaction, new properties emerge that were not obvious in examining the agents individually. Life thus appears as an emergent property of organic chemistry carried out under certain conditions of energy flow.

Indeed, it has been possible by means of a study of the metabolic core of metabolism, which is universal to all species and therefore presumably 4 billion years old, to begin to reconstruct the events leading to the universal ancestor. Within the constructs of organic chemistry, it seems likely that further analysis may even render the origin of metabolism comprehensible and deterministic.

With the problem of the origin or emergence of mind, the situation is less clear because our study in this area is much less mature. There is some feeling that we may come to understand mind as an emergent property of animal survival and evolution. There are no major discontinuities in the anatomy of hominid brains, and experiments on primate behavior show similarities in the minds of humans and great apes. This will remain one of the more speculative aspects of this essay. The 20th century nevertheless seems to be closing on a note of extreme loss of the uniqueness of human persona.

While the change of mankind's self-image was taking place, the relation of man to God and man's view of God were also undergoing deep changes. The God of Jerusalem was a very personal and national deity who related to the world through miracles, through revelations, and through intermediary angels. The God of Athens as presented by Aristotle was a much more abstract entity, the unmoved mover who motivated the universe. This divinity, bare of any quality, was also found in the views of Plato. The first attempt to reconcile these views of God was the work of Philo Judaeus of Alexandria (30 B.C.–40 A.D.). He tried to reconcile the God of Torah (Pentateuch) or God of faith with the Athenian God of reason. This is the beginning of an ongoing 2000-year unresolved philosophical theological issue between faith and reason.

The attempted reconciliation and its rejection resonates in the work of Islamic scholars from al-Kindi (9th century Arabia) to Averröes (1126–1198 Spain). The argument appears in the work of the Jewish scholar Maimonides (1135–1204) who was familiar with the works of the Islamic philosophers al-Farabi and Avicenna. With the subsequent introduction of Aristotelian thought into Christianity, we find the faith and reason argument emerging full blown in the writings of Thomas Aquinas (1225–1274). Thomas spent a lifetime wrestling with these matters. There is a continuous thread from Avicenna to Maimonides to Thomas. In organized religions, faith and reason arguments are always won by faith, for there is always a danger, clearly seen by al-Ghazzali (1058–1111), that a philosophy that could prove

God's existence could also disprove it. This professor at Naishapur wrote an attack on the theology of reason called *Incoherence of the Philosophers*.[6] For similar reasons, Thomas Aquinas was attacked by Etienne Tempier, bishop of Paris in 1277.

The battle between beliefs in the God of reason and the God of faith came to the fore in the works of Spinoza. The categories of faith and reason were shifted by his more precise and narrower understanding of the God of reason. The God of Spinoza was immanent, within all substance and coexistent with the laws of nature. This complete immanence left no room for the miracles, which are violations of the laws of nature. The God of the Old and New Testaments is clearly a God of miracles. This critical view of the miraculous, among other items, was involved in the heresy of Spinoza. He was excommunicated by the Jewish community of Amsterdam.

The traditional God of the Abrahamic religions is outside of nature but intervenes into nature in his interactions with mankind as in the parting of the waters of the Red Sea or in producing an abundance of loaves and fishes. This transcendence of God places him farther from the substance of the world (as viewed by Spinoza) but nearer to the affairs of mankind.

Following Spinoza, the God of nature and the God of miracles were matters of frequent discussion. The argument from design moved from the wonderful working of the world to its designer. An example of this approach is found in the *Bridgewater Treatises*,[7] a series of natural theology volumes from the mid-1830s. These monographs attempt to establish God's existence from the wonders of the scientific world. Intended as arguments for the God of faith, they can only reinforce the ideas of a God of reason. Following Darwin's theory, open warfare broke out between those who saw man's origin within the development of natural law and those who saw man as a miraculous creation of God. And so the 19th century ended with hostility between science and religion, and with mankind having undergone two millennia of decentralization.

Events of the 20th century are, I believe, leading to a possible new understanding of immanence and transcendence and toward the possibility of a unification of science and religion. We start in the seemingly unlikely domain of the mode of explanation that has characterized biology since the work of Claude Bernard (1813–1878). The entities of biology are hierarchical: molecule, organelle, cell, tissue, organ, organism, ecosystem, and so forth. Understanding in biology consists of explaining events at one hierarchical level in terms of the operation of entities at the next lower hierarchical level. This is the reductionism of the biologist and generally has no deeper metaphysical significance. Thus physiology looks to cell physiology, cell physiology looks to organelle physiology, organelle physiology looks to biochemistry, and biochemistry looks to molecular biology. This agenda has been remarkably successful and has led to a biotechnology that will alter the way in which we live.

Reductionism, however, has its limitations and does not allow us to predict from an understanding of one hierarchical level what will happen at the next higher level. This is because there are often many entities at a given level, and the laws that relate their interactions may be quite intricate; as a result, when we try to carry out our analysis, there is a bewildering combinatoric array of possibilities. This in the past provided a major and almost total block to theoretical mathematical biophysics.

Two related developments have pointed the way beyond these barriers to understanding. First, the invention of high-speed computers expands by many orders of magnitude the space of possibilities that may be analyzed, and second, a branch of complexity theory allows novelty to come out of theory in a natural way.

The problem at one level involves a number, often a large number, of entities or agents. These agents interact by a set of rules, which may be quite complicated and are nonlinear. This generates an array of possible outcomes that is so large as to be transcomputable, beyond the range of the largest and fastest imaginable computers. At this stage a new branch of complexity analysis was founded by John Holland[8,9] and others; techniques such as fitness criteria are introduced to prune or select in the space of possibilities. This leads to a new set of behaviors or a new set of relevant agents in the upper hierarchical level that are not simply predictable from the lower level but in no way contradict the laws at the lower level. Genetic algorithms introduced by Holland are an example of this search for emergent properties. These new properties are not totally random, nor are they totally determined. They result from an orderly exploration of a space of possibilities under a search algorithm with pre-agreed-on rules of selection. Emergence allows for ordered novelty without any violation of the reductionist program. This is a most exciting aspect of modern complexity theory.

This emergent feature of complex adaptive systems allows for a genuinely new way of pursuing scientific understanding. Novelty and innovation are now natural parts of the world of science. Thus, life emerges from organic geochemistry, not as a miracle as the fundamentalists would assert, not as a totally improbable event as the radical reductionist Jacques Monod asserts in *Chance and Necessity*,[10] but as an emergent property from the presence of atoms and their interactions. This idea was expressed by Lawrence Henderson in 1911 in his book *The Fitness of the Environment*.[11] For Henri Bergson in *Creative Evolution*,[12] the origin and evolution of life required a new principle, *élan vitale*. This vitalism becomes unnecessary if we replace *élan vitale* by the creative novelty of complex adaptive systems.

We next consider the evolution of cognition or perception. I don't wish to pretend that I understand these mental features of animals, but I am impressed by the arguments of Donald Griffen[13] that an animal in its lifetime will experience many more environmental conditions than the number of responses that can be preprogrammed into the genome. Therefore, a system with perception and choice of response will have an advantage for survival. Once an emergent system arises with this property, it will have a high degree of fitness. It will both persist and add new features. The human mind, as we know it, is a novel emergent feature of evolving animal species. Choice of response, which is the biological predecessor of free will, occurs very deep in the evolution of animals according to this view.

Now we come to a possible theological consequence of the new complexity theory. First, consider that miracles require a sophisticated enough human society to interpret an event as being outside of the usual order of things. It is meaningless to think of God carrying out miracles apart from the humans whom the miracles were meant to influence or help. Miracles can therefore be viewed as an emergent property of an immanent God that are related to the evolution of a human system of understanding. God's transcendence as inferred from the existence of miracles is a property of God in relation to humans. The human mind as an emergent property of

God's immanence becomes identified with God's transcendence. We *are* God's transcendence in its local manifestation.

Miracles are then either acts by humans or human interpretations of events that follow from the laws of nature. Free will, an emergent property of our biological selves, turns out to have enormous theological significance; it coexists with what we have regarded as God's transcendence.

A concept known as the anthropic principle states in a variety of forms that the only interesting universes are those in which the laws lead to minds that try to understand those universes. The laws of nature of such a universe can be regarded as God's immanence or as something that just happens. In such a universe, mind is an emergent gift of immanence and a great burden to the mentally endowed. For in either case, the theistic or the existential, the task is ours to discover an ethics and build a society to optimize the possibilities to move from mind to the next emergent level that some, such as Pierre Teilhard de Chardin, would call spirit. That is our task as humans. If one accepts mind as an emergent property of God's immanence, then our role is literally a partnership, because we are necessary for the transcendence of an immanent God.

The preceding three paragraphs are really quite startling. A few points have to be reviewed. The new finding from complexity theory is that novelty is possible within a world entirely governed by basic laws. This is due to selection for whatever reason among a huge number of possibilities. The persistent subset of emergent states thus constitutes a novel property. Among features that have arisen as emergent properties are life and mind. The latter evolves to a state where the laws of immanency may be contemplated and ethical choices can be posed. At that point, a transcendent property of nature emerges, and we are the agents of that transcendence. This is a simple and awesome conclusion.

It is impossible to describe the ideas under development here without using grandiose and imprecise language. We are just at the beginning of understanding. On a 4.6-billion-year-old planet, the level of hominid society that had to have been reached in order for the ideas of transcendence to be meaningful is at most 10,000, or a few 10,000s of years old. It might be as few as 5,000 years. The biblical time scale is more appropriate to God's transcendence than is the cosmological time scale which relates to the immanence. The vastness of the age of the universe is thus somehow reduced to human dimensions.

The biblical 5000 years, which went to 15 billion years since the Big Bang, dwarfing our human lifespan, is now reduced to 10,000 years of God's transcendence, or 300 generations of man. The vastness of the universe seen within the anthropic principle suggests that life and mind might emerge anywhere, but we can only deal in an ethical sense with the life and mind we know. The transcendent outcome of the immanence is very dependent on planetary details. This dependence on local detail also reduces the divine domain to a more human perspective.

The key to complex adaptive theology is the notion of emergence, the rise of novelty brought about by the operation of the laws of nature. The belief that God is coexistent with nature and its laws causes the story of the unfolding universe to be theology. To some this is just a matter of vocabulary. To the religious it is a cause for celebration; to the existentialists it just is. The evolution of *Homo sapiens*, with all its perceptive and cognitive aspects, is the central feature of this theology. Identifying the work of the human mind with divine transcendence impels us to a vigorous and deep study of the mind.

Our world and mind are given to us as the gift of an immanent God. We must do the right things with our lives and minds. That is the responsibility of the divine transcendence within us. The rest is history.

ACKNOWLEDGMENTS

Written in honor of Thelma Z. Lavine on the occasion of her retirement.

BIBLIOGRAPHY

1. DARWIN, CHARLES. 1858. The Origin of Species. J. Murray. London.
2. DARWIN, CHARLES. 1871. The Descent of Man. J. Murray. London.
3. GOSSE, PHILIP HENRY. 1857. Omphalos. Reprint, 1998. Ox Bow Press. Woodbridge, CT.
4. COWAN, G.A., D. PINES & D. MELTZER 1994. Complexity, Metaphor, Models, and Reality. Addision Wesley. Reading, MA.
5. HOLLAND, JOHN. 1998. Emergence. Addison Wesley. Reading, MA.
6. AL-GHAZZALI. 1977. Incoherence of Philosophers. Brigham Young University Press. Provo, UT.
7. 1830–1840. Bridgewater Treatises. Carey, Lea, and Blanshard. Philadelphia, PA.
8. HOLLAND, JOHN H. 1995. Hidden Order; How Adaptation Builds Complexity. Addision-Wesley. Reading, MA.
9. HOLLAND, JOHN H. 1998. Emergence; From Chaos to Order. Addison-Wesley. Reading, MA.
10. MONOD, JACQUES. 1971. Chance and Necessity. Alfred Knopf. New York.
11. HENDERSON, LAWRENCE. 1911. The Fitness of the Environment. Reprint, 1959. Beacon Press. Boston, MA.
12. BERGSON, HENRI. 1994. Creative Evolution. Random House. New York.
13. GRIFFEN, DONALD. 1992. Animal Minds. University of Chicago Press. Chicago.

Complexity and Evolutionary Law for Natural Systems

A "New Dialogue" with Nature: In Looking for a Language as a Means of Intercourse with Nature

S. F. TIMASHEV[a]

Karpov Institute of Physical Chemistry, 10, Vorontsovo Pole Street, Moscow, Russia

IS IT NECESSARY TO CARRY ON A DIALOGUE WITH NATURE?

"Then I saw a new heaven and a new earth; for the first heaven and the first earth had passed away, and the sea was no more" (Rev. 21:1). It may seem that the Revelation of Saint John the Divine (the Apocalypse), which reflects a variant view of the future, is addressed to our epoch. Indeed, for the first time anthropogenic forces have an effect on so great a scale that they may produce global changes, destroying not only separate ecosystems but even the biosphere as a whole. This is why the current discussions and heated arguments concerning "sustainable development" are well timed. The main problems may be yet to come. What are currently the most urgent problems? Which parameters, what paradigm, may be used for an adequate description of the state of the environment as well as the conditions of the gene banks of different natural communities and man? In what way do the different anthropogenic forces (chemical, ionizating radiation, physical fields—acoustic, electromagnetic) act on ecosystems and gene pools? What is the possible scale of the changes in biodiversity? The questions are endless. Is it possible, however, to find the answers? Is it possible to carry on a dialogue with nature, which is a very complex system—a "new dialogue," as defined by Prigogine and Stengers?[1] In the latter case we have to listen to the Earth.[2–5] "What is needed on our part is the capacity for listening to what the earth is telling us."[2] But this is not good enough: a good dialogue means that any questions, and as many questions as are necessary for finding the answers, must be put forth. It will be necessary to create a language to serve as a means of communication with nature.

Obviously, this "language" must provide the key for understanding the sense of the signals that natural objects produce while under observation. These signals, which are often measured with the aid of specialized equipment, are presented as a series over time of a dynamic variable $V(t)$ where t is time. We will have to learn to interpret this "book," this time series. Additionally, nature may use other forms of communication. Information about natural systems may be contained in the spacial configuration of natural structures. Measured spacial characteristics of these structures (heterogeneity, surface roughness, etc.) along some x-coordinate form the corresponding sequences of the dynamic variable $V(x)$, which I will call the "space

---

[a] Address for correspondence: (095) 975-2450 (fax); timashev@lmp.nifhi.ac.ru (e-mail), 007.

series." Therefore, the main question about the new dialogue with nature is, What information contained in these time series is adequate in itself, and what part of the information can only be obtained through time series processing. May this extracted information be presented as a sort of "data passport" of the systems under consideration? New possibilities in elaborating the new dialogue are associated with the ideas about nonlinear dissipate system dynamics and deterministic theory.[6–9] In this paper a new phenomenological method of analysis of nonlinear dissipate system dynamics, which may be called flicker-noise spectroscopy,[10–13] is presented. An algorithm is developed for determining enough phenomenological parameters—the data passport as it needed for describing the state of the complex dynamic system and the changes it undergoes during evolution. The developing approach may become a tool for undertaking new dialogue with nature. For the sake of precision, we will demonstrate the approach using time series analysis. Space structures may be examined in a similar manner.

EVOLUTION AS A TIME COLOR FRACTAL: METHOD OF ANALYSIS OF DYNAMIC DISSIPATE SYSTEM STATES

The main features of the evolution of nonlinear dynamic dissipate systems are complexity and nonregularity of changes in wide space–temporal intervals of measured dynamic variables of complex systems and structures of different origins, as well as self-similarity of these nonregularities. In the case of time series elaboration, the statement is manifested by observing very sharp jumps and other nonregularities of the observed dynamic variable $V(t)$ that demonstrate self-similarity in large temporal intervals (FIG. 1). This phenomenon raises the possibility of proposing a hypothesis that not *all* points, but a sequence of the discrete points in the time axis carry the main information about the state and the features of evolution of the system under consideration. The introduced hypothesis means that there is a set of shot-time, δ_i intervals, for every i space–temporal level of evolution, in which most of the information about the evolution $V(t)$ is kept. The time intervals between the short-time δ_i intervals also contain information about the evolution. The latter time intervals are not empty, they hold the information contained in the smaller δ_{i+1} intervals. There are definite correlation links between the individual δ_i-intervals and the introduced sequences of the δ_i-intervals (see below). These links are characterized by a set of parameters that carry information about the dynamic state of the system under consideration. These parameters will be introduced below.

The transition system from one δ_i interval to the adjacent one must be accompained by a change in the structure–energetic characteristics of the system state. The observed nonregularities of the dynamic variable $V(t)$ manifest this evolution regularity. According to the Triest theory of von Weizsacker,[14] an event happens when irreversibility is realized. This means that irreversibility must have occurred in the δ_i intervals, while the intervals between the adjacent δ_i intervals, which are "empty" for the i-level, may be designated as "now", according to Ruhnau.[15] The last image combines "the past," which has happened, and "the future," which potentially exists.

Let us idealize this image of the evolution and contract the δ_i intervals to the corresponding points. Let every new δ_i interval with zero duration carry the information

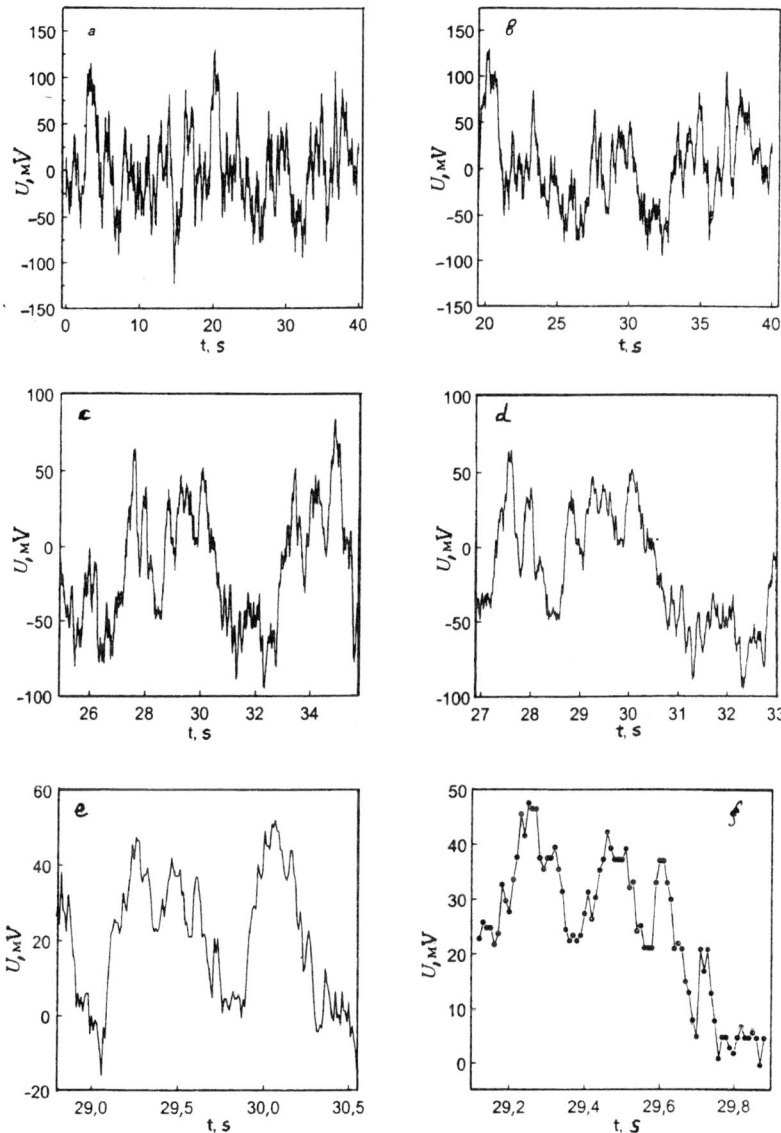

FIGURE 1. The electric votage fluctuation in the electromembrane system with cation exchange membrane under "over-limiting current" conditions when the relation, β, of the current density with the corresponding limiting current value is β = 7. (10 mM NaCl solution, membrane diameter is 0.1 cm).[29] (a) A votage fluctuation realization for a time of 40 sec (it contains 4000 readings); (b)–(f) fragments of the realization. The apparently discrete values given in (f) are the result of the equipment's resolving power. (Taken from Timashev[11]; used with permission from the *Mendeleev Chemistry Journal*.)

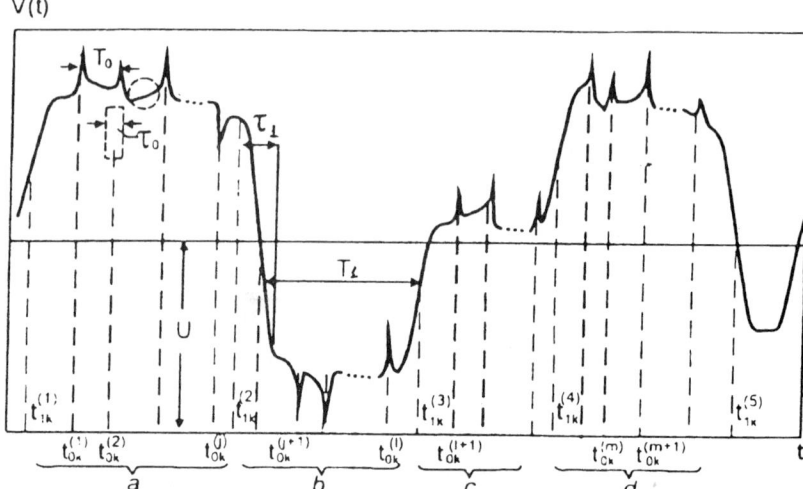

FIGURE 2. Intermittent evolution of a dynamic variable $V(t)$. (Taken from Timashev[10]; used with permission from the *Mendeleev Chemistry Journal*.)

about the structure–energy system state, which was previously contained in the "old δ_i intervals" that had finite durations. Zero duration of these new δ_i intervals means that the meaning of the dynamic variable $V(t)$ in the every point must contain "real" or "potential" singularities. "Real" singularity means that $V(t)$ is a generalized function of compact support in $\{0\}$ (supp $V(t) = \{0\}$).[16] "Potential" singularity means that $V(t)$ has shifts of meaning and discontinuities of the derivatives of different orders at indicated points. Obviously, the "potential" singularities transform to the "real" singularities after differentiation of $V(t)$.

The idealized $V(t)$ image that has been introduced (FIG. 2) is a type of "model-caricature" according to Frenkel's definition.[17] Frenkel wrote that a simple model-caricature had to be introduced for receiving simple equations that describe complicated real processes. Therefore, we postulate generalized intermittency dynamics[6,7] for i, the space–temporal level of the $V(t)$ signal. This dynamic is characterized by relatively weak variations of the value extended through long time intervals ("laminar" phases) with characteristic time T_0^i. Laminar phases are interrupted by "bursts" of short duration (τ_0^i; $\tau_0^i \ll T_0^i$) against the indicated background variations. We note t_{0k}^i as a time of the appearance of the burst of number $k(-\infty < t_{0k}^i < +\infty)$. Notice (see FIG. 2) that abrupt shifts of the $V(t)$ function are present at the same t_{0k}^i points. We call these shifts the "first type." In addition to this type of evolution, which was considered in Schuster[6] and Berge *et al.*,[7] we suppose that $V(t)$ value may drastically (in short time τ_1^i) change his background value (the mean value in the laminar region) just as the Heaviside function does. We note T_1^i as a characteristic time interval between these shifts ("the second type") as well as t_{1k}^i as the time of the appearance of the change in number $k(-\infty < t_{1k}^i < +\infty)$ and suppose $\tau_1^i \ll T_1^i$. We call the time moments t_{0k}^i and t_{1k}^i as "$0k$" and "$1k$" points, respectively. We also suppose that the nonregularities are characterized by self-similarity, that is, the image of the evolution

$V(t)$ presented in FIGURE 2, is reproduced in smaller scale into the dotted circle drawing in FIGURE 2. It allows us to present the evolution of a nonlinear dissipate system as a temporal fractal and receive the simple phenomenological equations that may be used as a base for studying the experimental time series and for getting the straight phenomenological information about the state of the system under consideration.

We suppose that the process under consideration is a stationary, stochastic one, whose statistical characteristics do not change in the course of time. It means[6,7] that autocorrelation function $\psi(\tau) = \langle V(t)V(t+\tau) \rangle$, where the brackets denote time averaging, depends only from τ and $\psi(\tau) = \psi(-\tau)$. In that case we derive the following equation for the Fourier transform $S(f)$ of the autocorrelation function or the power spectrum (according to the Wiener-Khinchin theorem):

$$S(f) = 2\int_0^\infty \cos(2\pi f\tau)\psi(\tau)d\tau \quad (1)$$

$$\psi(\tau) = 2\int_0^\infty \cos(2\pi f\tau)S(f)df$$

We also introduce the average difference moments $\Phi_{(p)}(\tau)$ (the so-called "structural function" in the theory of developed turbulence) of different order p:

$$\Phi_{(p)} = \langle |V(t) - V(t+\tau)|^p \rangle \quad (2)$$

where τ is a delayed time value. It is known that the spectral dependences $S(f)$ as well as the dependences $\Phi_{(p)}(\tau)$ from the parameter τ indicate the presence of correlating links between previous and future events. In the case of "white noise" the correlating links are absent, and $S(f)$ as well as $\Phi_{(p)}(\tau)$ is constant at $f > 0$ and $\tau > 0$, respectively. However, what is the sense of "the events" in every time series $V(t)$ considered? We shall demonstrate that in this approach, these events are marked by the introduced nonregularities of the signal $V(t)$.

At the first stage of the power spectrum $S(f)$ calculation, according to Timashev,[10] we decompose the signal $V(t)$ into two summands: the singular $V_S(t)$ one, which is formed by the bursts or spikes of the evolution (see FIG. 2), and the so-called "regular" $V_R(t)$ term, which is formed by $V(t)$ without the bursts:

$$V(t) = V_R(t) + V_S(t); \quad V_S(t) \equiv (V(t) - V_R(t)) \quad (3)$$

The second term in Equation (3) is nonzero in the regions of the bursts only. According to our hypothesis, we consider $V_S(t)$ as a generalized function of compact support in $\{0\}$ and use the well-known theorem of the theory of the generalized function[16]:

$$V_S(t) = \sum_{k,p} c_k^p \delta^{(p)}(t - t_{0k}) \quad (4)$$

where $\delta(t)$ and $\delta^{(p)}(t)$ are the Dirac delta-function and its time derivatives of p-order, which form the total set of the generalized functions of compact support in $\{0\}$; c_k^p

indicates the corresponding coefficients of the expansion of $V_S(t)$ in the complete system of the functions. Note that we have omitted in Equation (4) the i-index of the considered space–temporal level (scale). This is because we will determine the conditions when the calculated value of $S(f)$ does not depend on the space–temporal scale, which means that a self-similarity has been realized.

For calculating $S(f)$ it is necessary to substitute Equation (3) into Equation (1). Then $S(f)$ is expressed as a sum of the following four terms:

$$S_{RR}(f) = \langle V_R(t)V_R(t+\tau)\rangle, \qquad S_{SS}(f) = \langle V_S(t)V_S(t+\tau)\rangle,$$

$$S_{RS}(f) = \langle V_R(t)V_S(t+\tau)\rangle, \qquad S_{SR}(f) = \langle V_S(t)V_R(t+\tau)\rangle.$$

The simple transformation gives:

$$S(f) = S_{RR}(f) + 2 \sum_{n=-\infty}^{n=+\infty} C_n \cdot \cos(2\pi f t_n) \tag{5}$$

$$C_n = A_n + 2U_R B_n; \quad A_n \equiv \sum_{k=-\infty}^{k=+\infty} c_k^0 c_{k+n}^0 \qquad B_n \equiv \sum_{k=1}^{N_1} c_{k+n}^0 \tag{6}$$

$$U_R \equiv \frac{1}{N}(V_{R1} + V_{R2} + \ldots + V_{Rm}); \qquad N = N_1 + N_2 + \ldots + N_m$$

Here t_n, the time of the nth burst appearance, corresponds to the ith space–temporal level of $V(t)$ signal; T_0 is a characteristic interval between adjacent bursts; N_k is a number of the bursts at the kth ($k = 1, 2, \ldots, m$) interval with a characteristic duration, T_1, from the sequence a, b, c, d, …, for the signal $V(t)$ (see FIG. 2); m is the total amount of latter intervals during the time of observation; U_R is the mean value of the model signal $V_R(t)$ after averaging over all m intervals; V_{Rk} is a characteristic meaning of the $V_R(t)$ function at the kth interval.

Let us suppose for simplicity (it is possible to consider more general cases) that $U = 0$ and therefore, $C_n = A_n$. In the case $2\pi f T_0 \ll 1$, in the second term of the right part of Eq. (5) it is possible to replace the summation by integration over the dimensionless variable $\xi = 2\pi f T_0 n$ ($n = 1, 2, 3, \ldots$), so that $d\xi = 2\pi f T_0 \Delta n \ll 1$ if $\Delta n = 1$ for the indexes of adjacent bursts. Let us introduce the density probability function $\rho(n)$, which characterizes the spacing in the sequence of the bursts. In that case:

$$\begin{aligned}S(f) &= S_R(f) + \frac{1}{T_0^j f \pi} \int_0^\infty \Phi_0^i\left(\frac{\xi}{2T_0^j f \pi}\right) \cos\xi\, d\xi \\ &= S_R(f) + 2\int_0^\infty \Phi_0^i(\eta) \cdot \cos(2T_0^j f \eta \pi) d\eta\end{aligned} \tag{7}$$

where the function Φ_0^i characterizes the effective density of the burst sequence and takes into account the burst amplitudes.

According to the dimensions [sec] of the function $\Phi_0^i(z)$, we may present it in the form:

$$\Phi_0^i(\eta) = \frac{g}{4\pi} T_0^i \chi_0(b_0^i \eta) \tag{8}$$

where $\chi_0(b_0^i\eta)$ is a dimensionless function, b_0^i is a dimensionless scale parameter of the ith space–temporal level, g is a constant. Then we have:

$$S(f) = S_{RR}(f) + I_S(f)$$

$$I_S(f) \equiv \frac{g}{K_0 \pi} \int_0^\infty \chi_0(x)\cos Zx\, dx\,; \quad K_0 \equiv \frac{b_0^i}{T_0^i},\quad Z \equiv \frac{2f\pi}{K_0} \tag{9}$$

Here K_0 is the scale-invariant parameter which may be considered as an universal parameter of the system.

It is convenient to use various approximations of the function $\chi_0(b_0^i\eta)$. For instance, possible types of relaxation, including the retarded nonexponential relaxations of nonlinear systems with rearrangement of structure elements during relaxation under above-threshold excitation, are analyzed using the following approximations:

$$\chi(z) = z^{-\mu},\, (0<\mu<1) \text{—the "flicker-noise" approximation} \tag{10}$$

$$\chi(z) = \exp[-(\lambda_0 z)^s],\, (0<s<2) \text{—the Levy approximation} \tag{10a}$$

where μ, λ_0, and s are parameters. Approximations (10) are responsible for the self-similar equations of the functions $I(f)$, that is, $I(f)$ may be considered as a temporal fractal. We have:

(i) $I_S \equiv S(f) - S_{RR}(f) = \dfrac{g}{\pi K_0 z^{1-\mu}} \cdot \Gamma(1-\mu) \cdot \sin\dfrac{\mu\pi}{2} \sim \dfrac{1}{f^{1-\mu}}$

(ii) $Z_{01} \equiv Z/\lambda_0 \gg 1,\quad I_S = \dfrac{g}{\pi\lambda_0 K_0} \cdot \dfrac{1}{Z_{01}^{s+1}} \sim \dfrac{1}{f^{s+1}}$ (9b)

$Z_{01} \ll 1,\quad I_S = \dfrac{g}{\pi\lambda_0 K_0} \cdot \Gamma\!\left(1+\dfrac{1}{s}\right) = \text{const.}$

For the Levy approximation, the dependencies $I_S(f)$ may be found exactly if $s = 0.5$ and $s = 1$.[10] Specifically, in the case $s = 1$, we find:

$$I_S = \frac{4g(\lambda_0 K_0)^{-1} q^2}{\left(q^2 + \dfrac{1}{16}\right)} \tag{11}$$

where $q \equiv \lambda_0 K_0/(8\pi f)$.

From Equations (9)–(11), it follows that $I_S(f) \to 0$ if there are no correlations in the burst sequence (the Poisson distribution); that is, in the case: $\mu \to 0$, $s \to 0$, $\lambda_0 \to 0$. Obviously, the inequality $f \ll 1/(2\pi T_0^i)$ is realized for increasing scales if the ith space–temporal level rises as the frequency f decreases. The case (i) corresponds to the flicker-noise, or, in other words, to the realization of the "infinite memory" in the system. The Levy approximation (ii) corresponds to the "shot noise" case when the power spectrum does not depend on frequency in the limit of $f \ll \lambda_0 K_0/(2\pi)$ (the "losing memory" case at the time exeeding the value $t_M \sim [\lambda_0 K_0/(2\pi)]^{-1}$) and falls with frequency increasing as f^{-n}, where $n = s + 1$ and $1 < n < 3$ in the frequency interval: $\lambda_0 K_0/(2\pi) \ll f \ll K_0/(2\pi)$. The latter means that the parameter $\lambda_0 K_0$, which characterizes a nonlinear relaxation in the correlating sequence of bursts generated in the nonlinear dissipate system under action of outer energy sources, may be considered as a component of Kolmogorov K-entropy.[6] Note that dimensionless parameters $(1 - \mu)$ and $n \equiv (s + 1)$ characterize the content of the correlating links ("memory") in the correlating sequence of bursts in the relatively short ($t \ll t_M \sim [\lambda_0 K_0/(2\pi)]^{-1}$) time.

In the general case, the interpolation relation that characterizes the correlation links in the correlating sequence of the δ-like bursts of the "model-caricatures" $V(t)$, may be presented:

$$S(f) \approx g_0 \frac{\varepsilon^{a_1}(\lambda_0 K_0)^{a_2} U^{n-1}}{(\lambda_0 K_0/2\pi)^n + f^n} + S_{n0} \qquad (12)$$

Here, phenomenological parameters are introduced: ε is the specific energy dissipation rate; U is the mean velocity of the energy flow with which the wavenumber $k = f/U$ may be connected; n is above-mentioned parameter; g_0 is a dimensionless function that connects the physical parameters of the system under consideration with the introduced phenomenological parameters ε and $\lambda_0 K_0$; S_{n0} is a constant; a_1 and a_2 are indexes of a power that are found on the dimension ground.

Equation (12) may be used for describing different power spectra of the flicker-noise or shot-noise types, which are calculated on the base of various experimental time series data. Here it should be particularly emphasized that these dependencies are formed by the singular terms of the $V(t)$ evolution. One argument is to suppose that the nonspecific character of power laws in natural sciences as well as in human laws is a result of nonregularities of the corresponding evolutionary changes. It also means that the $S_{RR}(f)$ dependencies are changed more slowly at a low frequency limit and may manifest more specific features of the evolution in a region of higher frequencies. Of course, this is a hypothetical conclusion. Additional arguments in favor of this conclusion are needed. Among the possible experiments is research into the dependence of the observed power spectra from the "whole volume" υ (or its analogue) of the systems under consideration: $S(f) \sim \upsilon^{-m}$ ($m > 0$). Indeed, if the considered system has quite a large volume, the correlating sequence of the bursts will be generated at different points in this system, and there will not be any correlating links between the different sequences. Exactly the same effect takes place in semiconductor systems or thin metal films.[18] This is the so-called Hooge phenomenological law: $S(f) \sim \upsilon^{-m} f^{-n}$ ($n \sim 1$, $m \sim 1$). Probable, analogous dependencies may manifest themselves in other kinds of systems where evolutionary changes of the dynamic variable $V(t)$ are characterized by nonregular behavior, and correlation links between the dynamic fluctuations are realized.

Note that the numerical values of the introduced parameters $\lambda_0 K_0$ and n depend on the inner structure of the system under consideration and an ability of the structure to undergo local structure rearrangements because of inner links under the presence of inner and outer energy fluxes. The classic flicker-noise in electroconductive systems, which is expressed by the law $S(f) \sim 1/f$, corresponds to a partial case $n \to 1$ and $\lambda_0 K_0 \to 0$. This case means that system has a memory at "all times" and may be described (12), if $a_2 = n - 1 \to 0$ and ϵ has a dimension $[S(f) \text{Hz}]^{1/a_1}$. In the case of fully developed turbulence (Kolmogorov theory), when the parameter ϵ [cm^2/sec^3] has a sense of the mean energy dissipation rate, U is a mean velocity of the hydrodynamic flow, and the dimension of $S(f)$ is equal [cm^2/sec^3], we have[4]: $a_1 = (3 - n)/2$ and $a_2 = (3n - 5)/2$. It means that the classic flicker-noises, which are characterized by "infinite memory" in hydrodynamic flow are described by Equation (12) if $\lambda_0 K_0 \to 0$ and $n \to 5/3$ (the Kolmogorov-Obukhov law).

It was shown[10] that the dependencies $\Phi_{(p)}(\tau)$ are largely formed by shifts in the dynamic model-caricature $V(t)$, which are described by the Heaviside Θ function. This is easy to understand because only the changes due to jumps in the dynamic variable at the adjacent laminar regions contributed to $\Phi_{(p)}(\tau)$ at every point of $0k$ or $1k$ type. The values of the burst amplitudes do not contribute to $\Phi_{(p)}(\tau)$ values. To find the dependencies for $\Phi_{(p)}(\tau)$ we introduce a more general definition of the "difference moment of pth order" and rewrite Equation (2) in the form:

$$\Phi_p^{(1)}(\tau) = \left\langle \left| \int_t^{t+\tau} \frac{dV(z)}{dz} dz \right|^p \right\rangle \tag{2a}$$

where the upper index of the difference moments indicates that the integrand is the first derivative of the $V(t)$ function. Obviously, the definition provided by Equation (2a) coincides with Equation (2) only if $dV(t)/dt$ is a regular differentiable function in the whole range of the argument t variation. Let us pick out the "regular" and "singular" parts of the derivative $dV(t)/dt$. The corresponding regular part, which is noted as $[dV(z)/dz]_R$ includes the shifts of $dV(t)/dt$, which are determined by the above-mentioned discontinuities of the derivatives of the $V_R(t)$ function in the $0k$ points. The singular part of $(dV/dz)_S$ is defined by the sum of the Dirac delta functions and their time derivatives, which are the result of the differentiation of the $V(t)$ function in the regions of its "$1k$" and "$0k$" points, respectively.

The corresponding equations for the $\Phi_{(p)}(\tau)$ function were reported by Timashev.[10] Here, we present the interpolation equations for the second-order difference moment in the case when the contribution from the shifts at the $1k$ points is negligibly small:

$$\Phi_p^{(1)}(\tau) \approx g_1 \Gamma^{-2}(\nu_1 + 1) U^2 (\lambda_1 K_1 \tau)^{2\nu_1}, \quad \lambda_1 K_1 \tau \ll 1;$$

$$\Phi_p^{(1)}(\tau) \approx g_1 U^2 [1 - \Gamma^{-1}(\nu_1)(\lambda_1 K_1 \tau)^{\nu_1 - 1} \exp(-\lambda_1 K_1 \tau)]^2, \quad \lambda_1 K_1 \tau > 1 \tag{13}$$

Here K_1 is the second-scale invariant parameter of the process; ν_1 and λ_1 are parameters of nonlinear relaxation, which determines the rate of jump-like changes of the dynamic variable after perturbation at, respectively, short and long times; and $\Gamma(z)$ is the gamma function. Equations (13) and, more generally those of Timashev,[10] may be used for getting information about the phenomenological parameters ν_1 and $\lambda_1 K_1$ of the system under consideration by comparison of these equations with the corresponding expression derived on the basis of time series data. Note that the jumps at the 1k points contribute to $\Phi_{(p)}(\tau)$ dependence at longer time when $t \gg T_1$ (see FIG. 2). In that case the additional parameters ν_2 and $\lambda_2 K_2$ have to be introduced.[10]

Additional information about the evolution process is contained in the time series formed from the derivatives $d^m V/dt^m$ ($m \geq 1$). The latter time series may be received from the origin $V(t)$ time series by differentiation. Every new time series may produce new dependencies, namely, the Fourier spectra $I^{(m)}(f)$ of the corresponding autocorrelation function and the difference moments $\Phi_p^{(m)}(\tau)$. We do not discuss here the mathematical difficulties that arise with the realization of the differentiation procedures. We would like to highlight the possibility of finding as many parameters of the process under consideration as are necessary for the characterization of the dynamic system's state and the changes. All initial parameters are marked by nonregularities of different types at the 0k and 1k points. By this means, analysis of the time series formed by the first derivative dV/dt gives new information concerning the correlation of the jumps (by considering the $S^{(1)}(f)$ dependencies) as well as some new features of the evolution near the bursts (by considering the $\Phi_p^{(1)}(\tau)$ dependencies). The differentiation procedure has to converge as the order of the derivatives increases because the increase of the derivative order leads to identity ($0 \equiv 0$) in the process of differentiation.

THE MEANING OF THE INTRODUCED PARAMETERS

The basis of the presented methodology is a postulate about the meaning of the nonregularities of the measured dynamic variables (temporal or spatial). It corresponds to the abstract theory of information (ATI).[19] According to this theory,

> *knowledge is based on the possibility of distinction.* This remark reflects the fact that *distinquishability*—the possibility of making distinctions—presumably represents a precondition to the cognitive abilities of any rational living being. Without the possibility of distinction, we would not be able to perceive, to form concepts, or to speak. ... *Empirical knowledge presupposes temporality. The difference between the past and the future is a precondition of experience.* ... *Distinquishability and temporality are always interwoven. Any temporal transition can be looked upon as a change of distinquishabilities* .[19]

In the frame of our approach, all demands and the basic principles of the ATI are fulfilled. The introduced nonregularities are characterized by *distinquishability* (according to the types of the nonregularities) and *temporal fractal regulating*. The latter definition replaces and makes more particular the diffuse term *temporality* introduced in the ATI. In this case the idea of the fractal quality acquires a pragmatic meaning and illustrates a metaphysical image of the "realism of ideas."[20] In fact all the information corresponding to concrete evolutionary changes of each introduced "color nonregularity" may be extracted by means of concrete calculating procedures—from comparison of Equations (9) through (13) with the power spectra and the difference moments of different orders received from the time series data. In other words we realize the principal cognitive grasp (upon obtaining the corresponding numerical characteristics) of the distinquishable nonregularities of the observed dynamic variable.

In the frame of the proposed approach, two sequences of the introduced dynamic parameters may be formed—dimensionless (n, v, v_1, ...) and with dimension ($\lambda_0 K_0$, $\lambda_1 K_1$, $\lambda_2 K_2$, ...). These parameters may be considered the "passport data" of the dynamic system under consideration. The second set of dimension parameters conveys a sense of the components of the Kolmogorov K-entropy. Usually, the Kolmogorov entropy[6,7] is considered scalar. In our approach the introduced parameters characterize the rates of losing information, which depends on the type of color nonregularities. These parameters may differ for different types of nonregularities. We call the developed approach to time series analysis flicker-noise spectroscopy (FNS).[10-13] Obviously, FNS presents the possibility of getting complete information compared with the information contained in traditional Lyapunov exponents as well as in dynamic entropy. This conclusion corresponds with the opinions expressed in previous work (see Atmanspacher[21]). The developed phenomenological method of analysis is what we are calling flicker-noise spectroscopy.[10-13]

It is necessary to remark that the "polychromism" of the dynamic variable changes during evolution may be correlated with the image of "topo-chronology" introduced by D. Bohm.[22] "The choice of this word is intended to signify that not only must we emphasize spatial topology, i.e., the study of the order of placing one thing in relationship to another, but the study of how one event or moment acts physically in another."[22] In other words, according to our approach, the image of topo-chronology means that there may be correlation links realized between the structure-energetic states of the system, not only at adjacent time moments but also at different temporal intervals.

Obviously, the presented phenomenological approach, which may be called flicker-noise spectroscopy,[10-13] does not open the concrete physical essence of the introduced parameters, but it is responsible for the universality and nonspecificity of the method, which may be used for the analysis of different systems dynamics. As an illustration, time series corresponding to the fluctuations of hydrodynamic turbulent flows as well as to the electric voltage fluctuations in the electromembrane process (see FIG. 1) and in very thin polycrystal threads (whiskers) are analyzed[12] and the corresponding parameters (passport data) for characterizing these system states are found. The time series analysis may be used for studying various processes in geophysics and astrophysics, in ecology and medicine, in economics, and so forth. The approach under consideration may be used for discovering the main features of space structures or image (imprint) profiles of different origins (surface roughness, content of the imprint color and darkness, etc.) for the purpose of their description by introducing a set of the passport parameters. In these cases, we divide the space interval along the space axis (a Cartesian coordinate) of the structure or images (imprints) into small, equal intervals. Therefore, we have a set of points (the number of points must be quite large, more than several hundred) and the corresponding values of the measured dynamic variables under consideration (the heights of the roughness, the color nuance parameters, etc.). These "time series" which reflect a complexity of the nonlinear correlated processes of the structure formation ("evolution") may be examined for obtaining some parameters of evolution.

This method may be used for obtaining the parameters of various natural structures in geophysics as well as the passport data of medicinal images (tomography, mammography) and DNA sequences. This analysis may be fruitful in studying the correlation structures in music, novels, painting, and other art objects. Therefore, it gives us the potential to promote a cross contamination between the humanities and the sciences, to interlace the world of the arts with the world of natural laws and theories. FNS gives us a new possibilities for the analysis of literary and musical texts, in finding correlation links for fragments of painting canvases as well as for the temporal sequences of the historical events and for the sound signals of animals.[4,8,23,24] Note, that the introduction of "color" in FNS analysis makes it much more informative and gives us the possibility of reaching a high degree of subject personification.

ON THE METAPHYSICAL ESSENCE OF PHYSICAL LAWS

Obviously, the introduced "model-caricature" $V(t)$ is *a priori* a nonobservable, pure ideal (the compact support!) metaphysical image. It is necessary to point out that it is precisely the highest abstract images that are the basis of gnosiology.[14] Indeed, "hypothetic suppositions about nonobservable essences and about hidden mechanisms of natural phemomena"[25] form the basis for the general physical theories (quantum mechanics, the inflation model of forming the Universe, string theory, etc.). It is easy to understand that the secretiveness of the "first mechanisms," as well as the difference between the essence of the process and the process manifestation during the measuring procedure,[26] depends on the inertia of the real objects as well as the inevitability of experimental errors. In effect we have the situation described by Plato, in which a person in a cave has to make a conclusion about the essense of an object only by watching the shade of the object on the cave wall. Plato's image is a universal one. We have the similar situation in researching any phenomenon. That is why any scientific construction has to start from the metaphysical principles,[27] and the metaphysical arguments must be acknowledged as a legitimate tool for elaborating new ideas in physics and mathematics.[22] It is necessary to add that "every metaphysics is a realism of ideas."[20] It is hypothetical suppositions about the nonobservable essences that are defined as "basic principles of a theory from which the theory content is received deductively. These principles cannot be deduced using inductive generalization of the experimental data. These principles are always a result of the surmise and intuition which are shown by experiment. At this time, these principles cannot be the object of experimental research, and in this regard they belong to the metaphysical sphere."[25]

Quantum mechanics provides an unprecedented example in the history of physics. The basis of quantum mechanics is formed by aprioric, pure metaphysical constructions. "The quantum theory is highly abstract, and concepts derived from earlier, concrete experience are evidently inadequate to interpret it. … However it comprises all known basic ranges of physics (with the possible, but perhaps not final exception of general relativity), and no experimental results have been found that would recognizably contradict it."[14]

This example shows the adequateness of the basic metaphysical hypothesis for the approach developed for analysis of nonlinear dissipate system evolution. This method has to be tested by using it to solve practical problems. Among the various complex processes in natural sciences and humanities, environmental problems appear to have priority. It is important to develop the proposed approach for studying the unsolved problems of predicting the effects of dangerous natural phenomena.

This approach may be used to realize a new dialogue with nature, which will give us the necessary new knowledge for solving the urgent local and global environmental problems. Some problems that may be solved within the framework of the above-mentioned dialogue are presented in References 3–5 and 10–13.

ACKNOWLEDGMENTS

This work was supported by the Russian Foundation of Basic Research (Grants 96-03-33998 and 96-15-97608).

REFERENCES

1. PRIGOGINE, I. & I. STENGERS. 1984. Order Out of Chaos. Heinemann. London.
2. BERRY, T. 1990. The Dream of the Earth. Sierra Club Nature and Natural Philosophy Library. San Francisco.
3. TIMASHEV, S.F. 1991. The role of chemical factors in the evolution of natural systems (chemistry and ecology). Russ. Chem. Rev. **60:** 1183–1204.
4. TIMASHEV, S.F. 1996. Physicochemical principles of global ecology. Mendeleev Chem. J. **40:** 155–169.
5. TIMASHEV, S.F., S.P. PEROV & E.E. GUTMAN. 1994. Problems of the Physical Chemistry of the Ozone Layer. Russ. J. Phys. Chem. **68:** 1231–1242.
6. SCHUSTER, H.G. 1984. Deterministic Chaos. An Introduction. Physik-Verlag. Weinheim.
7. BERGE, P., Y. POMEAU & C. VIDAL. 1988. L'Ordre dans le Chaos. Nouvelle edition corrigee. Paris.
8. BAK, P. 1997. How Nature Works. The Science of Self-Organized Criticality. Oxford University Press. Oxford, England.
9. KURDYUMOV, S.P., G.G. MALINETSKII & A.B. POTAPOV. 1996. Nonstationary structures, dynamic chaos, cellular automata. In The Novel in Synergetics. Enigmas of the Non-equilibrium Structure World. 95–164. Nauka. Moscow.
10. TIMASHEV, S.F. 1997. Flicker-noise as an indicator of the "Time Arrow." Methodology of the time series analysis on the base of the deterministic chaos theory. Mendeleev Chem. J. **41**(3)**:** 17–29.
11. TIMASHEV, S.F. 1998. The principles of the nonlinear system evolution. Mendeleev Chem. J. **42**(3)**:** 18–35.
12. TIMASHEV, S.F. et al. 1998. The methodology of time series analysis on the base of the deterministic chaos theory. In Atlas of Temporal Variations of Natural, Anthropogenic and Social Processes. Vol. 2. Cuclical Dynamics in the Nature and Society: 386–397. Scientific World. Moscow.
13. KOSTUCHENKO, I.G. & S.F. TIMASHEV. 1998. Flicker-noise in processes of solar activity. Int. J. Bifurc. Chaos **8:** 4–5.
14. WEIZSACKER, C.F. VON. 1997. Time-emperical mathematics—quantum theory. In Time, Temporality, Now: Experiencing Time and Concept of Time in an Interdisciplinary Perspective. Harald Atmanspacher & Eva Ruhnau, Eds.: 91–104. Springer-Verlag. Berlin/Heidelberg/New York.

15. RUHNAU, E. 1997. The deconstruction of time and the emergence of temporality. *In* Time, Temporality, Now: Experiencing Time and Concept of Time in an Interdisciplinary Perspective. Harald Atmanspacher & Eva Ruhnau, Eds.: 53–69. Springer-Verlag. Berlin/Heidelberg/New York.
16. VLADIMIROV, V.S. 1979. Generalized Functions in Mathematical Physics. Mir. Moscow.
17. FRENKEL', YA.I. 1970. The Dawn of the Novel Physics. (in Russian). Nauka. Leningrad. pp. 307-308
18. TIMASHEV, S.F. 1995. The principles of the the flicker-noise spectroscopy. *In* Noise and Degradation Processes in the Semiconductor Devices. A.M. Gulyaev, Ed.: 5–19 (in Russian). A.S. Popov Society. Moscow.
19. LYRE, H. 1997. Time and information. *In* Time, Temporality, Now: Experiencing Time and Concept of Time in an Interdisciplinary Perspective. Harald Atmanspacher & Eva Ruhnau, Eds.: 81–89. Springer-Verlag. Berlin/Heidelberg/New York.
20. GESSEN, S.I. 1910. Mystics and metaphysics. *In* Logos. Book 1: 118–156. Musaget. Moscow.
21. ATMANSPACHER, H. 1997. Dynamic entropy in dynamical systems. *In* Time, Temporality, Now: Experiencing Time and Concept of Time in an Interdisciplinary Perspective. Harald Atmanspacher & Eva Ruhnau, Eds.: 327–346. Springer-Verlag. Berlin/Heidelberg/New York.
22. HILEY, B.J. & M. FERNANDES. 1997. Process and time. *In* Time, Temporality, Now: Experiencing Time and Concept of Time in an Interdisciplinary Perspective. Harald Atmanspacher & Eva Ruhnau, Eds.: 365–383. Springer-Verlag. Berlin/Heidelberg/New York.
23. BARROW, J.D. 1995. The Artful Universe. Penguin Books. London.
24. TRUBNIKOV, B.A. 1993. The Law of the Competitor Distribution. Nature (Russian). **11:** 3–13.
25. MAMCHUR, E.A. 1998. N'jan Yu. Cao. The Story of the XX Century's Field Conceptions. Probl. Philos. (Russia). **4:** 150–155.
26. KLOSE, J. 1997. Whitehead's Theory of Perception. *In* Time, Temporality, Now: Experiencing Time and Concept of Time in an Interdisciplinary Perspective. Harald Atmanspacher & Eva Ruhnau, Eds.: 23–42. Springer-Verlag. Berlin/Heidelberg/New York.
27. DALENOORT, G.J. 1997. Cognitive aspects of the representation of time. *In* Time, Temporality, Now. Experiencing Time and Concept of Time in an Interdisciplinary Perspective. Harald Atmanspacher & Eva Ruhnau, Eds.: 179–188. Springer-Verlag. Berlin/Heidelberg/New York.

A Dimly Perceived Horizon: The Complex Meeting Ground between Physical and Inner Time[a]

FRANCISCO J. VARELA[b]

LENA (Neurosciences Cognitives et Imagerie Cérébrale), CNRS URA 654, Hôpital de la Salpètrîere, 47 Blvd. de l'Hôpital, 75651 Paris Cedex 13, France

INTRODUCTION: PHYSICAL TIME AND INNER TEMPORALTY

The question I wish to address is the following. At face value, there are two kinds of time: *inner* time and *physical* time. The first is the linear sequence of moments given by the clock we live *by*, and the other is what we live *in*. Both are valid as sources of facts and of scientific investigation. The first gives rise to well-developed physical theories; the other, to human temporality, centered on the present and manifesting as a threefold unity of the just-past and the about-to-occur. Both can be developed in precise scientific detail. I will sketch some of my ideas in this regard concerning inner time in terms of modern cognitive neuroscience. More precisely my topic for this symposium is: What is the *meeting* ground in our discourse and understanding for these two kinds of time?

Analysis of the living present has early roots. Major contributions have been due to Augustine, William James, and Edmund Husserl. On the side of psychology the topic was taken up by a parallel stream within Gestalt psychology, then experimental psychology of time, and on to modern cognitive neuroscience. Instead of a long historical discussion, let me turn right away to a classic example that illustrates the situation very well. This is the chromatic phi-phenomenon, where a red light followed by a green light is shown at a specific temporal distance. The common perception is that of an "apparent" jumping of the lights, but interestingly the light changes color at midcourse! Unless we wish to assume a violation of the direction of causality, we are forced to conclude that perceived temporality is not simply isomorphic to linear time. Thus, this is a paradigmatic example of the needed distinction of not a single, but at least two kinds of time.

In order to address the question head-on, we need then to turn to the texture of the constitution of lived, inner time before we can link it with physical, linear flow. In this paper I will be using the thorough investigations of Edmund Husserl as my basis. More precisely, this paper is only concerned with a very specific topic: the dynamics of *retention*. Let me explain. Husserl's study of time from the early studies of 1905[1] on presented the constitution of internal time consciousness on the basis of a three-fold structure. To a first approximation, this texture can be described as follows:

---

[a]This paper is based on one section of a long study of neuroscience and the phenomenology of time published in 1999: "The Specious Present: the neurophenomenology of time consciousness." *In* Naturalized Phenomenology. J. Petitot, F. Varela, B. Pachoud & J.M. Roy, Eds. Stanford University Press. Palo Alto, CA.

[b]Address for correspondence: fv@ccr.jussieu.fr (e-mail).

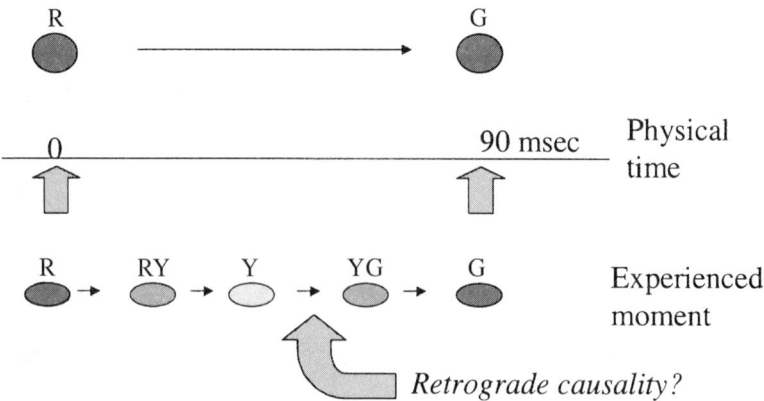

FIGURE 1. The color phi-phenomenon. R, red; G, green; Y, yellow.

There is always a *center*, the "now" moment with a focused intentional content (say, this room with my computer in front of me on which the letters I am typing are highlighted). This center is bounded by a horizon or fringe that is already past (I still hold the beginning of the sentence I just wrote), and it projects towards an intended next moment (this writing session is still unfinished).

These horizons are mobile: this very moment that was present (and hence was not merely described, but lived as such) slips towards an immediately past present. Then it plunges further out of view: I do not hold it just as immediately, and I need added depth to keep it at hand. This basic texture is the raw basis of what I will be discussing *in extenso* below. In its basic outline, I shall refer to it as the *three-part structure of temporality*. It represents one of the most remarkable results of Husserl's research as a result of phenomenological reduction.

Husserl introduced the terminology of *retention* and *protention* to designate this dynamics of impression. Retention is the attribute of a mental act that retains phases of the same perceptual act in a way that is distinguishable from the experience of the present (but that is not a re-presentation as we just saw). Retention is the key feature, for it is retention that provides direct contact with earlier perceptions, making perception at any given instant contain entities that show up as temporally extended. After application of the method of phenomenological reduction, duration has a spaciousness, it creates the space within which mental acts display their temporality. Similarly (but not symmetrically) another distinction seeks future threads or *protentions*. This is the three-part structure that transforms an intentional content into a temporal extension.

The question I wish to address is to what extent this retentional trail can be illuminated by the perspective of neurodynamics.

THE NEURODYNAMICS OF TEMPORAL APPEARANCE

I have to situate my presentation with regard to my overall approach to cognition, which is based on situated, embodied agents. I have introduced the name *enactive* to designate more precisely this approach, which comports two complementary aspects: (1) on the one hand, the ongoing coupling of the cognitive agent, a permanent coping that is fundamentally mediated by sensorimotor activities; (2) on the other hand, the autonomous activities of the agent whose identity is based on emerging, endogenous configurations (or self-organizing patterns) of neuronal activity. Enaction implies that sensorimotor coupling modulates (but does not determine) an ongoing endogenous activity that it configures into meaningful world items in an unceasing flow.

I cannot expand this overall framework more extensively,[2] but it is the background of my discussion of temporality as a neurocognitive process. Enaction is naturally framed in the tools derived from dynamic systems, in stark contrast to the cognitivist tradition that finds its natural expression in syntactic information-processing models. The debate between embodied-dynamics versus abstract-computational models as the basis for cognitive science is still very much alive. For some time I have argued for the first and against the second, and this choice justifies the extensive use of dynamic tools in this paper.

From an enactive viewpoint, any mental act is characterized by the concurrent participation of several functionally distinct and topographically distributed regions of the brain and their sensorimotor embodiment. It is the complex task of relating and integrating these different components that is the root of the temporality, as it is viewed from the perspective of the neuroscientist. A central idea pursued here is that these various components require a *frame or window of simultaneity that corresponds to the duration of the lived present*. In this view, the constant stream of sensory activation and motor consequence is incorporated on an endogenous dynamics (thus not informational–computational) framework which gives it its depth or incompressibility. This idea is not merely a theoretical abstraction: it is essential for the understanding of a vast array of evidence and experimental predictions.[3–6] These endogenously constituted integrative frameworks account for perceived time as discrete and not linear, because the nature of this discreteness is a horizon of integration rather than temporal "quanta."

At this point it is important to introduce three scales of duration to understand the temporal horizon as just introduced: (1) basic or elementary events, (the "1/10" scale); (2) relaxation time for large-scale integration (the "1" scale); (3) descriptive–narrative assessments (the "10" scale). This recursive structuring of temporal scales is surely a unified whole, and it only makes sense in relation to objects and events. It basically addresses how something that is temporally extended can show up as present but also reach far into my temporal horizon. The importance of keeping in mind this tri-level, recursive hierarchy will be apparent throughout this paper.

The first level is already apparent in the so-called fusion interval of various sensory systems: the minimum distance needed for two stimuli to be perceived as nonsimultaneous, a threshold that varies with each sensory modality (in Chapter XV of *Principles*, William James provides an elegant description of these data, which were extensively explored in the last century). These elementary events can be grounded today in the intrinsic cellular rhythms of neuronal discharges and in the temporal summation capacities of synaptic integration. These events fall within the range 10 msec (e.g., the rhythms of bursting interneurons) to 100 msec (e.g., the duration of an EPSP/IPSP in a cortical pyramidal neuron). These values are the basis for the 1/10 scale. Behaviorally, these elementary events give rise to microcognitive phenomena variously studied as perceptual moments, central oscillations, iconic memory, excitability cycles, and subjective time quanta. For instance, under minimum stationary conditions reaction time or oculomotor behavior displays a multimodal distribution with a 30-to-40-msec distance between peaks; in average daylight, apparent motion (or the psi-phenomenon) requires 100 msecs.

This leads us naturally to the second scale, that of long-range integration. Component processes already have a short duration on the order of 30–100 msec; how can such experimental psychological and neurobiological results be understood at the level of the fully constituted, normal cognitive operation? A long-standing tradition in neuroscience looks at the brain basis of cognitive acts (perception–action, memory, motivation and the like) in terms of *cell assemblies* or, synonymously, of *neuronal ensembles*. A cell assembly (CA) is a distributed subset of neurons with strong reciprocal connections.

CAs comprise distributed neuronal populations (certainly including neocortical pyramidal neurons, but not limited to them) requiring active connections. Because of their assumed strong interconnections, a CA can be activated or ignited from any of its smaller subsets, sensorimotor, or internal. Notice also that the term *reciprocal* is crucial here: one of the main results of modern neuroscience is to recognize that brain regions are indeed interconnected in a reciprocal fashion (what I like to refer to as the law of reciprocity). Thus, whatever the neural basis for interesting cognitive tasks turns out to be, it necessarily engages vast and geographically separated regions of the brain. Furthermore, these distinct regions cannot be seen as organized in some sequential arrangement: a cognitive act emerges from the gradual convergence of various sensory modalities into association or multimodal regions and into higher frontal areas for active decision and planning of behavioral acts. The traditional sequentialistic idea derives from the time of the dominance of the computer metaphor with its associated idea of information flow going in an upstream direction. Here, in contrast, I emphasize the strong dominance of dynamic network properties where sequentiality is replaced by reciprocal determination and relaxation time.

The genesis and determination of CAs can be seen as having three distinct causal and temporal levels of emergence: an ontogenetic level, which sets the anatomical architecture of a given brain into circuits and subcircuits; a second, developmental-learning level—sets of neurons that are frequently coactive strengthen their synaptic efficacies; and a third level of determination for CA constitution. The third level is the faster time scale of the experience of immediate daily coping, which manifests at the *perception–action* level, a duration on the order of seconds.

In the language of the dynamicist, the CA must have a *relaxation time* followed by a bifurcation or phase transition; that is, a time of emergence within which it arises, flourishes, and subsides, only to begin another cycle. This holding time is bounded by two simultaneous constraints: (1) it must be longer than the time of elementary events (the 1/10 scale); (2) it must be comparable to the time it takes for a cognitive act to be completed, on the order of a few seconds (the 1-scale).[3,5] In brief, as we said before, the relevant brain processes for ongoing cognitive activity are not only distributed in space, but also they are distributed in an expanse of time that cannot be compressed beyond a certain fraction of a second, the duration of integration of elementary events.

In view of the above, I will now introduce three interlinked (but logically independent) working hypotheses.

Hypothesis I: *A singular, specific cell assembly underlies the emergence and operation of every cognitive act.*

The emergence of a cognitive act is a matter of the coordination of many different regions allowing for different capacities: perception, memory, motivation, and so on. They must be bound together in specific groupings appropriate to the specifics of the current situation the animal is engaged in (and are thus necessarily transient) to constitute meaningful content in meaningful contexts for perception and actions. Notice that the Hypothesis I is strong in the sense that it predicts that only *one* dominant or major CA will be present during a cognitive act.

What kind of evidence is there to postulate that every cognitive act, from perceptuomotor behavior to human reasoning, arises by coherent activity of a subpopulation of neurons at multiple locations? And further how are such assemblies transiently self-selected for each specific task? Because this will be a recurrent topic for the remainder of this article, I will formulate it as the second part of my working hypothesis. The basic intuition to answer the problem just raised is that a specific CA emerges through a kind of temporal *resonance* or "glue." More specifically, the neural coherency-generating process can be understood as follows:

Hypothesis II: *A specific CA is selected through the fast, transient phase locking of activated neurons belonging to subthreshold, competing CAs.*

The key idea here is that ensembles arise because neural activity forms transient aggregates of *phase-locked* signals coming from multiple regions. Synchrony (via phase-locking) must *per force* occur at a rate sufficiently high so that there is enough time for the ensemble to "hold" together within the constraints of transmission times and cognitive frames of a fraction of a second. If, however, at a given moment several competing CAs are ignited, different spatio-temporal patterns will become manifest, and hence the dynamics of synchrony may be reflected in several frequency bands. The neuronal synchronization hypothesis postulates that it is the precise coincidence of the firing of the cells that brings about unity in mental–cognitive experience. If oscillatory activity promotes this conjunction mechanism, it has to be relatively fast to allow at least a few cycles before a perceptual process is completed (e.g., recognition of a face and head orientation).

This view has recently been supported by widespread findings of oscillations and sychronies in the gamma range (30–70 Hz) in neuronal groups during perceptual tasks. The experimental evidence now includes recordings during behavioral tasks at various levels, from various brain locations both cortical and subcortical, from animals ranging from birds to humans, and from signals spanning broad-band coherence from single units, local field potentials, and surface-evoked potentials (electric and magnetic).[7,8]

This notion of synchronous coupling of neuronal assemblies is of great importance for our interpretation of temporality, and we will return to it repeatedly below. This is where things get interesting, for we need to focus on an explanation to a view of cognition that is truly dynamic, making use of both recent advances in nonlinear mathematics and of neuroscientific observations.

Thus, we have neuronal-level constitutive events that have a duration in the 1/10-scale, forming aggregates that manifest as incompressible but complete cognitive acts in the 1-scale. This completion time is dynamically dependent on a number of dispersed assemblies and not on a fixed integration period; in other words, this represents the origin of duration without an external or internally ticking clock.[b] These are, then, new views about cognitive–mental functions, based on a large-scale, integrating brain mechanism, that have been slowly emerging with increasing plausibility.

But this entails that "nowness" seen from this perspective is pre-semantic in that it does not require a remembrance (or second-order presentification) for its emergence. The evidence for this important conclusion comes, again, from many sources. For instance, for up to 2–3 seconds, subjects can estimate duration quite precisely, but their performance decreases considerably for longer times; spontaneous speech in many languages is organized such that utterances last 2–3 sec; short, intentional movements (such as self-initiated arm motion) are embedded within windows of this same duration.

This brings to the fore the third duration, the 10-scale, appropriate to descriptive–narrative assessments. In fact, it is quite evident that this endogenous, dynamic emergence of nowness horizons can be, in turn, linked together in a broader temporal horizon. This temporal scale is inevitably linked to descriptive–narrative assessments and is inseparable from our linguistic capacities. It constitutes the "narrative center of gravity" in Dennett's metaphor, the flow of time related to personal identity.[9] It is the continuity of a self that breaks down under intoxication or in pathologic conditions such as schizophrenia or Korsakoff's syndrome.

I am now ready to advance the last step required to complete this part of my analysis:

Hypothesis III: *The integration-relaxation processes at the 1-scale are strict correlates of present-time consciousness.*

We have landed back on the experiential domain, which needs to be properly explored. This is why distinctions between the ongoing integrations in moments of nowness and how they may give rise to broader temporal horizons in re-membrance and narrative are also at the very core of the Husserlian analysis of intimate time, to which we now return.

RETENTION AS DYNAMIC TRAJECTORIES

Besides the substantial experimental support for these hypotheses, I need to make it clear that we are dealing with a *bona fide* candidate for the synthesis of a temporal space where cognitive events unfold. The discussion that follows requires at least some understanding of the dynamics of nonlinear phenomena as applied to cognitive events.

Let us be more specific. Large-scale phenomena seen in the nervous system as long-range integration via synchronization of ensembles cannot be dissociated from

---

[b]The precise timing is necessarily flexible (30–100 msec) because such events can naturally vary in their detailed timing depending on a number of factors: context, fatigue, type of sensorial mode used, age, and so on. This is why I speak of an order of magnitude, not absolute value.

intrinsic cellular properties of constituent neurons. Intracellular recordings studied *in vivo* and in slices of various brain regions *in vitro*, in both vertebrates and invertebrates, have shown the pervasive presence of intrinsic, slow rhythms mediated by specific ionic-gating mechanism.[10,11] In other words, what we are confronting is a large array of neural groups, which because of their intrinsic cellular properties qualify as complex nonlinear oscillators.

Why is this of importance here? For this reason: it leads us directly to an explicit view of the particular *kinds* of self-organization underlying the emergence of neural assemblies. These dynamical processes, in turn, illuminate emergence and temporality and the three-part structure of neural assemblies. Specifically, the emergence of assemblies at every moment of time arise from collections of a particular class of coupled, nonlinear oscillator, a very active field of research.[12,13] The key points to keep in mind are the following.

Self-organization arises from a *component* level, which in our case has already been identified as the 1/10-scale of duration, and reappearing here as single or groups of nonlinear oscillators. Second, we need to consider how these oscillators enter into synchrony (see Hypothesis II) as detected by a *collective* indicator or variable, in our case relative phase. Third, how such collective variable levels manifest at a global level as a cognitive *action* and behavior, which in our case corresponds to the emergence of a percept in multistability. This global level is not an abstract computation, but an embodied behavior subject to initial conditions (e.g., what I "aimed" at, what was the preceding percept) and nonspecific parameters (e.g., changes in viewing conditions, attentional modulation). The local–global interdependence is therefore quite explicit: neither can the emerging behavior be understood separate from the elementary components, nor can the components be rendered relevant separate from their global correlate.

What have we gained? Very simply that in the kinds of emergent processes involved above there is a natural account of the apparent discrepancy between what emerges and the presence of the past. The fact that an assembly of coupled oscillators attains a transient synchrony and that it takes time to do so is the explicit correlate of the origin of nowness (Hypothesis III). As the model (and the data, see below) show, the synchronization is dynamically *unstable* and thus will constantly give rise to new assemblies in succession. We may refer to these continuous jumps as the *trajectory* of the system. Each emergence bifurcates from the previous ones in its initial and boundary conditions. This makes the preceding emergence still present in the succeeding one.

To bring this idea closer to first-hand experience, consider the series of figures from Fisher,[14] in which a sitting girl transforms gradually over 20 steps into a man's face. Observers report that the positions over which the perceptual switch occurs do not coincide, but are lengthened in either direction along the image sequence.

When this is analyzed from a dynamic point of view, the following account is plausible. Looking at the extremes of the series ("girl," "man"), we see that the resonant assemblies are closer to a stable attractor basin corresponding to a single percept. As the parameter of ambiguity is increased (the place in the series), suddenly the emergence of a new percept is possible; that is, we have passed through a bifurcation or phase transition. However multistable, the system will still tend to stay close to the fixed point of origin, as it had left an active residue (hysteresis in technical jargon) for the trajectory, a remnant that is appended (to retake Husserl's term). The order parameters constrain the trajectories, the initial percept hovers around the stable percepts, but it may and does wander to different positions in phase space.

This embodies the important role of order parameters in a dynamic account. Two main aspects of order parameters can be described: (1) the current state of the oscillators and their coupling, or initial conditions; (2) the boundary conditions that shape the action at the global level: the contextual setting of the task performed and the independent modulations arising from the contextual setting where the action occurs (i.e., new stimuli or endogenous changes in motivation). The second visual task makes it clear that we are not dealing with an abstract purely syntactic description. Order parameters are defined by their embodiment and are unique to each case. The trajectories of this dynamics, then, enfolds both the current arising and its sources of origin in one synthetic whole as they appear phenomenally—a wooden iron indeed.

THE DYNAMICS OF MULTISTABILITY

The kind of specific dynamics we have brought to bear on the understanding of retention and the just-past are not simple. In particular, arrays of coupled oscillators are interesting because they in general do not behave according to the classical notion of stability that derives from a mechanical picture of the world. Stability here means that initial and boundary conditions lead to a trajectory concentrated into the small region of a point of the phase space where the system remains, a point attractor or a limit cycle. In contrast, the key role of phase transition in biological systems is that *in*stability is the base of *normal* functioning rather than a disturbance that needs to be avoided.

Lets us return once again to our experiential ground of visual multistability. As we saw in the example of Fisher's figures, the origin of multistability is due to properties that are *generic* to coupled oscillators and their phase relations; that is, their mode of appearance is an invariant under a number of conditions and subjects reporting. This is further clarified by recent experiments performed with a view to study the dynamics of multistability in visual perception. Kelso *et al.*[15] presented observers with *variant perspectives* of the classical Necker cube, a close relative of our first visual task. By asking the observer to push a button when the perceptual reversal occurs, one obtains a time series of reversal that obeys a stochastic distribution (a gamma function with a mean at around 1 sec). Kelso *et al.*, however, requested observers to perform the same task of time measurements but note separately the time series as a function of perspective, which is thus used as an order parameter.

The interesting observation is that, again, at the extremes of the images (the "hexagon" mode, and the "square" mode) the distribution of reversal intervals is considerably flattened; that is, the subjects are more likely to have sporadic figure reversal or to be "fixed" on one mode for a longer duration. In these extremes subjects report getting "blocked" by an interpretation. As before, one can think of these results as the manner of appearance of the coordination of a wide array of oscillators via a common variable of phase. By introducing perspectival variants, the location in phase space is accordingly modified, and new dynamic modes appear, in this case revealing a saddle instability.

That this dynamic interpretation is actually linked to neuronal ensembles as we have been assuming here is shown by recent experiments.[16] A monkey was rigorously trained to voluntarily reverse a set of ambiguous figures (binocular rivalry, a visual task known to be similar to Necker cube or Fisher figure reversal) and then to indicate the moment at which this reversal appeared for his perception. At the same time individual neurons were recorded from a number of visual cortical areas. The authors report that in motion sensitive area MT, a percentage of neurons correlate with reversal and can be modulated by the perceptual requirements of the tasks; this percentage is diminished in primary regions V1/V2. This kind of evidence strongly supports the notion that multistability arises through the large-scale coherence between neurons at many different places in the visual cortex and elsewhere in the brain, a concrete example of an emerging CA for a specific task having perceptual and phenomenal correlates.

CONCLUSION: MUTUAL CIRCULATION

In conclusion one can safely say that the retentional appearance uncovered by phenomenological analysis is in alignment with neuroscience if we take a dynamic stance for the emergence of neuronal assemblies. It is also clear that both the phenomenological account and the neuroscientific analysis are mutually enriched. However, the precise nature of this coemergence is not simple isomorphism, and this is discussed separately elsewhere.[17]

I want to conclude by drawing together the strands of what has been examined here. The main lesson is that the enterprise of neurophenomenology has taken us into the thicket of philosophical and methodological renewal. If this direction of research is to provide an answer to the otherwise unbridgeable explanatory gap, it cannot ignore the very constitutive basis for the mutual reciprocity that makes the mental and the physical hang together. It is thus evident that only from this renewed basis can neurophenomenology be other than a repetition of the past, in the form of searches for correspondence across the "mystery" line, be it through bridge locus or isomorphism between cognitive science and phenomenological data.

This mutual reciprocity without residue is the very nature of the region unique to the *Körperleib*. In the end, in this ontological region where reciprocity is manifest in all its vividness, three main threads need to be woven together on an equal footing to provide a seamless braid of continuity between the material and the experiential, the natural and the transcendental:

(I) the formal level because eidetic descriptive structures and implementation partake of the same mode of ideality and hence are effectively on common ground;

(II) the natural bodily process at the right level spanning across two levels of global emergence and local mechanisms that assure a direct relevance to both the psychological content if examined phenomenologically and to a detailed scientific examination;

(III) the pragmatic level of the *Leib/Körper* transition because it, and it alone, can have a situated bivalence, that excludes neither and provides the relevant

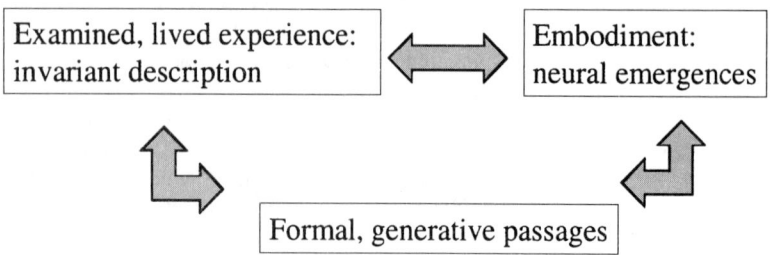

FIGURE 2. Mutual constraints: a triple braid. The strategy of mutual circulation.

basis for data for the preceding threads.

For the case at hand in the study of sound retention, the mutual constraints are by now quite explicit. If we take away any of these three threads of the braid, the entire enterprise is flawed on one side or the other. Take all three together, and the interpenetration of domains and reciprocity are apparent.

(I′) the dynamical analysis of trajectories arising from synchronous oscillators;

(II′) the emergence of a large-scale integration of the multiple cognitive dimension into a layered moment of cognition that is downwardly effective at the local level of subcircuits;

(III′) the phenomenology of nowness under reduction revealing a constitution that is both experientially meaningful as pure experience and structurally precise to as data for a dynamical description.

This triple braid provides, I believe, the complex meeting ground between physical and inner time which was my topic here. This kind of study is a renewal of the move toward nondual thinking in philosophy and science that is not declared or predefined by decree, but found at our very doorstep. Only a generative mutual reciprocity can replace the age-old friction of duality between matter and subjective life. We may glean the shape of such future thinking in the naturalization of phenomenology as the transcendence of nature.

REFERENCES

1. HUSSERL, E. 1966. Zur Phänomenologie des Inneren Zeitbewusstseins (1893–1917), Husserliana X. R. Bohm, Ed. M. Nijhoff. The Hague.
2. VARELA, F. 1996. Neurophenomenology: a methodological remedy for the hard problem, J. Consc. Studies **3**: 330–350.
3. VARELA, F., A. TORO, E.R. JOHN & E. SCHWARTZ. 1981. Perceptual framing and cortical alpha rhythms. Neuropsychologia **19**: 675–686.
4. DENNETT, D. & M. KINSBOURNE. 1991. Time and the observer: the where and when of time in the brain. Behav. Brain Sci. **15**: 183–247.
5. PÖPPEL, E. 1988. Mindworks: Time and conscious experience. Harcourt Brace Jovanovich. Boston.

6. PÖPPEL, E. & K. SCHILL. 1995. Time perception: Problems of representation and processing, *In* Handbook of Brain Theory and Neural Networks. M.A. Arbib, Ed.: 987–990. MIT Press. Cambridge, MA.
7. SINGER, W. 1995. Synchronization of cortical activity and its putative role in information processing and learning. Ann. Rev. Physiol. **55:** 349–374.
8. VARELA, F. 1995. Resonant cell assemblies: A new approach to cognitive functioning and neuronal synchrony. Biol. Res. **28:** 81–95.
9. KIRBY, A. P. 1991. Narrative and the Self. Indiana University Press. Bloomington, Indiana.
10. LLINAS, R. 1988. The intrinsic electrophysiological properties of mammalian neurons. Science **242:** 1654–1664.
11. NUNEZ, A., F. AMZICA & M. STERIADE. 1993. Electrophysiology of cat associatioon cortical cells in vivo: Intrinsic properties and synaptic responses. J. Neurophysiol. **70:** 418–430.
12. WINFREE, A. 1980. The Geometry of Biological Time. Springer Verlag. New York.
13. MACKEY, D. & L. GLASS. 1988. From Clocks to Chaos. Princeton University Press. Princeton, NJ.
14. FISHER, G.H. 1967. Measuring ambiguity. Am. J. Psychol. **80:** 541–547.
15. KELSO, J.A.S., P. CASE, T. HOLROYD, E. HORVATH, E. RACZASZEK, B. TULLER & M. DING. 1994. Multistability and metastability in perceptual and brain dynamics. *In* Multistability in Cognition. M. Staedler & P. Kruse, Eds. Springer Series on Synergetics. Berlin.
16. LEOPOLD, D. & N. LOGOTHETIS. 1996. Activity changes in early visual cortex reflect monkey's percepts during binocular rivalry. Nature **379:** 549–553.
17. VARELA, F. 1997 The Naturalization of Phenomenology as the Transcendence of Nature: Searching for generative mutual constraints. Alter: Revue de Phénoménologie No. 5. (Paris). pp. 355–385.

Analytical Setting: Time, Relation, and Complexity

GENOVINO FERRI[a,b] AND GIUSEPPE CIMINI[c]

Societá Italiana di Analisi Reichiana, via nazionale 396, Roseto degli Abruzzi, Teramo 64026, Italy
Societá Italiana de Analisi Reichiana, Via Parini 64, Giulianova Lido, Teramo 64022, Italy

The definition of the psychotherapeutic setting has never been entirely delineated on the one hand because of the various guidelines that tend to give preference to one element or the other in turn, and on the other hand because of the complexity of the topic. As a matter of fact, according to many well-informed opinions, the setting in classical analysis would be very different from the one that is generally created in psychotherapy, even though the setting represents one of the criteria for deciding what is analysis and what is not.

Freud faced this problem, but not systematically. The rules that he indicated are still used today, although not as rigidly, and they are represented by an internal and an external aspect; that is to say, a benevolent neutrality—the flowing attention that means not wanting to concentrate on anything specific and the rule of abstinence that means abstaining from giving the person being analyzed any indications that are not relevant to the purely psychoanalytical situation. On the other hand, the codified frequency of the sessions, the constancy of the timetable, the modality of payment, and the use of the analytical bed are still all part of the external conditions. As we pointed out initially, some differences exit in the settings, but we can still find some common elements: the psychoanalytic setting is regarded as a container in which the dynamics of the relation between therapist and patient are elaborated, the anxieties, the distances, the functions, and the way to perceive reality and unknown. But in the definitions of setting, whatever they are, it seems possible to establish rather a general idea involving a list of behaviors rather than a set of rules for functioning within a structure of relations that are absolutely decisive for the therapeutic goals.

Taking a step backward, we have to remember that analytical therapy, and therefore all psychotherapeutic treatments, was born and evolved from an overimplified deterministic conception of the world. The evolution of the complexity changed the background that was typical of Freud's period, and it introduced new ideas about the phenomena that psychiatrists and psychotherapists study: the elements that constitute the psyche, such as verbal and body language, depend on complex cognitive structures that are hardly reducible to simple dynamics.

In other words, we have witnessed a complication of the variables we meet in relational dynamics: the complexity of the world constituting of a bio-psycho social whole in many cases does not follow linear dynamics, and therefore complications and deterministic chaos can cause behaviors for which the output is not proportional

[b]Address for correspondence: 085/8944313 (voice); pepo@net-uno.it (e-mail).

to the input. It is clear that under these conditions it is possible to employ a new set of instruments that some use in a conceptual way, as a reference model, and others use to derive technical mathematical meaning, as has happened with schizophrenia and bipolar disturbances, and in the use of once unthinkable systems such as the transposition of models of research of dissipative systems, of the neural nets, of fractal mathematics, and of the theory of catastrophe. The study of illnesses in dynamic terms showed that the old paradigm of homeostasis may have to be reconsidered in light of the observations indicating that living systems have a higher possibility of vital success in the perspective of complexity than in a perspective of apparent stability.

In this view the definition of setting as explained above becomes inadequate: our starting point proposes that the setting forms a living structure, a complex system, a system of autopoiesis with the different phases and levels of organization that come from the contact of the analyst's fractals and the analysand's fractals. We are combining here the concept of the fractal figure with a basic concept in character analysis: the concept of trait. A fractal figure is a figure characterized by schemes that you continuously find in different orders—schemes, forms of the whole, that repeat themselves in similar ways, in every variety of orders and sizes.

The trait of character is a scheme, a figure, a form that we acquire at a particular stage of our evolution, in one of the developmental phases. This form, this scheme is by analogy a fractal figure: the structure of the character is based on traits that are etched by their own history and on them it "folds" and "unfolds" the form-character: these etched signs determine the building of the personality on a model that will organize itself on schemes that are similar to themselves and that will determine the fundamental order. This way, the character's order shows some elements that are constantly present in any observation carried out in every developmental phase that the model has: for example, the topic of acceptance—a trait that has an intrauterine origin—will always repeat itself in diachronic development, and its architecture will always be similar, even in the different phase explications. This fractal figure is imprinted in its specificity by the signs etched by its own history—*character* literally means etched sign—that remain in our personality's organization and that can be acted upon or recalled by the scenes of here and now.

If the setting is a living form, it has the capacity of a negentropic gradient. Negentropy is a negative variation of entropy starting from an original value; the birth of a the individual, the origin of life, the beginning of the biological evolution: this variation of entropy means the acquisition of greater order that is more evident with higher evolution. Negentropic development and character development can also be put together, and they form the evolution and stratification of the different organizations, of the different phases, and of the different phase shifts in the building of character, throughout the history of a person, up to the consciousness of the self. And there is a strong association with what Prigogine has said: away from equilibrium the matter becomes visible.

The setting is a living form that is born from the first contact between analyst and analyzed and it has the possibility of progressively developing a negative version of entropy from the original value, up to stratifications and specific forms, in a historical process belonging to the relation itself. We are saying that the setting will develop its own character, and the setting's character will be the connection between the character traits of the analyst and the character traits of the analysand, between these two fractalic figures, which will allow the possibility of a new complex system, its own self-organization, its autopoiesis, its developments, its phases. The character of the setting will allow an empathic contact, a being together with important consequences on the economy and negentropy of the self of the analysand and of the analyst and of the complex system of analyst and analyzed. To better clarify, the being together is a fractalic comprehension compatibility that probably takes place if the analyst has a fractalic figure larger than the one presented by the analysand.

The analyst's capacity to create contact determines a concept that has many analogies with the concept of flexibility, that is to say, the analyst's capacity to slide over his own positions, over the position of his own story's organizational levels, and to alight on the fractalic figure, able to find a resonance with the fractalic figure of the analysand. This move aims at creating a therapeutic alliance, a substratum of a possible coevolutive development of the relation in the setting.

We highlight coevolution, and we validate it in three forms: it is not sufficient to have negentropic evolution only in the analysand, and neither is it enough to also have the analyst's evolution, but it is fundamental to have an evolution of the *relation* between analyst and analyzed. It is implicit in the interdependence and in the maintenance of the diversity of the parts.

An analysis of the scene of the setting as described above involves a rereading of transference and countertransference. We can think of them as forces that flow from the person's structure, from the traits, from the fractalic figures of the analyst and of the analysand, flows of stage, trait, and specific fractalic figures that meet in interactions that correspond and find a place in specific stages of the new form: relation. The interactions among different trait transferences and countertransferences within the same relation recalls the concept of structural coupling, defined as recurrent interactions that begin structural modifications in the system. And accepting that a coupled structural system is an intelligent system that learns, an analytical setting has the capacity for intelligent negentropic development; it has the potential for intelligent structural couplings between the articulation of the number of the fractalic figures of the analyst and those of the analysand, reinforced by the fact that the setting is a privileged operational space, specific and protected. It is important here to highlight the incredible responsibility that the analyst has towards this relation; that it is not a simple relation, but an analytical therapeutic relation, a relation that has the fundamental aim to allow the analyzed person's self a greater vitality.

We introduce here a new equation between what we call phase shift—real critical points of instability—and the point of bifurcation. The "how" of the phase shift is determined by a sign etched in person's history, in the architecture of his character's development. To be clearer we cannot, for example, allow a person, who in his own history has the "how" of the bifurcation point etched at a point of high destructural risk, to go up and down the different organizational levels of the personality with rhythms higher than can be supported by a phase passage that is extremely fragile without a specific countertransference in form and time. We believe we can talk about a time arrow in the relation, in the setting, in the complex living system that determines the evolution of the structural couplings, of the possible negentropic stages. Inside this relation it is possible to manage certain times that sometimes have the connotation of egoic time, sometimes of chronological external time or of internal emotional time: time of the self; time of the complex system self–other; time beaten by an absolutely rooted condition in the person's history, made up of fears, or alarms of vigilance; time connected to the economic supportability of the possible fractalic evolutions of the trait and state shift, connected to the possible new configurations of the form of the relation; specific time of the individuality of the system that defines the complexity; one trait time, more trait times; a time that traces; a time that etches; a time that is traced; a time that irreversibly beats history; a time connected to organizational areas where the vital flow can become rarefied, the autopoiesis disintegrates, and a psychotic condition can explode; a time that defines form and is defined by the form itself.

The arrow of time indicates a direction for which past and future are not equivalent; and here we add that even more in the analytical therapeutic setting the states of time develop with the characteristic of irreversibility. Therefore, it is necessary to attentively revise the structure of all psychotherapy and to stop to reflect upon the key of optimistic reversibility contained in Freud's system: what is structured during the various stages of development, the different points of fixation, and the possible development of pathologies is not able to activate a regression that cannot be a reversibility in time, but that we can interpret as an actualization of the smaller fractals in the organization of that self. In the same way, it is necessary to reconsider how to carry on the therapy and the escapes into the atemporal or extratemporal unconscious.

We are requiring the analyst, a founding part of the new living form, a clear quality in the metacommunication, the capacity to talk about his own state and stage, an awareness of his self and of the other, and of the relation between self and other. This is a luxury that indicates negentropic stages quite far away from equilibrium, a creative touch that surrounds the form and continuously guarantees it. If we are allowed a parallel, the analysis of the character of the relationship, is the ego of the new living form, the awareness of the new living form that has reached the state of vision. And in this systemic fluctuant analysis, in this game of continuous cross references, we believe there is a characteristic of stage that also belongs to the new paradigm, in which the religion of uncertainty is one of its parameters, that does not have truth positions corresponding to negentropic necessities of minor stages, that has in the possibility of metacommunication an evolutive and constant point of bifurcation, a point of instability and fluctuation that gives the system the possibility of never closing.

The Consciousness of Time and the Conquest of Space

The Dawn of Civilization, Anthropology, and Archaeoastronomy

LUCIA G. GREGORI AND GIOVANNI P. GREGORI[a]

IFA - CNR, via Fosso del Cavaliere 100, 00133 Rome, Italy

The very first steps of civilization required an adequate mastery of territory, that is, the consciousness of space and time. Mankind is an effective recording instrument for monitoring proxy data related to long-range environmental variations. This instrument, however, needs suitable decoding and calibration. These different aspects are related to each other and are ultimately concerned with anthropology, which in any case requires "observational" evidence: archaeoastronomy is, perhaps, the best objective source for finding evidence of the early steps of human knowledge, a fascinating inquiry. The need for such knowledge can easily be focused by referring either (i) to the former nomadic societies, which continuously had to move and search for food, or (ii) to the settled societies which had to exploit a suitable mastery of irrigation, that is, they had to build aqueducts and to know surrounding territory for the purpose of organizing their defense and expanding their territory because of demographic increase. Moreover, trade and commerce soon became a tool for improving their standard of living. The knowledge of time and space became essential for defense and survival.

In early times, time measurement was based solely on observation of the sky, and space measurements were even more difficult. Only after the construction of the Roman network of roads did people become acquainted with space orientation based on local information, such as the presence of a road that could be recognized on a map, some landscape features, and so forth. Before such achievements, however, all landscapes appeared almost the same to a traveler, with no roads, no signs, no information available from the local (and generally hostile) inhabitants within very poorly inhabited (if inhabited at all) regions. Moreover, travelers had an almost total ignorance about geography. For the sake of clarity and simplicity, investigations can be distinguished into a few subcategories. The knowledge of time was concerned with the assessment either (i) of the seasonal cycle, (ii) of the lunar cycle, (iii) with time measurement during the day, (iv) with longer period variations such as equinox precession, or (v) with the introduction of a calendar as a real mass medium, an effective weapon for the control of society. Space orientation ought to be distinguished into four subcategories: (i) orientation and territory management, and the determination either (ii) of latitude or (iii) of Earth's radius or (iv) of longitude. Some con-

---

[a]Address for correspondence: +39 6 4993 4321 (voice) ; +39 6 20 660 291 (fax); gregori@atmos.ifa.rm.cnr.it (e-mail).

cise information is given here, but the interested reader is referred to a previous set of papers[1–4] for further discussion and references.

TIME MEASUREMENT

The Seasonal Cycle

Ancient Egyptians used temples like real optical benches. They managed *four* different calendars: lunar, solar, stellar, and Venusian. All other ancient peoples used only a lunar calendar, except the Mayas who, notwithstanding the fact that they only developed a Stone Age civilization, were capable of measuring the solar year with even a slightly better accuracy than the Egyptians. Egyptians are likely to have also used huge sundials with a gnomon obelisk, by which they could quickly determine the location of every day within the seasonal cycle. This skill was imported to Rome by Augustus, shortly after Julius Caesar's calendar reform, by which he got rid of the previous lunar calendar.

The Lunar Cycle

The lunar cycle was extensively investigated by almost every ancient society, being the real clock of every-day life, for seeding, harvesting, and traveling. Several archaeological relics support such an inference. Possibly, a large fraction of the prehistoric megaliths were concerned with lunar investigations.

Measuring Time during the Day

Various types of sundials were developed at different times by different people. The ancient Romans imported the knowhow directly from the Greeks. The most fascinating related object is the Pantheon in Rome, which seems to be the very first example of a half-spherical dome (*cupola*). It definitely appears to be an actual huge sundial, of the kind that Vitruvius called *antiboreum*, a few tens of examples of which are available in museums, that is, an empty half-sphere with a small hole through which a spot of sunlight is projected within the interior of the sphere, by which one can read the hour. The Pantheon has no window other than the central hole (*lumen*) at the center of its dome. The Pantheon's dome was later copied and changed by introducing a windowed drum for improving the illumination of the hall, and it is basically the ancestor of all dome architecture in the entire Western world.[5] Moreover, the light-spot inside the dome changes location according to season. Hence, Emperor Hadrian could instantaneously guess both the hour of the day and the date of the year at a glance. The Pantheon appears to be the most sophisticated technological tool still surviving after almost 20 centuries. It is not yet clear whether its internal decoration, mainly the floor, was related or not e.g. to the investigation of the Metonic cycle of the moon or to something else. Additional investigations are in progress.

Equinox Precession

This appears to be a difficult category to demonstrate. Its effect is quite small during a human lifetime, and the archaeological evidence is correspondingly very faint.

According to the authors' knowledge, four related pieces of evidence have to be mentioned. (i) De Santillana and von Dechend have analyzed popular tales and traditions from all over the world and claim that there is always a clear reference to a mythical turning structure, which can be identified with the night sky, and with a long period of regular motion that can be interpreted in terms of an apparent knowledge of equinox precession. (ii) Jean Richer referred to a Boeotian vase from approximately the VIII century B.C., that apparently shows four antipodal constellations corresponding to the celestial longitudes of the equinoxes and solstices as they appeared several centuries before the manufacture of that vase. (iii) Vlora and Tucci carried out a clever decoding of several symbols appearing within some ancient Egyptian tombs and claimed evidence of ancient knowledge of equinox precession. (iv) Moreover, in A.D. 1755 the astronomer Leonardo Ximenes discovered that the sundial within the Florence cathedral Santa Maria del Fiore, which had been built by Ser Filippo Brunelleschi in 1425–1436, displayed a discrepancy that demonstrates real measurement of equinox precession. It is up to the individual reader to decide whether these indications are sufficient evidence, or not for stating that our ancestors knew about equinox precession. Additional evidence will perhaps become available in the future.

The Calendar

All calendar reforms have involved long preparation and harsh debates. But, basically the calendars served as a tool for showing the entire world the power of their issuing institution. Moreover, the calendar was needed as a real clock for everyday life, for planning future actions, and so forth. Formerly, it had been important mainly for the timing of seeding and harvesting. Later on it became fundamental for every kind of social or economic or military activity. The present concept of date, however, is a comparatively recent cultural conquest. Calendars often contained trivial errors even until the last century, and European and other Christian people did not use a numerical date: all activities rather made reference to the dedication of each day to a saint. Several churches, chapels, and cathedrals contained some hole or small window through which on one specific day, at the local noon, a beam of sunlight dropped directly onto the image of the saint to which that day was dedicated. That was a mark suitable for all people who had neither printed calendar nor any equivalent. Recognizing a specific day was important for them, for example, for doing some particular agricultural operation. In this way, the Church used the calendar as a mass medium, and it ruled everyday life by providing a relevant service to society. Even the Gregorian reform of A.D. 1585 was a means of reaffirming the authority of the Pope as opposed to that of the Emperor's. Anglican and protestant countries adopted the Gregorian reform only after several years. The calendar was a real weapon for controlling society. Even the concept of time of day was much different from our present measurement. Ancient Romans divided either the entire timespan of daylight or the entire timespan of darkness, into 12 hours, with the result that the actual duration of one hour was different during the day from the duration of an hour during the night, and actual duration of these "hours" changed with season. An analogous difficulty even affected the scientific community in dealing with the age of Earth, which according to the Bible was only a few thousand years old, a ridiculous inference on the basis of the geological and paleontological evidence. But, when the Bible was written, authors did not have a modern consciousness of time; hence, its information just reflects the culture of that time. Nevertheless, religious doctrine severely interfered with the general acceptance of scientific evidence until a comparatively recent time; it was necessary first for people to become acquainted with the correct consciousness of time.

SPACE MEASUREMENT

Orientation and Territory Management

The ancient proto-Greeks seem to have developed an effective geodetic network composed of sacred temples and shrines. They used 12 symbols almost like an early alphabet, every symbol being associated with a 30° angle. Hence, the 12 symbols together spanned an entire circle of 360°. When these symbols were applied along the ecliptic plane for denoting celestial longitude, each symbol corresponded to a zodiac constellation, and in fact these symbols were the 12 zodiac signs. Ancient Greeks defined one "navel" (*omphalos*) of the world, located at a prechosen site within a given region. Every other site was associated with an azimuth with respect to such an omphalos and with a frame of reference fixed with Earth's surface; hence, every such site was denoted by one symbol, that is, by its azimuth. Its coins did not yet contain the name of the town, but contained such a zodiac sign. Each symbol was related to one god of the Greek Olympus; hence, that site was dedicated to that god, and every temple and shrine built at that site had to be dedicated to that god. Moreover, all temples and shrines were oriented either towards the omphalos, or towards each other. Even all the decorations sculpted on the metopes and pediments were not arbitrary: rather, they represented the azimuth reckoned with respect to the center of the temple. By this, every traveler in need of orientation had only to search for the closest temple or shrine and determine to what god it was dedicated and what was its orientation. By this means, the traveler immediately got the needed information. An accurate knowledge of mythology was the crucial mnemonic code for this purpose. All this resulted in an effective system of navigation for sailing within the Aegean Sea. Ancient Greeks developed four such *omphaloi* systems: Sardis, Delos, Delphi, and Ammonium, respectively. Each of them was the seat of an oracle: perhaps, the seat of every other oracle was also the omphalos of an additional reference frame? Ancient Egyptians are also likely to have developed some analogous geodetic network. Maybe they invented obelisks as improved versions of menhirs and used them together with pyramids for establishing their geodetic network. Their accurate cadastral charts were prepared by the *betamists*, who walked and counted their steps: but, they could hardly get rid of their inaccuracies unless they had some frame of reference for correcting error propagation. Such a network was also essential for building pyramids (as no plumb-line could be used). The ancient Greek savants were generally acknowledged in their time as the most authoritative geographers. They even acted as consultants for the Persian emperors, who wanted to plan new conquests. Alexander the Great moved eastward, not westward, and he went only as far as the eastern borders of the former Persian empire. He did not go further eastward, where he would have had to face India and China. The knowledge of geography and of territory, as well as the trade with the Chinese empire, apparently had already been highly effective for a long time.

The Determination of Latitude

Several sacred sites of ancient Greece were aligned along two parallels at a reciprocal distance of $\sim 1 \pm 0.05°$. It can be shown that they could obtain this degree of precision by using a wooden pall, say ~5 m tall, vertically located over a very flat platform (such as on top of a ziggurat), and measuring its shadow within a precision of a few millimeters.

The Determination of Earth's Radius

When the latitudes of any two given sites are known, which are believed to be at approximately the same longitude, you only need to measure the linear distance along Earth's surface between the two sites, and you can easily calculate the measurement of Earth's radius. The most crucial point, however, is that, in ancient times, they even lacked a suitable unit for measuring such long distances: they usually measured distance by the time spent for traveling, also factoring in obstacles and difficulties. Only the Greeks had a sense (that was later lost by the Romans, who relied only on their road system and on the travel times along them) of the need for a correct representation of every site on a spherical Earth. Because of this, they were the most renowned cartographers of antiquity. The impressively precise determination of Earth's radius by Eratosthenes is therefore likely to be the result of their high precision in measuring the distance between Syene (Aswan) and Alexandria by means of their accurate cadastral charts.

The Determination of Longitude

Sardis and Delphi were at the same latitude and were the base of two isosceles triangles, Delos and Ammonium being their vertices. That is, Delos and Ammonium were at the same longitude. How did they achieve this four-omphaloi pattern? They carried out triangulation among the islands of the Aegean Sea by means of large fires lit during clear nights, but it appears it was not feasible to measure the longitude of Delos and Ammonium in this way. Neither would it be appropriate to guess that their longitudes are equal just as a matter of a coincidence (a researcher should never appeal to chance). But in this respect, the Antikytherian machine envisages that we actually know too little about ancient technology. During Alexandrian times, sophisticated technology was available, but it was lost after the Arab destruction. Something, however, survived and was later brought to Europe by the Arabs in the form of mechanical watch technology. Ancient Greeks needed precision watches for carrying out such precise longitude determinations. But, on the other hand, the choice of the omphaloi dates back to protohistorical times—the puzzle is not solved.

CONCLUSION

Archaeoastronomy and ancient anthropology appear just like several other "exact" sciences, which are based on scanty and difficult observational evidence, inference, guess, trial-and-error, working hypotheses, and so forth. Moreover, mankind is an effective recorder of a special type of proxy data for investigating long-term environmental change.[6] We have a long way to go before we can infer either the fascinating origin of the very first understanding by mankind or the historical impact of mankind on its living environment.

REFERENCES

1. GREGORI, G.P. & L.G. GREGORI. 1996. Viewpoints in archaeoastronomy. An invitation to the prehistory of Earth's sciences. *In* Global Change and History of Geophysics, W. Schroeder & M. Colacino, Eds.: 12–75. IDCH of IAGA. Bremen-Roennebeck, Germany.
2. GREGORI, G.P. & L.G. GREGORI. 1998. Archaeoastronomy II. An invitation to the prehistory of environmental sciences. *In* Geomagnetism and Aeronomy. (with special case histories) W. Schroeder, Ed.: 8–64. IDCH of IAGA. Bremen-Roennebeck, Germany.
3. GREGORI, L.G. & G.P. GREGORI. 1996. The knowledge of territory in ancient civilisations. Temples and sacred sites as prehistoric geodetic networks? Archeologia e Calcolatori **7:** 193–212.
4. GREGORI, G.P. & L.G. GREGORI. 1999. Archaeoastronomy and the study of global environmental change. Mem. Ital. Astr. Soc. In press.
5. GREGORI, L.G. 1997. Renaissance and history of science. The case of construction theory. *In* Physics and Geophysics with Special Historical Case Studies (A Festschrift in honour of K.-H. Wiederkehr). W. Schroeder, Ed.: 346–378. Science Edition. Bremen-Roennebeck, Germany.
6. GREGORI, G.P. & L.G. GREGORI. 1998. Solar terrestrial relations. A historical reminder. Acta Geod. Geophys. Hungarica **33**(2–4): 391–459.

Time Complexity and Learning

ELDA GUALA[a] AND PAOLO BOERO

Department of Mathematics, Genoa University, Via Dodecaneso 35-16146, Genoa, Italy

INTRODUCTION

The debate about the physical existence of time[1–3] suggests the possibility that time could also be considered an intellectual construction in order to "treat" (that is, to describe/order/analyze) the flux of external events; in addition, it raises the problem of intellectual constructions suitable for "treating" the flux of internal events. On this point, we can speak about "mind times," metaphors that may help in "treating" mental processes, especially those intervening in complex problem solving. Bearing in mind our competencies (cognitive and epistemological aspects of teaching and learning mathematics), we will consider in a phenomenological manner the variety of "times" that the mind must manage in mathematical problem solving. We will also consider the intertwining among them, mentioning some examples[4,5] in which success or failure seems to depend on the capacity to manage such time complexity. Finally, we will consider the hypothesis that the analysis of "mind times" may be useful (in an "embodied cognition" perspective) for singling out some mental processes on which basic mathematical ideas and skills are founded.

PHENOMENOLOGY OF SOME MIND TIMES OCCURRING IN THE MENTAL DYNAMICS OF PROBLEM SOLVING

Our aim is to focus on some "time components" of "mental dynamics" that are relevant to mathematical problem-solving activities. By "mental dynamics" we mean the mental processes of creation, exploration, and transformation of space–time environments (in a proper or metaphoric sense). As for the focused "time components," they seem to be appropriate for describing some important mental activities and/or interpreting their misfunctioning.

Let us consider the following examples of times (the list is not exhaustive):

(a) *Time of past experience*: traced by means of memory tracks and supported by the ordering activity the mind performs on those tracks. How tracing works (quickly or slowly, in detail or not) can depend on the *involuntary* perception of the quality of an event (linked to the subject's experience and its emotional intensity) and/or on the *voluntary* reconstruction of past experience (performed by skipping episodes irrelevant to investigation and zooming in on relevant aspects). We may remark, however, that interesting hints may emerge from episodes considered irrelevant. In our opinion, both polarities (involuntary and voluntary) are important: time perception probably develops out of their dialectic relationship.

---

[a]Address for correspondence: +39-010-3536958 (voice); +39-10-3536752 (fax); guala@dima.unige.it (e-mail).

(b) *Contemporaneity times*: we can distinguish how the *observer* records the subject's observable behaviors on the time line and how the *subject* experiences contemporaneity while involved in a situation. This time can be evaluated by the subject (with estimations slowed down by wishes and accelerated by fears), who may graft on it virtual deplacements toward the future or past. We may note that these two contemporaneity times interact with each other (perhaps, it would be better to consider only one time of the observer–observer couple), and in particular intertwine during auto-observation.

(c) *Exploration times*: in open-ended tasks requiring the subject to find and concatenate suitable arithmetic operations, plan a geometric construction, build up a proof, and so forth, time projections can be realized from the past onward ("How will he have gone about solving. ...") or in the future and then towards the past ("I think up a solution and explore it in order to find the operations to perform, depending on available resources").

(d) *Synchronous connection time*: as a perception of coordinated functioning of the components of the real or virtual system under scrutiny, or as a discovery of the links among the variables of the problem; this can graft onto (c) as the solving moment. We may note that synchronous connection is nowadays a fascinating subject of investigation from different perspectives—including that of neuronal biology.[6]

For a person, the borders between one time and another are not clear, and intertwining and grafting of one time onto another are possible (see above; see also the following two sections).

One interesting research problem concerns the relationships between these "inner times" and the "external" *chronological time*, whose existence is subject to much debate.[2,7]

MIND TIMES AND DIFFICULTIES IN PROBLEM SOLVING

This section refers to episodes and situations taken from students' work in classes that are involved in the Genoa Group Projects for primary and lower secondary school. These projects include diagnostic and remedial activities for logic-linguistic and mathematical skills.

As concerns diagnosis and remedial work on some pupils' difficulties in the logic-linguistics and mathematics areas, "mind times" offer interesting interpretative keys.[4,5] For instance, some pupils still fail at grade IV level in subtraction problems where no easy, concrete analog model (e.g., thermometer, ruler, etc.) is available: particularly, in the mental calculation of "How much is needed to make .... ?" they do not succeed in coordinating (d) the onward counting process while underway and (b) control of its ending. Another example concerns the task of "drawing a small 23-cm plant with a 20-cm ruler" set for grade II pupils who are able to perform measurement within the length of the ruler. Some pupils are unable to project themselves into the time of the solved problem [see the drawn plant: (c)] in order to discover the operations to perform with the ruler. The teacher's intervention on the exploration time (c) ("with your ruler, extend the 20-cm segment already drawn") together with the pupil's intuition (d) in his contemporaneity time (b) ("I see 20 and 23 written close together ... I must add 3!") may allow the pupil to solve the problem.

We studied many cases like those reported above. Frequently, by taking into account collected information about the pupils' personalities and comparing their performances in different tasks, we arrived at the hypothesis that the pupils' difficulties depended less on cognitive resources than on serious problems related to the affective and emotional sphere (lack of hope for the future, lack of trust in themselves and/or in other people). In some cases positive changes were activated through the teacher's coordinated, repeated interventions in pupils' past experience times (a: reconstruction of experienced situations—also including emotional involvement—and discussion about the difficulties experienced in the process) and in exploration times (c) grafted on contemporaneity time (b) (immersion in present reality, acquiring a sense of security in moving backwards and forwards during the exploration of virtualities). In these cases the possibility of self-regulation of learning processes seemed to emerge gradually in students' behaviors.[3]

MIND TIMES AND EMBODIED COGNITION

From an embodied cognition perspective,[8] particular body experiences are "grounding metaphors" for important basic mathematics concepts.

Let us consider a gear system of coplanar wheels and analyze primary school pupils' behavior, partially described in Bartolini Bussi et al.[9]; we will try to find embodied mathematical contents and skills. The pupil's impulse starts the movement of one wheel and propagates through the gear, provoking clockwise and anticlockwise rotations (d). We can split the pupil's perceptions into *touch* and *view* senses, and formalize the chain of impulses, which the pupil perceives through touch as a causal chain, by means of differential relations $Fdt = Rdv$, where F indicates the physical effort and R the impression of inertia of the system; dt stands for the duration of the impulse, and dv the corresponding increase in rotation speed. When the pupil pays *visual* attention to rotation directions, he/she discovers alternance and may represent it by two different colors or letters; we may represent the direction of rotation of the Kth wheel in an algebraic manner by means of the formula: $(-1)^{K-1}$ (initial wheel: $K = 1$).

When started, the movement can be observed in the contemporaneity time (b). The system is harmonious, we might say Ptolemaic. Pupils discover differences in speeds and tempos, then count the number N of turns and grasp (d) the relationship with wheel diameters D. We know that by multiplying ratios we get the angular speed V_K of the Kth wheel: $V_K/V_1 = N_K/N_1 = D_1/D_K$. Proportion theory is the theoretical reference for the pupils' intuitions derived from their time experiences.

As another example, let us consider the production of a conjecture about the possibility that two nonparallel sticks produce parallel shadows.[4] Many VIIIth grade students imagine possible movements of the sun or perform/imagine possible movements of the sticks until stating "when sunrays ...," a sentence that then becomes, in the statement expressing the conjecture, "*if* sunrays ...." The conditionality of the statement is generated as a time section (d) in a process of dynamic exploration (c) in which the movement of the sun is accelerated, or virtual (or real) stick movements are performed by stopping the sun.

We believe that the analysis outlined above may lead to interesting developments for the following:

(I) helping teachers find reference situations for concept construction; here we refer to Vergnaud's definition of a concept[10] as "reference situations," "operational invariants" and "linguistic representation tools" related to that concept;

(II) interpreting mathematical behavior when (in difficult problem situations) we feel the need for heuristics related to "concrete" referents;

(III) suggesting further investigations about "embodiment" of important mathematical skills. For instance, the example concerning the genesis of conditionality of statements indicates only one possible root for conditionality. In Boero and Garuti,[11] another root is indicated for another statement. What are the roots of conditionality? Is it possible to characterize them in terms of mind times?

REFERENCES

1. COVENEY, P. & R. HIGHFIELD. 1991. La Freccia del tempo. Rizzoli. Milano.
2. KLEIN, E. & M. SPIRO, Eds. 1996. Le Temps et sa Fleche. Flammarion. Paris.
3. PRIGOGINE, I. 1996. La Fin des Certitudes. Editions Odile Jacob. Paris.
4. BOERO, P., R. GARUTI & M.A. MARIOTTI. 1996. Some dynamic processes underlying producing and proving conjectures. In Proceedings of PME-XX. Vol. 2: 121–128. Valencia, Italy.
5. BOERO, P. & E. SCALI. 1996. Il tempo (i tempi) nel lavoro mentale e alcune difficoltà in matematica. In Lo spazio e il Tempo. C. Caredda, B. Piochi & P. Vighi, Eds.: 59–65. Pitagora. Bologna.
6. VARELA, F., E. THOMPSON & E. ROSCH. 1991. The Embodied Mind: Cognitive Science and Human Experience. MIT Press. Cambridge, MA.
7. MACAR, F., V. POUTHAS & V. J. FRIEDMAN, Eds. 1992. Time, Action and Cognition. NATO-ASI Series. Kluwer Academic Publisher. Dordrecht, The Netherlands
8. LAKOFF, G. & R. NUNEZ. 1997. The metaphorical structure of mathematics. In Mathematical Reasoning. L. English, Ed.: 21-89. LEA. Hillsdale, NJ.
9. BARTOLINI BUSSI, M., M. BONI, F. FERRI & R. GARUTI. 1999. Early approach to theoretical thinking: Gears in primary school. Educ. Stud. Math. In press.
10. VERGNAUD, G. 1990. La thèorie des champs conceptuels. Rech. Did. Math. **10**: 133–170.
11. BOERO, P. & R. GARUTI. 1994. Approaching rational geometry: From physical relationships to conditional statements. In Proceedings of PME-XVIII. Vol. 2: 96–103. Lisbon.

Einstein's Twins

ENZO TIEZZI[a] AND NADIA MARCHETTINI

Department of Chemical and Biosystems Sciences, University of Siena, Pian dei Mantellini, 44, 53100 Siena, Italy

ENERGY VERSUS ENTROPY: THE LAST CATERPILLAR

If we are dealing with a highly organized system, rich in systemic relations, far from equilibrium and evolving in time (like a living cell), the details, time, and relations become crucial, and with them, the role of information. As Harold Morowitz[1] says, a methyl group in the wrong place can kill a whale. The entropy function and its relation to information, irrespective of the presence or absence of energy changes, now comes into the picture.

In general, if a system goes from a higher to a lower entropy state, there is a gain in information. Entropy is proportional to the lack of information on the microstate of the system. It is a maximum when all microstates have the same probability and cannot be distinguished. *Information is therefore equivalent to negentropy.*[2]

An energy flow can lead to destruction (increase in entropy, for example, a cannon ball) or organization (decrease in entropy, for example, photosynthesis). The same quantity of energy can destroy a wall or kill a man; obviously the loss of information and negentropy is much greater in the second case. Energy and information are never equivalent.

The classical example of the mixing of gases in an isolated system shows us that there can be an increase in entropy without energy input from outside. The point is that E and S are both functions of state, but energy is intrinsically reversible whereas entropy is not. Entropy has the broken time symmetry discussed by Prigogine.[3] In other words, entropy has an energy term plus a time term that energy does not have.

The singularity of an event also becomes of particular importance: if a certain quantity of energy is spent to kill a caterpillar, we lose the information embodied in the caterpillar. But were this the last caterpillar, we should lose its unique genetic information forever. The last caterpillar is different from the nth caterpillar.

TOWARD AN EVOLUTIVE THERMODYNAMICS

The role of thermodynamics in scientific thought boils down to defining relations and identifying constraints; thermodynamics is the science of what is possible and is to physics as logic is to philosophy. Entropy is the enigma of thermodynamics because it has the intrinsic properties of time irreversibility, quality, and information that other thermodynamic functions lack. This is why entropy is a central concept in biology and ecology.

---

[a]Address for correspondence; 0577-232012 (voice); 0577-232004 (fax); tiezzienzo@unisi.it (e-mail).

The evolutionary process is such that systems become more and more complex and organized. Biological diversity is the product of long-term interactions at a genealogical and ecological level: the genealogical interactions regard the dissipation of entropy by irreversible biological processes; the ecological interactions regard entropy gradients in the environment.

Far from equilibrium, we witness new states of matter having properties sharply at variance with those of equilibrium states. This suggests that irreversibility plays a fundamental role in nature. We must therefore introduce the foundations of irreversibility into our basic description of nature (*evolutive thermodynamics*).

Let us summarize the main statements concerning the relationship between entropy and the global evolutive biosphere:

a. The entropy of the universe is increasing (the second law of thermodynamics), whereas the entropy of Earth has decreased in the course of evolution by dispersing positive entropy into space.
b. If we think of the evolution of a system in terms of statistical mechanics, we think of evolution towards more and more disordered states (theorem H of Boltzmann) with high entropy values, whereas biological evolution goes towards more complex and ordered forms with low entropy values.
c. A living system feeds on negentropy, directly or indirectly via photosynthesis.
d. The thermodynamic equilibrium state (immobility, nondifferentiation, death) is characterized by an entropy maximum, whereas in the steady state (dynamicity, diversity, life) a flow of energy keeps the system as far as possible from equilibrium. The biosphere is a steady-state system.
e. The origin of life is related to the metabolic capacity to extract negative entropy from the environment.
f. Earth lives because of negentropy. Its life is associated with stochastic and irreversible processes occurring in the biosphere. Earth "plays dice," because this is an intrinsic characteristic of biological evolution, the entropy function, and the evolution of the energy–matter system. Obviously, this game of dice obeys certain rules determined by the constraints of the planet and by its history. An event occurs in a stochastic manner because it is preceded by others. There are historical, genetic and environmental constraints. Evolutionary events proceed in a manner that depends on time: they show a sense of direction of time; they are irreversible.

EINSTEIN'S TWINS

Time, in Einstein's view, is either mechanical and reversible, or a mere mental category. On this basis Einstein has developed the famous paradox of the twins.

According to Einstein's theory it is possible to show that the relationship between time and the velocity of light demonstrates that one of the twins, traveling at a speed close to c, remains younger than the other one. This is based on the following, well known equation:

$$t = t_0 / \sqrt{(1 - v^2/c^2)},$$

where t is the time and v the velocity of the twin.

In mechanics, velocity is the first derivative of space with respect to time (ds/dt). This algorithm is very useful for studying the trajectories of the planets, the motion of cars, trains, and so on. Of course, it is also very useful for comparing the velocity of a living system relative to a frame of refererence, for example, a runner on a track.

Nevertheless, in our opinion, it is not scientifically correct to compare the life span of a living organism (one of the twins) with the life span of another living organism (the second twin). These are two complex systems, both far from thermodynamic equilibrium, both with dissipative properties and capable of self-organization; that is, two evolving systems.

In this context space and time are categories belonging to different logical types, which should not be confused. By nature, time is evolutionary and irreversible, whereas space is conservative and reversible. A reversible quantity cannot be differentiated with respect to an irreversible one. It is not possible to compare evolving quantities, such as the life span of the twins, in the framework of reversible mechanics. Life, like all living properties, is irreversible and follows biological, evolutionary paths ($\pi\alpha\nu\tau\alpha$ $\rho\epsilon\iota$). So do the twins' lives: Einstein's paradox does not exist.

CONCLUSION

The equations of classical physics have no notion of the "time of events" or the "time of things." Prigogine and Stengers[4] write that although quantum mechanics and general relativity are revolutionary, as far as the concept of time is concerned, they are direct descendants of classical dynamics and carry a radical negation of the irreversibility of time.

The recognition that time is "real" leads to what Prigogine calls the *"time paradox."* He asks how it is possible that the basic equations of classical and quantum mechanics are reversible with respect to time at the microscopic level, whereas at the macroscopic level the arrow of time plays a fundamental role. How can time emerge from non time?

Prigogine and Stengers add that the theories of relativity, cosmology, and quantum mechanics have always sought separation from the time dimension, placing the laws of time in the dimension of eternity. However, life is not governed by atemporal and deterministic laws, but it is immersed in the flow of time, in constant relation with memory of the past and projection towards the future.

Boltzmann denied irreversibility, regarding it as a defeat; the physicists of Einstein's generation made this negation a scientific dogma. As Poincaré[5] pointed out in 1889, nothing more sophisticated than plain logic is needed to reveal the error of trying to explain irreversibility in terms of the reversible.

ACKNOWLEDGMENT

This paper has been supported by a COFIN.MURST 97 CFSIB grant.

REFERENCES

1. MOROWITZ, H. 1978. Foundations of Bioenergetics. Academic Press. New York.
2. SCHRÖDINGER, E. 1944. What Is Life? Cambridge University Press. Cambridge, England.
3. PRIGOGINE, I. 1991. *In* Ecological Physical Chemistry. C. Rossi & E. Tiezzi, Eds.: 1–24. Elsevier Science Publishers. Amsterdam.
4. PRIGOGINE, I. & I. STENGERS. 1988. Entre le Temps et l'Eternité. Fayard. Paris.
5. POINCARÉ, H. 1889. Sur les tentatives d'explication mécanique des principes de la Thermodynamique. C. R. Acad. Sci **CVIII**: 550–553.

Fractal Time and the Gift of Natural Constraints

SUSIE VROBEL[a]

Institute for Fractal Research (IF), Schachtenstrasse 5, 34130 Kassel, Germany

CONSTRAINTS

A constraint is something that limits the degrees of freedom a system may exploit. There are physical constraints such as the speed of light, social constraints such as our political and legal systems, linguistic constraints such as our language, and so forth. Constraints both limit a system's degrees of freedom and make it possible to describe and understand this system by means of its internal relations, thus rendering possible travel and trade, communication and science.

NATURAL CONSTRAINTS

Let us assume the system we are considering contains an observer who, on the one hand, profits from such constraints, but, on the other hand, wonders about their origin and whether they are inviolate. The term *inviolate* is used in D. R. Hofstadter's sense, referring to the fact that there is an inviolate level below every tangled hierarchy: "... in any system there is always some 'protected' level which is unassailable by the rules on other levels, no matter how tangled their interaction may be among themselves."[1] This observer may also wonder whether these constraints are generated by himself or his fellow-observers or whether they are simply "given" constraints. For some cases, the answer seems obvious: political, legal, and linguistic constraints are more or less arbitrarily fixed conventions. But what about the speed of light? Here, our observer may be tempted to talk about a "natural constraint" as opposed to convention. And if he considers biological constraints such as possible combinations of base pairs in nucleic acids, he may still think of them as being natural—as opposed to man-made. But as he proceeds to consider more recent developments in evolution such as possible recombinations of DNA segments, he may conclude that there is no clear boundary and that attaching the label "natural" to a constraint is a rather arbitrary decision. At this point, he may realize that, after all, conventions, too, are the result of an evolutionary process, as are behavioral patterns and habits. It seems to be not so much a question of how the constraint evolved but rather of how persistent it is. Therefore, it shall suffice, for my purposes, to denote as "natural" any constraint which has evolved and which we are now "stuck with" for the time being.

[a]Address for correspondence: (+49) 561 63226 (voice); (+49) 561 63229 (fax); Susanne.Vrobel@t-online.de (e-mail).

THE PHENOMENOLOGICAL APPROACH

Assuming our observer accepts this definition, he may still wonder whether these natural constraints are observer-induced or part of the outside world. This is a toughie. And there is an elegant way out: the phenomenological approach studies only the interface between the observer and the rest of the world. The concept of "interface reality" was developed by O. E. Rössler to describe the endophysical view of the world.[3,4] Interface reality is all our observer deals with, and it comprises all he sees and imagines, feels and thinks—in brief: his entire world. Thus, his field of study is that of natural constraints as they appear on his interface.

LODs AS NATURAL CONSTRAINTS

A different kind of constraint is *levels of description* (LODs). An LOD is an abstraction by means of which an observer scans reality in terms of x, in terms of $x + 1$, in terms of $x + n$, in terms of y, and so forth. These LODs may be chosen arbitrarily by an observer or may be at his disposal as a result of his experience of the world and himself. The number of LODs at an observer's disposal increases with every new experience this observer is able to embed into the LODs already available. New experiences can generate new LODs which, in turn, modify the observer's (interface) reality. LODs provide a means of quantifying internal spatial and temporal relations that make up reality (interface reality). For scale-invariant structures, LODs may be conveniently defined as the results of iterations or as the result of zooming into a structure at regular intervals, that is, with a constant scaling factor.

FIGURE 1. The Triadic Koch Island.

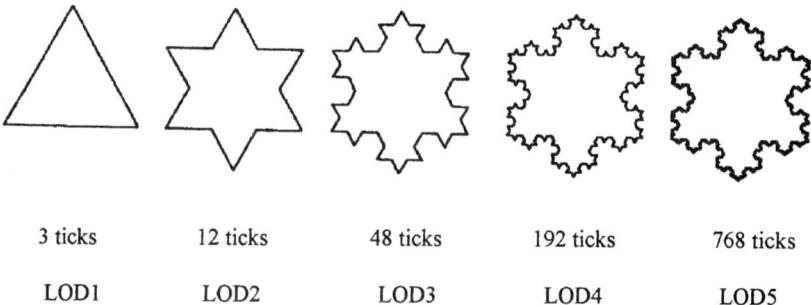

| 3 ticks | 12 ticks | 48 ticks | 192 ticks | 768 ticks |
| LOD1 | LOD2 | LOD3 | LOD4 | LOD5 |

FIGURE 2. Ticking away simultaneously on all levels of description.

For example, a unit of time may be defined as the period of time a tiny pointer following the circumference of the Triadic Koch Island (shown in FIG. 1) takes to complete one round. Depending on how much detail is accounted for by this pointer (and depending on the size of the pointer itself), this period of time is of differing duration for different LODs. As one zooms farther and farther into the edges of the Triadic Koch Island, more and more detail pops up and the length of the circumference—and, with it, the period of time the pointer takes to complete one round—approaches ∞. With every new zooming in (or out), a new LOD is generated. If one placed a pointer on each LOD and set these pointers to tick away the LOD-specific intervals simultaneously on all LODs, the resulting mechanism would be a fractal clock (see FIG. 2 for an early idea).

FRACTAL TIME

For scale-invariant systems such as the Koch-Curve (which is 1/3 of the circumference of the Triadic Koch Island), fractal clocks define their very own unit as the scale-invariant structure which can be found on all LODs. Imagine these scale-invariant structures arranged in a nesting cascade: the length of time of one unit varies from one LOD to the next and there are many parallelly arranged LODs, creating a simultaneity of scale-invariant structures on various LODs. These temporal extensions I have defined as Δt_{length} and Δt_{depth}.[6] Δt_{length} denotes the length of time which is defined as the number of incompatible temporal structures on one LOD. Δt_{depth} is the depth of time and is defined as the number of compatible temporal structures where one structure is nested into the same structure on the next level of description, and so forth. The theory of *fractal time* differentiates between Δt_{length}, Δt_{depth} and $\Delta t_{density}$. $\Delta t_{density}$ denotes the density of time.[6] It is defined by the ratio of the number of incompatible temporal structures per LOD and the scaling factor, which is the factor which describes the contraction of the scale from one LOD to the next. For scale-invariant systems, the internal relation between LODs of nested temporal structures may be described by the scaling factor s. In FIGURE 3, $s = 3$.

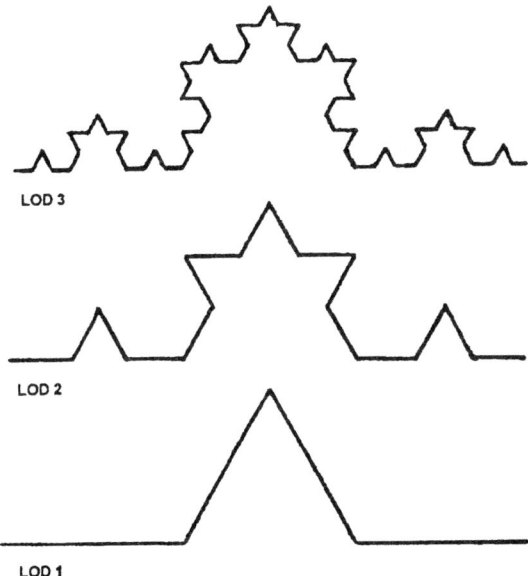

FIGURE 3. A nesting of scale-invariant structures.

THE PRIME AS A TEMPORAL NATURAL CONSTRAINT (TNC)

As a mathematically idealized structure, the Koch-Curve generates one nested structure upon the next, ad infinitum. In nature, these scale-invariant intervals usually have an upper and a lower limit. This means that there exists a structure at the top end of the cascade, which is itself part of and nested into the next-lower structure, but has no nesting potential itself anymore. I have denoted this structure as the prime[6] and defined it as the structure without nesting potential that is itself nested in a cascade of scale-invariant structures. This final structure at the top of the cascade of scale-invariant structures has a limiting effect: it is a nesting constraint. The limiting effect the prime has on the nesting potential and therefore the extention of Δt_{depth} makes it a *temporal natural constraint* (TNC).

THE PRIME STRUCTURE CONSTANT (PSC)

In a scale-invariant nesting cascade, the structure of the prime can be found on all LODs. The intervals delineated by the structure of the prime on all LODs serve as a reference scale that makes it possible to state the internal relations between the various LODs. An appropriate variation of scale would make the structure of the prime congruent on all LODs (as indicated in FIG. 4). If the structure of the prime is set as a constant—the *prime structure constant* (PSC)—a new kind of relativity can be said to result: time is "bent" with respect to the prime.

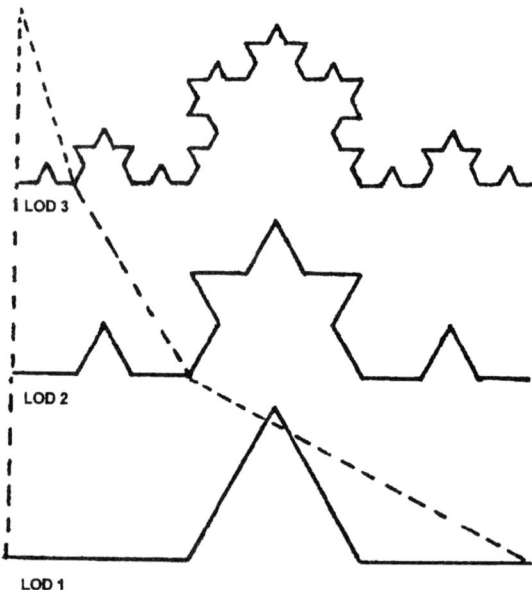

FIGURE 4. The condensation velocity v(c) for the Koch-Curve: LOD 1(4/3)/LOD 2 (4/9) = 3; LOD 2(4/9)/LOD 3 (4/27) = 3; ... etc.

CONDENSATION

The congruence required for such a "bending" of time in relation to the prime may be brought about by *condensation*. Condensation, a property (of time) generated by congruent, scale-invariant nestings, can be measured in condensation velocity $v(c)$ and condensation acceleration $a(c)$. The condensation velocity $v(c)$ for LOD2 ⊛ LOD1 (⊛ here denotes *nested in*) is determined by the quotient of Δt_{length} of LOD1 over Δt_{length} of LOD2. The condensation acceleration $a(c)$ is determined by the quotient of two condensation velocities of "neighboring" LODs. For scale-invariant structures, $v(c)$ is the scaling factor s (In FIG. 4, $s = 3$). Therefore, for the Koch-Curve, $v(c) = 3$.

NONTEMPORAL COGNITION OF TEMPORAL STRUCTURES

Condensation requires that scale-invariant nestings are detected and generated and that the observer is placed in a time interval that comprises the structure of the prime. Scale-invariant nestings are not scarce—they are abundant both in the world around us and inside us, the fractal observers.[5] Once time is "bent" with respect to the PSC, a glimpse beyond the immediate present will be possible for this observer. This glimpse will be of no duration in Δt_{length} for the observer: It is a nontemporal access to cognition of temporal structures that becomes possible for scale-invariant nested structures - an access which involves no duration in Δt_{length}.

OTHER CANDIDATES FOR TNCs

The prime as a first temporal natural constraint has the potential to be a powerful tool for modifying reality. With our perception sharpened by the introduction of this new concept, we can now scan the world for other TNCs. One promising candidate is the last period of a period-doubling cascade before the onset of chaos. *Period doubling* denotes "frequency-halving bifurcations (which) occur at smaller and smaller intervals of the control parameter."[2] The period-doubling scenario is governed by the *Feigenbaum number* and can be observed in the world around us as well as inside us, if we are fractal observers with a fractal interface generated by the LODs we have acquired. The Feigenbaum number is "the ratio of successive differences between period-doubling bifurcation parameters (which) approaches the number 4.669.... This property and the Feigenbaum number have been discovered in many physical systems in the prechaotic regime."[2]

The period-doubling scenario evolves as follows: "In the period-doubling phenomenon, one starts with a system with a fundamental periodic motion. Then, as some experimental parameter is varied, say λ, the motion undergoes a bifurcation of change to a periodic motion with twice the period of the original oscillation. As λ is changed further, the system bifurcates to periodic motions twice the period of the previous oscillation. One outstanding feature of this scenario is that the critical values of λ at which successive period doublings occur obey the following scaling rule (…):

$$\frac{\lambda_n - \lambda_{n-1}}{\lambda_{n+1} + \lambda_n} \to \delta = 4.6692016 \text{ as } n \to \infty."[2]$$

In the context of the theory of fractal time, period doubling may be interpreted as a dilation in the direction of Δt_{length}, as opposed to condensation in the direction of Δt_{depth}, which results in a "bending" of time with respect to the PSC described in this paper. Dilation in the direction of Δt_{length} requires congruent nestings in the direction of Δt_{length} that may be brought about by defining the structure of one periodic motion as a constant and thereby "bending" time with respect to this periodic motion constant (PMC).

In comparison to the PSC, the PMC at first sight seems to lack the dimension of Δt_{depth}. Therefore, it looks as if non-temporal cognition of temporal structures is not an option. Instead, a different option presents itself for the fractal observer: "bending" time with respect to the PMC results in a time dilation in the direction of Δt_{length} for the observer. At the same time, however, the doubling of a periodic motion to a period twice the period of the original is often accompanied by an emerging internal differentiation within these periods, for example, additional, though smaller, spikes may occur at "half time" and even smaller spikes may occur halfway within these half-periods, and so forth This internal differentiation within individual periods of the period-doubling scenario generates LODs and, therefore, Δt_{depth} and may thus provide the prerequisites for condensation in the direction of Δt_{depth}, in addition to a dilation in the direction of Δt_{length}. This analogy exceeds the topic of this paper, though, and will be discussed elsewhere. (The fractal-like structure of simultaneity may result, according to O. E. Rössler,[4] in further nonlinear distortions when different moments of external time are combined by the observer's interface.)

DISCUSSION

Natural constraints are a gift because, in the form of LODs and primes, they prompt us to identify and generate LODs and, at the same time, indicate and demarcate limits to our LODs: this enables us to study these LODs, in particular, their internal relations. It is conceivable that many of these constraints are not here to stay, but that we may modify our scale-invariant nesting cascades in such a way that we can still generate new LODs or initiate a condensation. In either case, we shall profit, it seems. TNCs may be selection effects: it is of advantage to define a limited temporal inertial system within which communication may be practiced without too many distortions caused by time condensation or dilation. TNCs set a limit to "structurability" of time by the observer.

Two modes of description appear to be available for two different kinds of observers: a non-fractal observer perceives reality by regarding the individual LODs for themselves, "unnested": To him, the prime structure of varying Δt_{length} on the individual LODs leads to the belief that this Prime structure is not congruent, but "bent," that is, condensed or dilated with respect to an absolute time. A fractal observer may perceive reality by "nested seeing" on several LODs simultaneously, i.e. through condensation: To him, the prime structure of constant Δt_{length} on the individual LODs leads to the belief that time is "bent" with respect to the constant prime structure.

A temporal natural constraint has been presented as a limitation to the observer's potential to generate temporal nestings and as a constraint to the "structurability" of the world. At the same time, it turns out to be a prerequisite for the observer to generate and access temporal structures that allow a nontemporal cognition of temporal structures. Furthermore, constraints not only define limits on the observer's side of the borderline, but also delineate what lies beyond, prompting us to try to put those limits behind us.

SUMMARY

Natural constraints are a gift. For a start, they make us wonder about whether the constraint encountered is observer induced or part of the outside world. As this question may prove hard to solve, it might be wise to start with a phenomenological approach, which allows us to study natural constraints as they appear on the interface of our world experience. One such natural constraint, the *prime*, is introduced as a *temporal natural constraint*, derived from a theory of fractal time, which differentiates between Δt_{length}, Δt_{depth}, and $\Delta t_{\text{density}}$. This differentiation allows an objective definition of duration that is independent of the individual time interval experienced. *Condensation*, a property generated by scale-invariant nestings, defines the structure of the prime as a reference scale for all levels of description. As a result, a nontemporal access to the cognition of temporal structures becomes possible. This access involves no duration in Δt_{length}. Thus, a temporal natural constaint is presented as a limitation of the observer's potential to generate temporal nestings and a constraint to the "structurability" of the world. At the same time, it is a prerequisite for the observer to generate and access temporal structures to allow for a nontemporal cognition of temporal structures.

ACKNOWLEDGMENTS

The author would like to thank Professor O. E. Rössler for helpful comments and his assistance in brushing up this paper.

REFERENCES

1. HOFSTADTER, D.R. 1979. Gödel, Escher, Bach: An Eternal Golden Braid. Harvester Press. U.K. (Penguin Edition: 1980).
2. MOON, F.C. 1987. Chaotic Vibrations. John Wiley. New York.
3. RÖSSLER, O.E. 1995. Intra-observer chaos: hidden root of quantum mechanics? *In* Quantum Mechanics, Diffusion and Chaotic Fractals, M.S. el Naschie, O.E. Rössler & I. Prigogine, Eds. Pergamon. Elsevier Science. Oxford, England.
4. RÖSSLER, O.E. 1998. Endophysics. World Scientific. Singapore.
5. VROBEL, S. 1997. Ice Cubes and Hot Water Bottles. *In* Fractals. An Interdisciplinary Journal on the Complex Geometry of Nature. Vol. 5 No. 1: 145–151. World Scientific. Singapore.
6. VROBEL, S. 1998. Fractal Time. The Institute for Advanced Interdisciplinary Research. Houston, Texas.

Alternating Oscillations and Chaos in a Model of Two Coupled Biochemical Oscillators Driving Successive Phases of the Cell Cycle

PIERRE-CHARLES ROMOND,[a] MAURO RUSTICI,[b] DIDIER GONZE, AND ALBERT GOLDBETER[c]

Faculté des Sciences, Université Libre de Bruxelles, Campus Plaine, C.P. 231, B-1050 Brussels, Belgium

INTRODUCTION

The cell division cycle is certainly one of the most important cellular processes in which nonlinear dynamics plays a major role. During the last decade, experimental evidence has accumulated to show that the onset of the M (mitosis) and S (DNA replication) phases of the embryonic and somatic cell cycles are controlled by the periodic activation of cyclin-dependent kinases (cdks) known as cdk1 and cdk2, respectively.[1-9] Various theoretical models have been proposed to account for the generation of sustained oscillations in the activity of these kinases, particularly the kinase cdc2 (cdk1) which controls the G2/M transition. The early theoretical models were either based on the positive feedback exerted by cdc2 on its own activation[10-12] (such positive feedback is now known to operate via cdc2 dephosphorylation by the cdc25 phosphatase which itself is activated by cdc2) or on the negative feedback involving the cdc2-induced degradation of cyclin which leads to cdc2 inactivation.[13-16]. More detailed models have since been proposed for both the fission yeast[17-20] and embryonic cell cycles.[19-23] The latter models take into account the various checkpoints that ensure the orderly progression through the successive phases of the cell cycle.

If the M and S phases are to follow each other during the cycle, rather than occurring concomitantly, and if each of the two phases is controlled by a biochemical oscillator involving, respectively, cdk1 and cdk2 with their associated cyclins, then it is necessary that the two oscillators be coupled through mutual control. Such control processes are part of the checkpoints mentioned above. There is evidence that this mutual control is of a negative nature. DNA replication is known to inhibit the triggering of mitosis until replication is completed.[3-5,24,25] Moreover, the kinase cdk1 that controls the initiation of mitosis inhibits the transition of the cell to a G1 replication-competent stage as long as the cell is in the S, G2, or M phases.[8,26]

---

[a]Present address: L.T.N.A., Faculté de Pharmacie, Université d'Auvergne, F-63000 Clermont-Ferrand, France.
[b]Present address: University of Sassari, Department of Chemistry, Via Vienna 2, 07100 Sassari, Italy.
[c]To whom correspondence should be addressed: (32 - 2) 650 5772 (voice); (32 - 2) 650 5767 (fax); agoldbet@ulb.ac.be (e-mail).

Coupling two oscillators may profoundly affect their dynamic behavior. We wish to explore here the consequences of a coupling through mutual inhibition of the biochemical oscillators controlling the onset of the M and S phases of the cell cycle. To keep the theoretical study relatively simple, we use for each of the two oscillators the model based on negative feedback previously proposed for sustained oscillations in cdk activity driven by cyclin synthesis and degradation.[13,15] In the present model each of the two oscillators contains three variables, that is, cdk1 or cdk2 with their associated cyclin and enzyme governing cyclin degradation. The coupling is assumed to follow from the direct inhibition of the synthesis of the cyclin of one oscillator by the cdk of the other oscillator. Similar results are obtained when assuming that inhibition occurs through mutual activation of cyclin degradation.

The present work does not aim at proposing a detailed model for the somatic cell cycle. Rather, we wish to investigate in a skeleton model how the mutual inhibition of two biochemical oscillators controlling different phases of the cell cycle may influence the oscillatory dynamics of such a coupled system. When studying the dynamic behavior of the coupled oscillators as a function of the strength of mutual inhibition, we recover, as most common behavior, alternating oscillations that likely correspond to the sequential activation of cdk1 and cdk2 and thus to the observed alternation between mitosis and the S phase. Such periodic alternation between the two kinases is, however, not the only mode of dynamic behavior predicted by the model. Thus, we also uncover the possibility, in slightly different conditions, of autonomous chaotic behavior.

In the next section we present the skeleton model of the double cdk1–cdk2 oscillator, as well as the kinetic equations that describe its time evolution. In the section ALTERNATING AND CHAOTIC OSCILLATIONS, we determine the various modes of periodic or chaotic oscillatory behavior as a function of the strength of the inhibitory coupling. The results are discussed in the DISCUSSION in regard to their physiological significance and to other examples of biological oscillators coupled through mutual inhibition.

SKELETON MODEL OF A DOUBLE OSCILLATOR CONTROLLING SUCCESSIVE PHASES OF THE CELL CYCLE

The model considered is schematized in FIGURE 1. It consists of two coupled minimal cascade models generating oscillations on the basis of cdk-induced cyclin degradation. The first oscillator, controlling the initiation of mitosis at the G2/M transition, involves the activation of cdk1 (M_1) by cyclin B (C_1), and the cdk1-induced degradation of cyclin B by an ubiquitin ligase (X_1) which is part of the ubiquitin-mediated proteolysis system. The second oscillator, controlling the initiation of DNA replication at the G1/S transition, is based on the activation of cdk2 (M_2) by cyclin E (C_2), and on the cdk2-induced degradation of cyclin E by another ubiquitin ligase (X_2). Note that, for simplicity, we do not consider the formation of a cyclin-cdk complex, but rather an activation of cdk by cyclin. Because the precise nature of the coupling is not yet fully characterized, we assume that the mutual inhibition of the two oscillators occurs as follows (see FIG. 1): cdk1 (M_1) inhibits the synthesis of cyclin E (C_2), while cdk2 (M_2) inhibits the synthesis of cyclin B (C_1).

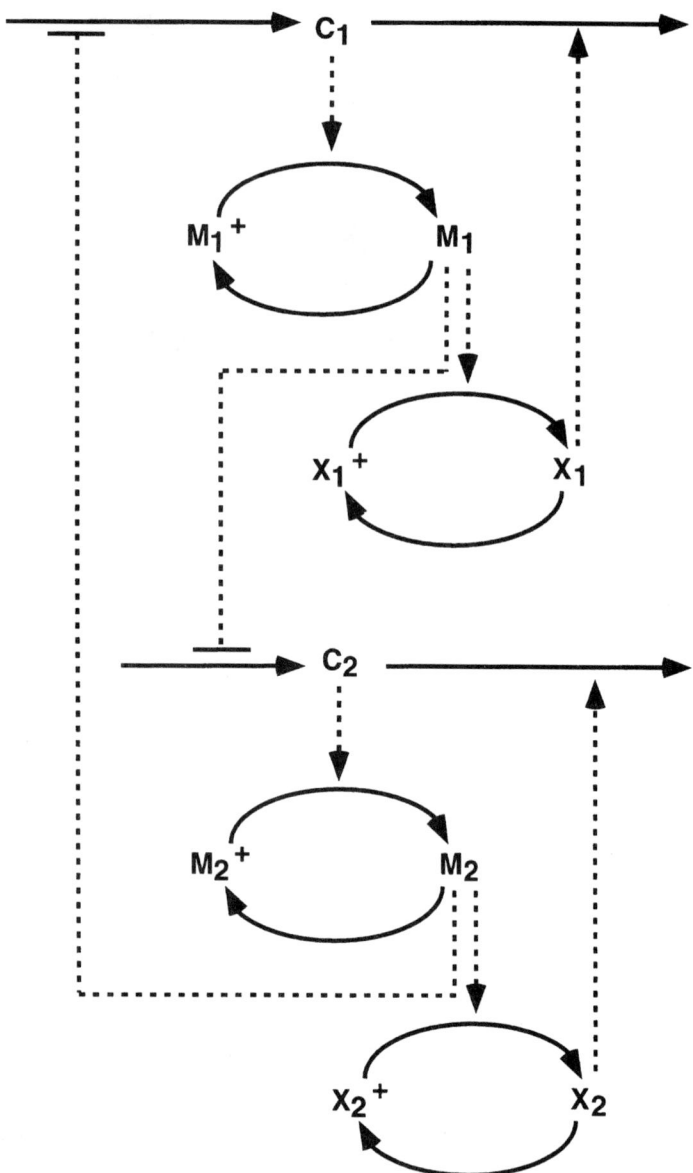

FIGURE 1. Skeleton model of two coupled biochemical oscillators controlling the M and S phases of the cell cycle. Each oscillator consists of a three-variable cascade involving a cyclin (C_1 or C_2), a cyclin-dependent kinase (cdk) (M_1 or M_2), and a cdk-activated ubiquitin ligase (X_1 or X_2) that controls cyclin degradation. The + sign indicates the inactive form of the enzymes. The dashed lines ending with a horizontal bar represent the inhibition exerted by M_1 and M_2 on the synthesis of C_1 and C_2, respectively.

The inhibitory action of cdk1 and cdk2 is described phenomenologically by treating the two kinases as inhibitors that directly modulate the rate of cyclin synthesis. Similar results are obtained if we assume that the coupling occurs through activation by M_1 (M_2) of cyclin C_2 (C_1) degradation (see below).

The time evolution of the system of two coupled biochemical oscillators is governed by the following system of kinetic equations[13,15]:

$$\frac{dC_1}{dt} = v_{i1}\frac{K_{im1}}{K_{im1} + M_2} - v_{d1}X_1\frac{C_1}{K_{d1} + C_1} - k_{d1}C_1 \tag{1a}$$

$$\frac{dM_1}{dt} = V_1\frac{(1 - M_1)}{K_1 + (1 - M_1)} - V_2\frac{M_1}{K_2 + M_1} \tag{1b}$$

$$\frac{dX_1}{dt} = V_3\frac{(1 - X_1)}{K_3 + (1 - X_1)} - V_4\frac{X_1}{K_4 + X_1} \tag{1c}$$

$$\frac{dC_2}{dt} = v_{i2}\frac{K_{im2}}{K_{im2} + M_1} - v_{d2}X_2\frac{C_2}{K_{d2} + C_2} - k_{d2}C_2 \tag{1d}$$

$$\frac{dM_2}{dt} = U_1\frac{(1 - M_2)}{H_1 + (1 - M_2)} - U_2\frac{M_2}{H_2 + M_2} \tag{1e}$$

$$\frac{dX_2}{dt} = U_3\frac{(1 - X_2)}{H_3 + (1 - X_2)} - U_4\frac{X_2}{H_4 + X_2} \tag{1f}$$

where

$$V_1 = \frac{C_1}{K_{c1} + C_1}V_{M1}, \quad V_3 = M_1 \cdot V_{M3} \tag{2a,b}$$

$$U_1 = \frac{C_2}{K_{c2} + C_2}U_{M1}, \quad U_3 = M_2 \cdot U_{M3} \tag{2c,d}$$

In the above equations, C_1 and C_2 denote the concentrations of cyclins B and E, while M_1, M_2 and X_1, X_2 refer to the fractions of activated cdk1 and cdk2 or of enzymes X_1 and X_2. Moreover, v_{ij} and v_{dj} ($j = 1, 2$) denote the constant rate of cyclin synthesis and the maximum rate of cyclin degradation by enzyme X_j reached for $X_j = 1$ for the first ($j = 1$) and second ($j = 2$) oscillator, respectively; K_{d1} and K_{c1} (K_{d2} and K_{c2}) denote the Michaelis constants for cyclin degradation and for cyclin activation of the phosphatase acting on the phosphorylated (inactive) form of the kinase cdk1 (cdk2); k_{d1} (k_{d2}) represents an apparent first-order rate constant related to nonspecific degradation of cyclin. Moreover, V_i (U_i) and K_i (H_i) ($i = 1, \ldots, 4$) denote the effective maximum rate and the Michaelis constant for each of the four enzymes involved in the two cycles of the cascade for each oscillator, namely, on one hand, the phosphatase and the kinase acting on cdk1 (cdk2), and on the other hand, the kinase cdk1 (cdk2) and phosphatase acting on the enzyme governing cyclin B (cyclin E) proteolysis (see FIG. 1). For each converter enzyme, the two parameters V_i (U_i) and K_i (H_i) are divided by the total amount of relevant target protein, that is, the total amount M_{1T} (M_{2T}) of cdk1 (cdk2) or the total amount X_{1T} (X_{2T}) of ubiquitin-conjugating enzyme acting on cyclin B (cyclin E); M_{1T}, M_{2T}, X_{1T}, and X_{2T} are considered as constant throughout the cell cycle.

For each oscillator the coupling between the two cycles of the cascade arises from the expressions for the effective, maximum rates V_1 (U_1) and V_3 (U_3) given by Equations (2a–d). Expression (2a) reflects the assumption that cyclin B activates in a Michaelian manner the phosphatase that acts on cdk1; V_{M1} denotes the maximum rate of that enzyme reached at saturating cyclin levels. On the other hand, Equation (2b) expresses the proportionality of the effective maximum rate of cdk1 to the fraction M_1 of active enzyme; V_{M3} denotes the maximum velocity of the kinase reached for $M_1 = 1$. Whereas Equations (2a) and (2b) pertain to the cdk1 oscillator controlling the G2/M transition, Equations (2c) and (2d) yield the expressions for the maximum rates U_1 and U_3 of the corresponding enzymes in the second oscillator, which controls the G1/S transition through the periodic activation of cdk2.

The coupling between the two oscillators is introduced via the first term in the kinetic equations for C_1 and C_2. This term reflects the Michaelian inhibition of C_1 synthesis by M_2 and of C_2 synthesis by M_1. Parameter K_{im1} (K_{im2}) denotes the inhibition constant divided by the total amount of cdk2 (cdk1). The smaller K_{im1} and K_{im2}, the stronger the inhibition.

We shall restrict the present analysis to the symmetric case in which corresponding parameter values are identical for each of the two oscillators. In particular, the maximum rate of cyclin synthesis ($v_{i1} = v_{i2} = v_i$) and the inhibition constant ($K_{im1} = K_{im2} = K_{im}$) have the same values for both oscillators. In the following we shall determine the dynamic behavior of the system of Equations (1a–f) as a function of the strength of the mutual inhibition measured by parameter K_{im}. We shall briefly discuss at the end of this paper the effect of introducing asymmetries in parameter values between the two oscillators. Focusing on the symmetrical case provides a convenient reference situation, since asymmetries in parameter values may be treated, in a second stage, as perturbations from such a reference state.

ALTERNATING AND CHAOTIC OSCILLATIONS

The most natural parameter for studying the effect of a coupling between the two enzymatic cascades controlling the periodic activation of cdk1 and cdk2 is K_{im}, which measures the strength of mutual inhibition of the two oscillators. In FIGURE 2 we present a series of phase portraits obtained by projecting the trajectory of the full, six-variable system, on the C_1–C_2 plane, for decreasing values of the inhibition constant K_{im}. The data show the existence of a rich spectrum of dynamic behavior. The actual sequence of oscillatory behavioral modes depends on the value of parameter v_i. The influence of this and other parameters will be examined in a subsequent publication. Here, we select the value of v_i so as to show the variety of oscillatory phenomena that the coupling of the two oscillators may bring about.

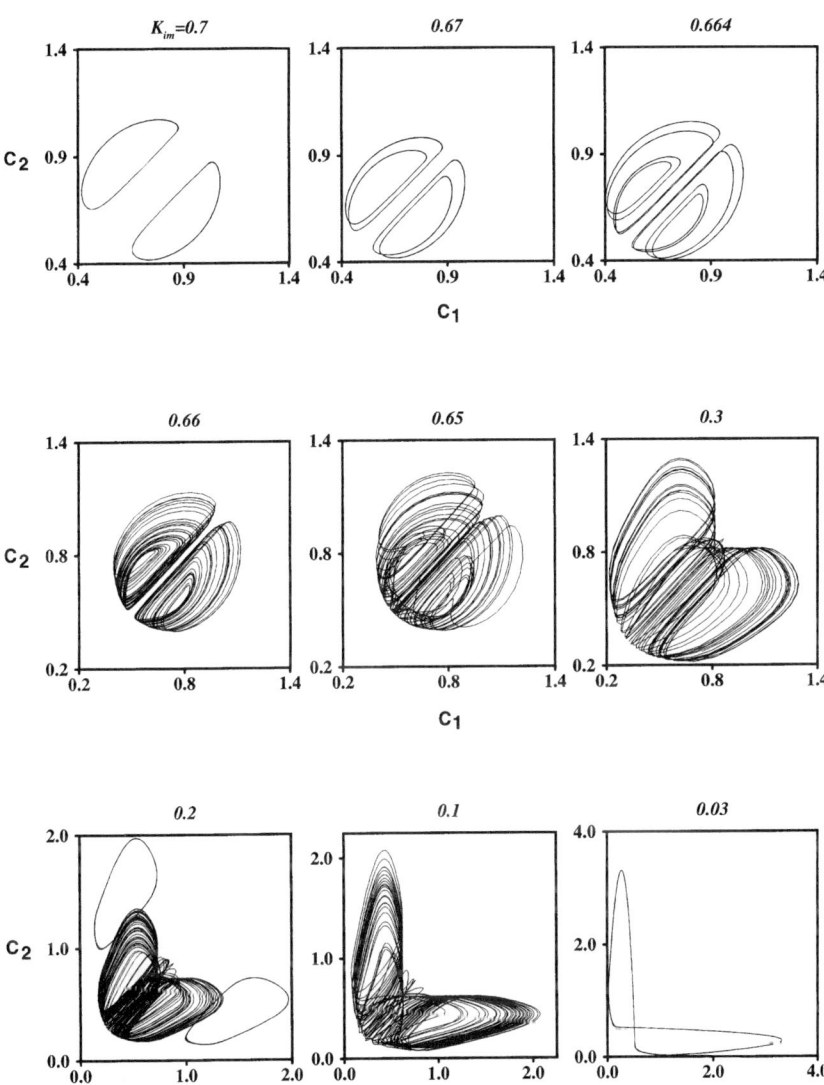

FIGURE 2. Dynamic behavior of the double cdk1-cdk2 oscillator model as a function of the inhibition constant K_{im}. Shown is the trajectory followed by the six-variable system (1a–1f) projected onto the C_1–C_2 plane (concentrations C_1 and C_2 are both expressed in µM). The sequence of diagrams is obtained at a constant value of the rate of cyclin synthesis ($v_i = 0.05$ µM min^{-1}). When $K_{im} = 0.7$ a pair of nonsymmetric limit cycles arise. In the range $0.7 < K_{im} < 0.664$ a period doubling cascade is observed. For $K_{im} = 0.66$ a pair of nonsymmetric chaotic attractors are found, and when K_{im} reaches the value of 0.65 they merge into a single antisymmetric, chaotic attractor. In the range $0.3 < K_{im} < 0.2$ a coexistence between an antisymmetric chaotic attractor and a pair of nonsymmetric limit cycles is observed. For $K_{im} = 0.1$ the two nonsymmetric limit cycles disappear and only the antisymmetric chaotic

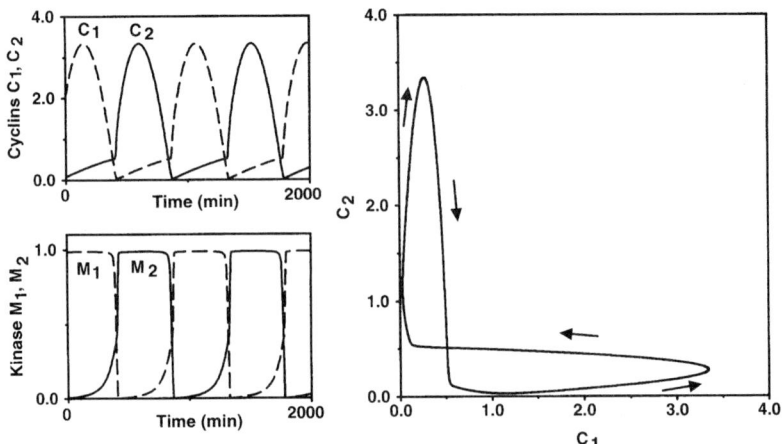

FIGURE 3. Alternating oscillations of cyclins C_1 and C_2 (*left upper panel*) and of kinases M_1 and M_2 (*left lower panel*) in the double cdk1-cdk2 oscillator model. These oscillations correspond to the antisymmetric limit cycle (*right panel*) obtained in FIGURE 2 for $K_{im} = 0.03$. The time evolution is obtained by numerical integration of Equations (1a)–(1f) using a routine for solving stiff differential equations.

Starting with weak inhibition corresponding to a relatively large value of the inhibition constant ($K_{im} = 0.7$), we observe the presence of two limit cycles that are symmetric with respect to each other. In each cycle, however, the variations of C_1 and C_2 are not symmetrical, so that the two coexisting trajectories can be described as nonsymmetric limit cycles. Upon decreasing K_{im} down to 0.67 and 0.664, we observe for each cycle a sequence of period doubling, which eventually leads to the coexistence of two strange attractors ($K_{im} = 0.66$). Further decrease of K_{im} to 0.65 causes the fusion of the two chaotic attractors. The resulting, unique strange attractor remains present at smaller values of K_{im} (e.g., $K_{im} = 0.3$), and later is seen to coexist with two nonsymmetric limit cycles ($K_{im} = 0.2$). For $K_{im} = 0.1$, the trajectory is again unique and takes the form of an antisymmetric chaotic attractor. Finally, for very low values of the inhibition constant ($K_{im} = 0.03$), that is, when the mutual inhibition is strong, we observe an antisymmetric limit cycle.

In the present paper we wish to focus on two of the behaviors shown in the sequence of FIGURE 2 that appear to be of particular biological significance. The first is the case of antisymmetric oscillations which correspond to the trajectory obtained for $K_{im} = 0.03$. This trajectory in the C_1–C_2 plane is again shown in FIGURE 3, together with the corresponding oscillations in C_1, C_2, M_1, and M_2. We see that in the presence of strong mutual inhibition, the two oscillators operate out of phase, so that

---

attractor remains (see also FIG. 4). For very low values ($K_{im} = 0.03$) of the inhibition constant an antisymmetric limit cycle is obtained (see also FIG. 3). The other parameters for the first and second oscillators are: H_i ($i = 1, ..., 4$) = K_i ($i = 1, ..., 4$) = 0.01; $V_{m1} = U_{m1} = 0.3$ min^{-1}; $V_2 = U_2 = 0.15$ min^{-1}; $V_{m3} = U_{m3} = 0.1$ min^{-1}; $V_4 = U_4 = 0.05$ min^{-1}; $K_{c1} = K_{c2} = 0.5$ μM; $v_{d1} = v_{d2} = 0.025$ μM min^{-1}; $K_{d1} = K_{d2} = 0.02$ μM; $k_{d1} = k_{d2} = 0.001$ min^{-1}.

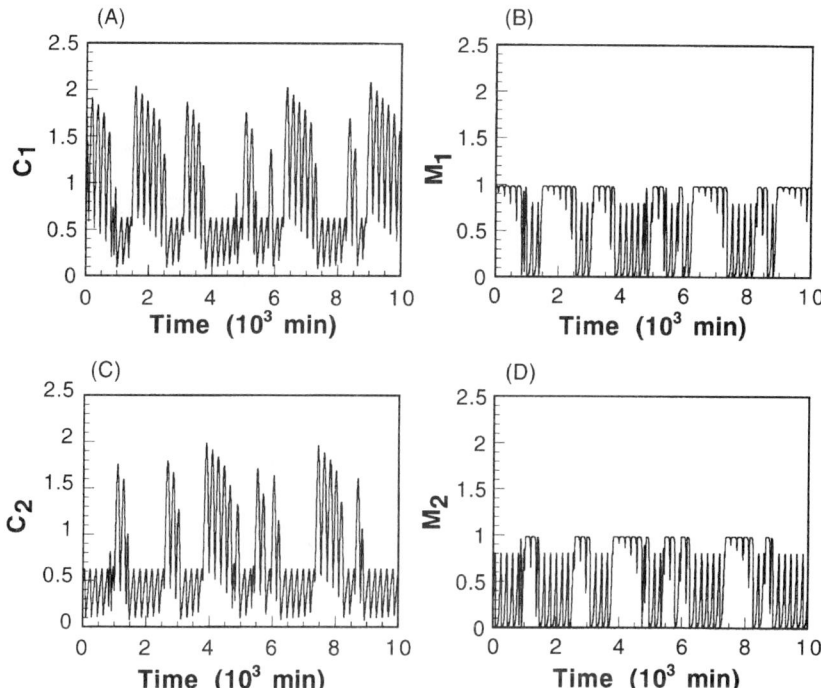

FIGURE 4. Autonomous chaos in the double cdk1-cdk2 oscillator model. The aperiodic oscillations correspond to the strange attractor shown for $K_{im} = 0.1$ in FIGURE 2. Panels (A) and (C) show the aperiodic evolution of the cyclins C_1 and C_2 (in µM). Panels (B) and (D) show the aperiodic evolution of the kinases M_1 and M_2. The curves are obtained as described in FIGURE 3.

cdk1 (M_1) reaches its maximum when cdk2 (M_2) is close to its minimum, and vice versa. This situation of alternating oscillations in cdk1 and cdk2 likely corresponds to the physiological case (see DISCUSSION). Because this study is of a qualitative rather than quantitative nature, the values of the parameters in FIGURE 3 have been selected arbitrarily so as to yield a period of mitotic oscillations of the order of somatic cell cycle lengths.

A second case of particular interest is that in which aperiodic oscillations result from the coupling between the cdk1 and cdk2 oscillators. Shown in FIGURE 4 are the chaotic oscillations corresponding to the strange attractor obtained in FIGURE 2 for $K_{im} = 0.1$. The variations in M_1 and M_2 are again out of phase, but have lost their periodic nature.

The question arises as to whether the results depend on the precise form of the coupling through mutual inhibition. To address this point we have studied another version of the model in which mutual inhibition is achieved through activation of C_1 (C_2) degradation by M_2 (M_1). For simplicity, we treat this putative regulation as a direct activation process. Then, Equations (1a) and (1d) for C_1 and C_2 are replaced by Equations (3a) and (3b):

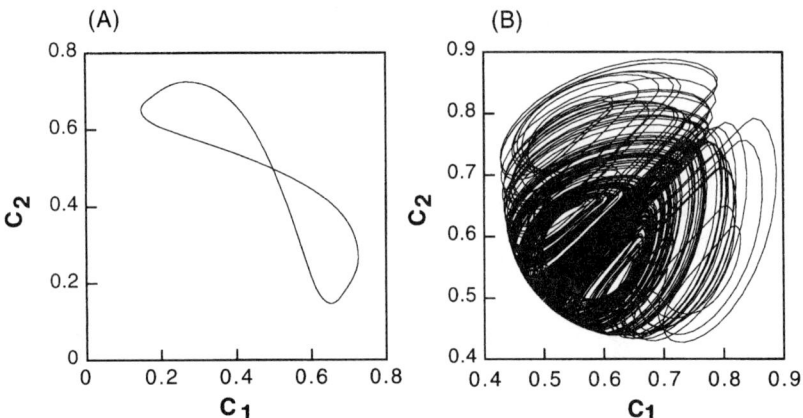

FIGURE 5. Alternating oscillations (A) and chaos (B) obtained in the model when the mutual inhibition of the two oscillators occurs through the putative activation of cyclin degradation by the cyclin-dependent kinases. The evolution of the coupled oscillators is then governed by eqs. (1b), (1c), (1e), (1f), (3a), and (3b). Shown are the projections of the trajectories followed by the system onto the C_1–C_2 plane; C_1 and C_2 are expressed in µM. Parameter values are: H_i ($i = 1, \ldots, 4$) = K_i ($i = 1, \ldots 4$) = 0.01; $V_{m1} = U_{m1} = 0.3$ min^{-1}; $V_2 = U_2 = 0.15$ min^{-1}; $V_{m3} = U_{m3} = 0.1$ min^{-1}; $V_4 = U_4 = 0.05$ min^{-1}; $K_{c1} = K_{c2} = 0.5$ µM; $v_{d1} = v_{d2} = 0.025$ µM min^{-1}; $K_{d1} = K_{d2} = 0.02$ µM; $k_{d1} = k_{d2} = 0.001$ min^{-1}; moreover, for (A) $K_{a1} = K_{a2} = 0.1$ µM, $v_{i1} = v_{i2} = 0.005$ µM min^{-1}, and for (B) $K_{a1} = K_{a2} = 1.0$ µM, $v_{i1} = v_{i2} = 0.01$ µM min^{-1}.

$$\frac{dC_1}{dt} = v_{i1} - v_{d1} X_1 \frac{C_1}{K_{d1} + C_1} \frac{M_2}{K_{a1} + M_2} - k_{d1} C_1 \tag{3a}$$

$$\frac{dC_2}{dt} = v_{i2} - v_{d2} X_2 \frac{C_2}{K_{d2} + C_2} \frac{M_1}{K_{a2} + M_1} - k_{d2} C_2 \tag{3b}$$

The time evolution of variables M_1, X_1, M_2 and X_2 remains given by Equations (1b), (1c), (1e), and (1f), respectively. The results presented in FIGURE 5 show that alternating oscillations as well as chaos are recovered in that version of the model.

DISCUSSION

Nonlinear phenomena play a key role in the dynamics of the cell cycle. In animal cells, there is evidence for autonomous oscillations in the activity of the cyclin-dependent kinases cdk1 and cdk2, the rise of which initiates mitosis (M phase) and DNA replication (S phase), respectively. The avoidance of concomitancy, that is, the ordering of these events is achieved through checkpoint mechanisms that are partly based on the mutual inhibition of mitosis and DNA replication.[3–5,8,24–26]

Here we have explored theoretically the effects of coupling through mutual inhibition the two biochemical oscillators that control the M and S phases of the cell cycle. To determine the effects of such a coupling, we resorted, for each of the two oscillators, to a simple model of a phosphorylation–dephosphorylation enzymatic cascade regulated by negative feedback. This model was previously shown to provide a simple mechanism for sustained oscillations in the activity of a cyclin-dependent kinase. Moreover, it fits very simply with the fact that cyclin and the active form of cdk, respectively, are slow and fast variables. We coupled the two oscillators by considering that the synthesis of cyclin in each oscillator is inhibited by the cdk of the other oscillator. The strength of this mutual inhibition is measured by the inhibition constant K_{im}. This skeleton model of two coupled oscillators can be viewed as a simple, cdk1–cdk2 double oscillator model for the cell cycle.[13,15] Obviously, it would be desirable to consider a more realistic model, but the number of variables and parameters would rapidly increase. Such a model could, for example, incorporate an inhibitor of both cdk1 and cdk2 that keeps them turned off in G1 phase; release from this inhibition would be connected with cell growth. It appears, indeed, that mitosis is prevented as long as the cell has not reached a critical size.

Our analysis shows that the dynamic behavior of the double oscillator model markedly depends on the strength of mutual inhibition. When this inhibition is strong, antisymmetric (i.e., alternating) oscillations are observed, in which one cdk reaches its maximum while the other is at a minimum. Such an alternation likely corresponds to the physiological situation as it would ensure the ordered progression of the cell through the S and M phases. In FIGURE 3 which illustrates this situation, the fractions of active cdk1 (M_1) and cdk2 (M_2) are seen to reach a plateau during their active phase. This situation with rapid alternations of replication and mitosis without intervening G1 and G2 gaps is most relevant to the embryonic cell cycle. For smaller values of the rate of cyclin synthesis v_i, however, M_1 and M_2 do not reach a prolonged plateau during the active phase, and the latter occupies a shorter portion of the period. The value of v_i selected for FIGURES 2 and 3 was chosen so as to maximize the variety of dynamic phenomena that can be observed in the model.

When the inhibition is weaker, chaos is found over a relatively large range of parameter values (see FIGURE 2). The question remains as to whether this phenomenon is of physiological significance. Previous theoretical studies have raised the possibility that the variability of the cell cycle duration might be due to the chaotic nature of its underlying dynamics.[27–30] In this context some theoretical models have been proposed in which chaos originates from the periodic forcing of a model biochemical oscillator.[31] Here, in contrast, chaos is autonomous as it occurs in the absence of any periodic forcing. The study of the cdk1–cdk2 double oscillator model shows that aperiodic oscillations can naturally arise from the coupling between the two cdk oscillators when the strength of mutual inhibition is in the appropriate range, that is, neither too strong (in which case periodic, alternating oscillations would occur) nor too weak (in which case two nonsymmetric limit cycles may coexist, as shown in FIGURE 2 for $K_{im} = 0.7$). These results provide a first theoretical indication as to how autonomous chaos may occur in the cell cycle dynamics.

The study of the double oscillator model also shows the possibility of a coexistence between multiple periodic or chaotic attractors (see FIG. 2), as well as other types of oscillations (symmetric or nonsymmetric) not shown there. The very rich repertoire of dynamic behavior will be studied in further detail in a subsequent publication where a bifurcation diagram in the v_i–K_{im} parameter plane will be presented. Such a diagram shows that the sequence of temporal patterns in FIGURE 2 depends on the value of parameter v_i. For the value of v_i considered in this figure, the system evolves toward a stable steady state in the absence of mutual inhibition, that is, at very large K_{im} values. Thus, in this case, the coupling is necessary for oscillations to occur. Oscillations can occur in the absence of coupling, however, at smaller values of v_i.

Admittedly, the two-oscillator model is a much simplified caricature of the dynamics of the animal cell cycle. More detailed models have been proposed for both the fission yeast and embryonic cell cycles.[17–23] In yeast there is a single cdk that controls the onset of both the M and S phases in conjunction with different cyclins and a cdk inhibitor.[18,19] The question arises whether a detailed exploration of these more realistic models in parameter space could also produce results similar to those presented here, in regard to alternating oscillations and/or chaos in the time evolution of the different cdk–cyclins complexes. A virtue of the present skeleton model is that it brings to light the sorts of dynamic behavior that are expected to occur as a result of the coupling between two oscillators controlling the S and M phases.

We have focused here on the case where the mutual coupling is achieved through inhibition of cyclin synthesis. As shown in FIGURE 5, similar results, including antisymmetric and chaotic oscillations, are also obtained when the coupling is exerted through the activation of cyclin degradation in one oscillator by the cdk belonging to the other oscillator. The results are also recovered, to a large extent, when a limited degree of asymmetry is introduced for the parameter values of the two oscillators.

In contrast to the present approach, some authors have suggested that the cell cycle is not described by a limit cycle oscillator but rather by a bistability phenomenon, each steady state corresponding to a particular phase (M or S) of the cell cycle.[20,32–34] We do not think, however, that the latter view necessarily contradicts the limit cycle description. Indeed, to account for the succession of cell cycle phases, the approach based on bistability corresponds to a sequence of "frozen frames." Including the variation of control parameters, so as to link these frames continuously with one another, should confer a recurrent (i.e., oscillatory) nature on the sequence of events which, each on its own, can be interpreted in terms of bistability.

Our results can be related to those obtained in the study of other biological oscillators coupled through mutual inhibition. Of particular interest in this respect is the analysis of a model of two mutually inhibiting pacemaker neurons carried out by Borisyuk et al.[35] These authors also obtained in that different context evidence for both antisymmetric and chaotic oscillations. The present study shows that if the cell cycle is governed by two biochemical oscillators coupled through mutual inhibition, the most prevalent mode of dynamic behavior will be that of periodic, alternating oscillations corresponding to the ordered sequence of the cell cycle phases. Our results indicate, however, that besides this physiological situation, depending on the strength of mutual inhibition, such a coupling may also give rise to more complex dynamic phenomena including chaos.

SUMMARY

The animal cell cycle is controlled by the periodic variation of two cyclin-dependent protein kinases, cdk1 and cdk2, which govern the entry into the M (mitosis) and

S (DNA replication) phases, respectively. The ordered progression between these phases is achieved thanks to the existence of checkpoint mechanisms based on mutual inhibition of these processes. Here we study a simple theoretical model for oscillations in cdk1 and cdk2 activity, involving mutual inhibition of the two oscillators. Each minimal oscillator is described by a three-variable cascade involving a cdk, together with the associated cyclin and cyclin-degrading enzyme. The dynamics of this skeleton model of coupled oscillators is determined as a function of the strength of their mutual inhibition. The most common mode of dynamic behavior, obtained under conditions of strong mutual inhibition, is that of alternating oscillations in cdk1 and cdk2, which correspond to the physiological situation of the ordered recurrence of the M and S phases. In addition, for weaker inhibition we obtain evidence for a variety of dynamic phenomena such as complex periodic oscillations, chaos, and the coexistence between multiple periodic or chaotic attractors. We discuss the conditions of occurrence of these various modes of oscillatory behavior, as well as their possible physiological significance.

ACKNOWLEDGMENTS

This work was supported by the programme "Actions de Recherche Concertée" (ARC 94-99/180) launched by the Division of Scientific Research, Ministry of Science and Education, French Community of Belgium. One of the authors (M.R.) was supported by a short-term EMBO fellowship.

REFERENCES

1. FELIX, M. A., J. PINES, T. HUNT & E. KARSENTI. 1989. A post-ribosomal supernatant from activated *Xenopus* eggs that displays post-translationally regulated oscillation of its cdc2+ mitotic kinase activity. EMBO J. **8:** 3059-3069.
2. NURSE, P. 1990. Universal control mechanism regulating onset of M-phase. Nature **344:** 503-508.
3. NURSE, P. 1994. Ordering S phase and M phase in the cell cycle. Cell **79:** 547–550.
4. MURRAY, A. W. & T. HUNT. 1993. The Cell Cycle: An Introduction. Oxford University Press. Oxford, U.K.
5. HUTCHISON, C. & D. M. GLOVER. 1995. Cell Cycle Control. Oxford University Press. Oxford, U.K.
6. DULIC, V., E. LESS & S. I. REED. 1992. Association of human cyclin E with a periodic G1-S phase protein kinase. Science **257:** 1958–1961.
7. JACKSON, P. K., S. CHEVALIER, M. PHILIPPE & M. W. KIRSCHNER. 1995. Early events in DNA replication require cyclin E and are blocked by p21CIP1. J. Cell Biol. **130:** 755–769.
8. STILLMAN, B. 1996. Cell cycle control of DNA replication. Science **274:** 1659–1664.
9. KRUDE, T., M. JACKMAN, J. PINES & R.A. LASKEY. 1997. Cyclin/Cdk-dependent initiation of DNA replication in a human cell-free system. Cell **88:** 109–119.
10. HYVER, C. & H. LE GUYADER. 1990. MPF and cyclin: modelling of the cell cycle minimum oscillator. Biosystems **24:** 82–50.
11. NOREL, R. & Z. AGUR. 1991. A model for the adjustment of the mitotic clock by cyclin and MPF levels. Science **251:** 1076–1078.
12. TYSON, J. J. 1991. Modeling the cell division cycle: cdc2 and cyclin interactions. Proc. Natl. Acad. Sci. USA **88:** 7328–7332.

13. GOLDBETER, A. 1991. A minimal cascade model for the mitotic oscillator involving cyclin and cdc2 kinase. Proc. Natl. Acad. Sci. USA **88:** 9107–9111.
14. GOLDBETER, A. 1993. Modeling the mitotic oscillator driving the cell division cycle. Comments Theor. Biol. **3:** 75–107.
15. GOLDBETER, A. 1996. Biochemical oscillations and cellular rhythms: the molecular bases of periodic and chaotic behaviour. Cambridge University Press. Cambridge, U.K.
16. ROMOND, P. C., J. M. GUILMOT & A. GOLDBETER. 1994. The mitotic oscillator: temporal self-organisation in a phosphorylation enzymatic cascade. Ber. Bunsenges. Phys. Chem. **98:** 1152–1159.
17. NOVAK, B. & J. J. TYSON. 1995. Quantitative analysis of a molecular model of mitotic control in fission yeast. J. Theor. Biol. **173:** 283–305.
18. NOVAK, B. & J. J. TYSON. 1997. Modeling the control of DNA replication in fission yeast. Proc. Natl. Acad. Sci. USA **94:** 9147–9152.
19. MARLOVITS, C. J., C. J. TYSON, B. NOVAK & J. J. TYSON. 1998. Modeling M-phase control in *Xenopus* oocyte extracts: the surveillance mechanism for unreplicated DNA. Biophys. Chem. **72:** 169–184.
20. NOVAK, B., A. CSIKASZ-NAGY, B. GYORFFY, K. CHEN & J. J. TYSON. 1998. Mathematical model of the fission yeast cell cycle with checkpoint controls at the G1/S, G2/M and metaphase/anaphase transitions. Biophys. Chem. **72:** 185–200.
21. NOVAK, B. & J. J. TYSON. 1993. Modeling the cell division cycle: M-phase trigger, oscillations, and size control. J. Theor. Biol. **165:** 101–134.
22. NOVAK, B. & J. J. TYSON. 1993. Numerical analysis of a comprehensive model of M-phase control in *Xenopus* oocyte extracts and intact embryos. J. Cell. Sci. **106:** 1153–1168.
23. BORISUK, M. T. & J. J. TYSON. 1998. Bifurcation analysis of a model of mitotic control in frog eggs. J. Theor. Biol. **195:** 69–85.
24. DASSO, M. & J. W. NEWPORT. 1990. Completion of DNA replication is monitored by a feedback system that controls the initiation of mitosis in vitro: studies in *Xenopus*. Cell **61:** 811–823.
25. ELLEDGE, S. J. 1996. Cell cycle checkpoints: preventing an identity crisis. Science **274:** 1664–1672.
26. DAHMANN, C., J. F. DIFFLEY & K. A. NASMYTH. 1995. S-phase-promoting cyclin-dependent kinases prevent re-replication by inhibiting the transition of replication origins to a pre-replicative state. Curr. Biol. **5:** 1257–1269.
27. ENGELBERG, J. 1968. On deterministic origins of mitotic variability. J. Theor. Biol. **20:** 249–259.
28. MACKEY, M. C. 1985. A deterministic cell cycle model with transition probability-like behavior. *In* Temporal Order. L. Rensing & N. I. Jaeger, Eds.: 315–320. Springer. Berlin.
29. MACKEY, M. C., M. SANTAVY & P. SELEPOVA. 1986. A mitotic oscillator with a strange attractor and distributions of cell cycle times. *In* Nonlinear Oscillation in Biology and Chemistry. H. Othmer, Ed.: 34–45. Springer-Verlag. Berlin.
30. GRASMAN, J. 1990. A deterministic model of the cell cycle. Bull. Math. Biol. **52:** 535–547.
31. LLOYD, D., A. L. LLOYD & L. F. OLSEN. 1992. The cell division cycle: a physiologically plausible dynamic model can exhibit chaotic solutions. Biosystems **27:** 17–24.
32. TYSON, J. J., B. NOVAK, G. M. ODELL, K. CHEN & C. D. THRON. 1996. Chemical kinetic theory: understanding cell-cycle regulation. Trends Biochem. Sci. **21:** 89–96.

33. THRON, C.D. 1996. A model for a bistable biochemical trigger of mitosis. Biophys. Chem. **57:** 239–251.
34. THRON, C.D. 1997. Bistable biochemical switching and the control of the events of the cell cycle. Oncogene **15:** 317–325.
35. BORISYUK, G. N., R. M BORISYUK, A. I. KHIBNIK & D. ROOSE. 1995. Dynamics and bifurcations of two coupled neural oscillators with different connection types. Bull. Math. Biol. **57:** 809–840.

Self Organization in Synthetic Polymeric Systems

JOHN A. POJMAN[a]

Department of Chemistry and Biochemistry, University of Southern Mississippi, Hattiesburg, Mississippi 39406, USA

Temporal and spatial self organization occurs in many polymeric systems that are industrially important. Temporal oscillations occur in industrial processing and can interfere with manufacturing. Spatial pattern formation can occur on the microscopic level to form structures reminiscent of those seen in the Belousov-Zhabotinsky reaction and in biological systems. Beautifully regular spirals and bubble formations can be formed in self-propagating polymerization reactions that interfere with novel approaches to preparing materials. How these patterns emerge and how that knowledge can be used to prepare new materials will be discussed.

INTRODUCTION

Self organization is essential to living systems and not surprisingly occurs in polymeric systems. Natural polymers comprise most living things, from the genetic code in the polymer DNA to the proteins that catalyze reactions to the structural polymers such as chitin that make up the exoskeletons of crustaceans. Plastics made from synthetic polymers are ubiquitous, from Tupperware to artificial hearts. About half the world's chemists work in polymer related industries. Still, relatively little work has been done on self organization in nonequilibrium, synthetic polymeric systems.

In this paper we survey some of the work that has been done in self organization in polymerization processes. We will consider polymerization reactions in a continuous-stirred flow tank reactor (CSTR) that exhibit oscillations through the coupling of temperature-dependent viscosity and viscosity-dependent rate constants. Emulsion polymerization, which produces small polymer particles dispersed in water, can also oscillate in a CSTR. Both types of systems are important industrially, and their stabilities have been studied by engineers with the goal of eliminating their time-dependent behavior. Our favorite oscillating system, the Belousov-Zhabotinsky reaction, can be used to create an isothermal periodic polymerization reaction in a batch or continuous system through the periodic termination of the polymerization process.

In most industrial processes, nonlinear behavior is seen not as an advantage but as something to be avoided. However, frontal polymerization may be useful for making new materials, and they are interesting because of the rich array of nonlinear phenomena they show, with pulsations, convection, and spinning fronts.

---

[a]Address for correspondence: 601-266-6075 (fax); john.pojman@usm.edu (e-mail).

WHAT ARE POLYMERS?

We can only touch on the immense field of polymer chemistry but will emphasize free-radical addition polymerization, which is one of the most important industrial methods. A polymer is a large molecule composed of many, many subunits, much as a train is composed of many different cars hooked end to end. If the subunits are all the same, then we call it a homopolymer. (This would be like a freight train made of many boxcars.) Important industrial polymers such poly(ethylene) and poly(methyl methacrylate) are in this class. If two different monomers react, they form a copolymer.

Free-radical polymerization with a thermal initiator can be approximately represented by a three-step mechanism. First, an unstable compound I, usually a peroxide or a nitrile, decomposes to produce radicals:

$$I \rightarrow 2fR\bullet$$

where f is the efficiency, which depends on the initiator type and the solvent. A radical can then add to a monomer to initiate a growing polymer chain:

$$R\bullet + M \rightarrow P_1\bullet$$

$$P_n\bullet + M \rightarrow P_{n+1}\bullet$$

The propagation step continues until a chain terminates by reacting with another chain (or with an initiator radical):

$$P_n\bullet + P_m\bullet \rightarrow P_n + P_m \text{ (or } P_{n+m}\text{)}$$

The major heat release in the polymerization reaction occurs in the propagation step.

The number of units in a polymer molecule may vary from tens to millions, depending on the reaction conditions and the monomers themselves. Synthetic polymers are never produced with a single molecular weight—only living organisms can claim this. Instead polymers always occur as a distribution, which in the best case is a Poisson distribution but usually is much broader and reflects the equilibrium between polymerization and depolymerization. For more detailed information on polymer chemistry, we refer the reader to several texts.[1,2]

EQUILIBRIUM SELF ORGANIZATION

Many striking patterns can form at equilibrium because of phase separation. We will only mention two because our interest is in nonequilibrium self organization. Almost all polymers are immiscible with each other and even with different molecular weights of themselves. If two immiscible polymers are dissolved in a common solvent, which is then removed by evaporation, phase separation will occur. If the process is done slowly then an equilibrium structure will result, completely determined by the free energy of mixing. If the solvent is removed rapidly, then nonequilibrium patterns may result.[3]

One interesting area of equilibrium pattern formation is with block copolymers, which consist of two distinct sections of immiscible polymers that are linearly connected. Because the immiscible components are connected, interesting patterns can form.[4]

SOURCES OF FEEDBACK

Self organization in any system requires some type of feedback. The most obvious source of feedback in polymerization reactions is thermal autocatalysis. The heat released by the reaction increases the rate of reaction, which increases the rate of heat release, and so on. This phenomenon can occur in almost any reaction and will be important when we consider frontal polymerization.

Free radical polymerizations often exhibit autoacceleration at high conversion via the isothermal "gel effect" or "Trommsdorff effect."[5,6] These reactions occur by means of the creation of a radical that attacks an unsaturated monomer, converting it to a radical, which can add to another monomer, propagating the chain. The chain growth terminates when two radical chains ends encounter each other and terminate. Each chain grows briefly and then becomes unreactive. As the degree of polymerization becomes high, the viscosity increases. The diffusion-limited termination reactions are slowed, increasing the overall rate of reaction, even though the rate constant of the propagation step is unaffected. The propagation step involves a monomer diffusing to a polymer chain. The addition of a radical to a double bond is a relatively slow process, having a rate constant of $\approx 10^4$ (Msec)$^{-1}$. The termination process requires two polymer chains to diffuse together. As the concentration of polymer molecules increases, the rate of diffusion drops dramatically for both chains. The rate of termination decreases because the polymer chains entangle, but monomers can diffuse through the chains to maintain propagation.

INDUSTRIAL REACTORS

Industrial reactors are prone to instabilities because of the slow rate of heat loss for large systems due to the low surface to volume ratio. Because the consequences of an unstable reactor can be disastrous, industrial plants are often operated under far from optimal conditions to minimize the chance of unstable behavior. Teymour and Ray studied vinyl acetate polymerization in a CSTR.[7] The monomer and initiator were flowed into the reactor, which was maintained at a sufficiently elevated temperature for the initiation of polymerization. As the degree of conversion increased, the rate of polymerization increased. The higher rate of reaction meant that the heat produced had less time to dissipate, so the temperature rose. The reaction might have reached a new steady state with higher conversion had the higher temperature not lowered the viscosity. The decrease in viscosity increased the rate of termination. Because these competing processes occurred on different time scales, the system did not reach a steady state, but exhibited temporal oscillations in temperature and conversion. FIGURE 1 shows a time series and the experimental phase plot for vinyl acetate. The period of oscillation is long, about 200 minutes, which is typical for polymerization in a CSTR.

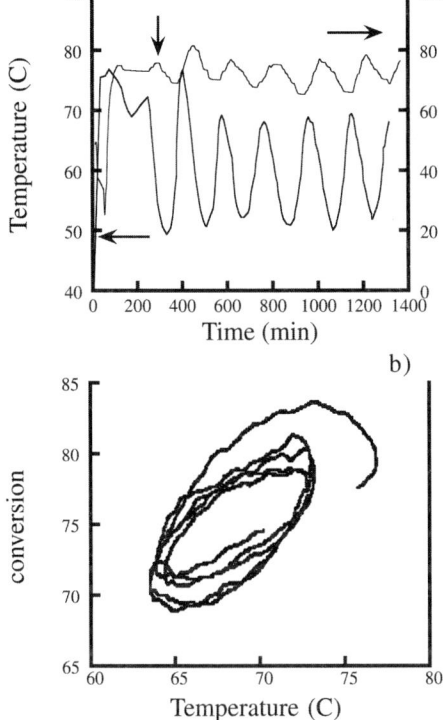

FIGURE 1. (a) Time series for conversion and temperature for vinyl acetate polymerization in a CSTR. Vertical arrow indicates change in residence time from 60 to 90 minutes. (b) Experimental phase plot for data in (a). Adapted from Teymour et al.[7]

Another class of oscillating polymerization reactions in a CSTR was observed in emulsion polymerization.[8,9] In this process, a water-insoluble monomer/polymer is dispersed throughout the aqueous phase with the aid of a surfactant. Schork and Ray demonstrated slow oscillatory behavior in the conversion and in the surface tension of the aqueous phase.

POLYMERIZATION COUPLED TO OSCILLATING REACTIONS

Váradi and Beck[10] observed in 1973 that acrylonitrile, a very reactive water soluble monomer, inhibits oscillations in the ferroin-catalyzed Belousov-Zhabotinsky (BZ) reaction while producing a white precipitate, indicating the formation of free radicals. Pojman studied the cerium-catalyzed BZ reaction (ferroin is a poor catalyst because it can form a complex with poly(acrylonitrile) in a batch reactor).[11] Because poly(acrylonitrile) is insoluble in water, the qualitative progress of the polymerization was monitored by measuring the relative decrease in transmitted light due to scattering of an incandescent light beam passed through the solution. ESR data show the oscillations in the malonyl radical concentration.[12] Oscillations in the bromine dioxide concentration have a greater amplitude ($\approx 10^{-6}$ M) and are out of phase with those of malonyl radical ($\approx 10^{-8}$ M).

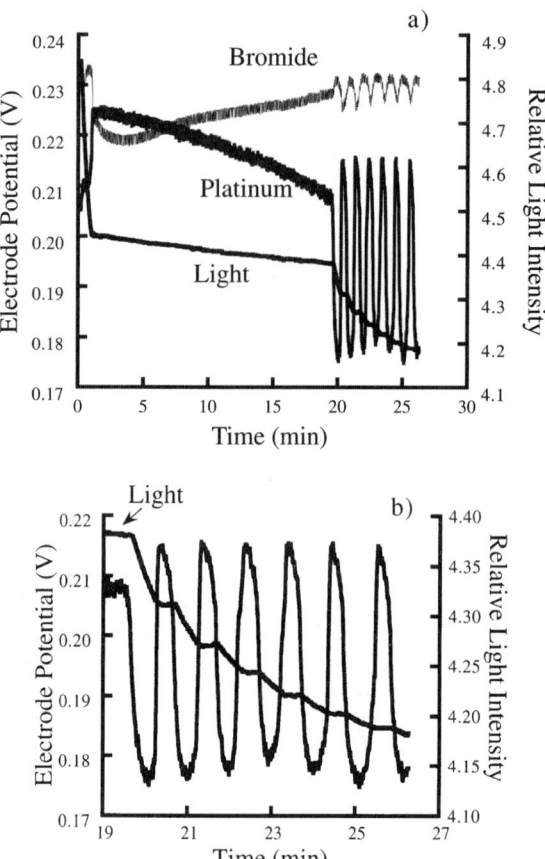

FIGURE 2. (a) The evolution of a BZ reaction in which 1.0 ml acrylonitrile was present before the Ce(IV)/H_2SO_4 solution was added. $[NaBrO_3]_0$ = 0.077 M; [malonic acid]$_0$ = 0.10 M; $[Ce(IV)]_0$ = 0.0063 M; $[H_2SO_4]_0$ = 0.90 M. (b) An enlargement of the oscillatory region. The light intensity is inversely related to the turbidity of the solution. Adapted from Pojman et al.[11]

Acrylonitrile halts oscillations for a period of time proportional to the amount added. However, no polymer precipitates until oscillations in both the Pt electrode potential and the bromide concentration return. Then, a white precipitate forms continuously during the oscillations. Even if acrylonitrile in excess of its solubility limit is added, oscillations continue.

Oscillations and polymerization occur in both batch and flow reactors into which acrylonitrile is continuously flowed along with the other BZ reactants. Polymerization occurs periodically, in phase with the oscillations (FIG. 2). Washington et al. determined that it is not periodic initiation by malonyl acid radicals that causes the periodic polymerization but periodic termination by bromine dioxide.[13]

Although these experiments are interesting, it remains to be seen if using an oscillating reaction to initiate polymerization can be more useful than current approaches.

FRONTAL POLYMERIZATION

Frontal polymerization is a mode of converting monomer into polymer via a localized reaction zone that propagates, most often through the coupling of thermal diffusion and Arrhenius reaction kinetics. Frontal polymerization reactions were first discovered in Russia by Chechilo and Enikolopyan in 1972.[14] They studied methyl methacrylate polymerization to determine the effect of initiator type and concentration on front velocity[15] and the effect of pressure.[16] A great deal of work on the theory of frontal polymerization was performed.[17–22] The literature up to 1984 has been reviewed by Davtyan *et al.*[23]

Pojman and his co-workers demonstrated the feasibility of propagating fronts of polymerization in solutions of thermal free-radical initiators in a variety of high-boiling-point monomers[24–26] and with a solid monomer.[27] Frontal copolymerization has also been considered.[28] The macrokinetics and dynamics of frontal polymerization have been examined in detail and applications for materials synthesis considered.[30]

BASIC PHENOMENA

Frontal polymerization is relatively easy to study. In the simplest case, a test tube is filled with the reactants. The front is ignited by applying heat to one end of the tube with an electric heater. The position of the front is obvious because of the difference in the optical properties of polymer and monomer. FIGURE 3 shows a typical front. The experiments are so simple to perform that they are routinely done in the physical chemistry lab at the University of Southern Mississippi.[31]

Under most cases, a plot of the front position versus time produces a straight line whose slope is the front velocity. The velocity can be affected by the initiator type and concentration but is on the order of a one centimeter per minute.

The defining feature of frontal polymerization is the sharp temperature gradient present in the front. FIGURE 4 shows three different temperature profiles for methacrylic acid for different initiators. Notice that the temperature jumps about 200°C over as little as a few millimeters, which corresponds to polymerization in a few seconds at that point.

ADVANTAGES AND POTENTIAL APPLICATIONS

Most examples of frontal polymerization are done *without solvent*, which means the final product does not need to be separated from the solvent and that no contaminated solvent requires disposal. Such bulk polymerization of reactive monomers like acrylates cannot normally be accomplished safely, which is a polite way of say-

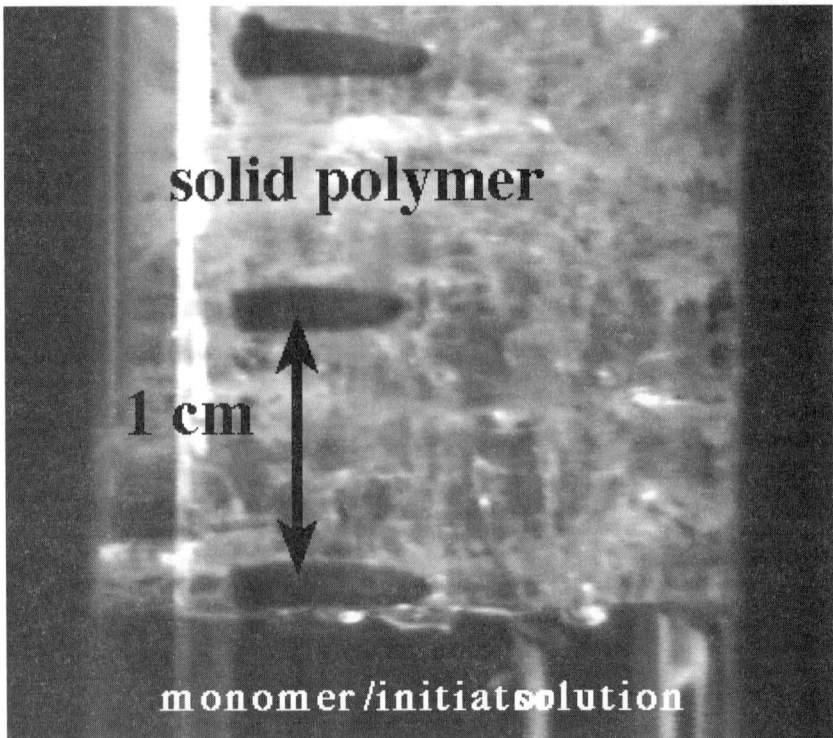

FIGURE 3. A descending case of frontal polymerization with triethylene glycol dimethacrylate and benzoyl peroxide as the initiator.

ing a tube of methacrylic acid and peroxide will explode if it adiabatically polymerized.

Why then does frontal polymerization work? Because the reaction is localized in a narrow region. Consider an analogy from combustion. We do not heat our homes by filling them with natural gas and igniting it. We safely heat them by localizing the combustion in a burner, that is, we use a flame. Frontal polymerization works well because it is like a "liquid flame" of polymerization.

Other advantages include rapid curing of thick epoxy composites without an autoclave. White has shown it is possible to have a frontal curing of thick layers of a commercial epoxy prepreg with superior properties compared to homogeneous curing.[32,33] Chekanov et al. has shown that standard epoxy/amine systems can be cured an order of magnitude faster than batch methods while still achieving 90% of the mechanical properties.[34] Pojman et al. have shown it is possible to produce polymer dispersed liquid crystals (PDLCs) via frontal curing of epoxies in which a liquid crystal is initially dissolved in the resin.[35] Pojman et al. have also demonstrated Interpenetrating Polymer Network (IPN) formation via frontal polymerization.[36]

Producing uniform materials from materials that will sediment during batch curing is very promising. Nagy et al. showed that a thermochromic composite could be

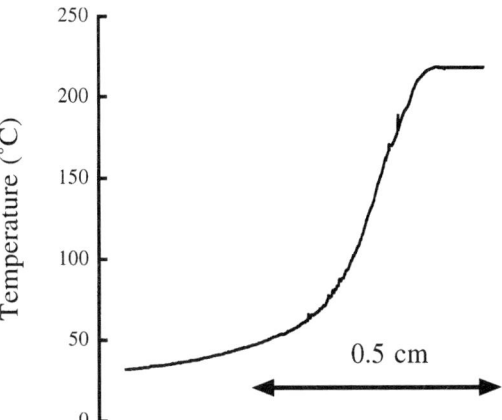

FIGURE 4. Temperature profiles for a methacrylic acid polymerization front with 1.5% vol/vol *t*-butyl peroxide initiator.

prepared better frontally than through batch curing.[37] Szalay *et al*. prepared a novel conducting composite frontally.[38] Pojman *et al*. have developed a process for producing Functionally Graded Polymeric materials using frontal polymerization.[39]

PERIODIC MODES

In stirred reactions, a steady state can lose its stability as a bifurcation parameter is varied, leading to oscillations.[40] Propagating thermal fronts can show analogous behavior. The bifurcation parameter for a thermal front is the Zeldovich number.[41] He assumed that the reaction occurs in an infinitely narrow region in a single step with activation energy E_{eff}, initial temperature T_0, and maximum temperature T_m:

$$Z = \frac{T_m - T_0}{T_m} \frac{E_{\text{eff}}}{RT_m}$$

A great deal of theoretical work has been devoted to determining the modes of propagation that occur.[42–46] In a one-dimensional system, the constant velocity front becomes unstable as Z is increased. A period-doubling route to chaos has been shown numerically.[47] A wide array of modes has been observed in self-propagating high-temperature synthesis (thermite) reactions.[48]

Decreasing the initial temperature of the reactants increases the Zeldovich number (if the conversion does not change) and drives the system away from the planar front propagation mode into nonstationary regimes. The stability analysis of the reaction–diffusion equations describing the propagation of reaction waves along the cylindrical sample predicts the existence of spin modes with different numbers of heads.[46] Depending on the Zeldovich number and the tube diameter (provided the kinetic parameters are kept constant), the number of spinning heads can vary from one to infinity as the tube diameter increases.

Begishev et al. studied anionic polymerization fronts with ε-caprolactam.[49] Two aspects of this system were interesting. First, the polymer crystallizes as it cools, which releases heat. Thus, a front of crystallization follows behind the main front. Volpert and coworkers investigated this two-wave system.[50] Second, a "hot spot" moved around the front as it propagated down the tube, leaving a spiral pattern in the product. The entire front propagated with a velocity on the order of 0.5 cm/min, which was a function of the concentrations of activator and catalyst. The hot spot circulated around the outside of the 6-cm (i.d.) front 16 times as rapidly as the front propagated. Similar behavior has been observed in the frontal polymerization transition metal nitrate acrylamide complexes.[51]

Experimental study of frontal polymerization of methacrylic acid has shown the existence of a rich variety of periodic regimes.[29,52] At ambient initial temperature with heat exchange to room temperature air, only a stable planar front mode exists. Decreasing the initial temperature of the reactants, as well as increasing the rate of heat loss, leads to the occurrence of periodic modes. The most impressive pattern (a large translucent spiral in FIG. 5) has a single-head spin mode that can be realized

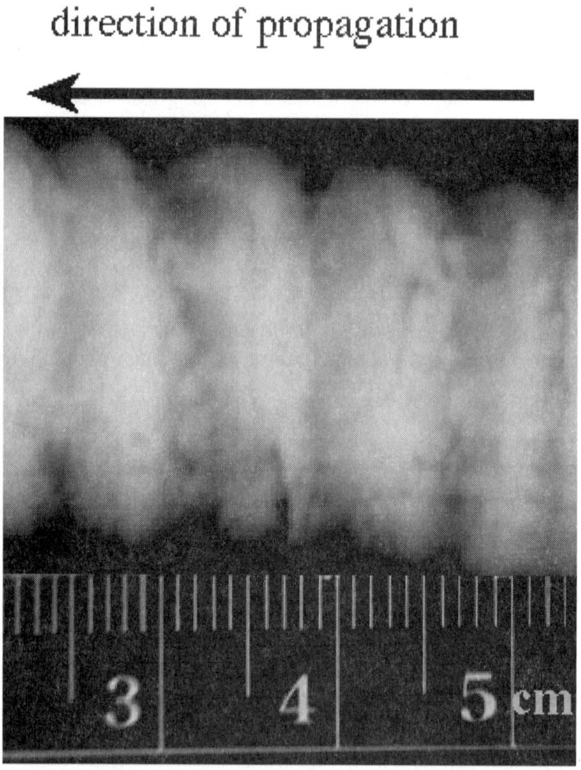

FIGURE 5. The pattern left in poly(methacrylic acid) by a single-head-spin mode. The methacrylic acid was initially at 0°C, and benzoyl peroxide (2% wt/vol) was used as the initiator.

with a clearance of 0.5 cm between the front and an ice-water bath. The single "hot spot" propagates around the front (itself moving with a velocity of 0.66 cm/min) with a period of 1.3 min. The direction of the single-head spinning is arbitrary and can change during front propagation.

What is the physical mechanism for these instabilities? First we consider the one-dimensional case of uniform pulsations. For a given activation energy, if the initial temperature is "high enough" heat diffuses ahead of the front and initiates polymerization at a temperature near the initial temperature T_i. A constant front velocity is observed with a front temperature equal to $T_i + |\Delta H/C_p|$ (assuming 100% conversion). If the initial temperature is low, the heat diffuses and accumulates until a critical temperature is reached (the ignition temperature for the homogeneous polymerization); the polymerization then occurs very rapidly, consuming all of the monomer. Again, the front temperature is $T_i + |\Delta H/C_p|$. Because all of the monomer in the reaction zone has been consumed, the rate of reaction plummets, and the velocity drops. Heat then diffuses into the unreacted region ahead of the front, repeating the cycle.

Another way of looking at the problem is to compare the rates of heat production and heat transport. When the initial temperature is low, the rate of heat production can exceed the rate of heat transport along the direction of propagation, causing an overshoot in the temperature. When all the monomer has been consumed, the rate of heat transport exceeds the rate of heat production, causing a drop in the temperature and the propagation velocity. This lack of a balance is reflected in the changing temperature gradient. When the initial temperature is high, the heat production is balanced by diffusion, and a constant velocity and temperature profile can exist.

The spinning mode is more complicated. Let us consider the plane front and a localized temperature perturbation on it. As we discussed above, at the spot of the perturbation there are two competing processes, heat production and heat transfer. Again, if the energy of activation is sufficiently large or the initial temperature suf-

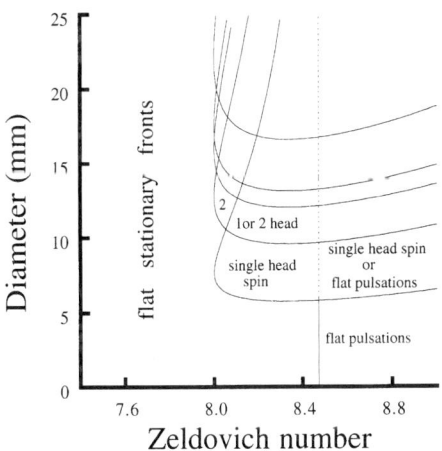

FIGURE 6. Calculated stability diagram for methacrylic acid fronts as a function of the Zeldovich number and the tube diameter. Ea = 18 kcal mol^{-1}; κ = 0.001 cm^2 sec^{-1}; $T_m - T_0$ = 200°C. Adapted from Ilyashenko and Pojman.[53]

ficiently low, the temperature can begin to increase. As it increases reactants are consumed more rapidly. From where? There are two possibilities: (a) to take the reactants from ahead of the front, that is, to move in the direction perpendicular to the front; and (b) to take the reactants from the left or from the right of the perturbed spot, that is, to move along the front to the left or right. In some cases it is more effective to move along the front because the reactants there are preheated, while in the perpendicular direction they are cold. This motion of the temperature perturbation along the front gives a two-dimensional instability that can be a spinning mode or some other mode, depending on geometry and the values of the parameters.

Ilyashenko and Pojman studied the single-head spin mode in detail[53] and used the previous results of Sivashinsky[46] to build a mode map in the tube diameter-Zeldovich number plane using kinetic data for methacrylic acid fronts (FIG. 6). The map shows the expected mode appearance as a function of test tube diameter. They were able to calculate the spiral pitch to be 0.5 cm, in good agreement with the experiments. We see that for typical polymerization parameters the only way to achieve periodic modes is to lower the initial temperature because the energy of activation is fixed by the kinetic parameters. Therefore, spin modes are not a problem at room temperature. Or are they?

THE PLOT THICKENS...

In studying frontal propagation with triethylene glycol diacrylate (TGDMA) at room temperature, we observed spin modes! We found that adding bromophenol blue dye made the spin modes more visible without the use of an IR camera.[54] We undertook a detailed study with hexane diol diacrylate (HDDA) using a plasticizer as a diluent. The following spin-head doubling sequence is shown in FIGURE 7. At HDDA and plasticizer (diethyl phthalate, DEP) percentages of 60% (vol/vol) and 40% (vol/vol), respectively, a single-headed polymerization spin was observed (FIG. 8). Using the bromophenol blue indicator, the spin can be observed as a pale yellow line moving into the green reactant solution in a spiral fashion. A helical green spiral was observed in the zone between a spiral and its predecessor. The zone is associated with low polymer conversion and poor mechanical properties. Upon breaking the polymer rod along its axis, the traces of the spin were found within as alternating

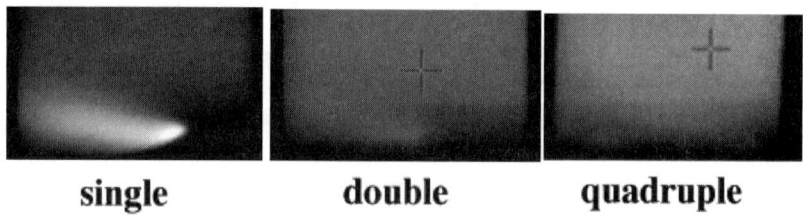

FIGURE 7. IR images of spin-head doubling with 1,6-hexanediol diacrylate. For the double and quadruple, only half the heads can be seen at a time, but notice that each head is smaller the greater their number. (Image courtesy of J. Masere.)

FIGURE 8. A single head spin mode pattern in an HDDA/DMSO descending front with bromophenol blue added to enhance contrast. Tube diameter is 1.5 cm. (Image courtesy of J. Masere.)

translucent and pale sections. This clearly demonstrates that the reaction occurs not only on the surface but also within the bulk of the specimen. The single-head mode polymerization regime becomes double-headed when the percentage of HDDA is increased into the 42–46% range. The two spins' heads become equidistant to each other as the front propagates, and the size of the hot spots are smaller in comparison to the single-head hot spots. This is different from the images of single-head spots where a spot can be observed as it moves along the surface closer to the camera and takes some time to reappear as it travels along the surface away from the camera view. When the free radical indicator is used, the two spin heads can easily be visualized. Four spin heads are observed upon a further increase in the HDDA percentage (46–49% range). Using an IR camera, only two of these spots can be observed at a time because the other two are are hidden. A multispot mode is observed as the range of HDDA percentage is increased. It is difficult to count the number of hot spots because each of the spots changes direction as the reaction progresses. A reversal of direction is observed every time spin heads collide. The number of these spots in-

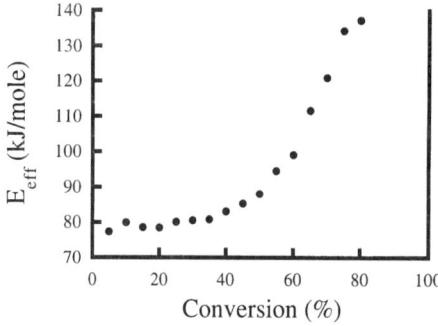

FIGURE 9. The activation energy dependence on conversion for a diacrylate (HDDA) with a peroxide initiator (Lupersol 231). Adapted from Gray.[55]

creases with an increase in HDDA percentage and some spots are observed along the entire reaction zone plane. The front behavior appears to be random and may be chaotic. There seems to be a narrow region between the four-head spins and multiheaded mode in which the front appears to move in a pulsating fashion. Spinning hot spots are also evident. It is possible that there is an overlap between these modes. Traces remaining on the surface are annular and almost parallel. However, no pure pulsating fronts were observed.

The question is, Why do spin modes appear in diacrylates at room temperature when a monoacrylate, such as methacrylic acid, only exhibits spin modes at low temperatures? Difunctional systems will cross-link at low conversion, forming a gel that drastically affects the kinetic parameters. Gray studied the photopolymerization of HDDA and determined the effective energy of activation as a function of conversion.[55] FIGURE 9 shows the overall energy of activation of HDDA including the value from a peroxide initiator (Lupersol 231). Using data from Gray[55] at 80% conversion, we estimate a Zeldovich number of 14. This is far beyond the stability boundary ($Z = 8$) and explains how spin modes are observed at room temperature.

When DEP was replaced by benzyl acrylate as the diluent, T_{max} was found to be independent of the benzyl acrylate percentage. Although a change in the front behavior was observed as the percentage of the reactive diluent varied, that change occurred within a broad range of percentage composition as opposed to the rather narrow ranges associated with passive diluents.

By maintaining the initial concentration of vinyl moieties constant, we can alter the degree of cross-linking without changing T_{max}. By maintaining the total vinyl concentration at 0.11 M, single-headed spin behavior was observed below 22% HDDA and 36.8% benzyl acrylate. Two spin heads were observed between 22.4% and 44.8% HDDA and between 36.8% and 4.4% benzyl acrylate. When the HDDA and benzyl acrylate ranges were shifted to 45.2%–48% and 4%–0.0%, respectively, four-headed spin modes were observed. It is worth noting that Lupersol was kept constant at 4%.

Thus, varying the relative amounts of a monoacrylate and multifunctional acrylate, the dynamics of frontal polymerization can be controlled. This is an extremely convenient experimental system for the study of nonlinear phenomena in thermal fronts.

CONVECTION

Because of the large temperature and compositional gradients created in a front, large density gradients are created that can cause convection. The conditions for the onset of convection have been studied theoretically and experimentally.[25,56–58] Convection can be so severe with liquid monomers that produce a liquid polymer that experiments can only be performed by adding silica gel to increase the viscosity or in simulated microgravity.[59]

NONRANDOM BUBBLE DISTRIBUTIONS

Bubbles are produced in most polymer fronts because nitrogen or carbon dioxide is released as a product of the thermal decomposition of nitriles and peroxides.[29] An experiment aboard the *Conquest I* sounding rocket showed that bubbles form periodic structures in a thermoplastic (polymer that is a liquid at high temperature).[59] From a considerable body of literature (reviewed by Wozniak *et al.*) and from exper-

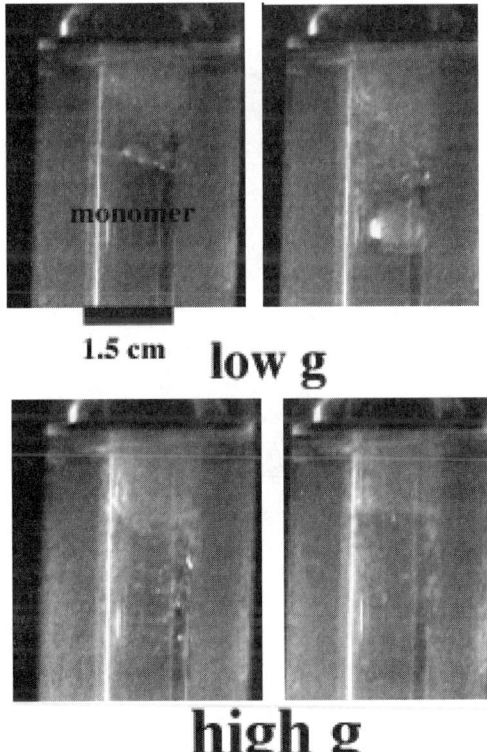

FIGURE 10. Fronts of benzyl acrylate polymerization in low and high *g*. The fronts are propagating from the top of the images on down. Time between images is 12 seconds.

iments on the space shuttle, we know that bubbles migrate toward the higher temperature in a gradient.[61] Bubbles will tend to migrate toward the center of the tube where the temperature is highest. How then does the front propagate? Given that the bubbles are collected into a string of large bubbles in a periodic pattern, does the front propagate periodically? This is a complicated situation because the bubbles are being produced *in situ* at the front from the initiator decomposition. The interaction of bubbles in a gradient is also complicated.

We performed experiments on NASA's KC-135 aircraft, which provides 20 sec of $0.01g$ followed by one minute of $1-1.8g$. FIGURE 10 shows how in weightlessness, the bubbles aggregate in the poly(benzyl acrylate) with Lupersol 231 as the initiator. When the high g arrives, the bubbles rise to the top of the tube, and the molten polymer sinks. We propose that the bubbles move toward the hotter interior of the tube via surface tension–induced convection and aggregate as they do so.

We repeated the experiment using triethylene glycol dimethacrylate, which forms a rigid cross-linked product. We deliberately left a bubble (about 0.5 ml) when filling

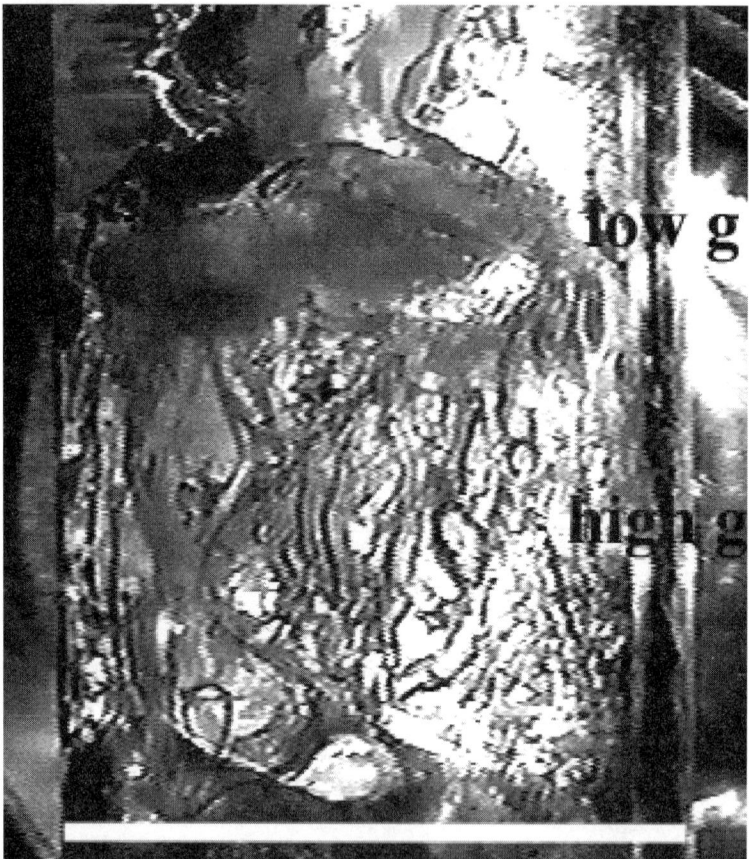

FIGURE 11. Image of TGDMA sample (from front in FIG. 10) showing bubble aggregation under low g.

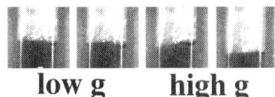
low g   high g

FIGURE 12. Images of TGDMA frontal polymerization under low and high g. Time between images is approximately 10 seconds. Lupersol 231 was the initiator.

the tube to allow copious bubble formation. As bubbles are produced at the front in $1g$, they form long strings of small bubbles (FIG. 11). It seems that existing bubbles acts as a nucleation sites for further bubbles formation. However, under low g conditions, a single large bubble forms that can impede the front propagation, as can be seen in FIGURE 12. As the buoyant force is restored, the front grows around the bubbles, and only small bubbles are seen. We propose this mechanism: Under $1g$, the buoyant force holds a bubble up against the front, and so bubbles grow in long chains. However, in low g the bubble is free to move via surface tension–induced convection. This type of motion allows bubbles to aggregate into a large bubble.

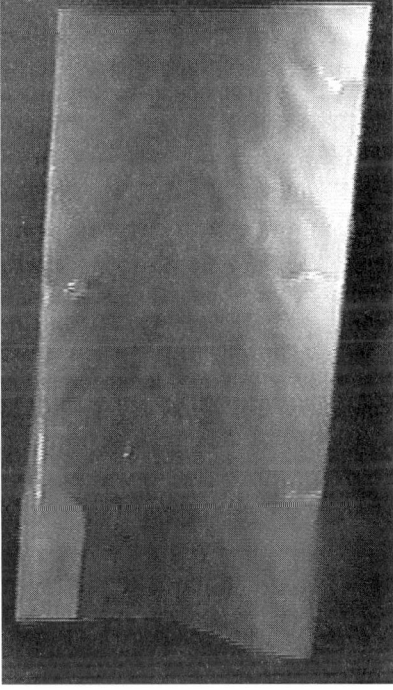

FIGURE 13. (*Left*) Periodic bubble structures in the frontal copolymerization of acryloyloxyethyltrimethylammonium chloride (AETMA) (80% in water) and acrylamide with 10% w/v ammonium persulfate initiator with respect to the total volume of solution. (*Right*) Periodic phase separation in the binary frontal polymerization of triethylene glycol dimethacrylate and EPON 862 (diglycidyl ether of bisphenol A, DGEBA) cured with a BCl_3-amine complex (Accelerator DY 9577, Ciba). Tube diameter in both images is 1.5 cm.

Bubbles also form nonrandom patterns on earth. While studying frontal copolymerization of acrylamide and acryloyloxyethyltrimethylammonium chloride (AETMA) with ultrafine silica gel added to prevent convection, we observed a regular bubble pattern formed from the water vapor. FIGURE 13a shows the splayed bubble pattern. Interestingly, a similar pattern is observed in a system that exhibits macroscopic phase separation. Figure 13b shows an image of a frontally produced interpenetrating polymer network (IPN) of triethylene glycol dimethacrylate and an epoxy. In this system two miscible monomers polymerize only with themselves but not with each other. They also cross-link and become immiscible at higher conversion.[36] In FIGURE 13B we can see a similar pattern of phase-separated regions radiating from the center.

We propose the following mechanism: The temperature in the center is higher because of heat loss from the walls. The temperature difference can be as large as 50°C. There are two essential features to the explanation: (1) There is a radial temperature gradient with higher temperature in the center. (2) The phase separation is temperature dependent.

Consider bubble formation. Vapor pressure increases and gas solubility decreases with temperature. At the front, a bubble will form first in the center because the solubility is lower there and thus the probability of nucleation is greater. As the front propagates, the region underneath the bubble is depleted of gas because of diffusion to the existing bubble so a new bubble will not form. The bubble will grow to the right or left because those regions will be higher in gas that has not nucleated because of the lower temperature. When the bubble has grown a sufficient distance from the center such that diffusion can not deplete the gas concentration, a new bubble will nucleate. Thus, the distance between bubbles horizontally and vertically will be approximately the same.

For the TGDMA/DGEBA system, something similar should happen, but it is harder to predict which phase separates at the higher temperature. And the high temperature also affects the rate of cross-linking, which also affects the rate of phase separation. Nonetheless, the nonrandom phase separation in both cases is a result of the front propagation with heat loss.

SUMMARY

Polymer systems exhibit all the dynamic instabilities we have studied: oscillations, propagating fronts, and pattern formation. Some of the instabilities, such as those in a CSTR have been studied with the goal of eliminating them from industrial processes. A new trend is developing to harness the instabilities to create new materials or to create old materials in new ways.

ACKNOWLEDGMENT

This project was supported by the National Science Foundation (Grant No. CTS-9319175) and by NASA (NAG8-1466).

REFERENCES

1. ODIAN, G. 1991. Principles of Polymerization. Wiley. New York.
2. ALLCOCK, H.R. & F.W. LAMPE. 1981. Contemporary Polymer Chemistry. Prentice-Hall. Englewood Cliffs, NJ.
3. EDEL, V. 1995. Early and late stage phase separation dynamics of poly styrene/poly(methyl methacrylate-stat-cyclohexyl methacrylate) blends macromolecules. **28:** 6219–6228.
4. LAURER, J.H., S.D. SMITH, J. SAMSETH, K. MORTENSEN & R.J. SPONTAK. 1998. Interfacial modification as a route to novels bilayered morphologies in binary block copolymer/homopolymer blends. Macromolecules **31:** 4975–4985.
5. NORRISH, R.G.W. & R.R. SMITH. 1942. Catalyzed polymerization of methyl methacrylate in the liquid phase. Nature **150:** 336–337.
6. TROMMSDORFF, E., H. KBHLE & P. LAGALLY. 1948. Zur polymerisation des methacrylsduremethylesters. Makromol. Chem. **1:** 169–198.
7. TEYMOUR, F. & W.H. RAY. 1992. The dynamic behavior of continuous polymerization reactors–v. experimental investigation of limit-cycle behavior for vinyl acetate polymerization. Chem. Eng. Sci. **47:** 4121–4132.
8. SCHORK, F.J. & W.H. RAY. 1987. The dynamics of the continuous emulsion polymerization of methylmethacrylate. J. Appl. Poly. Sci. (Chem.) **34:** 1259–1276.
9. RAWLINGS, J.B. & WH. RAY. 1987. Stability of continuous emulsion polymerization reactors: a detailed model analysis. Chem. Eng. Sci. **42:** 2767–2777.
10. VÁRADI, Z. & M.T. BECK. 1973. Inhibition of a homogeneous periodic reaction by radical scavengers. J. Chem. Soc. Chem. Comm.: 30–31.
11. POJMAN, J.A., D.C. LEARD & W. WEST. 1992. The periodic polymerization of acrylonitrile in the cerium-catalyzed Belousov-Zhabotinskii reaction. J. Am. Chem. Soc. **114:** 8298–8299.
12. VENKATARAMAN, B. & P.G. SORENSEN. 1991. ESR studies of the oscillations of the malonyl radical in the Belousov-Zhabotinsky reaction in a CSTR. J. Phys. Chem. **95:** 5707–5712.
13. WASHINGTON, R.P., G.P. MISRA, W.W. WEST & J.A. POJMAN. 1999. Polymerization Coupled to Oscillating Reactions: 1. A Mechanistic Investigation and Numerical Simulation of Acrylonitrile Polymerization in the Belousov-Zhabotinsky reaction in a batch reactor. J. Am. Chem. Soc. In press.
14. CHECHILO, N.M., R.J. KHVILIVITSKII & N.S. ENIKOLOPYAN. 1972. On the phenomenon of polymerization reaction spreading. Dokl. Akad. Nauk SSSR **204:** 1180–1181.
15. CHECHILO, N.M. & N.S. ENIKOLOPYAN. 1975. Effect of the concentration and nature of initiators on the propagation process in polymerization. Dokl. Phys. Chem. **221:** 392–394.
16. CHECHILO, N.M. & N.S. ENIKOLOPYAN. 1976. Effect of pressure and initial tempera ture of the reaction mixture during propagation of a polymerization reaction. Dokl. Phys. Chem. **230:** 840–843.
17. DAVTYAN, S.P., N.F. SURKOV, B.A. ROZENBERG & N.S. ENIKOLOPYAN. 1977. Influence of the gel effect on the kinetics of radical polymerization under the conditions of the polymerization front propagation. Dokl. Phys. Chem. **232:** 64–67.
18. DAVTYAN, S.P., E.A. GEL'MAN, A.A. KARYAN, A. TONOYAN & N.S. ENIKOLOPYAN. 1980. Applicability of the principle of quasistationary concentrations to the radical polymerization of vinyl monomers under adiabatic conditions. Dokl. Phys. Chem. **253:** 579–582.
19. ENIKOLOPYAN, N.S., M.A. KOZHUSHNER & B.B. KHANUKAEV. 1974. Molecular weight distribution during isothermal and frontal polymerization. Dokl. Phys. Chem. **217:** 676–678.

20. KHANUKAEV, B.B., M.A. KOZHUSHNER & N.S. ENIKOLOPYAN. 1974. Theory of the propagation of a polymerization front. Dokl. Phys. Chem. **214:** 84–87.
21. KHANUKAEV, B.B., M.A. KOZHUSHNER & N.S. ENIKOLOPYAN. 1974. Theory of polymerization-front propagation combust. Explos. Shock Waves **10:** 562–568.
22. SURKOV, N.F., S.P. DAVTYAN, B.A. ROZENBERG & N.S. ENIKOLOPYAN. 1976. Calculation of the steady velocity of the reaction front during hardening of epoxy oligomers by diamines. Dokl. Phys. Chem. **228:** 435–438.
23. DAVTYAN, S.P., P.V. ZHIRKOV & S.A. VOL'FSON. 1984. Problems of non-isothermal character in polymerisation processes. Russ. Chem. Rev. **53:** 150–163.
24. POJMAN, J.A. 1991. Traveling Fronts of Methacrylic Acid Polymerization J. Am. Chem. Soc. **113:** 6284–6286.
25. POJMAN, J.A., R. CRAVEN, A. KHAN & W. WEST. 1992. Convective Instabilities In Traveling Fronts of Addition Polymerization J. Phys. Chem. **96:** 7466–7472.
26. POJMAN, J.A., J. WILLIS, D. FORTENBERRY, V. ILYASHENKO & A. KHAN. 1995. Factors affecting propagating fronts of addition polymerization: velocity, front curvature, temperature profile, conversion and molecular weight distribution. J. Polym. Sci. Part A: Polym Chem. **33:** 643–652.
27. POJMAN, J.A., I.P. NAGY & C. SALTER. 1993. Traveling fronts of addition polymerization with a solid monomer. J. Am. Chem. Soc. **115:** 11044–11045.
28. TREDICI, A., R. PECCHINI & M. MORBIDELLI. 1998. Self-propagating frontal copolymerization. J. Polym. Sci. Part A: Polym. Chem. **36:** 1117–1126.
29. POJMAN, J.A., V.M. ILYASHENKC & A.M. KHAN. 1996. Free-radical frontal polymerization: self-propagating thermal reaction waves J. Chem. Soc. Faraday Trans. **92:** 2825–2837.
30. KHAN, A.M. & J.A. POJMAN. 1996. The use of frontal polymerization in polymer synthesis trends. Polym. Sci. (Cambridge, U.K.) **4:** 253–257.
31. POJMAN, J.A., W.W. WEST & J. SIMMONS. 1997. Propagating fronts of polymerization in the physical chemistry laboratory. J. Chem. Ed. **74:** 727–730.
32. KIM, C., H. TENG, C.L. TUCKER & S.R. WHITE. 1995. The continuous curing process for thermoset polymer composites. Part 1: Modeling and demonstration. J. Comp. Mater. **29:** 1222–1253.
33. WHITE, S.R. & C. KIM. 1993. A simultaneous lay-up and in situ cure process for thick composites. J. Reinf. Plast. Comp. **12:** 520–535.
34. CHEKANOV, Y., D. ARRINGTON, G. BRUST & J.A. POJMAN. 1997. Frontal curing of epoxy resin: comparison of mechanical and thermal properties to batch cured materials. J. Appl. Polym. Sci. **66:** 1209–1216.
35. POJMAN, J., N. GILL, J. WILLIS & J.B. WHITHEAD. 1996. Polymer dispersed liquid crystal (pdlc materials produced via frontal epoxy curing. Polym. Mater. Sci. Eng. Prep. **75:** 20–21.
36. POJMAN, J.A., W. ELCAN, A.M. KHAN & L. MATHIAS. 1997. Binary polymerization fronts: a new method to produce simultaneous interpenetrating polymer networks (SINs). J. Polym. Sci. Part A: Polym. Chem. **35:** 227–230.
37. NAGY, I.P., L. SIKE & J.A. POJMAN. 1995. Thermochromic composite prepared via a propagating polymerization front. J. Am. Chem. Soc. **117:** 3611–3612.
38. SZALAY, J., I.P. NAGY, I. BARKAI & M. ZSUGA. 1996. Conductive composites prepared via a propagating polymerization front die. Ang. Makr. Chem. **236:** 97–109.
39. POJMAN, J.A., Y.A. CHEKANOV, C. CASE & T. MCCARDLE. 1998. Functionally-graded polymeric materials prepared via frontal polymerization. Polym. Mater. Sci. Eng. Prep. Div. Polym. Mater. Sci. Eng. **79:** 82–83.
40. EPSTEIN, I.R. & J.A. POJMAN. 1998. An introduction to nonlinear chemical dynamics: oscillations, waves, patterns and chaos. Oxford University Press. New York.

41. ZELDOVICH, Y.B., G.I. BARENBLATT, V.B. LIBROVICH & G.M. MAKHVILADZE. 1985. The Mathematical Theory of Combustion and Explosions. Consultants Bureau. New York.
42. SHKADINSKY, K.G., B.I. KHAIKIN & A.G. MERZHANOV. 1971. Propagation of pulsating exothermic reaction front in the condensed phase. Combust. Explos. Shock Waves **1**: 15–22.
43. BAYLISS, A. & B.J. MATKOWSKY. 1990. Two routes to chaos in condensed phase combustion. SIAM J. Appl. Math. **50**: 437–459.
44. SHCHERBAK, S.B. 1983. Unstable combustion of samples of gas-free compositions in the forms of rods of square and circular cross section combust. Explos. Shock Waves **19**: 542
45. MATKOWSKY, B.J. & G.I. SIVASHINSKY. 1978. Propagation of a pulsating reaction front in solid fuel combustion. SIAM J. Appl. Math. **35**: 465–478.
46. SIVASHINSKY, G.I. 1981. On spinning propagation of combustion. SIAM J. Appl. Math. **40**: 432–438.
47. BAYLISS, A. & B.J. MATKOWSKY. 1987. Fronts, relaxation oscillations and period doubling in solid fuel combustion. J. Comput. Phys. **71**: 147–168.
48. STRUNINA, A.G., A.V. DVORYANKIN & A.G. MERZHANOV. 1983. Unstable regimes of thermite system combustion. Combust. Explos. Shock Waves **19**: 158–163.
49. BEGISHEV, V.P., V.A. VOLPERT, S.P. DAVTYAN & A.Y. MALKIN. 1985. On some features of the anionic activated ε-caprolactam polymerization process under wave propagation conditions. Dokl. Phys. Chem. **279**: 1075–1077.
50. VOLPERT, V.A., I.N. MERGABOVA, S.P. DAVTYAN & V.P. BEGISHEV. 1986. Propagation of the caprolactam polymerization wave. Combust. Explos. Shock Waves **21**: 443–447.
51. SAVOSTYANOV, V.S., D.A. KRITSKAYA, A.N. PONOMAREV & A.D. POMOGAILO. 1994. Thermally initiated frontal polymerization of transition metal nitrate acrylamide complexes. J. Poly. Sci. Part A: Poly. Chem. **32**: 1201–1212.
52. POJMAN, J.A., V.M. ILYASHENKO & A.M. KHAN. 1995. Spin mode instabilities in propagating fronts of polymerization. Physica D **84**: 260–268.
53. ILYASHENKO, V.M. & J.A. POJMAN. 1998. Single Head Spin Modes in Frontal Polymerization. Chaos **8**: 285–287.
54. Masere, J. & J.A. Pojman. 1998. Free radical-scavenging dyes as indicators of frontal polymerization dynamics. J. Chem. Soc. Faraday Trans. **94**: 919–922.
55. GRAY, K.N. 1988. Photopolymerization kinetics of multifunctional acrylates. Master's thesis, University of Southern Mississippi, Halliesburg, MS.
56. BOWDEN, G., M. GARBEY, V.M. ILYASHENKO, J.A. POJMAN, S. SOLOVYOV, A. TAIK & V. VOLPERT. 1997. The effect of convection on a propagating front with a solid product: comparison of theory and experiments. J. Phys. Chem. B, **101**: 678–686.
57. MCCAUGHEY, B., J.A. POJMAN, C. SIMMONS & V.A. VOLPERT. 1998. The effect of convection on a propagating front with a liquid product: comparison of theory and experiments. Chaos **8**: 520–529.
58. VOLPERT, V., V.A. VOLPERT, J.A. POJMAN & S.E. SOLOVYOV. 1996. Hydrodynamic stability of a polymerization front. Eur. J. Appl. Math. **7**: 303–320.
59. POJMAN, J.A., A.M. KHAN & L.J. MATHIAS. 1997. Frontal polymerization in microgravity: results from the *Conquest I* sounding rocket flight. Microg. Sci. Technol. **X**: 36–40.
60. WOZNIAK, G., J. SIEKMANN & J. SRULIJES. 1988. Thermocapillary bubble and drop dynamics under reduced gravity—survey and prospects. Z. Flugwiss. Weltraumforsch **12**: 137.

61. BALASUBRAMANIAM, R., C.E. LACY, G. WONIAK & R.S. SUBRAMANIAN. 1996. Thermocapillary migration of bubbles and drops at moderate values of the Marangoni number in reduced gravity. Phys. Fluids **8:** 872–880.

Relevance of the Protein–Lipid Interaction on the Functioning of the Bacterial Reaction Center

A. AGOSTIANO[a] AND M. DELLA MONICA

Dipartimento di Chimica, Università di Bari, Via Orabona No. 4, 70126 Bari, Italy

The secret of the molecular logic of photosynthetic processes resides in the specific organization of the components the photosynthetic apparatus into protein membrane complexes. These highly organized complexes are built up by several molecular classes, cooperating in complementary roles: architectonic support, light absorption, energy and electron transfer, and products separation.

The smallest photosynthetic unit able to perform primary charge separation is a pigment–protein complex, called a reaction center (RC), a multisubunit complex that spans the lipid membrane.[1,2] It consists of three subunits, L, M, and H, and of several cofactors: four bacteriochlorophylls (Bchl), two bacteriopheophytins (Bphe), two quinones (Q), and an iron atom (Fe). The L and M subunits form a heterodimer that binds all the cofactors.

Light absorption results in excitation of the primary donor (P) constituted of two of the bacteriochlorophylls. An electron is then transferred through a molecule of Bphe to the primary quinone (Q_A), which is ubiquinone-10 in *Rhodobacter sphaeroides*. Stabilization of the primary charge separated state $P^+Q_A^-$ is achieved by replacement of the electron on P^+ by reduced cytochrome (Cyt) c_2 and by a forward electron transfer to a second molecule of ubiquinone-10 (Q_B) which works as a two-electron acceptor and binds to a relatively polar domain of the protein. Q_B is loosely bound so that *in vivo* the free exchange of quinone between the Q_B binding site of RC and membrane pool is possible.[3] In absence of an exogenous electron donor to P^+, the charge recombination between P^+ and Q_B^- is observed.

As is mentioned above, the RC is an integral membrane protein that can be solubilized in water by substitution of membrane phospholipids with detergent molecules. The role played by the detergent goes farther than the mere solubilization of the protein. Comparative EPR and ENDOR studies[4] on the electronic structure of the radical cation of the primary electron donor in four different species of purple bacteria clearly indicate that the surfactant environment of the reaction centers influences both the electronic structure of the cofactors and the electron transfer rate. Two distinct conformations of oxidized primary donor can be obtained, depending on the detergent properties and the detergent/RC ratio. The detergent interacts with the channel through which the loosely bound UQ_{10} at the Q_B site is uptaken or released[5] in its reduced form during the photocycle of the RC.

---

[a]Address for correspondence: 39-80-5442060 (voice); 39-80-5442128 (fax); agostiano@area.ba.cnr.it (e-mail).

This experimental evidence supports the assumption that the biological membrane not only provides the structural matrix for proteins, but also affects the activity of a large number of membrane-associated enzymes and transport systems.[6,7] For this reason the use of a phopspholipid-based membrane model in which the RC is incorporated should be a major step toward a better understanding of the energy stabilization in the reaction centers.

In this paper the kinetic behavior of charge recombination of RC in proteoliposomes, in hexane phospholipids reverse micelles, and in organogels has been analyzed. In particular the influence of the nature of the phospholipids and of their form

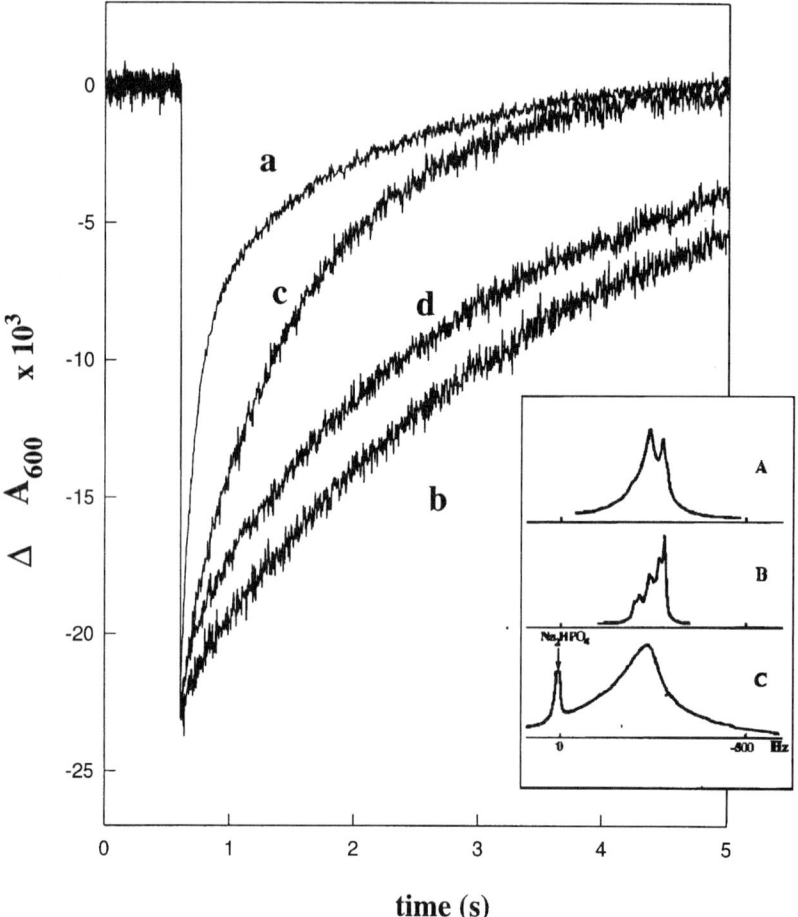

FIGURE 1. Charge recombination kinetics of RC in the different hosting systems. **(a)** RC in TLE buffer; **(b)** RC in liposomes ([PS]/[PC] = 2.2); **(c)** RC in n-hexane phospholipid reverse micelles **(d)** RC in organogel ([PS]/[PC] = 2.2, W0 = 3.5). All the traces were normalizrd to trace **(a)**. *Inset*: Phospholipid 31P NMR spectra of liposomes **(A)**, reverse micelles **(B)**, organogel **(C)**.

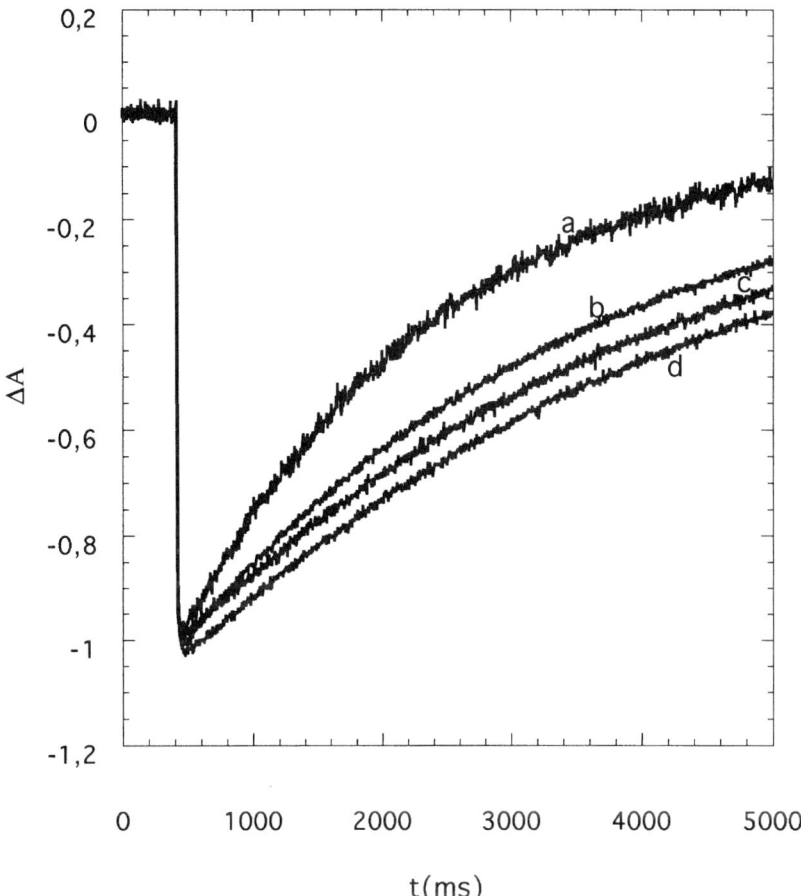

FIGURE 2. Charge recombination kinetics in RC proteoliposomes made by lipids with different net charge. (a) PC; (b) PS; (c) PG; (d) PI.

of aggregation on the reduction kinetics of the primary donor has been reported in order to obtain information on the role played by the protein–lipid interaction on the quinone exchange modality at the Q_B site. The RC photoactivity has been investigated by flash spectroscopy, in the absence of an exogenous electron donor for P^+, observing the charge recombination from the state $P^+Q_B^-$ at 600 nm (FIG. 1).

In aqueous detergent solution, the charge recombination from the $P^+Q_B^-$ state is characterized by a half-time of approximately 700 msec. The biphasicity of the charge recombination kinetics can be attributed to a RC fraction lacking the secondary quinone Q_B. In lipid systems the situation appears to be more complex.

A dramatic difference in the charge recombination kinetics from $P^+Q_B^-$ is observed, together with multiphasic P^+ decay. The diffusion of quinone through heterogeneous microphases could play a role in the multiphasic character of the kinetics by means of a rate limiting effect on the binding equilibrium at Q_B site.[8] Moreover an increase in the viscosity of the environment in which the protein is reconstituted could influence the diffusion process and consequently the charge recombination kinetics. The reduced rates obtained in proteoliposomes and organogel compared to that observed in reverse micelles, indicate, in fact, that a decrease in lipid mobility increases the charge-separated state stabilization according to similar results obtained in proteoliposomes at different temperatures.[9]

The lipid mobility in the membrane model system used was studied by NMR spectroscopy ($[^{31}P]NMR$). The inset of FIGURE 1 shows the $[^{31}P]NMR$ spectrum of the liposomial system whose main components are PC, PE, and PS in the concentration ratio 1:1:2. The peaks relative to the main phospholipid components, PC and PS, are enlarged, indicating reduced mobility of the lipid polar head in liposomal bilayer. The lipid mobility increases for the reverse micellar system, as suggested from the narrow and clearly resolved ^{31}P resonance bands of the phospholipids in this system. Finally, a unique broad phosholipid NMR signal is observed in organogel, suggesting that not only should the PC and PS molecules be confined in the same chemical environment, presumably the wall of the cylindrical reverse micelles, but also there should be a high reduction of the lipid mobility in such systems.[10]

The presence of a negative charge on the lipids greatly influences the stability of the charge-separated states, as evidenced by the data reported in FIGURE 2. In liposomes prepared with negative lipids, the process is always slower than that of a RC reconstituted in liposomes formed by zwitterionic lipids.

In aqueous solution the negatively charged phospholipids form vesicles characterized by a large configuration of their polar head group because of the electrostatic repulsion. In addition, the negatively charged polar head can interact with the positively charged zone of the RC located between the hydrophilic moieties of the protein, stabilizing the charge-separated state. The difference in microstructure of the host systems, electrostatic interaction between protein and lipid, and the geometric parameter of the phospholipid molecules seem to dramatically influence the kinetics of the process. The presence of multiple conformational states of the RC could account equally for the results observed in our lipid systems, in which local heterogeneity could play a role in modulating the interaction of Q_B with its surroundings as well as the energy gap between $P^+Q_A^-$ and $P^+Q_B^-$.

In conclusion, the charge recombination turned out resulted to be strongly dependent on the characteristics of the lipid system employed. The charge of the lipids and the factors affecting their packing in the double layer seem to play a relevant role in the stabilization of the charge-separated state. The presence of a net negative charge on the lipids in liposomes results in a stabilization of the reduced ubiquinone and can also be related to different conformations of the protein in the dark and in the light and to the translocation mechanisms of the Q_B from its binding site.

REFERENCES

1. FEHER, G. & M.Y. OKAMURA. 1978. Photosynthetic Bacteria. Plenum Press. New York.
2. BACCARINI-MELANDRI, A., G.A. HAUSKA, B.A. MELANDRI & A.R. CROFTS. 1975. Asymmetry of an energy transducing membrane the location of cytochrome c2 in

Rhodopseudomonas spheroides and *Rhodopseudomonas capsulata*. Biochim. Biophys. Acta **387**: 212–217.
3. OKAMURA, M.Y., G. FEHER & N. NELSON. 1982. Photosynthesis, Energy Conversion by Plants and Bacteria. Academic Press. New York.
4. RAUTTER, J., F. LENDZIAN, S. WANG, J.P. ALLEN & W. LUBITZ. 1994. Comparative study of reaction centers from photosynthetic purple bactiria:electron paramagnetic resonance and electron nuclear double resonance spectroscopy. Biochemistry **33**: 12077–12084.
5. BEEKMAN, L.M., R.W. VISSCHERS, R. MONSHOUWER, M. HEER-DAWSON, T.A. MATTIOLI, P. MCGLYNN, C.N. HUNTER, B. ROBERT, I.H. VAN STOKKUM, R. VAN GRONDELLE, et al. 1995. Time-resolved and steady-state spectroscopyc analysis of membrane-bound reaction centres from rhodobacter sphaeroide: comparisons with detergent-solubilized complexes. Biochemistry **34**: 14712–14721.
6. JAIN, M. 1988. Introduction to Biological Membranes. Second edit. Wiley. New York.
7. GENNIS, R.G. 1984. Biomembranes: Molecular Structure and Function. Springer-Verlag, New York.
8. AGOSTIANO, A., L. CATUCCI, M. DELLA MONICA, A. MALLARDI, G. PALAZZO & G. VENTUROLI. 1995. Charge recombination kinetics of photosynthetic reaction centers in water-in-oil phospholipids organogels. Bioelectrochem. Bioenerg. **38**: 25–33.
9. BACIOU, L., T. GULIK-KRYWICKI & P. SEBBAN. 1991. Biochemistry 1991: Involvement of the protein–protein interactions in the thermodynamics of the electron-transfer process in the reaction centers from *Rhodopseudomonas viridis*. Biochemistry **30**: 1298–1302.
10. AGOSTIANO, A., L. CATUCCI, G. COLAFEMMINA, M. DELLA MONICA, G. PALAZZO, M. GIUSTINI & A. MALLARDI. 1995. Charge recombination kinetics of photosynthetic reaction centers in different membrane models. Gazz. Chim. Ital. **125**: 615–620.

Time-Dependent Properties of Isotactic Polypropylene Fibers

G. BARRA,[a] C. D'ANIELLO,[a] R. RUSSO,[b] AND V. VITTORIA[a,c]

[a]*Dipartimento di Ingegneria Chimica e Alimentare, Università di Salerno, 84084 Fisciano, Salerno, Italy*
[b]*Istituto di Ricerca e Tecnologia dei Polimeri, Via Toiano, 6 Arco Felice, Naples, Italy*

INTRODUCTION

Many physical properties of polymers exhibit appreciable changes with the time. Some of these changes are due to the phenomenon reported as "physical aging." Physical aging of polymers, depending only on time, occurs in amorphous polymers stored at a temperature below the glass transition temperature (Tg): when the system is rapidly cooled through Tg, it contains an excess of free volume and is not in thermodynamic equilibrium. The achievement of equilibrium in the glassy state is hindered by kinetic phenomena. The free volume excess decreases with time, with consequent decrease of mobility, which produces a slowing down of the overall process that is therefore defined as "self retarding." The slow approaching of the thermodynamic equilibrium determines the change of the physical properties of the glassy polymers.[1] In the case of semicrystalline polymers, either unoriented or cold drawn, changes of properties can be observed even in systems stored at temperatures above the Tg, due to a multiplicity of mechanisms. Stiffening occurs, for example, in fibers of low-density polyethylene and polypropylene, during the time the sample is left at room temperature after the mechanical treatment.[2] The changes occurring in a material with aging can affect its application, performance, and lifetime.

In this paper we analyze samples of isotactic polypropylene (iPP) drawn at 110°C and stored at room temperature up to two months: in this period of time, we followed many properties, to correlate change of structural organization and physical properties of the drawn samples. Fibers of isotactic polypropylene are widely used for many applications, and a study of the aging phenomenon, occurring after drawing, can help improve their applications and lifetime.

EXPERIMENTAL

The isotactic polypropylene was a RAPRA (GB) product, $M_W = 307,000$ and $Mn = 15,600$. Quenched films were drawn at 110°C using a dynamometric apparatus Instron 4301. The stretching rate was 1 cm/min with an initial gauge length of 1 cm. The sample was drawn up to draw ratio $l/l_0 = 10$, where l_0 is the initial length of the sample (Sample D10). The elastic modulus E of the drawn sample D10 was measured at room temperature, as soon as possible and after different time intervals. It

---

[c]To whom correspondence may be addressed: vittoria@post.dica.unisa.it (e-mail).

was obtained from the stress–strain curve for a deformation less than 1% and averaged over many measurements. The density d, g/cm^3, was measured after different intervals of time by floating the sample in a mixture of liquids with different densities and weighing with a picnometer. Sorption and diffusion of dichloromethane were measured at 25°C, immediately after drawing and again after two months, with a microgravimetric method. Equilibrium concentration of vapor was measured as a function of the activity of the vapor $a = p/p_0$, where p is the pressure to which the sample was exposed and p_0 the saturation pressure at 25°C. Diffusion values were determined by the initial slope of the sorption as a function of the square root of time. The transport properties are related to the fraction and to the thermodynamic state of the amorphous component.[3]

RESULTS

In FIGURE 1 the density (g/cm^3) and the elastic modulus (GPa) of sample D10 are reported as a function of the logarithm of the aging time. We observe an increase of density linear with $\log(t)$, whereas a steeper increase of the elastic modulus is observed for short times. The increase of density is due either to a densification of the amorphous component, or to a secondary crystallization phenomenon. This second effect could also be responsible for the increase of the elastic modulus.

An important contribution to the experimental description of the aging phenomenon can be given by the transport properties that, as is generally known, are very sensitive to any microstructural change, particularly in the amorphous component.[3] The zero concentration diffusion coefficient, D_0, is affected by the fractional free volume and is related to the morphology of the amorphous component.

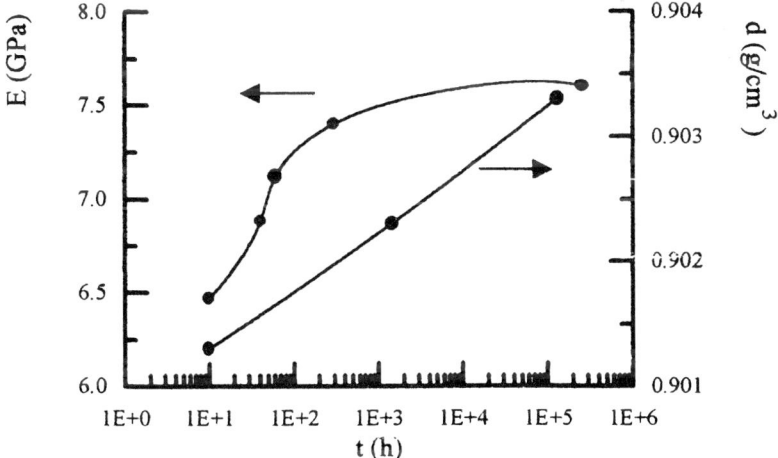

FIGURE 1. Density (g/cm^3) and elastic modulus (GPa) of sample D10 as a function of the logarithm of aging time.

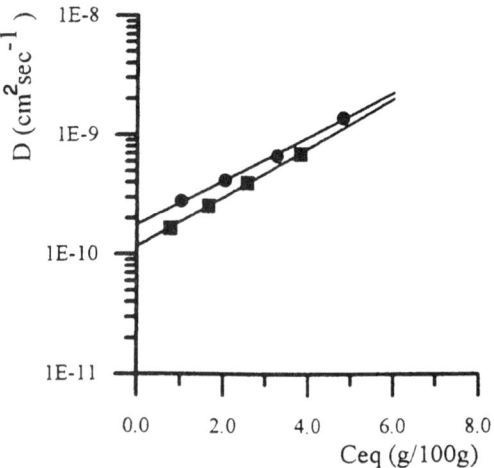

FIGURE 2. The diffusion coefficient, D, as a function of the equilibrium concentration of vapor, C_{eq}, for sample D10 analyzed as soon as possible (●) and again after two months (■).

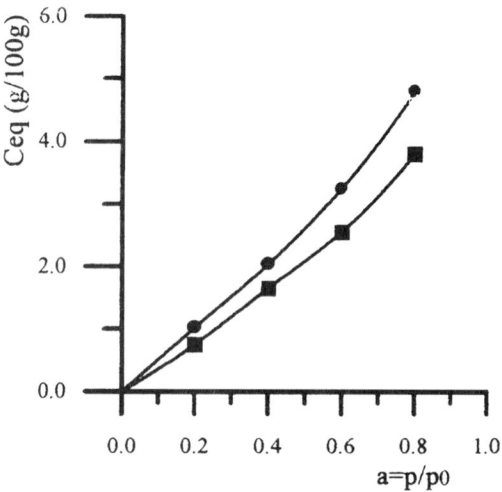

FIGURE 3. Equilibrium concentration of vapor as a function of activity, $a = p/p_0$, for the fresh sample (●) and the sample aged for two months (■).

In FIGURE 2 we report the diffusion coefficient, D, as a function of the equilibrium concentration of vapor, C_{eq}, for the sample D10 analyzed as soon as possible and again after two months. It is evident that aging induces a double effect: D_0 (the value of D at $C_{eq} = 0$) decreases, whereas the slope increases upon aging. The two effects are indicative of a reduced fractional free volume in the amorphous component and, therefore, of a reduced mobility.

In FIGURE 3 the equilibrium concentration of vapor is reported as a function of activity, $a = p/p_0$, for the fresh sample and for the sample aged for two months. From the sorption at low activity, it is possible to derive the amorphous fraction by comparing the equilibrium concentration of the sample with that of a completely amorphous sample, since the reduction of sorption is due to the fraction of the crystalline and/or impermeable phase present in the sample. We calculated an amorphous fraction of 0.32 for the fresh sample and 0.22 for the aged sample. From the density we can obtain a crystallinity of 0.57 for the fresh sample and 0.62 for the aged sample. The sorption reveals, therefore, that the fraction of the impermeable phase (not amorphous), derived from sorption, is consistently higher than the fraction of the crystalline phase, as derived from density. This impermeable phase, already present in the fresh-drawn sample, increases more during the aging.

These results seem to confirm that aging is a complex phenomenon and is connected with the concept of metastability consequent to thermal and/or mechanical treatments. Here, thermal and mechanical treatments are superimposed, and the aging mechanism is more complex; the aging must be connected to the molecular and structural reorganization induced by drawing. Drawing generates a new morphological unit, the microfibril, which is absent in the initial film. In the microfibril, crystalline blocks and amorphous layers are arranged with a regular alternating distribution along the draw axis; different blocks are axially connected by tie molecules that cross the amorphous layers. The tie molecules play a crucial role in determining the properties of the fiber. During the drawing the tie molecules tend to assume the extended conformation and become taut. The taut tie molecules are responsible for the high axial modulus and determine a reduction of the molecular mobility by compression of the amorphous layers.[4]

Our results support the hypothesis that the increase of modulus is due to a secondary crystallization and to the increase of the taut tie molecules. This is suggested by the difference between the crystallinity derived from the density and the fraction of the impermeable phase derived by sorption. In particular, the new molecular topology generated by the drawing process could permit the inclusion of oriented chains in crystalline blocks by an epitaxial mechanism, in addition to a more and more reduced mobility of the taut tie molecules, due to the decrease of the fractional free volume with time. This reduced mobility could determine the impermeability to the vapors of this fraction of molecules, making it similar to the crystalline phase.

REFERENCES

1. STRUIK, L.C.E. 1978. Physical Aging of Amorphous Polymers and Other Materials. Elsevier. New York and Amsterdam.
2. ARRIDGE, R.G., P.J. BARHAM & A. KELLER. 1977. J. Polym. Sci. Polym. Physic. Ed. **15:** 388.
3. A. PETERLIN. 1975. J. Macromol. Sci. Phys. Ed. **11:** 57.
4. A. PETERLIN. 1975. Polym. Eng. Sci. **17:** 183.

Suspended Life in Biological Systems

Fragility and Complexity

C. BRANCA, A. FARAONE, S. MAGAZÙ,[a] G. MAISANO, P. MIGLIARDO, AND V. VILLARI

Dipartimento di Fisica and INFM, Università di Messina, P.O. Box 55, 98166 Messina, Italy

Various cellular systems—plant seeds, spores, certain crustaceans, some species of soil-dwelling organisms, and so forth—under stress conditions (drying and freezing), pass into a vitreous state of suspended animation, succeeding in surviving for long periods (up to 120 years) and recovering full viability upon rehydration.[1] It's now well established that this phenomenon involves the synthesis of trehalose (α-D-glucopyranosyl α-D-glucopyranoside), a nonreducing disaccharide of glucose. Although the role of trehalose as a bioprotector has been supported by plenty of experimental studies,[2,3] science has not so far been able to unravel the connection between the overwhelming complexity of the above-mentioned biosystems and the elemental molecular processes involved by trehalose. The behavioral properties of the system depend on the $(3N + 1)$-dimensional potential energy hypersurface in the configuration space. The complexity of the energy landscape, explored by the system, can be correlated with the density of the minima of the hypersurface (degeneracy) and with the distribution of the barrier heights between them. These features determine the structural sensitivity of a system to temperature changes, namely its *fragility* (m). In this contribution we propose to connect this qualitative scenario to a quantitative approach by connecting complexity with fragility. The latter, in fact, is operatively defined as

$$m = \Delta C_p(T_g)/\Delta\mu = \left.\frac{d\log\eta}{d(T_g/T)}\right|_{T=T_g}$$

where $\Delta C_p(T_g)$ is the heat capacity at T_g, $\Delta\mu$ the energy barrier height and η the shear viscosity.[4] In this frame Angell's classification[5] of glass-forming systems as "strong" and "fragile" takes particular relevance. In such a classification, based on the choice of an invariant viscosity at the scaling temperature T_g ($\eta(T_g) = 10^{13}$ poise), the departure from the Arrhenius law is taken as a signature of the degree of fragility of the system. FIGURE 1 shows that an Arrhenius behavior of viscosity in the T_g-scaled plot and a small heat capacity variation at T_g characterize the strongest systems, whereas a large departure from Arrhenius law and a large ΔC_p characterize the most fragile ones. Between these two limiting cases, intermediate behaviors are

---

[a]To whom correspondence may be addressed: Salvatore Magazù, Dipartimento di Fisica and INFM dell'Università di Messina, P.O. Box 55, C.da Papardo s.ta Sperone 31, 98166 Messina, Italy. +39-090-391478 (voice); +39-090-395004 (fax); magazu@dsme01.messina.infm.it (e-mail).

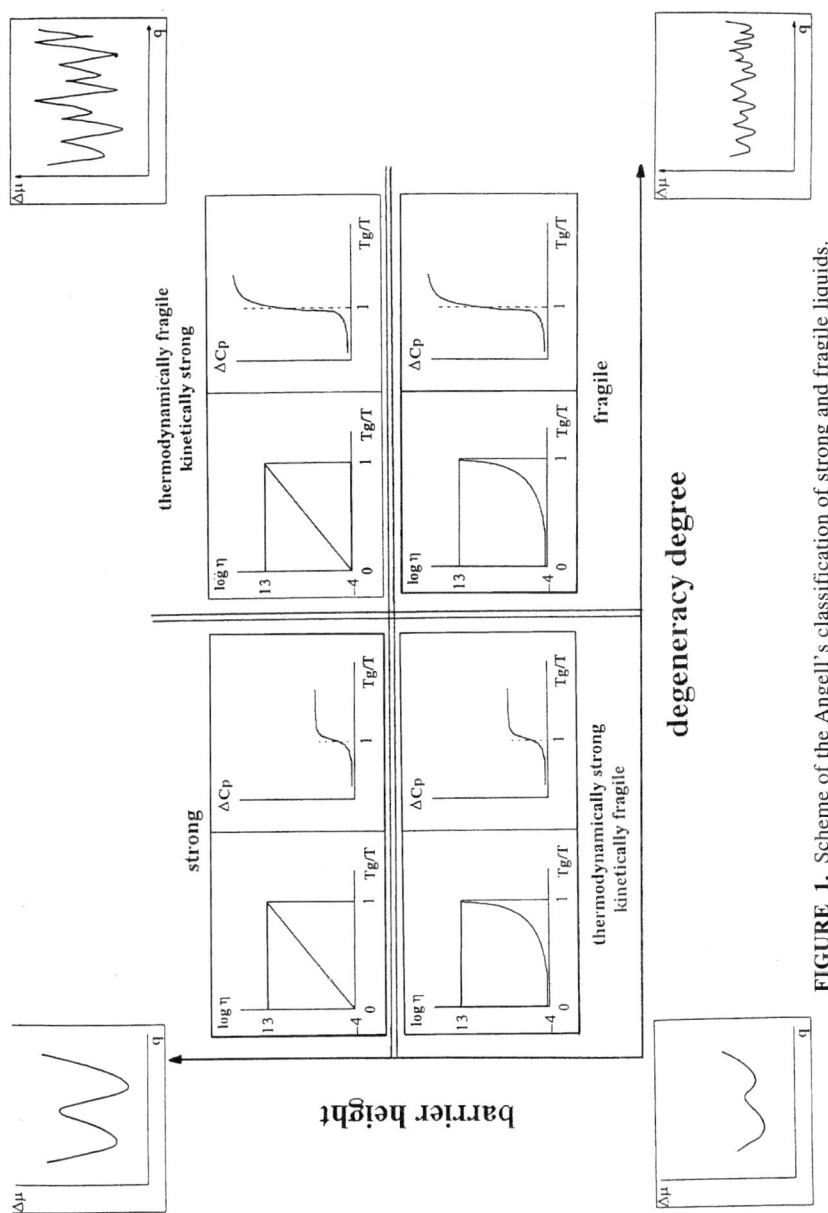

FIGURE 1. Scheme of the Angell's classification of strong and fragile liquids.

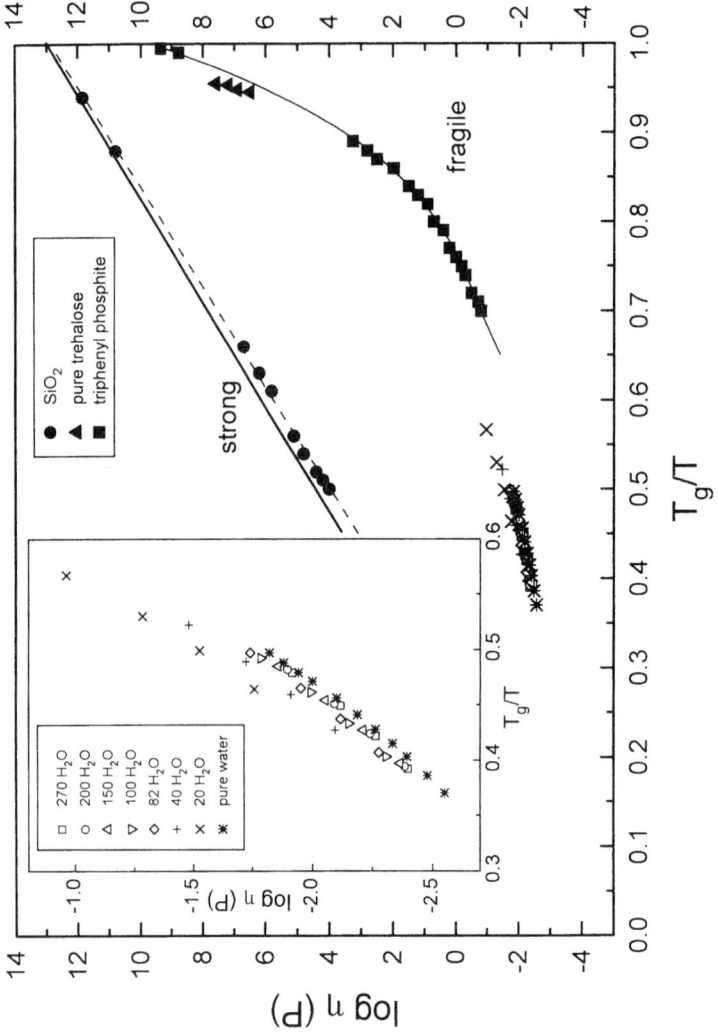

FIGURE 2. Strong-fragile pattern of viscosity-temperature relations for T_g-based normalization of trehalose aqueous solutions.

interpreted in terms of different kinetic and thermodynamic contributions: thermodynamically strong (fragile) and kinetically fragile (strong) systems are characterized by a low (high) configurational degeneracy and small (large) barrier heights. Viscosity measurements provide a simple criterion capable of measuring the fragility of a system. In the trehalose–water system viscosity shows strongly non-Arrhenius behavior (see FIG. 2), with a fragility that decreases with the water content from $m = 118.4$ at a weight fraction of 0.066 to $m = 113.7$ at a weight fraction of 0.322. In addition, the temperature dependence of the self-diffusion coefficient, as obtained by NMR measurements,[6] indicates that the trehalose molecule, in the diluted concentration range, is characterized by a structural flexibility that gives rise to internal motions of the two glucose rings around the glycosidic oxygen.

Taken together these experimental findings suggest a great ability of the trehalose molecule to conform to the irregular surfaces of biostructures at high dilution and to pack them, preserving them from damage, at low water contents (cryptobiotic effect).

In conclusion, transition from a state characterized by exceptional complexity, which can be associated with the state of living organisms, toward a state characterized by less fragility at low dilution, correlated with the state of suspended life, could constitute the exceptional feature of the trehalose-water system.

REFERENCES

1. VEGIS, A. 1964. Annu. Rev. Plant. Physiol. **15:** 185
2. GREEN, J. L. & C. A. ANGELL. 1989. J. Phys. Chem. **93:** 2880
3. FOX, K. C. 1995. Science. **267:** 1922
4. ANGELL, C. A. 1997. Prog. Theoret. Phys. Suppl. No. **126:** 1
5. ANGELL, C. A., P. H. POOLE & J. SHAO. 1994. Il Nuovo Cimento D. **16:** 993
6. MAGAZÙ, S., G. MAISANO, P. MIGLIARDO, E. TETTAMANTI & V. VILLARI. 1999. Mol. Phys. **96**(3): 381–387.

Coupling of Erratic Belousov-Zhabotinsky Oscillators Observed by Spectrophotometric Measure

M. BRANCA,[a,b] A. BRUNETTI,[c] C. CARAVATI,[a] AND M. RUSTICI[a]

[a]Università di Sassari, Dipartimento di Chimica, Via Vienna 2, 07100 Sassari, Italy
[b]Università di Sassari, Istituto di Fisica, Via Vienna 2, 07100 Sassari, Italy

INTRODUCTION

One reason for the current interest in oscillating reactions is the numerous periodic phenomena that occur in spatial and temporal control of processes, including phenomena in living organisms such as reactions catalyzed by enzymes and reactions across membranes. The study of the interaction of two or more oscillating systems is directly connected with biological problems and with spatial pattern formation. For this reason, the coupling of chemical oscillators has been the subject of a number of studies. Experiments have been carried out using batch reactors, as well as CSTR reactors.[1–8] The simplest system in which the coupling can be studied is a single solution. The coupling that occurs inside the same solution is very interesting; as in a closed reactor, the different parts of the solution are obviously physically coupled.

We used the Belousov-Zhabotinsky (BZ) reaction, a metal ion–catalyzed oxidation and bromination of an organic substrate, this being the most frequently used reaction for experiments in chemical oscillations.[9]

In unstirred batch reactors, diffusion and convection may be as important for a system's dynamics as for its local kinetics. In particular, the coupling between a temporal oscillation and a spatial dishomogeneity may produce chaotic behavior.[10,11] Epstein and Showalter recently dedicated a review to nonlinear chemical dynamics, which contains an exhaustive discussion on coupled systems.[12] The phenomenon most frequently observed in coupled systems is the adaptation of the oscillators of one to the other so that the frequencies are rationally related.[13]

In this paper the coupling of the chaotic or regular oscillating phenomena that occurs in different parts of the same well-premixed oscillating solution and the effect of splitting the solution into two different portions are investigated. In order to avoid ions transport resistance phenomena and other complications that are present in electrical coupling experiments, we used the spectrophotometric method[14] instead of the electrochemical method.

bAddress for correspondence: 39-079-229555 (voice); branca@ssmain.uniss.it (e-mail).

TABLE 1. Concentrations of starting solutions

	1	2	3
Erratic oscillations	Malonic acid, 0.30 M H_2SO_4, 1 M	$KBrO_3$, 9.27 10^{-2} M H_2SO_4, 1 M	$Ce(SO_4)_2$, 4.51 10^{-3} H_2SO_4, 1 M
Regular oscillations	Malonic acid, 0.15 M H_2SO_4, 1 M	$KBrO_3$, 4.64 10^{-2} M H_2SO_4, 1 M	$Ce(SO_4)_2$, 2.14 10^{-3} H_2SO_4, 1 M

MATERIALS AND METHODS

The experiments were performed isothermally at room temperature (~20°C) in a spectrophotometric cuvette used as a batch reactor. All chemicals were of analytical quality and were used without further purification. Stock solutions of reagents were prepared in appropriate sulfuric acid solutions (TABLE 1).

All the experiments were performed in the following way: the oscillator began by mixing equal quantities of reagents in an Erlenmeyer flask under strong stirring in the following order: 1, 2, 3 (TABLE 1). After 10 minutes, this premixed solution was sampled and transferred in the cuvettes for spectrophotometric measurement.

Oscillations were recorded using a Varian spectrophotometer, series 634, operating either in a single- or a double-beam mode. The dynamics were monitored by

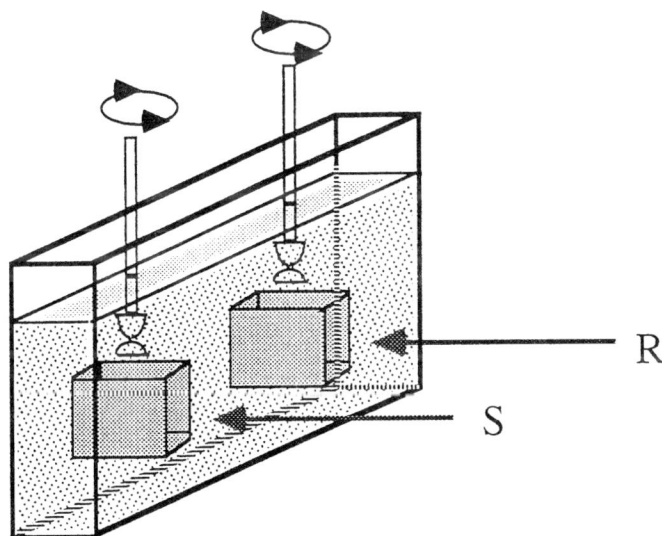

FIGURE 1. Sketch of the handmade cuvette not drawn to scale. This cuvette consists of a parallelepiped with the edges of 1 × 4 × 20 cm so that the same solution acts both as sample and as reference. The cross-area of each beam of the photometer is 30 mm². The volume, indicated in gray, spanned by the each beam is 300 mm³ and was located 2 cm away from the interface liquid air, 1 cm away from the bottom of the cuvette. The distance between S and R beam is 11.0 cm. The volume of solution is 80 ml. Two driven glass propellers can stir the reaction solutions in the cuvettes.

means of the solutions' absorption at 320 nm using two commercial cuvettes of 1-cm pathlength or one handmade, large glass cuvette of 1 cm pathlength that consisted of a parallelepiped with the edges of $1 \times 4 \times 20$ cm so that the same solution acts both as sample and as reference (FIG. 1). The cross-sectional area of the spectrophotometer beam was 30 mm^2. The volume spanned by the beam was 300 mm^3 and was located 2 cm away from the liquid–air interface and 1 cm away from the bottom of the cuvette. The distance between S and R beams was 11.0 cm. The used volumes were 80 ml in the case of the large cuvette and 3 ml for the small one.

Inside the cuvette two driven glass propellers stirred the reaction solution. We designed the stirring device so as to ensure the same stirring speed for the two propellers. All the experiments show high reproducibility, and neither depend on the starting stock solutions nor on the kind of experiment performed. Nevertheless, it is crucial to start from the indicated concentrations.

RESULTS AND DISCUSSION

FIGURE 2 shows oscillatory behavior at the specific concentrations (TABLE 1) in the presence or absence of stirring. The photometric signal was detected after the reactive solution was premixed in an Erlenmeyer flask for 10 minutes and using a sulfuric acid solution as reference. After a transient phase of about 10 minutes, the signal associated with the unstirred but pre-mixed solution did not show any regular periodic pattern (FIG. 2); nevertheless, regular periodic oscillations appeared (FIG.

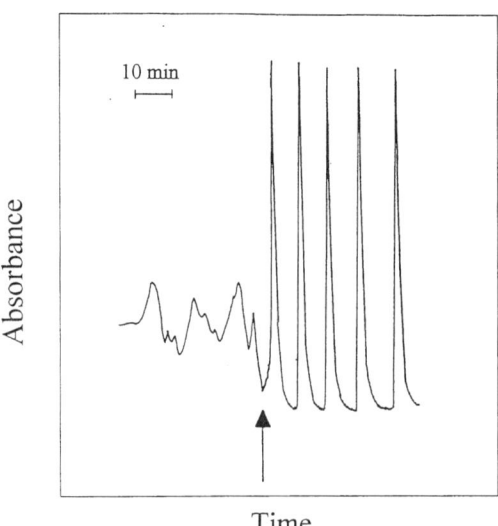

FIGURE 2. Spectrophotometric recordings at room temperature of cerium(IV) concentration in a solution initially containing 0.1 M malonic acid, 0.031 M KBrO$_3$, 0.0015 M Ce(SO$_4$)$_2$, and 1 M sulfuric acid. Wavelength, 320 nm. Reference: 1 M sulfuric acid. The typical oscillatory behavior is aperiodic in the absence of stirring and periodic in the presence of stirring. The *arrow* indicates the start of stirring.

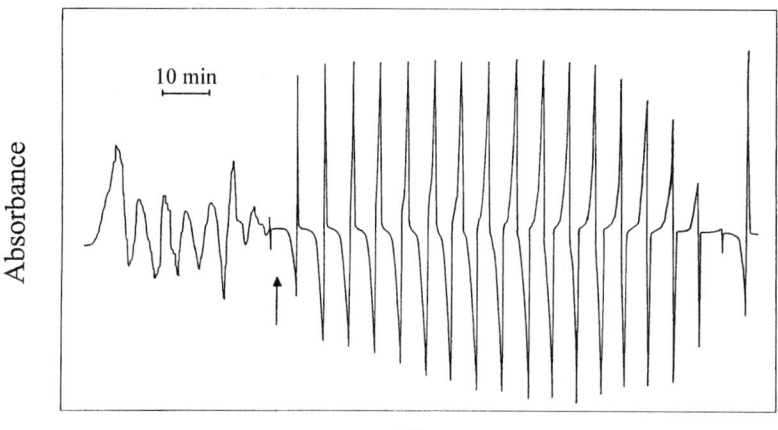

FIGURE 3. Spectrophotometric recordings at room temperature of cerium(IV) concentration in a solution initially containing 0.1 M malonic acid, 0.031 M $KBrO_3$, 0.0015 M $Ce(SO_4)_2$, and 1 M sulfuric acid. Wavelength, 320 nm. Reference and sample cuvettes are filled with the same solution. The oscillatory behavior is aperiodic in the absence of stirring and periodic in the presence of stirring. The *arrow* indicates the start of stirring.

2) when stirring occurred. The disappearance of chaos during stirring can be explained because in this case the dynamic is regulated only by kinetics. Recently, we showed that the observed aperiodic signal is a manifestation of chaos and is due to an interplay between chemical kinetics and convection-diffusion.[14]

When the same reacting solution is divided into two portions that are used to fill both the reference and the sample cuvettes, the spectrum difference in FIGURE 3 is obtained in the absence of stirring. The solution's separation entails a perturbation on the two "new" solutions, which start from different initial conditions. Therefore, the aperiodic pattern mirrors the sensibility on initial conditions (i.e. chaotic behavior) in the absence of stirring.

When the stirring starts simultaneously in the R and S cuvettes, the erratic behavior disappears, and the periodic signal shown in FIGURE 3 appears. This signal is due to the difference of two uncoupled regular oscillators that oscillate with a small frequency difference ($f_1/f_2 = 1.05$, where the f_1 and f_2 are the frequencies of the two oscillators). This result shows that the division of the reactive parent solution into two new solutions introduces some perturbations, and the new systems will oscillate independently. As a result of the fact that there is no coupling between the two new "un-identical" systems, they will start to oscillate in a regular way with a very similar but nonidentical frequency.

In order, to observe whether the detected chaotic behavior is only temporal behavior, we constructed a big cuvette so that the reference beam and the sample beam analyze different portions of the same solution at the same time. When we measured the difference spectrum in the absence of stirring of the reactive solution contained in the big cuvette, a new erratic spectrum was obtained (see FIG. 4). The signal observed is related to the difference of two erratic oscillations. Each part of the solution

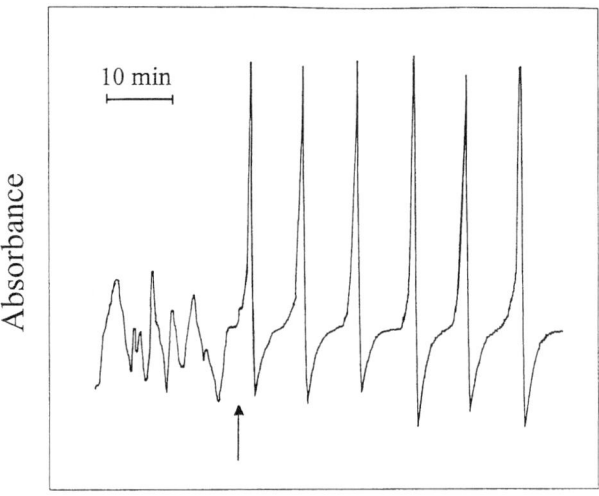

FIGURE 4. Spectrophotometric recordings at room temperature of cerium(IV) concentration in a solution initially containing 0.1 M malonic acid, 0.031 M $KBrO_3$, 0.0015 M $Ce(SO_4)_2$, and 1 M sulfuric acid. Using a single big "cuvette" (FIG. 1), reference and sample beams analyze different portions of the same solution at the same time. Wavelength, 320 nm. The oscillatory behavior is aperiodic in the absence of stirring and periodic in the presence of stirring. The *arrow* indicates the start of stirring.

noticed by the instrument oscillates in time and space in a chaotic and an unconstrained way. As soon as the stirring occurred, we expected a signal constant at zero absorbance because this process would destroy the etherogeneities zones and generate a single homogeneous solution. Instead, FIGURE 4 shows the appearance of a periodic signal, and this means that the solution even under stirring is not homogeneous. The shape of this signal is due to the difference of two regular periodic oscillations with the same frequency but different phase. To verify if the behavior exhibited in FIGURES 2 and 4 is characteristic and exclusive of the solution that originates the chaotic behavior, we repeated the experiments using reagent's concentrations that give place to regular oscillation sustained in the time (TABLE 1). FIGURE 5a shows the regular oscillation obtained in the absence of stirring using sulfuric acid as reference solution. FIGURE 5b shows the signal obtained when the two different cuvettes S and R are filled with the same well-premixed solution. In this case, like the chaotic solution under stirring, the signal obtained is due to the difference of two uncoupled periodic oscillators with a slight difference in the phase and frequency.

When the S and R solutions are coupled, only the difference in the frequency disappears. In fact, when different portions of the same solutions are analyzed at the same time, using a big cuvette, the signal shown in FIGURE 5c appears. The signal could be reconstructed by subtraction of two oscillations with the same frequency but different phase.

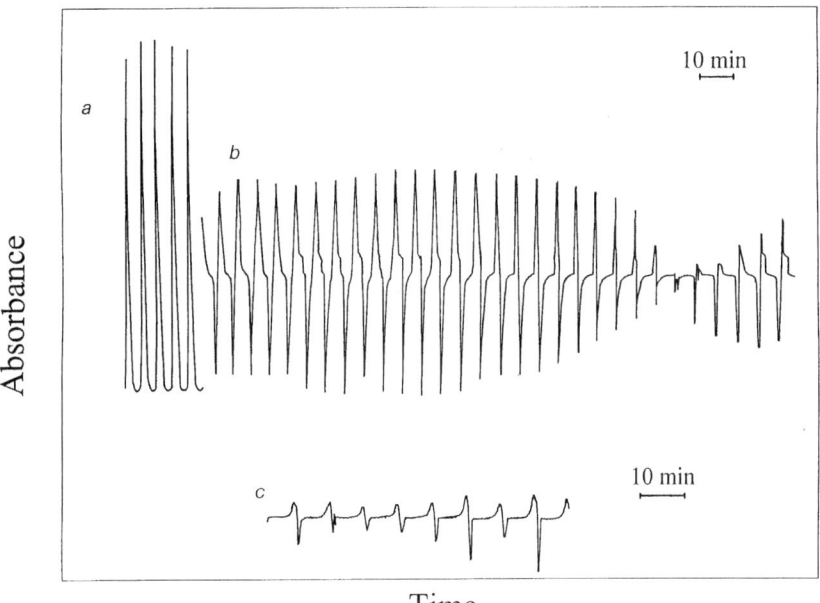

FIGURE 5. Spectrophotometric recordings at room temperature of cerium(IV) concentration in a solution initially containing 0.05 M malonic acid, 0.015 M $KBrO_3$, 0.00075 M $Ce(SO_4)_2$, and 1 M sulfuric acid. Wavelength, 320 nm. (**a**) Reference: 1 M sulfuric acid. (**b**) Reference and Sample cuvettes are filled with the same solution. (**c**) Reference and sample beams analyze different portions of the same solution at the same time using a single big "cuvette" (FIG. 1).

CONCLUSIONS

We have analyzed the BZ solution in terms of two oscillating subsystems having a volume of 300 mm³ and a distance between them of 11 cm. When the well-premixed solution is divided into two portions, the starting conditions are nearly identical. Therefore, the perturbations induced by the division lead to a very small difference in the two new solutions so that a different time for each solution, which are in a pseudo steady-state, will be necessary to reach the periodic "attractor." In the case of a regular oscillating solution, this little temporal difference would be responsible for the observed difference not only in the phase but also in the frequency of the oscillations.

It is well known that in the case of a periodic attractor, like a limit cycle, regulating the dynamics, every perturbation induces transience and the system after a variable time, will oscillate at the same initial frequency. Closed systems far from equilibrium do not have a real attractor because the reagents' concentrations slowly change over time. Therefore, the trajectory should be similar to a spiral fragment.

When the behavior of the solution is erratic, the two separated subsystems oscillate independently, the perturbation induced by the division produces in the two new solutions a divergence in the dynamic, and chaos in the space and time is observed. When different portions of the same BZ undivided solution are analyzed, that is, when different parts of the same solution are looked at at the same time, the subsystems are chemically and physically connected. In a solution oscillating in a regular way, the oscillations are entrained in the frequency, regular chemical waves will appear, and a phase difference in the oscillation is observed.

When the behavior of the solution is erratic, each part of the solution noticed by the instrument, oscillate in time and space in a chaotic way so that chaos in space and time again is observed.

ACKNOWLEDGMENTS

This project was carried out with the support of MURST 97 CFSIB (Italian Ministry of University and Scientific Research).

REFERENCES

1. MAREK, M. & I. STUCHL. 1975. Synchronization in two interacting oscillatory systems. Biophys. Chem. **3:** 241–248.
2. FUJII, H. & Y. SAWADA. 1978. Phase-difference locking of coupled oscillating chemical systems. J. Chem. Phys. **69:** 3830–3832.
3. NAKAJIMA, K. & Y. SAWADA. 1980. Experimental studies on the weak coupling of oscillatory chemical reaction systems. J. Chem. Phys **72:** 2231–2234.
4. STUCHL, I. & M. MAREK. 1982. Dissipative structures in coupled cells: experiments. J. Chem. Phys. **77:** 2956–2963.
5. BAR-ELI, K. 1984. Coupling of chemical oscillators. J. Phys. Chem. **88:** 3616–3622.
6. BAR-ELI, K. & S. REUVENI. 1985. Stable stationary states of coupled chemical oscillators. Experimental evidence. J. Phys. Chem. **89:** 1329–1330.
7. CROWLEY, M.F. & R.J. FIELD. 1986. Electrically coupled Belousov-Zhabotinskii oscillators. 1. Experiments and simulations. J. Phys. Chem. **90:** 1907–1915.
8. DOUMBOUYA, S.I., A.F. MÜNSTER, C.J. DOONA & F.W. SCHNEIDER. 1993. Deterministic chaos in serially coupled chemical oscillators. J. Phys. Chem. **97:** 1025–1031.
9. FIELD, R.J., E. KOROS & R. M. NOYES. 1972. Oscillations in chemical systems. II. Thorough analysis of temporal oscillations in the bromate–cerium–malonic acid system. J. Am. Chem. Soc. **94:** 8649–8664.
10. ZHABOTINSKY, A.M. 1991. A history of chemical oscillations and waves. CHAOS **1:** 379–386.
11. SCOTT, S.K. 1991. Chemical Chaos. Oxford University Press. Oxford, England.
12. EPSTEIN, I.R. & K. SHOWALTER. 1996. Nonlinear chemical dynamics: oscillations, patterns and chaos. J. Phys. Chem. **100:** 13132–13147.
13. CROWLEY, M.F. & I.R. EPSTEIN. 1989. Experimental and theoretical studies of a coupled chemical oscillator: phase death, multistability, and in-phase and out-of-phase entrainment. J. Phys. Chem. **93:** 2496–2502.
14. RUSTICI, M., M. BRANCA, C. CARAVATI & N. MARCHETTINI. 1996. Evidence of a chaotic transient in a closed unstirred cerium catalyzed Belousov-Zhabotinsky system. Chem. Phys. Lett. **263:** 429–434.

Complexity at the Molecular Level

A Dynamic View of Biomolecules Obtained by NMR

ALESSANDRO DONATI,[a] LYDIA BELLIK, MARIA PIA PICCHI, AND CLAUDIA BONECHI

Department of Chemical & Biosystem Sciences, University of Siena, Pian dei Mantellini, 44 - 53100 Siena, Italy

INTRODUCTION

Dominant processes in natural systems are usually controlled by feedback mechanisms that are directly related to self-organization and complex behavior. Much work has been done to understand the biophysical basis of complexity in ecology and biology. Information at the molecular level, such as macromolecular flexibility, is fundamental for understanding this feedback regulation.[1]

Conformational changes can cause variations in the activity of enzymes by interaction with reaction products at the regulating site. Here we analyze the problem of molecular flexibility in order to understand how it affects the activity of biomolecules from another point of view. The molecule studied was the pain-producing nonapeptide hormone bradykinin (Arg^1-Pro^2-Pro^3-Gly^4-Phe^5-Ser^6-Pro^7-Phe^8-Arg^9), which is formed by proteolysis of kininogen during inflammation.[2] Small peptides, like bradykinin, usually do not have any single stable conformation in solution. The conformation that causes the biological effect is adopted only at the active site. Also in this case the flexibility of the biomolecule determines the mechanism of action.

We are interested in studying whether the molecule has a random-coil behavior in solution and adopts the active conformation only when the interaction occurs at the active site, or whether the molecule rapidly exchanges between a few low-energy conformations and the interaction just stabilizes one of them.

We have determined the conformation of bradykinin in solution by nuclear magnetic resonance (NMR) coupled with theoretical calculations based on dipolar relaxation theory and molecular mechanics. The experimental results are compatible neither with a single static conformation nor with a "random-coil" molecule in solution.

RESULTS AND DISCUSSION

Small polypeptides do not usually adopt any definite conformation in solution,[3] though this rule does not apply to all sequences. On the other hand, proteins that usually have a stable structure may have parts of the chain with high flexibility in solution.[4] Bradykinin was found to be an intermediate case. In fact, NMR parameters

---

[a]Address for correspondence: 39-0577-232006 (voice); 39-0577-232004 (fax); alex@spectrum.chim.unisi.it (e-mail).

TABLE 1. Differences in ^{1}H chemical shift between experimental and random coil values for residues of the bradykinin in DMSO at 298K[5]

Residue	Δ Chemical Shift (ppm)			
	NH	H_α	$H_{\beta 2}$	$H_{\beta 3}$
Arg1	—	—	0.066	0.105
Gly4	0.081	0.049	—	—
Phe5	−0.080	0.318	0.127	0.118
Ser6	0.435	0.106	0.121	0.097
Phe8	−0.305	0.046	0.106	0.108
Arg9	−0.147	−0.179	0.103	0.109

NOTE: Values for Pro2, Pro3, and Pro7 were not obtained.

that are sensitive to conformational changes, namely the ^{1}H-^{1}H nuclear Overhauser effect (NOE), ^{1}H chemical shift (δ) and ^{1}H-^{1}H vicinal coupling constant (^{3}J), were only in part compatible with a structure in which any single bond freely rotates in solution.

TABLE 1 shows the differences in chemical shift between the observable protons of bradykinin and those reported for the random coil situation.[5] The four types of protons analyzed showed non-homogeneous behavior: for the $H_{\beta 2}$ and $H_{\beta 3}$ small differences were found, whereas differences were significant for the amide and H_α protons, the especially for the Ser6$_{NH}$ and Phe8$_{NH}$ and Phe5$_{H\alpha}$ which were at the upper limit of the boundaries coded for a single stable structure.

The $H_{\beta 2,3}$ chemical shift showed values similar to those of the random coil because these atoms are not located along the backbone of the peptide, and a certain degree of insensitivity to conformational restriction is conceivable. It is interesting that the resonances shifted from the random coil value range in the inner part of the molecule, probably because of the lower degree of freedom with respect to the outer residues and because of restriction of the motion around the Phe5 and Phe8 residues that allows a deshielding chemical shift anisotropy effect on Ser6$_{NH}$ and Phe8$_{NH}$, respectively.

FIGURE 1 shows an expanded region of the 2D-NOESY spectrum with the interaction between the amide protons and C_α and C_β protons, from which we can observe different cross-peak intensities for those proton couples that should have the same behavior in a random coil system. In fact the coupling between NH and the diastereotopic protons $H_{\beta 2}$ and $H_{\beta 3}$ should not differ when there is free rotation around the backbone and/or lateral single bonds. Significantly, in the case of the interaction Phe5$_{H\beta 3}$–Ser6$_{NH}$, it is not possible to obtain the corresponding Phe5$_{H\beta 2}$–Ser6$_{NH}$, which is perfectly consistent with the above analysis of chemical shift.

It is also important that the coupling $^3J_{H\alpha\text{-NH}}$ results were not consistent with a molecule spending most of its time in a single stable conformation. In fact the values of the five observable $^3J_{H\alpha\text{-NH}}$ were very close to the random coil value (≅7 Hz).

These experimental results are not consistent with a pure random coil or a single static structure. A compromise between this two extremes is a type of motion constrained between two or more rapidly exchanging conformations.

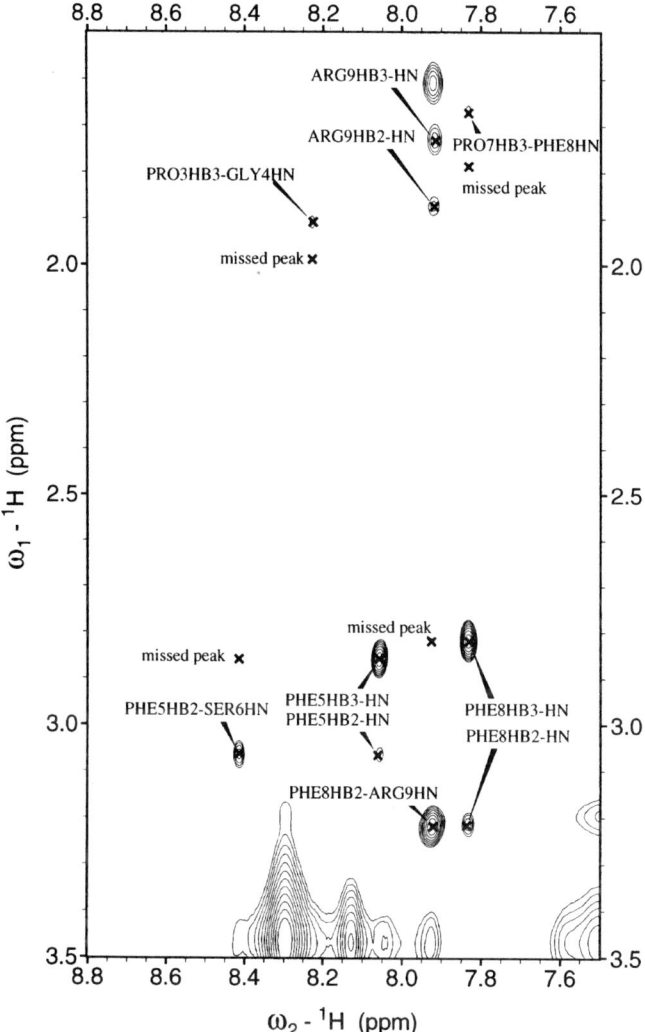

FIGURE 1. Portion of the 2D-NOESY spectrum showing the dipolar couplings between the amide protons and the diastereotopic protons $H_{\beta 2,3}$ of the bradykinin. The spectrum was recorded with 200 msec mixing time at 300K.

DETERMINATION OF THE STRUCTURAL ENSEMBLE

The structure of a macromolecule is usually resolved by using the geometric parameters extracted from the experimental NMR data as constraints for theoretical simulations of molecular mechanics (MM) and molecular dynamics (MD).[6] Here, it involved the following steps: (i) calculation of ^1H–^1H distances from NOESY inten-

TABLE 2. R_{NMR}-factors for the four approaches used to obtain structural ensembles to resolve bradykinin structure in solution

	Single Minimized Structure	Full MC Ensemble	rMD Ensemble	Restricted rMC Ensemble
R_{NMR}-factor	0.48	0.43	0.41	0.26

sities by complete relaxation matrix analysis using the program MARDIGRAS[7]; (ii) application of restrained molecular mechanics (rMM),[8] using a Monte Carlo (rMC) approach for sampling the conformational space or molecular dynamic (rMD)[9] simulations; (iii) statistical analysis of the pool of conformers obtained by the rMC and rMD simulations, using the program XCluster[10]; (iv) calculation of the theoretical

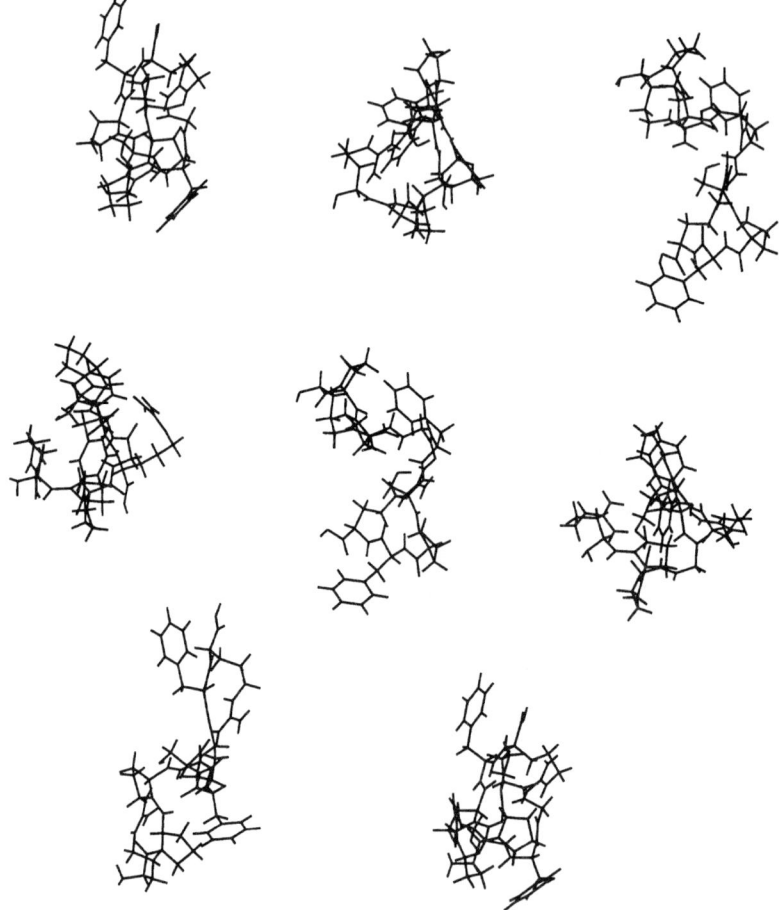

FIGURE 2. The eight representative structures from the rMC simulation followed by cluster analysis. The whole ensemble of conformers was clustered in eight subsets of structures by R.M.S. atomic displacement.

NOESY intensities by complete relaxation matrix analysis using the program CORMA[11]; (v) comparison of the experimental and theoretical data using the crystallographic-type R_{NMR}-factor[12] :

$$R_{NMR} = \Sigma |I_0(i) - I_c(i)| / \Sigma I_0(i)$$

Using this procedure we were able to compare the goodness of the fit between the experimental (I_0) and theoretical (I_c) data for several cases simulating the different behaviours of the molecule in solution.

TABLE 2 gives the R_{NMR}-factor for the following four cases: (i) a single lowest energy structure obtained by the full minimization of a set of conformers from a rMC simulation; (ii) an ensemble of conformers obtained from an unrestrained Monte Carlo simulation; (iii) an ensemble of conformers obtained from constant temperature (300K) rMD; (iv) a restricted ensemble of structures obtained by cluster analysis of the whole pool of structures obtained from the rMC simulation. The lower is the R_{NMR}-factor, the value in better accordance between experimental and theoretical data. Higher values of the R_{NMR}-factor were obtained for the single fully minimized structure and for the complete ensembles; the lowest value is obtained for the restricted set of conformers from cluster analysis (FIG. 2). These data are in line with the experimental data reported above. It is interesting that the best R_{NMR}-factor was obtained with the small ensemble of structures (iv) which represent a type of behavior of bradykinin in solution intermediate between a single static structure and a random coil, in which a few conformers rapidly exchange.

This evidence is consistent for bradykinin adopting an active form in solution and interaction with the receptor stabilizing it at the active site. The results encourage us to continue research into the implications of molecular flexibility for the mechanism of action of complex biological macromolecules.

ACKNOWLEDGMENTS

This paper has been supported by the Cofin. MURST 97 CFSIB Grant and the Consorzio CSGI.

REFERENCES

1. GOLDBETER, A 1996. Biochemical Oscillations and Cellular Rhythms. Cambridge University Press. Cambridge, U.K.
2. FRAMER, S.G. & R.M. BURCH. 1991. In Bradykinin Antagonists: Basic and Clinical Research. R.M. Burch, Ed.: 1–31. Marcel Dekker. New York.
3. WÜTHRICH, K. 1986. NMR of Proteins and Nucleic Acids. John Wiley and Sons. Wiley Interscience Publication. New York.
4. BONVIN, A.M.J.S. & A.T. BRÜNGER. 1995. Conformational variability of solution nuclear magnetic resonance structures. J. Mol. Biol. 250: 80–93.
5. BUNDI, A., C. GRATHWOHL, J. HOCHMANN, R.M. KELLER, G. WAGNER & K. WUTHRICH. 1975. Proton NMR of the protected tetrapeptides TFA-Gly-Gly-L-X-L-Ala-OCH3, where X stands for one of the 20 common amino acids. J. Magn. Reson. 18: 191–198.
6. LEACH, A.R. 1996. Molecular Modelling. Principles and Applications. Longman Press. Harlow Essex. England.

7. BORGIAS B.A. & T.L. JAMES. 1990. MARDIGRAS: A procedure for matrix analysis of relaxation for discerning geometry of an acqueous structure. J. Magn. Reson. **87:** 475–487.
8. STILL, C.A. 1995. MacroModel, Interactive Molecular Modeling System 5.0. Columbia University. New York.
9. PEARLMAN, D.A. D.A. CASE, J.C. CALDWELL, W.S. ROSS, T.E. CHEATHAM III, D.N. FERGUSON, G.L. SEIBEL, U.C. SINGH, P.K. WEINER & P.A. KOLLMAN. 1996. AMBER 5.0 University of California San Francisco, San Francisco.
10. STILL, C.A. 1995. XCluster 1.2. Columbia University. New York.
11. BORGIAS, B.A., P.D. THOMAS & T.L. JAMES. 1989. Complete Relaxation Matrix Analisys (CORMA). University of California, San Francisco. San Francisco.
12. GONZALES C., J.A.C. RULLMAN, A.M.J.J. BONVIN, R. BOELENS & R. KAPTEIN. 1991. J. Magn. Reson. **91:** 659–664.

Thermodynamics of Extremely Diluted Aqueous Solutions

VITTORIO ELIA[a] AND MARCELLA NICCOLI

Department of Chemistry, University "Federico II" of Naples, Via Mezzocannone 4, 80134 Naples, Italy

An extensive thermodynamic study has been carried out on aqueous solutions obtained through successive dilutions and succussions of 1% wt/vol of some solutes up to extremely diluted solutions, (less than 1×10^{-5} mol kg^{-1}) obtained via several 1/100 successive dilution processes. The interaction of acids or bases with the extremely diluted solutions has been studied calorimetrically at 25°C. Measurements have been performed of the heat produced by the mixing of acid or basic solutions of different concentrations, with bidistilled water or with the extremely diluted solutions. Despite the extreme dilution of the solutions, an exothermic heat of mixing in excess has been found in about the 92% of the cases, compared to the corresponding heat of mixing with the untreated solvent. Here, we show that successive dilutions and succussions may permanently alter the physical–chemical properties of the solvent water. The nature of the phenomenon here described still remains unexplained, but significant experimental results are obtained.

A thermodynamic study on aqueous solutions gives interesting information about the behavior of solutes and their interactions with the solvent. The interaction of acids or bases with the extremely diluted solutions has been studied calorimetrically at 25°C. The extremely diluted solution is obtained starting from a solution at 1% wt/vol. After succussion, that solution is named 1CH preceded by the name or formula of the solute. The succussion process consists of vertical shakings of the solution by means of a mechanical apparatus. In a simple succussion process, 100 vertical strokes in six seconds are given to the glass vessel containing the solution. To prepare the successive dilution, 1 g of this solution is added to 99 g of water that again gets succussed, obtaining the 2CH solution. The iteration of this process produces the extremely diluted solutions studied. Measurements have been performed of the heats of mixing of acid or basic solutions of different concentrations with bidistilled water or with solutions, at a concentration of 0.01 mol kg^{-1}, used as reagent, whereas the concentrations of the extremely diluted solutions or with extremely diluted solutions. Procedures for the calorimetric determination of the heat of dilution or mixing are well developed.[1] The experimental results are treated according to the MacMillan-Mayer approach,[2] modified by Friedman and Krishnan.[3] The enthalpies of mixing two solutions are given by the following equations:

$$\Delta H^{mix}(\text{J kg}^{-1}) = h_{xx}m_x^f(m_x^f - m_x^i) + 2h_{xy}m_x^f m_y^f + h_{yy}m_y^f(m_y^f - m_y^i) + \text{higher terms}$$

---

[a]Address for correspondence: +39 081 5476517 (voice); +39 081 5527771 (fax); Elia@chemna.dichi.unina.it (e-mail).

where in m_x^f, m_y^f, m_x^i, m_y^i are the molalities (mol kg^{-1}) after and before the mixing process, respectively, and h_{xx}, h_{yy}, and h_{xy} the enthalpy interaction coefficients, are adjustable parameters. Their values fall in the range 1×10^2–1×10^4 J kg mol^{-2}. Consequently, when the concentration of the solute y of an extremely diluted solution is of the order of 10^{-5} mol kg^{-1} or less, while the concentration of the solute x is a finite one—1×10^4–1×10^{-2} mol kg^{-1}—the previous equation reduces to the sole contribution of x. This actually happens on the third successive 1/100 dilution. Then, the extremely diluted solutions, as those described before, because of the practical absence of the solute, cannot produce any contribution to the heat of mixing via the y component. For electrolyte solutions, the powers of the molalities in the previous equation are fractionary, but the conclusions stay absolutely the same. By an extensive study it was assessed that, using aqueous solutions of acids or bases as reagent, it is possible to distinguish qualitatively the behavior of pure solvent from that of the extremely diluted solutions, whose chemical composition is the same as that of the solvent used. The interaction between acids or bases with the extremely diluted solutions has been studied calorimetrically, determining the heats of mixing at 25°C by means of a thermal activity monitor (TAM) from Thermometric (Sweden).

The heats of mixing with the solvent (bidistilled water) and those with the extremely diluted solutions were determined. Sodium hydroxide (NaOH) or hydrochloric acid (HCl) aqueous solutions, at a concentration of 0.01 mol kg^{-1}, used as reagent, whereas the concentrations of the extremely diluted solutions were less than 1×10^{-5} mol kg^{-1}. Despite the extreme dilution of the solutions used, an excess exothermic heat of mixing has been found in nearly all the cases, with respect to the heat of mixing of the same reagents with the solvent. The excess heat of mixing, namely, the difference between the heat of mixing of the reagent (a solution at finite concentration) with the extremely diluted solutions and the heat of mixing of the same reagent with the solvent, is of the same order of magnitude or higher than the heat of mixing of the reagent with the solvent. To explain this heat in excess, we are forced to focus our attention on the solvent and, in particular, on possible chemical–physical changes induced by the procedure employed in preparing the solutions.

The excess heats of mixing of about 300 experimental measurements, using as reagent aqueous solutions of NaOH or HCl 0.01 mol kg^{-1}, are reported in columns 1–5 of TABLE 1, together with the heats of mixing of the same reagent with the solvent. The reported heats in excess are detectable for some weeks. From this table it clearly appears that a new phenomenon occurs and, because of the absence of solute, it can be inferred that the physical–chemical properties of the solvent must be permanently altered by the procedure of successive dilutions (1/100) and succussions used to prepare the extremely diluted solutions. Thus, we can firmly state that it is now easily possible to measure a chemical–physical property, the heat of mixing with acids or bases, characterizing this new state of the water,[4–6] using a commercial microcalorimeter.

To confirm these very surprising findings, but otherwise "objective" instrumental responses, and to get a deeper insight into this new behavior, a calorimetric titration procedure was adopted. The excess heats of mixing , thus produced in about 300 experimental measurements are reported in columns 6–13 of TABLE 1. A "titration" of the extremely diluted solutions implies the determination of the heat of mixing in ex-

TABLE 1. Excess heats of the mixing for extremely diluted solutions with sodium hydroxide and hydrochloric acid solutions

System	$-Q^g$ NaOH 5×10^{-3} m^e	N^d	$-Q^g$ HCl 5×10^{-3} m^e	N^d	$-Q^g$ NaOH 2.5×10^{-3} m^e	N^d	$-Q^g$ NaOH 1×10^{-3} m^e	N^d	$-Q^g$ NaOH 5×10^{-4} m^e	N^d	$-Q^g$ NaOH 2×10^{-3} m^e	N^d
H$_2$O bidist^a	2.1±0.1^a	30	0.85±0.01^a	30	1.7±0.1^a	30	1.4±0.1^a	30	1.0±0.1^a	30	0.5±0.2^a	30
H$_2$O 1 CH^{b,c}	0^f–10	21	0.4–0.6	12	0^f–9.8	11	0^f–7.9	11	0^f–3.3	11	0^f–2.1	11
H$_2$O 3 CH^{b,c}	0^f–38	52	0^f–3.2	35	0^f–35	18	0^f–11	18	0^f–5.9	18	0^f–3.6	18
H$_2$O 30 H^{b,c}	0.8–16	35	0^f–5.8	26	1.5–3.4	4	1.4–3.1	4	1.1–2.9	4	0.5–1.7	4
NaCl 3 CH^c	0^f–17	31	0^f–3.3	19	1.3	1	1.3	1	1.2	1	0.8	1
NaCl 30 CH^c	0^f–16	25	0^f–1.5	19	—	—	—	—	—	—	—	—
IAA 7 CH^c	2.4–3.8	2	—	—	2–3.6	2	2–3.1	2	1.8–2.7	2	1–1.6	2
IAA 8 CH^c	3.1–4.5	2	—	—	3–4	2	2.7–3.7	2	2.4–3.1	2	1.4–1.6	2
IAA 9 CH^c	1.4–3.1	2	—	—	1.4–3	2	1.3–2.6	2	1–2.2	2	0.4–1.2	2
IAA 10 CH^c	8.3–9.3	2	—	—	8–9	2	7.1–7.9	2	5.7–6.2	2	2.7–2.9	2
IAA 11 CH^c	4.5–7.9	2	—	—	4.3–7.6	2	3.8–6.6	2	3.2–5.5	2	1.6–2.9	2
IAA 12 CH^c	0^f–4.9	12	—	—	3.8–4.8	12	3.3–4.1	12	2.9–3.4	12	1.7–1.8	12

ABBREVIATIONS: Sodium chloride, NaCl; Indole-3-acetic acid, IAA; 2,4-dichloro phenoxyacetic acid, 2,4-D; N-(phosphonomethyl)-glycine, GLP.
^{a}Heats of dilution (mean + SD) of sodium hydroxide solutions.
^{b}In these cases the procedure of preparation starts with pure solvent. Succussion and dilution is performed just as in the cases of 1% solutions.
^{c}Because of the quantitative variability of the excess heats of mixing for these systems, the range of values obtained is reported.
^{d}Number of experiments performed.
^{e}Concentration (mol kg^{-1}) of the reagent after the mixing process. In these experiments the final concentration is half of the initial one.
^{f}Percentage of experiments that give null excess: 8%.
^{g}Excess heats of mixing (J/kg) of the extremely diluted solution.

TABLE 1 — continued

System	$-Q^g$ NaOH 5×10^{-3} m^e	N^d	$-Q^g$ HCl 5×10^{-3} m^e	N^d	$-Q^g$ NaOH 2.5×10^{-3} m^e	N^d	$-Q^g$ NaOH 1×10^{-3} m^e	N^d	$-Q^g$ NaOH 5×10^{-4} m^e	N^d	$-Q^g$ NaOH 2×10^{-3} m^e	N^d
H$_2$O bidisi^a	2.1±0.1^a	30	0.85±0.01^a	30	1.7±0.1^a	30	1.4±0.1^a	30	1.0±0.1^a	30	0.5±0.2^a	30
2,4-D 3 CH^c	2.6	1	—	—	2.5	1	2.1	1	1.7	1	1	1
2,4-D 5 CH^c	0^f	1	—	—	0^f	1	0^f	1	0^f	1	0^f	1
2,4-D 7 CH^c	1.6	1	—	—	1.2	1	1.2	1	0.8	1	0.3	1
2,4-D 8 CH^c	1.6	1	—	—	1.5	1	1.2	1	0.9	1	0.6	1
2,4-D 12CH^c	0.5–2.6	6	—	—	0.2–2.4	6	0.2–2.4	6	0.2–1.8	6	0.2–1.1	6
GLP 4 CH^c	0^f	1	—	—	0^f	1	0^f	1	0^f	1	0^f	1
GLP 5 CH^c	0.6	1	—	—	0.6	1	0.4	1	0.3	1	0.1	1
GLP 6 CH^c	0.3	1	—	—	0.3	1	0.2	1	0.1	1	0.1	1
GLP 7 CH^c	0^f	1	—	—	0^f	1	0^f	1	0^f	1	0^f	1
GLP 8 CH^c	0.4	1	—	—	0.4	1	0.4	1	0.3	1	0.1	1
GLP 9 CH^c	1.1	1	—	—	1	1	1	1	0.8	1	0.4	1
GLP 10 CH^c	4.9	1	—	—	4.6	1	4.2	1	3.4	1	1.8	1
GLP 11 CH^c	3.5	1	—	—	3.3	1	3.1	1	2.5	1	1.2	1
GLP 12 CH^c	0^f	1	—	—	0^f	1	0^f	1	0^f	1	0^f	1

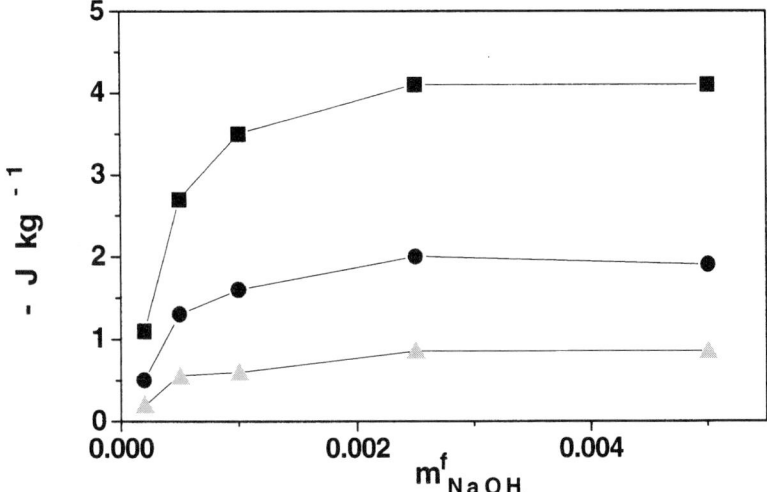

FIGURE 1. Calorimetric titration curves. Extremely diluted solutions of IAA 12CH (■) and its 1:1 (●) and 1:2 (▲) normally diluted solutions with bidistilled water.

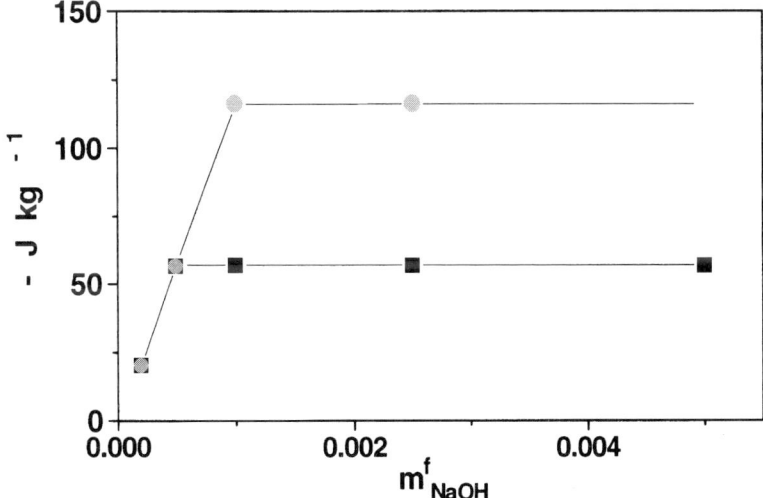

FIGURE 2. An example of calorimetric titration of at 2×10^{-3} (●) mol kg^{-1} HCl solution and its 1:1 (■) diluted solution with bidistilled water.

cess, compared to the heat are obtained with the reference bidistilled water, whereas solutions of NaOH at different concentrations are mixed with the samples under examination. About 60 titrations have been performed with about 40 different samples. These titration curves present two peculiar features (see FIG. 1). First, a plateau appears at the highest concentrations of the titrant, and, second, there's a "break" point

TABLE 2. Excess heats of mixing in the titration of the extremely diluted solutions with sodium hydroxide

$M_{NaOH}{}^a$	$-Q^b$ 1AA^c 12 CH	$-Q^b$ 1AA^c 12 CH diluted 1:1	$-Q^b$ 1AA^c 12 CH diluted 1:2
2×10^{-4}	1.1	0.5	0.2
5×10^{-4}	2.7	1.3	0.55
1×10^{-3}	3.5	1.6	0.6
2.5×10^{-3}	4.1	2.0	0.85
5×10^{-3}	4.1	1.9	0.85

^{a}Concentration (mol kg^{-1}) of the reagent after the mixing process. In these experiments the final concentration is half of the initial one.
^{b}Excess heats of mixing (J/kg) of the extremely diluted solution.
^{c}Indole-3-acetic acid, IAA.

at a concentration of about 0.001 mol kg^{-1} of the reagent used (NaOH) in the final solution. This latter feature, particularly the fact that it appears exactly at the same concentration, is common to all experiments (with different samples of the extremely diluted solutions used). On the other hand, the magnitude of the excess heat, characterizing the plateau, depends on the nature of the solutions.

To test the stability of these chemical–physical changes in water "structure," the extremely diluted solutions were further diluted (without the succussion procedure) in different proportions (e.g., 1:1, 1:2, ...) with bidistilled water. These "simply diluted" solutions were "titrated" with the NaOH solutions. The resulting curves are characterized by plateaux that are proportional to the degree of the "simple" dilution (i.e. for the proportions just cited as examples, 1/2, 1/3 of the plateaux obtained with the original samples of the extremely diluted solutions), but showing the "break" points at the same concentration of the reagent used (TABLE 2, FIG. 1). This means that the modifications in water "structure" induced by the preparation procedure are stable with respect to a normal dilution process and that the reagent interacts via a destroying mechanism, revealing a pHdependent phenomenon. FIGURE 1 reports titration curves for a highly diluted sample. For the sake of comparison, TABLE 3 and FIGURE 2 show typical calorimetric titration curves for an acid–base reaction, obtained by titrating two solutions of the acid, whose concentrations are in the 1:2 ratio. As can be seen, two different plateaux are reached, and two different equivalent points are identified, both in the 1:2 ratio, thus showing a sharp difference, either in the amount of heat, slope of the curves, and equivalent point positions, with respect to the behavior of the extremely diluted solutions for which each "titration" curve reaches its own plateau at the same "break" point. It is very interesting to look at the heat of mixing versus pH diagrams (see FIG. 3): they reveal an extraordinary similarity with the normally reported ones for a two-state, pH-induced denaturation process of proteins.[7]

The exact nature of the phenomena here described still remains unexplained, but significant experimental results have been obtained. The mixing process of acids or bases reveals a statistically significant exothermic excess heat with respect to the same process carried on the untreated solvent, bidistilled water, despite the physical absence of solute molecules in the solution obtained after just a few dilution/succus-

TABLE 3. Excess heats of mixing in the titration of hydrochloric acid with sodium hydroxide

$M_{NaOH}{}^a$	$-Q^b$ HCl 1×10^{-3} mol kg^{-1}	$-Q^b$ HCl 2×10^{-3} mol kg^{-1}
2×1^{-4}	19.7	19.7
5×10^{-4}	57	57
1×10^{-3}	57	116.2
2.5×10^{-3}	57	116.2
5×10^{-3}	57	116.2

[a]Concentration (mol kg^{-1}) of the reagent after the mixing process. In these experiments the final concentration is half of the initial one.
[b]Excess heats of mixing (J/kg) of the extremely diluted solution.

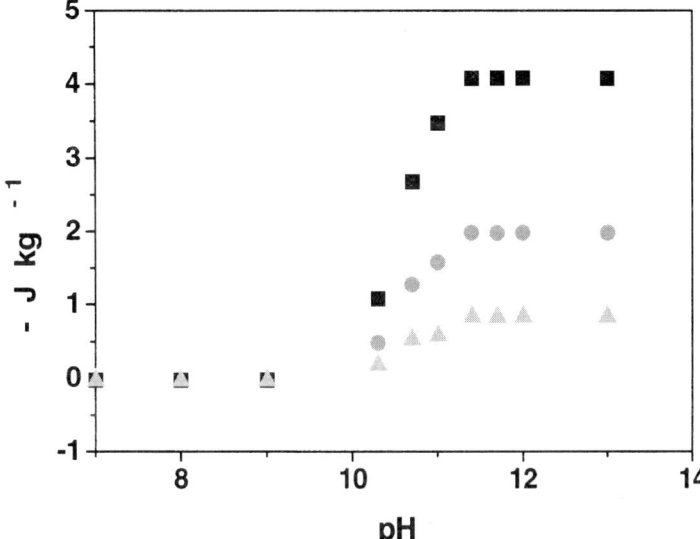

FIGURE 3. Calorimetric titrations curves: Extremely diluted solutions of IAA 12CH (■) and its 1:1 (●) and 1:2 (▲) normally diluted solutions with bidistilled water, as a function of the pH in the final solutions.

sion procedures. All that is not, at present, in agreement with current theories concerning the properties of liquid water at room temperature,[8] and consequently, the need for appropriate new theoretical studies is urgent.

A hypothesis of disorder–order transition could be proposed, based on the exothermic excess heat and its pH dependence, induced by the addition of acid and/or basic reagents.

ACKNOWLEDGMENTS

We thank Dr. Filomena Velleca for help with experimental measurements and Prof. Liberato Ciavatta for helpful suggestions and discussions. This work was supported by the Ministry of University and Scientific Research (MURST), Rome, Italy Cofin.MURST 97 CFSIB.

REFERENCES

1. CASTRONUOVO, G., V. ELIA & F. VELLECA. 1997. Hydrophilic interactions determine cooperativity of hydrophobic interactions and molecular recognition in aqueous solutions of non electrolytes. The preferential configuration model. Curr. Top. Solut. Chem. **2:** 125–142.
2. MACMILLAN, W.G. & J.E. MAYER. 1945. The statistical thermodynamics of multicomponent systems. J. Chem. Phys. **13:** 276–305.
3. FRIEDMAN, H.L. & C.V. KRISHNANN. 1973. Studies of hydrophobic bonding in aqueous alcohols: enthalpy measurements and model calculations. J. Solut. Chem. **2:** 119–140.
4. DAVENAS, E. et al. 1988. Human basophil degranulation triggered by very dilute antiserum against IgE. Nature **333:** 816–818.
5. LO, S.Y. 1996. Anomalous state of ice. Mod. Phys. Lett. B **10:** 909–919.
6. LO, S.Y. et al. 1996. Physical Properties of Water with I_E Structures. Mod. Phys. Lett. B **10:** 921–930.
7. PRIVALOV, P.L. 1973. Stability of proteins. Small globular proteins. Adv. Prot. Chem. **33:** 167–241.
8. FRANKS, F. 1976. Water. A Comprehensive Treatise. Plenum. New York.

Linear and Nonlinear Properties of Heart Rate Variability: Influence of Obesity

A. GASTALDELLI, R. MAMMOLITI, E. MUSCELLI, S. CAMASTRA, L. LANDINI, E. FERRANNINI, AND M. EMDIN

C.N.R. Institute of Clinical Physiology, Department of Internal Medicine and Department of Informatic Engineering, University of Pisa, 56126 Pisa, Italy

INTRODUCTION

Physiological systems are best characterized as time-varying processes exhibiting rhythmic and complex behavior. The interaction among system variables, external noise, and state changes modulates the overall variability of physiological signals such as heart rate, arterial pressure, and respiration, which may therefore present both linear and nonlinear patterns. To describe the complex and periodic dynamics of living systems, various analytical tools have been employed, especially in the cardiovascular field.[1] Among them, power spectral analysis (PSA)[2] and recurrence quantification analysis (RQA)[3,4] have been used to describe, respectively, linear and nonlinear dynamics of heart rate variability (HRV). PSA is a validated method that quantifies autonomic nervous modulation of cardiac activity by describing the fluctuations of HR linked to vasomotion and respiration. RQA evaluates complexity and determinism in time series by detecting state changes in drifting or exciting dynamical systems. RQA can be easily applied to cardiovascular signals because it does not require any *a priori* mathematical assumption, such as stationarity or linearity; parameters introduced by RQA, based on distance, recurrence, and entropy of recurrence plots (RP),[5] may be related to different physiological states. Nevertheless, no correlation has been shown between RQA parameters and autonomic nervous activity.

It has recently been shown that obesity is a state of reduced sensitivity of the sinoatrial node to both sympathetic and vagal influences.[6] Data from obese and lean subjects were therefore analyzed by PSA and RQA, and parameters derived by the two methods were compared for the two groups of subjects.

METHODS

PSA and RQA were applied to the R-wave peak interval (RR interval) time series as derived by continuous electrocardiographic (ECG) monitoring (250-Hz frequency sampling). We analyzed 21 ECG tracings recorded during 60 min of quiet, supine rest. Subjects were divided into two groups, 13 obese and 8 lean, on the basis of their body mass index (BMI > 28 $kg \cdot m^{-2}$). The characteristics of the subjects are shown in TABLE 1.

TABLE 1. Characteristics of the study subjects

	Obese	Lean
Age (years)	37 ± 2	33 ± 2
Height (cm)	163 ± 3	173 ± 4
Weight (kg)	93 ± 5[a]	65 ± 5
BMI (kg·m^{-2})	35 ± 1[a]	21 ± 1

[a]Significantly different from the lean group.

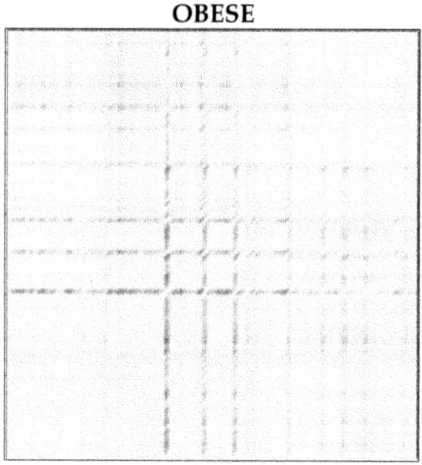

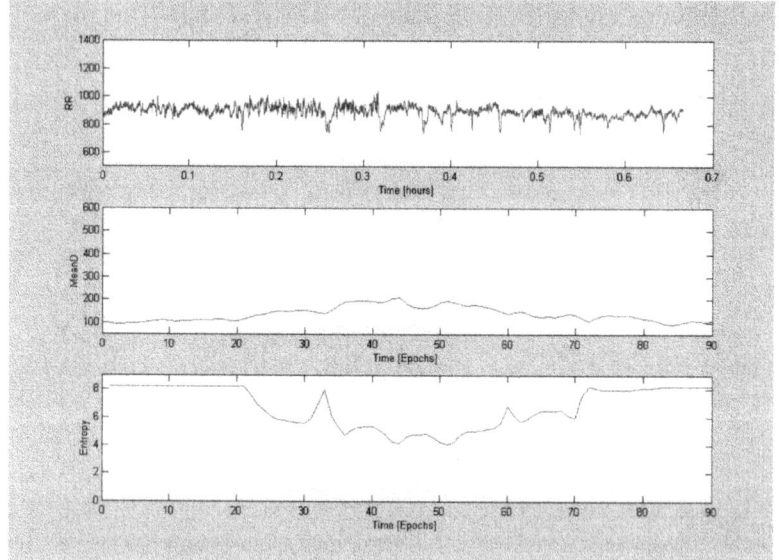

FIGURE 1. Panel A (*top to bottom*). Recurrence plot (RP) with embedding = 8 and delay = 4 of an obese patient, relevant tachogram during a basal session, and two RQA indices (meanD and Entropy). Data points referring to RQA anlysis are evaluated on 300-beat epoch length.

Power Spectral Analysis

Autoregressive PSA was performed on a 12-order model using the Levinson-Durbin recursive algorithm over consecutive 256 data-point intervals.[1,6] Parameters obtained with this approach were: mean RR interval (MEANRR), power of the low-frequency component (LF, 0.04–0.15 Hz, which reflects sinus node baroreflex responsiveness) and of the high-frequency component (HF, 0.15–0.40 Hz,

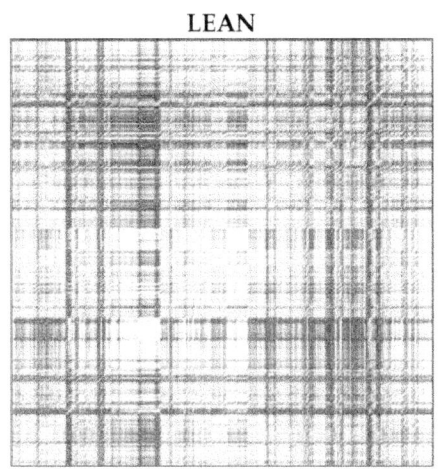

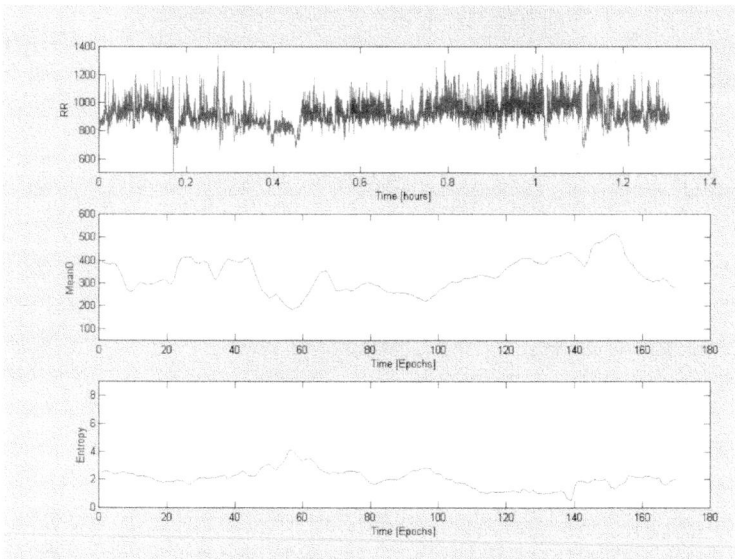

FIGURE 1. Panel B (*top to bottom*). Recurrence plot (RP) with embedding = 8 and delay = 4 of an obese patient, relevant tachogram during a basal session, and two RQA indices (meanD and Entropy). Data points referring to RQA anlysis are evaluated on 300-beat epoch length.

TABLE 2. PSA and RQA parameters during quiet supine rest

	Obese	Lean
Mean RR (ms)	894 ± 31	848 ± 19
P (msec^2)	1590 ± 338[a]	2901 ± 553
LF (msec^2)	851 ± 193	1326 ± 234
HF (msec^2)	645 ± 150[a]	1299 ± 323
LF/HF	1.95 ± 0.37	1.22 ± 0.22
MEAND (msec)	203 ± 16[a]	260 ± 21
%REC	83.4 ± 3.5	73.0 ± 5.4
%DET	92.5 ± 2.9	88.4 ± 5.0
Entropy (bits/bin)	4.5 ± 0.4[a]	3.3 ± 0.4
MaxL (au)	295.5 ± 1.9[a]	278.5 ± 10.1

[a]Significantly different from the lean group.

reflecting vagally modulated respiratory sinus arrhythmia); total spectral power (P); the LF/HF ratio, which is an index of sympatho-vagal balance.

Recurrence Quantification Analysis

For RQA, RR was projected into an appropriate multidimensional space by embedding procedures: the rows of the embedding matrix correspond to consecutive vectors of length m (embedding dimension), while the lag ($T = 4$) was selected by minimizing a nonlinear correlation function. The choice of dimensionality ($m = 8$) was obtained by singular value decomposition of the embedding matrix, consistent with the presence of no more than three independent oscillators.[8] The quantitative descriptors used in the present analysis were: the mean of normalized vector distances (MEAND), the plot point recurrence percentage (%REC), the RP line distribution Shannon entropy (ENT), and the longest diagonal line segment (MAXL), whose reciprocal measures the divergence of near trajectories and is correlated with the first Lyapunov exponent. From a mathematical point of view, chaotic behavior is characterized by high values of MEAND and low values of %REC, ENT, and MAXL.[3,8]

Statistical Analysis

Data are given as mean ± SE. Mean group values were compared by analysis of variance. Linear regression was carried out by standard methods.

RESULTS

Although RR was not significantly different between the two groups, the obese subjects showed lower overall HRV (as reflected by the total spectral power). In addition, obesity was associated with depressed vagal tone (HF) (TABLE 2). By RQA, the obese group showed greater periodicity and lower complexity properties, as reflected by the mean values on a 300-beat epoch length of MEAND, ENT, and MAXL

TABLE 3. Correlation between PSA and RQA parameters

	Mean RR	P	LF	HF	LF/HF
MEAND	0.49[a]	0.87[a]	0.76[a]	0.85[a]	−0.46[a]
%REC	−0.54[a]	−0.71[a]	−0.62[a]	−0.59[a]	0.33
%DET	−0.60[a]	−0.67[a]	−0.56[a]	−0.60[a]	0.28
Entropy	−0.50[a]	−0.86[a]	−0.78[a]	−0.82[a]	0.54[a]
MaxL	−0.48[a]	−0.51[a]	−0.60[a]	−0.26	0.10

[a]Statistically significant at $p < 0.05$ or less.

(TABLE 2), the time-course of relevant parameters and by the original recurrence plot (FIG. 1).

On the pooled data, statistically significant correlations were found between linear and nonlinear parameters. In particular, MEAND correlated positively with MEANRR ($p < 0.02$), P ($p = 0.0001$), LF ($p = 0.0001$), and HF ($p = 0.0001$), and negatively with LF/HF ($p < 0.04$). %REC and %DET correlated negatively with MEANRR ($p < 0.03$), P ($p < 0.002$), LF ($p < 0.02$), and HF ($p < 0.02$). ENT correlated negatively with MEANRR ($p<0.02$), P ($p = 0.0001$), LF ($p = 0.0001$), and HF ($p = 0.0001$), and positively with LF/HF ($p < 0.02$). MAXL correlated negatively with P ($p < 0.05$) and LF ($p < 0.02$) (TABLE 3).

DISCUSSION

The obese subjects displayed lower heart rate variability, mainly in the vagally modulated frequency component, with a relative sympathetic dominance. Furthermore, they showed greater periodicity and lower complexity properties than lean subjects (TABLE 2, FIG. 1). The PSA parameters related to periodicity were well correlated with the RQA parameters related to complexity. In the obese, loss of chaotic properties seems to be strictly linked with a reduced vagal response, as shown by the correlation between chaotic parameters, such as ENT and MEAND, and the LF/HF ratio. RQA parameters evaluated in the two groups were significantly different whereas LF/HF ratio, although higher in the obese group, did not reach statististical significance presumably because of the small sample size. In larger groups of subejcts, Muscelli et al.[6] have reported significantly higher values of LF/HF in obese than in lean nondiabetic subjects.

Our findings indicate that (1) RQA is a valuable tool to study heart rate varibility; (2) autonomic outflow modulates both linear responses to endogenous stimuli and nonlinear properties of heartbeat; (3) obesity, a clinical condition with an excess of cardiac morbidity and mortality[7] possibly due to autonomic dysfunction,[6] is characterized by consensual changes in both linear and nonlinear characteristics of heart rate variability.

REFERENCES

1. TASK FORCE OF THE EUROPEAN SOCIETY CARDIOLOGY AND THE NORTH AMERICAN SOCIETY OF PACING AND ELECTROPHYSIOLOGY. 1966. Circulation **93**(5): 1043–1065.

2. PAGANI, M. *et al.* 1986. Circ. Res. **59:** 178–193.
3. WEBBER, JR., C.L. *et al.* 1994. J. Appl. Physiol. 76: 965–973.
4. TRULLA, L.L. *et al.* 1996. Phys. Lett. A **223:** 255–260.
5. ECKMANN, J.P. *et al.* 1987. Europhys. Lett. **4:** 973.
6. MUSCELLI, E. *et al.* 1998. J. Clin. Endocrinol. Metab. **83:** 2084–2090.
7. LISSNER, L. *et al.* 1991. N. Engl. J. Med. **315:** 1839–1844.
8. MAMMOLITI, R. *et al.* 1998. Comp. Cardiol. **25:** 145–148.

Fractal Analysis in Human Pathology

P. LUZI,[a] G. BIANCIARDI, C. MIRACCO, M.M. DE SANTI, M. T. DEL VECCHIO, L. ALIA, AND P. TOSI

Institute of Pathological Anatomy and Histology, University of Siena, Via delle Scotte 6, 53100 Siena, Italy

INTRODUCTION

Living structures may be described as being in a self-organizing, fluctuating steady-state far from equilibrium.[1] Self-organization and a state far from equilibrium are characteristics of chaotic structures. Chaotic structures present fractal geometry, so is not too astonishing that the branching pattern of the airways in the lung or the arterial vascular pattern of the cardiovascular system have been described with fractal properties.[2–4]

Like coastlines, a tumor examined by light microscopy has a complex, irregular border and retains a similar level of complexity over a range of magnifications.[5–7] Euclidean morphometric measurements were found to be invalid outside precisely defined conditions of resolution and magnification.[8]

In our Institute we are applying fractal dimension analysis to study human tumors at light and ultrastructural levels. Here, we present data obtained studying the epithelial–connective tissue interface in basal cell carcinoma of the skin, the boundaries of invasive bladder carcinomas (urothelial neoplasia), and the lymphocytic nuclear membrane in mycosis fungoides and chronic dermatitis.

MATERIAL AND METHODS

At the level of light microscopy, samples of basal cell carcinoma (BCC) of the skin, of different diagnostic classes, and of invasive bladder carcinomas (IBC) were analyzed. In particular, 147 cases of BCC were selected from the files of the Institute of Pathological Anatomy and Histology of the University of Siena. Hematoxylin-and eosin-stained paraffin sections of each BCC were reviewed by a single pathologist and assigned to three diagnostic categories: (1) Circumscribed BCC (CBCC) are tumors composed of large islands (one or more) of basaloid cells, aggregated in cohesive clusters bound together by a fibrovascular stroma. The tumor margins are convex and the neoplasms grows expansively with a regular front of invasion. (2) Infiltrative BCC (IBCC) are tumors that lack a central cohesive mass of basal cell islands as seen in solid BCCs, consisting of elongated islands and cords that are widely separated spatially. (3) Mixed BCC (MBCC) are tumors that have a mixture of solid and infiltrative growth patterns. Five-micron-thick sections were stained with monoclonal antibodies against human cytokeratins and used for image analysis.

---

[a]Address for correspondence: Anatomia Patologica, Via delle Scotte 6, 53 100 Siena, Italy. +39 577 263234 (voice); +39 577 263235 (fax); luzi@unisi.it (e-mail).

TABLE 1. Fractal dimension of histological outlines of basal cell carcinoma tissue, confusion matrix between actual and predicted group membership, after linear discriminant analysis

Group[a]	Number of Cases	Predicted Group Membership[b]		
		1	2	3
CBCC	60	51(85%)	9 (15%)	0(0%)
MBCC	39	12(31%)	21(54%)	6(16%)
IBCC	48	3(6%)	9(19%)	36(75%)

[a]CBCC, circumscribed basal cell carcinoma; MBCC, mixed basal cell carcinoma; IBCC, infiltrative basal cell carcinoma.
[b]Percent of cases correctly classified = 74%; $p < 0.001$.

Twenty-seven cases of IBC taken from the records of the Institute were reviewed by a single pathologist and assigned to diagnostic categories (histological grades). Twelve cases were classified as low-grade (well-preserved cellular pattern, cells with increased and uniform size, fine and even chromatin, papillary and loose cluster cell arrangement) and 15 as high-grade tumors (low-preserved cellular pattern, cells with increased and pleomorphic size, coarse and uneven chromatin, isolated and loose cluster arrangement).[9] Images of keratine-stained sections were caught by a CCD camera.

At the level of transmission electron microscopy, 30 samples of the skin were taken. Diagnosis was performed by a dermatopathologist on the basis of histological and clinical aspects. In particular, nuclei of T-cells from 15 cases of chronic dermatitis and from 15 cases of a malignant lesion of the skin (early stages of mycosis fungoides) were grabbed, and the nuclear contour used for image analysis.

Fractal analysis was performed by using the box-counting method, using software developed by us (Visual Basic 3.0). Briefly, each image was covered with a net of square boxes (from 4 to 100 pixels), and the number of boxes containing any part of the outline was counted. A log–log plot of the reciprocal side length of the square against the number of outline-containing squares was performed. The slope of the best linear segment of the graph represented the fractal dimension of the image.[4,5]

RESULTS

Linear discriminant analysis was applied by a Wilks' lambda statistic, and the results summarized in a confusion matrix, in order to evaluate the predictive significance of the fractal dimension with respect to the qualitative classification of the basal cell carcinomas: none of the cases of Circumscribed BCC was placed in the Infiltrative BCC group. The percent of correct classification was 74% (TABLE 1).

Variance analysis was used to analyze the samples of invasive bladder carcinomas (urothelial neoplasia) of different histological grades. The lowest histological grade resulted in the lowest value of fractal dimension (TABLE 2, $p < 0.05$).

The Mann-Whitney test was applied to verify the mean differences between the data obtained from mycosis fungoides and chronic dermatitis: the fractal dimension of the former was higher than the latter (TABLE 2, $p < 0.001$).

TABLE 2. Fractal dimension of histological outlines in two different histological types of invasive bladder carcinoma (IBC) (*top*) and of the nuclear contour in chronic dermatitis (CD) and early mycosis fungoides (EMF) (*bottom*)[a]

Low-grade	High-grade	
IBC = 1.35 + 0.20, $n = 12$	IBC = 1.50 + 0.17, $n = 15$	$p < 0.05$
CD = 1.12 + 0.01, $n = 15$	EMF = 1.20 + 0.02, $n = 15$	$p < 0.001$

NOTE: Mean values ± standard devation.
[a]IBC, invasive bladder carcinoma; CD, chronic dermatitis; EMF, early mycosis fungoides.

DISCUSSION

Image analysis is a powerful tool for isolating important discriminating factors useful in diagnostic decisions. Several different morphometric measurements have been performed to discriminate between different diagnostic classes; examples include the measure of nuclear diameter, perimeter, area, and respective form factors. However, all these Euclidean parameters present many problems when applied to the highly irregular biological structure, for example, all are dependent on scale or, in other words, on the magnification. In the recent years, in the field of pathology some works have highlighted the power of a fractal approach in obtaining useful information concerning diagnosis and prognosis in humans.[4–7]

Here we present our original data concerning the great capacity of fractal analysis to distinguish among diagnostic classes in the study of tumors. Fractal analysis, performed with the box-counting method, has revealed the ability of this approach to distinguish between circumscribed and infiltrative patterns of basal cell carcinoma of the skin, thus originating a nonsubjective method to analyze that type of tumor. Moreover, fractal analysis has allowed us to obtain an index that distinguishes between tumors of different histological types, both low and high grade, in the urothelial neoplasia. Finally, at the ultrastructural level, fractal analysis of the outline of the lymphocyte nucleus allows us to distinguish between chronic dermatitis and early mycosis fungoides.

REFERENCES

1. EIGEN, M. 1987. Stufen zum leben. R. Piper GMBH & Co. KG, München.
2. MAINSTER, M.A. 1990. The fractal properties of retinal vessels: embryological and clinical implication. Eye **4:** 235–241.
3. LONG, C.A. & J. E. LONG. 1992. Fractal dimension of cranial sutures and waveforms. Acta Anat. **145:** 201–206.
4. CROSS, S.S. 1994. The application of fractal geometry analysis to microscopic images. Micron **25**(1): 101–113.
5. CROSS, S.S. *et al.* 1994. Fractal geometry analysis of colorectal polyps. J. Pathol. **172:** 317–323.
6. LOSA, G.A. & F. NONNENMACHER. 1996. Self-similarity and fractal irregularity in the pathologic tissues. Mod. Pathol. **9:** 174–182.
7. CROSS, S.S. 1997. Fractals in the pathology. J. Pathol. **182:** 1–8.
8. PAUMGARTNER, D., G. LOSA & E.R. WEIBEL. 1981. Resolution effect on the stereological estimation of surface and volume and its interpretation in terms of fractal dimensions. J. Microsc. **121:** 51–63.
9. MURPHY, W.M. *et al.* 1984. Urinary cytology and bladder cancer. Cancer **53:** 1555–1565.

Recurrence Quantification Analysis in Molecular Dynamics

CESARE MANETTI,[a,b] MARC-ANTOINE CERUSO,[a] ALESSANDRO GIULIANI,[c] CHARLES L. WEBBER, JR.,[d] AND JOSEPH P. ZBILUT[e]

[a]*Department of Chemistry, University of Rome "La Sapienza," Piazzale Aldo Moro, 5-00185 Rome, Italy*
[c]*Istituto Superiore di Sanitá, TCE Lab, Rome 00161, Italy*
[d]*Department of Physiology, Loyola Univ. Medical Center, 2160 South 1st Avenue, Maywood, Illinois 60153, USA*
[e]*Department of Molecular Biophysics and Physiology, Rush University, 1653 W. Congress, Chicago, Illinois 60612, USA*

INTRODUCTION

The quantitative analysis of molecular dynamics (MD) trajectories implies the need for the individuation of salient phenomena embodied in time series data: unique patterns in the dynamics require taxonomies. This need has engendered, as a consequence, the use of classical multivariate data analysis techniques such as principal components analysis (PCA)[1,2] and cluster analysis (CA).[3]

An important criterion for the choice of analysis method for MD trajectories is the method's dependence on dynamical components of the data set, and its relative independence from purely statistical characteristics. This requirement is expressible in terms of "phase information'" sensitivity and is broadly defined as having properties that are destroyed by random shuffling of the series itself (shuffling sensitive information). From this perspective, the usual statistical descriptors (e.g., mean, rms) do not carry any phase information (shuffling resistant), while both PCA and CA retain some information about the dynamics of the system (shuffling sensitive).

The development of nonlinear dynamics has enlarged the range of possible descriptors of time series. One of most promising methods that has been found to be diagnostically useful in the quantitative assessment of time series structure in fields ranging from molecular dynamics to physiology[4–6] is recurrence quantification analysis (RQA). RQA is a technique originally developed by Eckmann, Kamphorst, and Ruelle[7] as a purely graphical descriptive tool to analyze dynamic processes without any stationarity constraint. Recently, Webber and Zbilut demonstrated the high sensitivity of this technique in the detection of subtle state changes of the studied dynamics.[8,9]

RQA is based on the computation of the distance matrix between the rows of the embedding matrix correspondent to the studied series. The distance matrix gives a general picture of the autocorrelation structure of the series and permits us to derive useful information about the dynamics. The local character of the distance function,

---

[b]Address for correspondence: manetti@caspur.it (e-mail).

upon which RQA is based, makes the technique particularly suited for detecting fast transients in the dynamics.

In this work, RQA was applied to the potential energy time series of conformational space explored during MD simulations. The main goal was to demonstrate the ability of RQA to discriminate between the dynamics of a simple system [a Lennard-Jones (LJ) fluid, which does not carry any phase information since it is a purely statistical, shuffling-resistant system] and those of a complex system, such as the MD trajectory of the B1 immunoglobulin G-binding domain of streptococcal protein G (B1-IgG) simulated in water (a protein that demonstrates shuffling-sensitive phase information because of the existence of structured paths between its microstates).[10]

The testing strategy is straightforward: whereas in the case of an LJ fluid the RQA measures must remain invariant after shuffling, they should change significantly in the case of the protein. We will try to sketch a physical characterization of RQA measures relative to MD.

MATERIALS AND METHODS

LJ System MD Simulations

For the LJ simulation, we considered a system of 125 particles, enclosed in a cube of side L, with periodic boundary conditions interacting through a two-body potential of the LJ type:

$$V(r) = 4[(\sigma/r)^{12} - (\sigma/r)^6]$$

with the parameter, $\sigma = 3.405$Å, corresponding to argon, so the energies are expressed in units of ε ($\varepsilon = 119.8$K)[11]. The simulation were performed at different temperatures and varying L at different density. We used the same protocol for both simulations: as a first step, the initial velocities were taken from a Maxwellian distribution to perform 200 psec of simulation with only the last 100 psec being used for analysis. The potential energy time series was sampled at 0.05 psec.

Protein MD Simulation

All simulations of the protein were performed with the GROMACS simulation package[12]. A modification[13] of the GROMOS87[14] force field was used with additional terms for aromatic hydrogens[15] and improved carbon—oxygen interaction parameters.[13] SHAKE[16] was used to constrain bond lengths, allowing a time step of 2 fsec.

The initial protein configuration was taken from the protein databank (1pga)[17]. The protein was immersed in a pre-equilibrated box of SPC water,[18] while four water molecules with the highest electrostatic potential were replaced by sodium ions, resulting in an electrically neutral cubic box ($a \cong 4.1$ nm) containing 1790 water molecules and four sodium counter ions for a total of 5936 atoms. Care was taken that all crystallographic water molecules be conserved.

In order to prepare the solvated system for molecular dynamics, a three-step procedure was followed. Using a restraining harmonic potential, all heavy atoms of the protein and the crystallographic water oxygens were constrained to their initial positions while surrounding SPC water molecules were first minimized and then submitted to 5 psec of constant volume MD at 300°K. The resulting system was then minimized, without any constraints, before starting constant temperature and constant volume MD. A nonbonded cutoff of 1.2 nm was used for both LJ and coulomb potentials. The pair lists were updated every 10 steps. A constant temperature of 300 K was maintained by coupling to an external bath[19] using a coupling constant ($\tau = 0.002$) equal to the integration time step (2 fsec). A total of 1.9 nsec of simulation were produced in this manner. The potential energy of the protein was sampled every 0.1 psec.

Data Analysis

The potential energy series of both LJ fluid and protein was submitted to a 10 dimensional embedding.

The embedding space was constructed by the method of delays[20]: the space is generated by the construction of a multivariate matrix having as columns the original series shifted by a fixed lag consecutively applied to the series. The embedding dimension equals the number of columns of the matrix. Example: given the series: 10, 11, 21, 32, 41, ... the corresponding three-dimensional embedding space at lag = 1 is:

$$
\begin{array}{ccc}
10 & 11 & 21 \\
11 & 21 & 32 \\
21 & 32 & 41 \\
32 & 41 & . \\
41 & . & . \\
. & . & .
\end{array}
$$

The rows of the embedding matrix correspond to subsequent epochs of length n (embedding dimension). In our case the lag was equal to one and corresponded to 0.05 psec. The choice of 10-dimensional embedding was dictated by the need of having a sufficient dimensionality to highlight even minor oscillators.[21]

On the embedding matrix we applied the RQA. The RQA is based on the computation of a distance matrix between the rows (epochs) of the embedding matrix of a time series. This distance matrix is displayed by darkening the pixel located at specific (i, j) coordinates that correspond to a distance value between i and j rows (epochs) lower than a predetermined cutoff. The features of the distance function make the plot symmetric ($D_{i,j} = D_{j,i}$) and with a darkened main diagonal corresponding to the identity line ($D_{i,j} = 0, i = j$). The darkened points individuate the recurrencies (recurrent points) of the dynamic process and the plot can be considered to be a global picture of the autocorrelation structure of the system under study[7,8] (FIG. 1).

Besides the global impression given by the graphic appearance of the plot, Webber and Zbilut[8] developed five quantitative descriptors of the recurrence plot that were demonstrated to be very useful in quantifying the evolution of the dynamic process studied.

The RQA descriptors are: recurrence (REC), which quantifies the percentage of the plot occupied by recurrent points. It corresponds to the proportion of recurrent

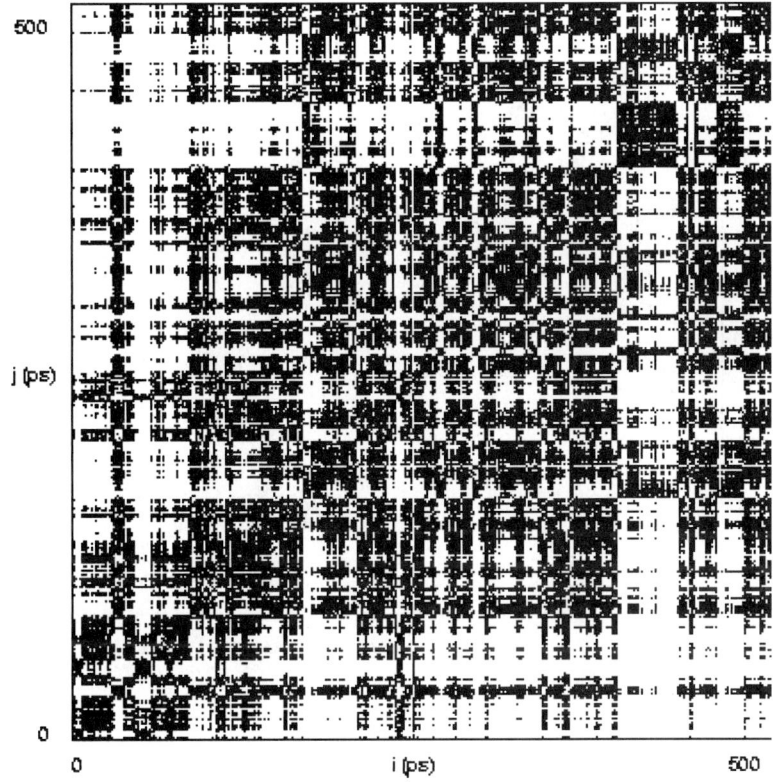

FIGURE 1. Protein recurrence plot.

pairs over all the possible pairs of epochs or, equivalently, the proportion of pairwise distances below the chosen radius among all the computed distances. Determinism (DET) is the percentage of recurrent points that appear in sequence, forming diagonal line structures in the distance matrix. DET corresponds to the amount of patches of recurrent behavior in the studied series, that is, to portions of the state space in which the system resides for a time longer than expected by chance alone (see Refs. 22 and 23). This is a crucial point: a recurrence can, in principle, be observed by chance whenever the system explores two nearby points of its state space, on the contrary, the observation of recurrent points consecutive in time (and then forming lines parallel to the main diagonal) is an important signature of deterministic structuring.[7,24] The superposition between determinism and Lyapunov exponents is proof of this point.[7] Entropy (ENT), which is defined in terms of the Shannon-Weaver formula for information entropy,[8,25] is computed over the distribution length of the lines of recurrent points and measures the richness of deterministic structuring of the series. LYAP is simply the length (in terms of consecutive points) of the longest recurrent line in the plot. LYAP was found to accurately predict ($r = 0.93$) the value of

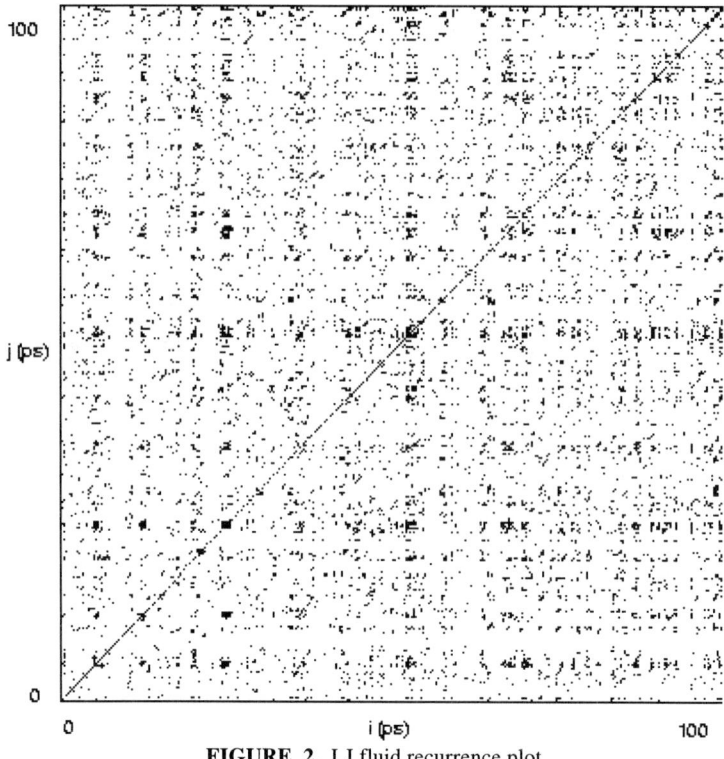

FIGURE 2. LJ fluid recurrence plot.

the maximum Lyapunov exponent in a logistic map going from a regular to a chaotic regime.[26] Finally, TREND is the regression coefficient of the relation between time (in terms of distance from the main diagonal) and the amount of recurrence. TREND quantifies the fading away of recurrence going forward in time and represents a measure of stationarity.[8] Additionally, a time series of any of the RQA descriptors can be produced by windowing the originally embedded scalar series and overlapping in a manner similar to time-varying spectral plots.

RESULTS

All RQA descriptors were computed for both the LJ fluid and the protein MD potential energy time series. In particular, for the fluid, data from a simulation at $T = 0.8$ (usual reduced unit) were used. To test for the null hypothesis that the MD series are stochastic, the original trajectories were randomly shuffled to obtain 30 copies of each series (TABLE 1). The 95% confidence intervals for the RQA descriptors were computed, and the position of the original series relative to the confidence intervals checked. Except for REC (and here it is noted that the value for the MD sim-

TABLE 1. RQA results: shuffled versus original series

LJ fluid	MD Simulation	Shuffled Mean	95% Conf. Int.	Range
REC	0.688	0.76	0.74–0.78	0.66–0.87
DET	39.03	39.47	38.85–40.08	36.57–42.63
ENT	2.347	2.33	2.31–2.35	2.25–2.50
LYAP	16	17	16–18	14–21
TREND	−0.005	0.009	−0.03–0.05	−0.23–0.24
Protein				
REC	5.12	0.72	0.69–0.75	0.58–0.86
DET	69.48	39.20	38.08–40.31	33.20–44.88
ENT	3.25	2.31	2.27–2.36	2.06–2.62
LYAP	31	14	13–15	10–21
TREND	-2.27	0.033	−0.08–0.15	−0.45–0.81

ulation falls within the range of obtained shufflings), the null hypothesis for LJ fluid could not be rejected, pointing to the stochastic character of the fluid simulation. For the protein MD, however, the RQA values were well beyond the confidence limits of the shuffled series, thus demonstrating the presence of strong "phase information" for the protein dynamics. These features are qualitatively evident when looking at the recurrence plots of the protein (FIG. 1) and the LJ fluid (FIG. 2): while protein shows a very rich and intermingled texture, the LJ plot is much more homogeneous.

DISCUSSION

A relevant portion of the theoretical work on MD was based on LJ fluid simulations performed by Rahman.[27] These trajectories can be defined as recurrent, Hamiltonian, mixing, and K-flow, or according to some authors, "Lyapunov unstable."[28] The simulated LJ system evolves toward an equilibrium state, and the constant energy surface defined by the initial conditions is accessible to the system itself. The motion of such a system is at least mixing so as to sample the entire explored surface.

In the case of the LJ fluid, the result obtained, in Eckmann, Kamphorst, and Ruelle terms,[7] can be defined as "autonomous," that is, typical of a system evolving according to time-independent equations: this corresponds to our operational definition of a "shuffling-resistant" potential energy time series. In fact, the RQA measurements of the shuffled series are not statistically different from the original series. This behavior corresponds to a randomlike sampling of the phase space of the system, even if the sampling is driven by a deterministic "engine" such as MD. These kinds of "experiments" were used by Verlet[11,29] to compute thermodynamic properties of fluids by means of a formalization introduced by Birkhoff,[30,31] and based on Boltzmann's view of ergodicity assumptions.

The recurrence plot of the protein simulation (FIG. 1) immediately shows the impossibility of direct averaging of the data. The simple visual inspection of the plot highlights abrupt changes in the texture pointing to multiple minima in the trajectory (rugged landscape as opposed to flat surface). More importantly, the shuffling procedure significantly alters the numerical values of the RQA descriptors, thus demonstrating the "shuffling sensitivity" of the underlying trajectory. The "ergodic" constraints of complete accessibility and mixing are not sufficient to make the system evolve to an equilibrium situation given the finite time of the simulation, and as a result, the trajectory is trapped in a limited portion of the energy surface. In such situations we can speak of metastable states,[32,33] which obviate the possibility of computing direct averages. In order to compute physical measures on such simulations, the local minima of the phase space must be revealed and their relative depth estimated.

The thermalization algorithms used in MD are not guaranteed to preserve the microcanonic properties of the system[19]; nevertheless, we think that the quantitative RQA measures can be correlated with the thermodynamic properties of the system under investigation. In any case, these measures allow us to derive some useful information about the shape of the energy landscape of the simulation. As a matter of fact, the basic algorithm of recurrence plots was developed by Eckmann and Ruelle[34] with the aim of reconstructing the dynamics relative to a time series in a finite dimensional space and of generating a tangent map of the reconstructed dynamics in order to calculate Lyapunov exponents.

In the recurrence plot of the LJ fluid (FIG. 2), the loss of any preferential directionality of the system (quantitatively proved by the shuffling invariance) is clear. The protein recurrence plot (FIG. 1) has a preferential directionality in time (shuffling sensitivity) that allows us to appreciate the effective dimensionality of the explored conformational space. With adequate sampling time, we can resolve both the Frauenfelder substates[10,35] in terms of large-scale typology of the plot and the features of the single substate in terms of texture.

It is important to note that, in practice, several different variables could have been chosen for the MD simulations, such as coulomb energy, or van der Waals' energy. Potential energy was chosen because it is a global descriptor of the physical motions that can be adequately sampled[36] relative to long relaxation times as compared to anharmonic motions.[37] Thus potential energy is suited for the studied time scale. In a related matter, while the generalized ergodic measure (GEM) of Straub and Thirumalai[38] can compute a distribution of energy barriers between substates; currently, this is not possible with RQA. The advantage of RQA over GEM is its sensitivity to local fast transients, thus being much more suited to dynamically individuate phase transitions. In this sense, the two methods are complementary: GEM is better for energy characterization; RQA is better for time resolution.

Finally, it should be remarked that RQA allows for the identification of putatively important events along the studied dynamics using only one variable (potential energy) instead of many (e.g., single dihedral angles). The meaning of these events may, in fact, require specific analyses using variables such as vibrational times, ring flips of amino acids, folding times, protonation times, or diffusion times of water, and require a case-by-case approach.

Our analysis is particularly useful in the calculation of the free energy difference between reactant and product in which various MD simulations can reduce uncertainties in free-energy calculations[39]; although, as with MD studies in general, it is still limited by the general time scale of physiologically important substates which are on the order of tens of micro- to milliseconds.

In summary RQA seems to constitute a very promising tool for the characterization of conformational substates in MD simulations.

REFERENCES

1. GARCIA, A.E. 1992. Large-amplitude nonlinear motions in proteins. Phys. Rev. Lett. **68:** 2696–2699.
2. AMADEI, A., A.B.M LINSSEN & H.J.C. BERENDSEN. 1993. Essential dynamics of proteins. Proteins **17:** 412–425.
3. KARPEN, M.E., D.J. TOBIAS & C.L. BROOKS, III. 1993. Statistical clustering techniques for the analysis of long molecular dynamics trajectories: analysis of 2.2-ns trajectories of YPGDV. Biochemistry **32:** 412–420.
4. GIULIANI, A. & C. MANETTI. 1996. Hidden peculiarities in the potential energy time series of a tripeptide highlighted by a recurrence plot analysis: a molecular dynamics simulation. Phys. Rev. E **53:** 6336–6340.
5. MESTIVIER, D., N.P. CHAU, X. CHANUDET, P. BAUDECEAU & P. LARROQUE. 1997. Relationship between diabetic autonomic dysfunction and heart rate variability assessed by recurrence plot. Am. J. Physiol. **272:** H1094–1099.
6. FAURE, P. & H. KORN. 1997. A nonrandom dynamic component in the synaptic noise of a central neuron. Proc. Natl. Acad. Sci. USA **94:** 6506–6511.
7. ECKMANN, J.-P., S.O. KAMPHORST & D. RUELLE. 1987. Recurrence plots of dynamical systems. Europhys. Lett. **4:** 973–977.
8. WEBBER, C.L., JR. & J.P. ZBILUT. 1994. Dynamical assessment of physiological systems and states using recurrence plot strategies. J. Appl. Physiol. **76:** 965–973.
9. WEBBER, C.L., M.A. SCHMIDT & J.M. WALSH. 1995. Influence of isometric loading on biceps EMG dynamics as assessed by linear and nonlinear tools. J. Appl. Physiol. **78:** 814–822.
10. NIENHAUS, G.U., J.D. MÜLLER, B.H. MCMAHON & H. FRAUENFELDER. 1997. Exploring the conformational energy landscape of proteins. Physica D **107:** 297–311.
11. VERLET, L. 1967. Computer "experiments" on classical fluids. I. Thermodynamical properties of Lennard-Jones molecules. Phys. Rev. **159:** 98–103.
12. VAN DER SPOEL, D., H.J.C. BERENDSEN, A.R. VAN BUUREN, E. APOL, P.J. MEULENHOFF, A.L.T.M. SIJBERS & R. VAN DRUNEN. 1995. Gromacs User Manual. Nijenborgh 4, 9747 AG Groningen, The Netherlands. Internet:http://rugmd0.chem.rug.nl,gmx.
13. VAN BUUREN, A.R., S.-J. MARRINK & H.J.C. BERENDSEN. 1993. A molecular dynamics study of the decane/water interface. J. Phys. Chem. **97:** 9206–9212.
14. VAN GUNSTEREN, W.F. & H.J.C. BERENDSEN. 1987. Gromos Manual. BIOMOS, Biomolecular Software, Laboratory of Physical Chemistry, University of Groningen, The Netherlands.
15. VAN GUNSTEREN, W.F., S.R. BILLETER, A.A. EISING, P.H. HÜNENBERGER, P. KRÜGER, A.E. MARK, W.R.P. SCOTT & I.G. TIRONI. 1996. Biomolecular Simulation: The GROMOS96 Manual and User's Guide. Biomos b.v. Zürich, Groningen.
16. RYCKAERT, J.-P., G. CICCOTTI & H.J.C. BERENDSEN. 1977. Numerical integration of the cartesian equations of motion of a system with constraints: molecular dynamics of n-alkanes. J. Comp. Phys. **23:** 327–341.
17. GALLAGHER, P.T., P.B. ALEXANDER & G.L. GILLIAND. 1994. Two crystal structures of the B1 immunoglobulin-binding domain of streptococcal protein G and comparison with NMR. Biochemistry **33:** 4721–4729.
18. BERENDSEN, H.J.C., J.P.M. POSTMA, W.F. VAN GUNSTEREN & J. HERMANS. 1981. *In* Intermolecular Forces. B. Pullman, Ed.: 331–342. D. Reidel Publishing Company. Dordrecht, The Netherlands.

19. BERENDSEN, H.J.C., J.P.M. POSTMA, W.F. VAN GUNSTEREN, A. DI NOLA & J.R. HAAK. 1984. Molecular dynamics with coupling to an external bath. J. Chem. Phys. **81:** 3684–3690.
20. BROOMHEAD, D.S. & G.P. KING. 1986. Extracting qualitative dynamics from experimental data. Physica D **20:** 217–236.
21. TAKENS, F. 1980. *In* Dynamical Systems and Turbulence. D.A. Rand & L.-S. Young, Eds.: 366–381. Springer-Verlag. New York, Heidelberg, Berlin.
22. ZAK, M., J.P. ZBILUT & R.E. MEYERS. 1997. *In* From Instability to Intelligence: Lecture Notes in Physics, M49. Springer. Heidelberg, Germany.
23. ZBILUT, J.P., M. ZAK & R.E. MEYERS. 1996. A terminal dynamics of heartbeat. Biol. Cybern. **75:** 277–280.
24. ZBILUT, J.P., A. GIULIANI A. & C.L. WEBBER, JR. 1998. Recurrence quantification analysis and principal components in the detection of short complex signals. Phys. Lett. A **237:** 131–135.
25. SHANNON, C.E. 1948. A mathematical theory of communication. Bell. Syst. Tech. J. **27:** 379–423.
26. LLIGONA-TRULLA, L., A. GIULIANI, J.P. ZBILUT & C.L. WEBBER, JR. 1996. Recurrence quantification analysis of the logistic equation with transients. Phys. Lett. A **223:** 255–260.
27. RAHMAN, A. 1964. Correlation in the motion of atoms in liquid argon. Phys. Rev. **136:** A405–411.
28. HAILE, J.M. 1992. *In* Molecular Dynamics Simulation: Elementary Methods: 53. John Wiley. New York.
29. LEBOWITZ, J.L., J.K. PERCUS & L. VERLET. 1967. Ensemble dependence of fluctuations with application to machine computations. Phys. Rev. **153:** 250–254.
30. BIRKHOFF, G.D. 1931. Proof of the ergodic theorem. Proc. Natl. Acad. Sci. USA **17:** 656–659.
31. V. NEUMANN, J. 1932. Physical Applications of the Ergodic Hypothesis. Proc. Natl. Acad. Sci. USA **18:** 263–266.
32. FORD, J. 1973. The transition from analytic dynamics to statistical mechanics, Adv. Chem. Phys. **24:** 155–185.
33. HONEYCUTT, J.D. & D. THIRUMALAI. 1990. Metastability of the folded states of globular proteins. Proc. Nat. Acad. Sci. USA. **87:** 3526–3529.
34. ECKMANN, J.-P., S.O. KAMPHORST, D. RUELLE & S. CILIBERTO. 1986. Liapunov exponents from time series. Phys. Rev. A **34:** 4971–4979.
35. ANSARI, A., J. BERENDZEN, S.F. BOWNE, H. FRAUENFELDER, I.E.T. IBEN, T.B. SAUKE, E. SHYAMSUNDER & R.D. YOUNG. 1985. Protein states and proteinquakes. Proc. Natl. Acad. Sci. USA **82:** 5000–5004.
36. CLARAGE, J.B., T. ROMO, B.K. ANDREWS, B.M. PETTITT & G.N. PHILLIPS, JR. 1995. A sampling problem in molecular dynamics simulations of macromolecules. Proc. Natl. Acad. Sci. USA **92:** 3288–3292.
37. STEINBACH, P.J. & B.R. BROOKS. 1996. Hydrated myoglobin's anharmonic fluctuations are not primarily due to dihedral transitions. Proc. Natl. Acad. Sci. USA **93:** 55–59.
38. STRAUB, J.E. & D. THIRUMALAI. 1993. Theoretical probes of conformational fluctuations in S-peptide and RNase A/3'-UMP enzyme product complex. Proteins **15:** 360–373.
39. HODEL, A., T. SIMONSON, R.O. FOX & A.T. BRUNGER. 1993. Conformational substates and uncertainty in macromolecular free energy calculations. J. Phys. Chem. **97:** 3409–3417.

Photodegradation of Organic Waste Coupling Hydrogenase and Titanium Dioxide

G. M. MURA,[a] M. L. GANADU, G. LUBINU, AND V. MAIDA

Department of Chemistry, University of Sassari, via Vienna, 2, 07100 Sassari, Italy

INTRODUCTION

Dairy-farming waste is a major environmental polluting agent because of its high biological (BOD) and chemical oxygen demand (COD). This is mainly due to the high content of lactose in dairy waste. Thus, the major problem in dairy waste disposal or reuse is the lactose component. It may be possible to reuse the lactose and proteins present in dairy waste by means of separation systems, which allow them to be purified from the dairy waste.[1]

The use of the separation techniques is environmentally advantageous, although relatively expensive especially for small producers. In small cheese factories scattered over a large area (e.g., Sardinia or South Italy), the collection of dairy waste and its disposal is a difficult and expensive procedure. Even when conditions are favorable, only 10% of dairy waste is re-utilized in this way.

Dairy waste may also be degraded, and the main aim of this work was to evaluate the possibility of lactose degradation using the photocatalytic process based on TiO_2.

Recently there have been many studies on the photooxidation of the pollutants in the liquid and gas phases.[2,3] TiO_2 seems to be the most widely studied photocatalyst, and it is considered to be the most interesting one because of its stability, photocatalytic efficiency, environmental friendliness, availability, and low cost.[4]

SCHEME 1 presents the degradation mechanism using TiO_2. When the photon hits the TiO_2, an electron is excited from the valence band to the conduction band of TiO_2, thus inducing a positive charge (hole, h^+) and a negative charge (electron, e^-) on the semiconductor surface. The h^+ site oxidizes an electron donor, and the electron reduces an electron acceptor.[5,6] The electron may be transferred either directly to the electron acceptor or by means of an electron carrier (methylviologen (MV) in the scheme 1) used as an intermediate. Molecular oxygen is the electron acceptor commonly used in the environmental application of photocatalysis, because of the involvement of its intermediates in the next oxidation step of the catalytic process.[7] The photocatalytic oxidation of organic compounds is of great use in dealing with environmental problems, and their efficient degradation is the theme of many studies. For example, TiO_2 has been used to degrade highly toxic dyes,[8] purify drinking water,[9] and in the reductive removal of metal species.[10] However, the use of TiO_2 in recycling foodstuffs, in which concentrations of pollutants are normally higher than in other wastes has not been extensively studied.

---

[a]Address for correspondence: +39 79 229542 (voice); +39 79 229559 (fax); billia @ssmain.uniss.it (e-mail).

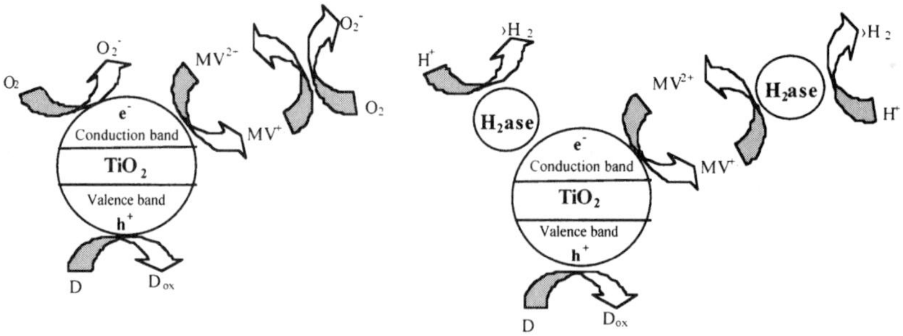

SCHEME 1

The technologies using enzymes, highly efficient and specific biocatalysts, are one of the most promising lines of research in this field. Enzymes coupled to photo-induced semiconductors such as TiO_2, SnO, or CdS have been used to catalyze a large number of redox reactions.[11,12] For example, the complete degradation of formic acid by means of CdS coupled to *Thiocapsa roseopersicina* hydrogenase has been reported recently.[13]

In this work, the degradation of dairy waste was checked in two ways: (a) measuring the disappearance of O_2 used as electron acceptor; and (b) measuring the H_2 production, using a hydrogenase extract from the hyperthermophilic bacterium *P. furiosus* as electron acceptor. The use of the latter system leads to the production of a clean fuel as the final product of the reduction process, that is, H_2. The coupling between TiO_2 and hydrogenase is more efficient if an electron carrier is present and many hydrogenases show no reactivity if one is not present.[14] The most widely used electron carrier with hydrogenases is methylviologen and this compound was used in the experiments described in this work.

MATERIALS AND METHODS

Chemicals

TiO_2 powder (P25, 50 m^2/g) was obtained from Degussa. Sodium ditionite was obtained from BDH. Tris, methylviologen (MV), and synthetic sea water were purchased from Sigma. Phosphate salts were obtained from Merk. Lactose, freeze-dried dairy waste, and protein fraction were kindly provided by Alimenta S.r.l. (Macomer, Italy). The chromatographic media and columns were obtained from Pharmacia.

Growth of Pyrococcus furiosus

Pyrococcus furiosus (DSM 3638) was cultured at 85°C in a 5-liter vessel, inoculated with 0.5 liter of preculture and mechanically stirred under an argon flow (40 ml/min). The growth medium composition was: yeast extract (1 g/l), triptone (5 g/l), cysteine (0.5 g/l) and synthetic sea water (30 g/l). The pH was adjusted to about 6.2 before inoculum. The cells were collected at an OD of about 0.5 at 600 nm after 24 hours, centrifuged, and stored at −80°C.

Hydrogenase Purification

Hydrogenase purification and activity assay were performed as previously described.[15]

Gas-Chromatographic Measures of Hydrogen Production and Oxygen Consumption

Hydrogen and oxygen were measured on a Perkin Elmer 8500 gaschromatograph with a molecular sieve column (1 m, 1/16″). Gas carrier: argon (15 ml/min). Detector: HWD range 1, injector temperature: 200°C, oven temperature: 35°C.

The photoreactions were performed by irradiating the reaction vials with an 80-watt high-pressure Hg lamp (Philips, XPL-N), in 14-ml pyrex vials, jacketed and thermostated at 70°C with water, magnetically stirred, and closed with silicone rubber stoppers.

The reaction mixtures (7 ml) for the hydrogen photoproduction were composed of the following: 0.4 U/ml hydrogenase (200 µl from a stock with specific activity of 40 U/mg); 2 mM methylviologen, 20 mM phosphate buffer, 50 mM lactose, or dairy waste (5 or 50 mM in lactose); protein fraction (when present) (1, 5, or 10 mg/ml), adjusted with HCl or NaOH to the desired pH; 4 mg/ml TiO_2; Hydrogenase was added after the reaction mixture had been degassed for 10 minutes with Argon. Dairy waste was diluted to obtain a concentration in lactose of 5 or 50 mM, on the basis of its composition. The protein fraction was derived from the separation process of dairy waste. The vials were thermostated at 70°C, and the reaction started by exposure to the prewarmed lamp.

The reaction mixtures to measure oxygen consumption were prepared in the same way, with no hydrogenase added and without the degassing procedure. For the gas analysis, 100-µl samples were extracted with a gas syringe from the headspace of the reactor.

Oxygraphic Measures of Oxygen Consumption

A 80-W, high-pressure Hg lamp (Philips, HPLN) was located behind a box that had a tightly closed cylindrical photoreactor. The photoreactor consisted of a 16-cm^3 Pyrex glass vial surrounded by a 1-cm-thick, thermostated jacket. The radial distance from the UV lamp to the reactor was 10 cm. Solutions were magnetically stirred and thermostated at 40°C.

A YSI model 5300 oxygen monitor with a 5331 standard oxygen probe (Clark electrode) was used to measure the oxygen consumption. The electrode was inserted into the reactor to expell all the air of the head space through the access slot. When the system was operating correctly, there was a linear relationship between dissolved oxygen and probe current.

Calibration in air-saturated water solutions at known temperature and pressure and of known composition was necessary. The actual oxygen consumption rate was determined by setting as 100% an air-saturated sample of aqueous solution, which contained 4.83 ml O_2/ml at pH 7.5, 1 atm, and 40°C. When the display read 100%, an output of 1 VDC +/− 0.002 was recorded. The output signal was sent to a personal computer to accurately elaborate the data by means of an I/O board (National Instruments, PC-LPM-16/PnP). About 0.13% error accrued for every 15 minutes of operation, due to the oxygen consumption by the polarographic probe.

Water solutions of lactose (α-lactose, Sigma) or dairy waste in the presence 20 mM of phosphate buffer, pH 7.5, were used as electron donors The photoreactor was filled with an aqueous suspension containing 4 mg/ml of TiO_2 (P25 anatase, 50 m^2/g, Degussa) and the appropriate donor concentrations (0–50 mM). The reaction volume was 14 ml for all experiments. Before the photocatalytic reaction started, the reaction medium was saturated with air. Suspensions were irradiated with the prewarmed lamp, and reactions were followed until the oxygen completely disappeared. All the experiments were performed at least twice.

RESULTS AND DISCUSSION

To study the photodegradation of dairy waste and lactose two systems were used: the classic one, with TiO_2 and O_2 as final electron acceptor, and TiO_2 with hydrogenase. We measured the O_2 disappearance rate in the first system and the H_2 production rate in anaerobic conditions in the second system. The consumption of O_2 and the production of H_2 in tightly closed and thermostated vials was tested to measure the electron donor photodegradation. Because *P. furiosus* hydrogenase is active above 70°C, a temperature of 72°C was chosen for the experiments with hydrogenase as well as for those with oxygen.

To study the lactose and the dairy waste photodegradation, a lactose concentration of 50 mM and a lactose concentration in dairy waste of 50 mM were used. Becuse a significant decrease of pH, from 8.0 to about 4.0, was observed during the photoreduction process, buffered solutions were used in the experiments with both hydrogenase and oxygen. The considerable decrease in pH could render the enzyme inactive, as it only functions in a narrow pH range (approximately from pH 6 to 8.5). A buffer with 20 mM of Na_2HPO_4/NaH_2PO_4 was used. This concentration was chosen to minimize the inhibitory effects of phosphate on TiO_2 reported earlier.[3] The results obtained at pH values 6.5, 7.5, and 8.5 for two different electron donors, lactose

TABLE 1. Effect of pH on H_2 production rate (*upper*) and O_2 consumption rate (*lower*)

Donor	rH_2 μmol/(l·min)		
	pH 6.5	pH 7.5	pH 8.5
Lactose 50 mM	28 ± 3	27 ± 5	25 ± 12
Dairy waste 50 mM	Traces	Traces	Traces
Blank	0	0	0
	rO_2 μmol/(l·min)		
	6.5 pH	7.5 pH	8.5 pH
Lactose 50 mM	23 ± 10	27 ± 12	8 ± 8
Dairy waste 50 mM	Traces	Traces	0
Blank	0	0	0

CONDITIONS: electron acceptor: hydrogenase 0.4 U/ml (upper) or oxygen (lower); electron donor: lactose 50 mM or dairy waste 50 mM in lactose. TiO_2 4 mg/ml, MV 2 mM, phosphate buffer 20 mM.

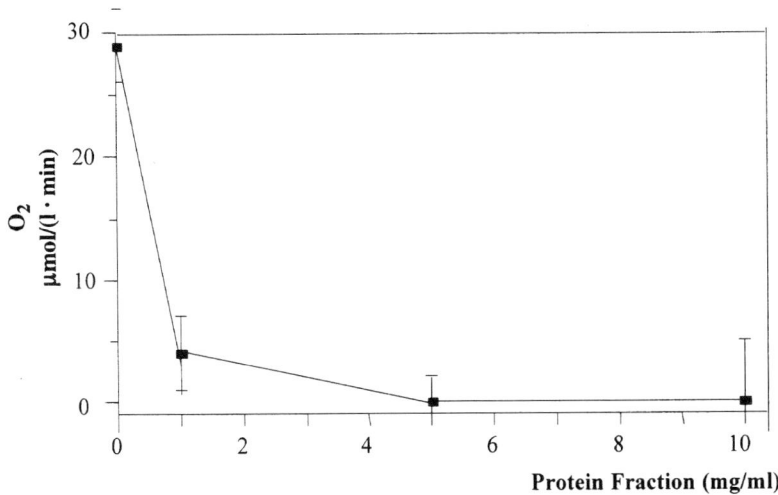

FIGURE 1. Effect of protein fraction addition on O_2 consumption rate. Conditions: 4 mg/ml TiO_2, 50 mM lactose, 2 mM MV, 20 mM phosphate buffer.

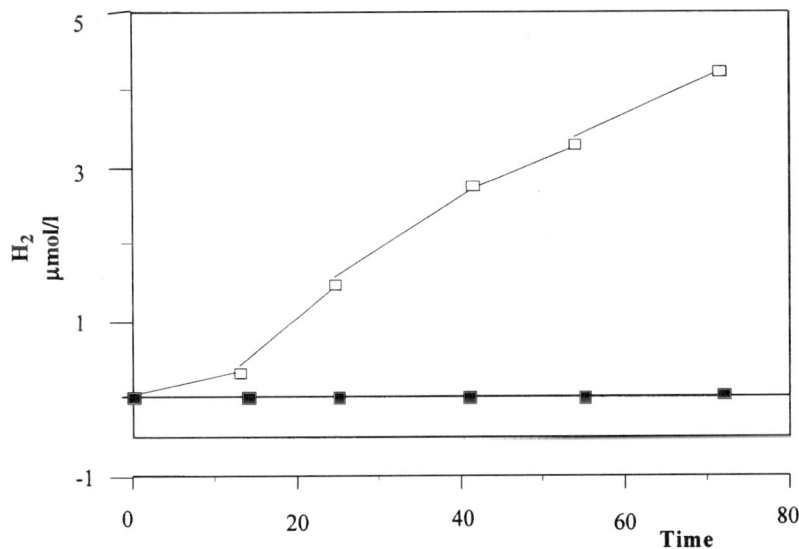

FIGURE 2. Effect of protein fraction addition of protein on H_2 production rate. Conditions: TiO_2 4 mg/ml, , 2 mM lactose, 10 mg/ml protein, 2 mM MV, 20 mM phosphate buffer, 0.4 U/ml hydrogenase. □ without protein, ■ with protein: (10 mg/ml).

and dairy waste are shown in TABLE 1. The H_2 production rate as well as the O_2 disappearance rates are similar for both the electron acceptors (hydrogenase and oxy-

TABLE 2. Effect of dairy waste dilution

Donor	MV	H_2 μmol/(l·min)
No donor	+	0
Dairy waste: 50 mM, pH 7.5	+	0
Dairy waste: 5 mM, pH 8	+	3 ± 0.7
Dairy waste: 5 mM, pH 6	+	14 ± 1.2
Dairy waste: 5 mM, pH 8	−	0
Dairy waste: 5 mM, pH 6	−	0

H_2 production rate in the presence or absence of an electron carrier and under different pH conditions. Conditions: 4 mg/ml TiO_2, 0.4 U/ml Hydrogenase, dairy waste, 5 mM in lactose; 2 mM MV, 20 mM phosphate buffer.

gen) at pH 6.5 and 7.5, while some decrease in the hydrogen production rate was observed at pH 8.5.

When dairy waste is used as an electron donor, neither H_2 production nor O_2 consumption are observed. Because lactose is the principal component of the dairy waste studied in this experiment, the protein component seems likely to be responsible for the inhibition of the process observed in our experiment.

To verify this hypothesis, a protein solution obtained from the dairy waste was added to the pure lactose solution. With the addition of increasing quantities of protein to the 50 mM lactose solution, using O_2 as an electron acceptor, a decrease in the reaction rate was already observed at a protein concentration of 0.5 mg/ml (FIG. 1). A similar effect was observed for the reaction in which hydrogenase was used as an electron acceptor (FIG. 2). As in dairy waste, if the protein concentration exceeds 5 mg/ml, its inhibitory effect on the dairy waste degradation could be a serious problem. This effect is due to the adsorption of the protein on the surface of the TiO_2; thus, diluting the dairy waste could result in its effective degradation.

The results showing the effect of the dilution on the degradation process are reported in TABLE 2. It is clear that, in the system containing hydrogenase, the 10-fold dilution of the solution considerably increases the reaction rate. Thus, the degradation of dairy waste under the experimental conditions used in this experiment is feasible when electron donor concentrations are low, always in the presence of the electron carrier MV. Inasmuch as the adsorption of the reacting species on the TiO_2 surface may be influenced by the net charge of the surface,[16] the effect of the pH on the H_2 production rate was also observed. Indeed at pH 6 an increase in H_2 production rate was measured compared to that with pH 8 (TABLE 2).

Among the main components of dairy waste, that is, lactose and proteins, only lactose acts as an efficient electron donor, and pH variation has no effect on its reaction rate. Thus the pH effect is probably due to the inhibiting effect of the proteins. The overall charge of protein molecules is strongly pH-dependent and their absorption on the TiO_2 surface, which is related to the net charge of the surface,[16] is strongly influenced by the pH. By fixing the proper pH, one can decrease protein absorption on the surface of the photocatalyst, thus increasing the lactose degradation rate.

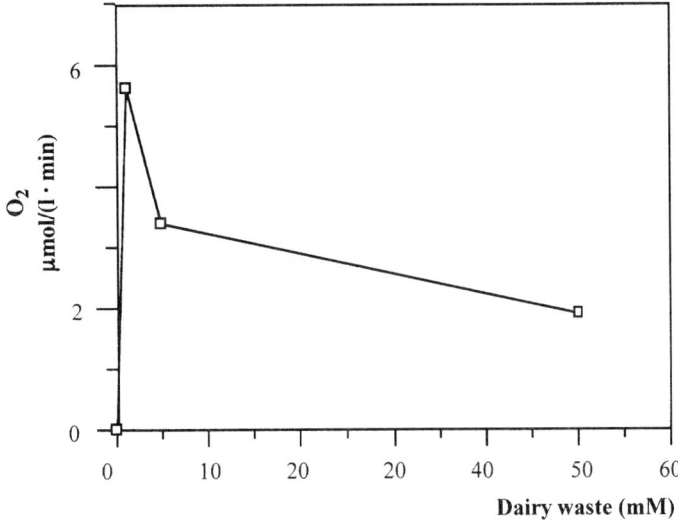

FIGURE 3. Effect of dairy waste dilution on O_2 consumption rate: oxymetric data. Conditions: 4 mg/ml TiO_2, dairy waste (1, 5, or 50 mM in lactose), 2 mM MV, phosphate buffer 20 mM.

The use of the electron carrier, MV, for dairy waste degradation with hydrogenase seems to be necessary. When MV is not used, no H_2 production is observed (TABLE 2).

In the case in which oxygen is the electron acceptor, the gas-chromatographic data shows no reaction rate if dairy waste is the electron donor, even if the dairy waste is diluted 10 times (data not shown). The experiments were repeated using the more sensitive oxymeter technique. The oxymetric data indicates (FIG. 3) that, even in the absence of MV, diluting the dairy waste from 50 mM to 1 mM (with respect to the lactose) causes the reaction rate to increases.

These results indicate that it is possible to degrade dairy waste using either hydrogenase or oxygen. A relatively low concentration of dairy waste is needed to make the degradation process effective. In the case of oxygen as electron acceptor, degradation is likely even in the absence of the electron carrier methylviologen, whereas in the case of hydrogenase, without MV the degradation reaction is negligible.

SUMMARY

In this work the results of the study on the degradation of dairy waste and lactose with use of titanium dioxide (TiO_2), as photocatalyst, are presented. The ability of TiO_2 to degrade dairy waste is compared either in the presence of molecular oxygen or of a bacterial hydrogenase as electron acceptor. The enzyme-mediated H_2 production rates or the O_2 consumption rates are used to measure electron donor degradation.

The results obtained clearly indicate that dairy waste can be degraded by the TiO_2 photocatalyst. Proteins present in the dairy waste have a strong inhibitory effect on the degradation process. Indeed lactose alone can easily be degraded. When the protein extract obtained from the dairy waste was added to the lactose solution, the reaction for both electron acceptors, hydrogenase and oxygen, was inhibited. When the dairy waste solutions were diluted, there was a positive effect on the reaction rate. This was particularly true in the case of hydrogenase, and to a lesser extent in the case of oxygen acceptor. The reduction of the pH from 8 to 6 also increased H_2 production when the enzyme was used.

REFERENCES

1. KESSLER, H.G. 1981. Food engineering and dairy technology. Verlag. Freising, Germany.
2. HOFFMANN, M.R., M.T. MARTIN, W. CHOI & D.F. BAHNEMANN. 1995. Environmental applications of semiconductor photocatalysis. Chem. Rev. **95:** 69–96.
3. MILLS, A. & S. LE HUNTE. 1997. An overview of semiconductor photocatalysis. J. Photochem. Photobiol. A: Chem. **108:** 1–35.
4. RAJESHWAR, K. & J.G. IBANEZ. 1995. Electrochemical aspects of photocatalysis: application to detoxification and disinfection scenarios. J. Chem. Educ. **72:** 1044–1049.
5. LINSEBIGLER, A.L., G. LU & J.T. YATES, JR. 1995. Photocatalysis on TiO_2 surfaces: principles, mechanisms and selected results. Chem. Rev. **95:** 735–758.
6. RABANI, J., K. YAMASHITA, K. USHIDA, J. STARK & A. KIRA. 1998. Fundamentals reactions in illuminated titanium dioxide nanocrystallite layers studied by pulsed laser. J. Phys. Chem. B. **102:** 1689–1695.
7. PELIZZETTI, E. & C. MINERO. 1993. Mechanism of the photo-oxidative degradation of organic pollutants over TiO_2 particles. Electrochim. Acta **38:** 47–55
8. RICHARDSON, S.D., A.D. THRUSTON, JR., T.W. COLLETTE, K.S. PATTERSON, B.W. LYKINS, JR. & J.C. IRELAND. 1996. Identification of TiO_2/UV disinfection byproducts in drinking water. Environ. Sci. Technol. **30:** 3327–3334.
9. VINODGOPAL, K., D.E. WYNKOOP & P.V. KAMAT. 1996. Environmental photochemistry on semiconductor surfaces: photosensitized degradation of a textile azo dye, acid orange 7, on TiO_2 particles using visible light. Environ. Sci. Technol. **30:** 1660–1666.
10. LAU, L.D., R. RODRIGUEZ, S. HENERY, D. MANUEL & L. SCHWENDIMAN. 1998. Photoreduction of mercuric salt solutions at high pH. Environ. Sci. Technol. **32:** 670–675.
11. WILLNER, I. & D. MANDLER. 1989. Enzyme-catalysed biotransformation through photochemical regeneration of nicotinamide cofactors. Enzyme Microb. Technol. **11:** 467–483.
12. PARKINSON, B.A. & P.F. WEAVER. 1984. Photoelectrochemical pumping of enzymatic CO_2 reduction. Nature **309:** 148–149.
13. NEDOLUZHKO, A.I., I.A. SHUMILIN & V.V. NIKANDROV. 1996. Coupled action of cadmium metal and hydrogenase in formate photodecomposition sensitized by CdS. J. Phys. Chem. **100:** 17544–17550.
14. MURA, G.M., G. GALLI, C. PRATESI, G. FRASCOTTI, L. SERBOLISCA, P. PEDRONI & G. GRANDI. 1996. Photoinduced hydrogen production using titanium dioxide coupled to thermostable hydrogenases. J. Mar. Biotechnol. **4:** 68–74.
15. PEDRONI, P., A. DELLA VOLPE, G. GALLI, G.M. MURA, C. PRATESI & G. GRANDI. 1995. Characterization of the locus encoding the (Ni-Fe) sulfhydrogenase from the

archaeon *Pyrococcus furiosus*: evidence for a relationship to bacterial sulfite reductases. Microbiology **141:** 449–458.
16. KORMANN, C., D.W. BAHNEMANN & M.R. HOFFMANN. 1991. Photolysis of chloroform and other organic molecules in aqueous TiO_2 suspensions. Environ. Sci. Technol. **25:** 494–500.

Complex Rotational Dynamics of the Cu(II)–Bleomycin System in Mobile and Viscous Media

REBECCA POGNI, ELENA BUSI, GIOVANNI DELLA LUNGA, AND
RICCARDO BASOSI[a]

Department of Chemistry, University of Siena, Pian dei Mantellini 44, 53100 Siena, Italy

INTRODUCTION

Metal-free bleomycin is an antitumor antibiotic. Studies on the biological activity of bleomycin indicate that the metal-free bleomycin is activated through its binding of available metal ions, such as Cu(II) and Fe(II). The Cu(II)–bleomycin [Cu(II)Blm] complex is the transport form found in blood and can be considered a "magnetopharmaceutical" compound. Its final target is cancerous tissue where it becomes localized.[1,2] Living systems present a wide range of environments that can alter the dynamics of the nominally isotropically tumbling agents. Rotational correlation time, τ_R, is important in determining proton relaxation enhancement.[3,4] A previous study in the mobilized phase showed a difference in the electron paramagnetic resonance (EPR) parameters for cupric bleomycin in the liquid as opposed to the solid state.[5] Aqueous and 46% wt/vol sucrose solutions have been studied from near physiological temperature to near freezing to determine the dominant rotational correlation time in aqueous and viscous environments.

RESULTS AND DISCUSSION

From a mathematical point of view, the rotation of the complex plays an important role in determining the effectiveness of proton relaxation enhancement in first-sphere, second-sphere, and rotationally modulated outer-sphere relaxation. EPR spectra often are extremely sensitive to rotational motion; thus, τ_R may be determined by simulation of the EPR spectral line shape whenever suitable programs are available.[6]

Rigid limit spectra taken at 100 K were used first to derive the hyperfine coupling constants A ($A_\parallel$ = 175 G) and the g values ($g_\parallel$ = 2.21 and $g_\perp$ = 2.04) for the Cu(II)Blm complex. These values are consistent with those expected for Cu(II) in a square planar array of ligands. Simulations were performed considering the coordination of four nitrogens (A_N = 14.5 G) with the same magnetic parameters. The excellent agreement between the simulated and experimental spectra (FIG. 1) leads us to be confident that the isotropic Brownian diffusional model used for simulating the slow

---

[a]To whom correspondence may be addressed: ++39 577 263545 (voice); ++39 577 263546 (fax); basosi@unisi.it (e-mail).

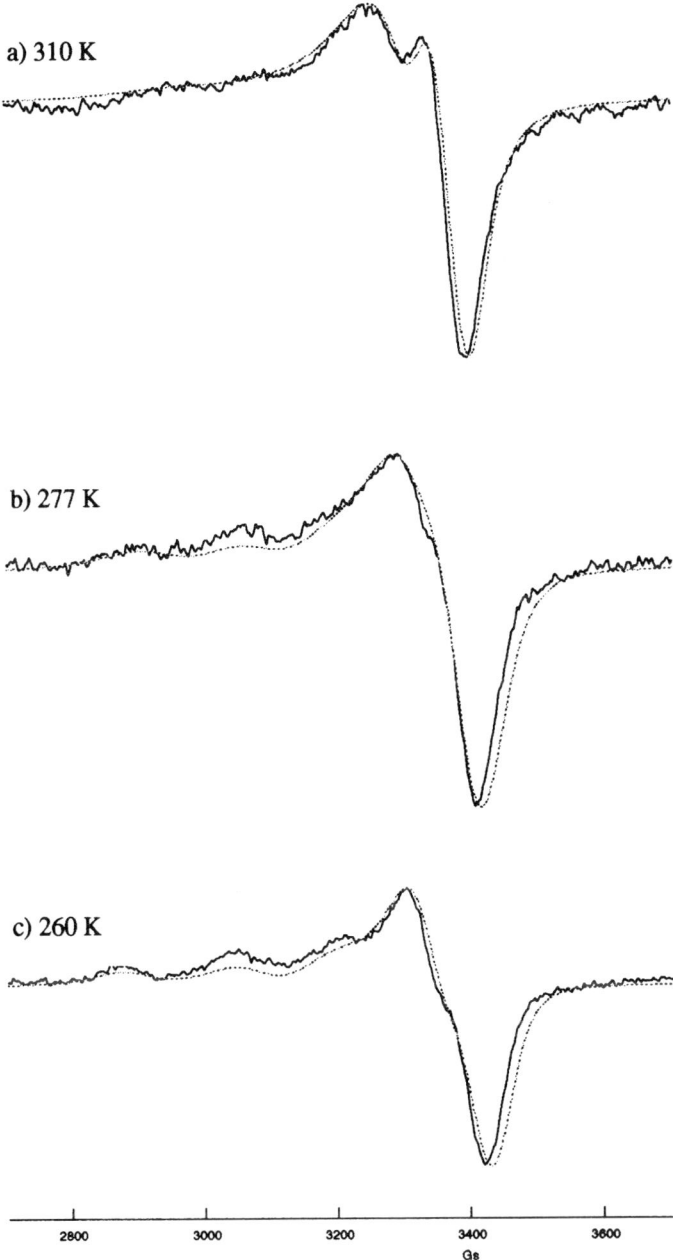

FIGURE 1. Experimental EPR spectra of the Cu(II)Blm complex in aqueous solution pH 7.4, at (**a**) 310 K, (**b**) 277 K, (**c**) 260 K paired with the simulations that eventually gave the best fit.

TABLE 1. Best-fit rotational correlation times (τ_R) from EPR spectra

T(K)	τ_R(s)
310	3.51×10^{-10}
310 + sucrose	8.97×10^{-10}
277	1.06×10^{-09}
260	2.17×10^{-09}

motional EPR spectra is valid. The rotational correlation time τ_R was computed from the isotropic rotational diffusion rate constant, D_R, derived from simulated spectra using the following relationship:

$$D_R = 1/6\ \tau_R$$

In FIGURE 1 the experimental EPR spectra of the Cu(II)Blm complex in aqueous solution from physiological temperature (310 K) to near freezing (260 K) are report-

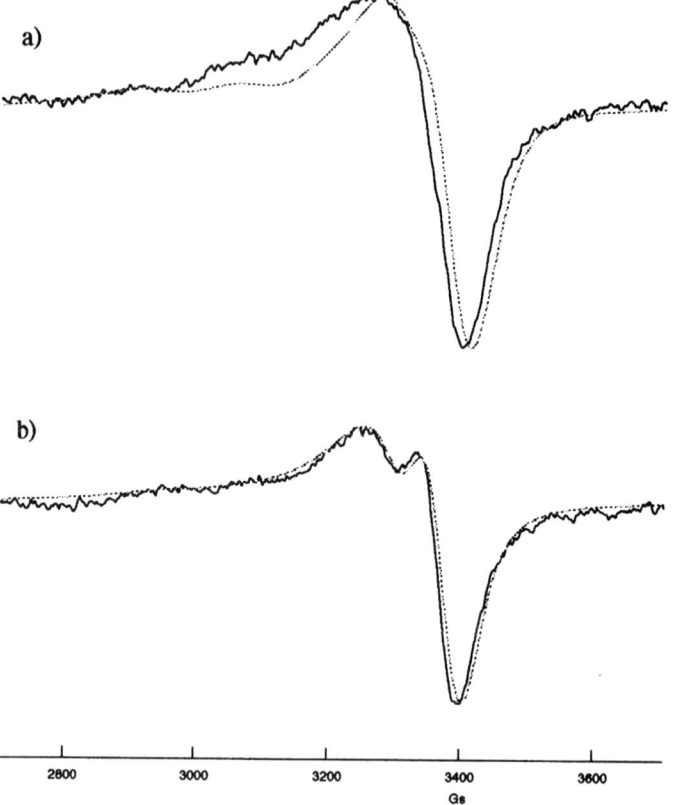

FIGURE 2. (a) Experimental EPR spectrum in 46% wt/vol sucrose solution of the Cu(II)Blm complex. (b) Same EPR spectrum as in (a) but recorded in aqueous solution. All the spectra are paired with their best simulation. pH 7.4, $T = 310$ K.

ed. Spectra are paired with the simulations that eventually gave the best fit. The relative rotational correlation times are reported in TABLE 1. As can be inferred from a lineshape analysis of the EPR spectra at physiological temperature (310 K), the appearance of the spectrum is that of an incipient slow-motion spectrum consistent with the molecular weight of the complex (MW ≈ 1700). As the temperature decreases, the lineshape of the spectrum varies continuously from that reported in FIGURE 1a until the temperature of 277 K is reached, at which point the spectral anisotropy starts to become evident. At 260 K the solution freezes and a dramatic increase in the rotational correlation times occurs (see TABLE 1). A linear relationship between temperature and rotational correlation times is evident. Living systems present a wide range of environments that can alter the dynamics of the nominally isotropically tumbling agents. FIGURE 1 shows how EPR is sensitive to changes in the environment. A 46% wt/vol sucrose solution was used to study the rotational dynamics in a viscous mediium. In fact the sucrose solution simulates environments that may exist *in vivo*. In FIGURE 2a the EPR spectrum of the complex Cu(II)Blm in 46% sucrose solution at 310 K, paired with its best simulation, is reported. In FIGURE 2b the spectrum is compared with that for the same experimental conditions but without sucrose. As can be seen from FIGURE 2, the effect of sucrose reduces the motion of the complex. This is evident from the appearance of anisotropy in the spectrum of FIGURE 2a. The lineshape of this spectrum is very similar to that reported in FIGURE 1b at 277 K, even if in the former case the high field part of the spectrum is broader. Fitting in FIGURE 2a is worse than the previous one, probably due to a more complex rotation.

ACKNOWLEDGMENTS

Blenoxane was generously supplied by Bristol Myers Squibb Co. This study was funded by Cofin.MURST 97 CFSIB

REFERENCES

1. ANTHOLINE, W.E., J.S. HYDE, R.C. SEALY & D.H. PETERING. 1984. Structure of cupric bleomycin. J. Biol. Chem. **259**(7): 4437–4440.
2. PRON, G., J. BELEHRADEK, JR. & L.M. MIR. 1993. Identification of a plasma membrane protein that specifically binds bleomycin. Biochem. Biophys. Res. Commun. **194**(1): 333–337.
3. LAUFFER, R.B. 1987. Paramagnetic metal complexes as water proton relaxation agents for NMR imaging: theory and design. Chem. Rev. **87**: 901–927.
4. CHEN, J.W., F.P. AUTERI, D.E. BUDIL, R.L. BELFORD & R.B. CLARKSON. 1994. Use of EPR to investigate rotational dynamics of paramagnetic contrast agents. J. Phys. Chem. **98**: 13452–13459.
5. ANTHOLINE, W.E., G. RIEDY, J.S. HYDE, R. BASOSI & D.H. PETERING. 1984. ESR parameters for cupric bleomycin in the mobilized state. J. Biomol. Struct. Dynam. **2**(2): 469–480.
6. DELLA LUNGA, G., R. POGNI & R. BASOSI. 1994. Computer simulation of EPR spectra in the slow-motion region for copper complexes with nitrogen ligands. J. Phys. Chem. **98**(15): 3937–3942.

On Crystallization Kinetics of Polytetrafluoroethylene

RACHELE PUCCIARIELLO[a] AND CLAUDIA MANCUSI

Dipartimento di Chimica, Universitá della Basilicata, Via N. Sauro 85, 85100 Potenza, Italy

INTRODUCTION

The kinetics of melt crystallization of polymers have been extensively studied. The major part of the work concerns isothermal crystallization, since this process is easily accessible through the experimental determination of the macroscopic crystallinity as a function of time, for example, through calorimetric measurements.

The description of the macroscopic development of crystallinity in terms of nucleation and linear crystal growth is similar to the Poisson's raindrops on a pond problem.[1] In fact, the crystalline growth of two-dimensional spherulites is analogous to expanding waves generated by raindrops falling on a surface of water. In the framework of this model, the overall crystallization rate, as the development of crystallinity versus time, can be described by the Avrami equation,[2] derived on the basis of the following hypotheses:

1. Primary nucleation is completely random; that is, in the supercooled polymer melt, crystallization nuclei appear at any point of the melt bulk, independently of the development of the process.
2. The number of nuclei increases during the whole process (thermal nucleation, new crystals started growing throughout the crystallization), or all the nuclei act at the same time and instantaneously as the temperature reaches values lower than the melting temperature (athermal nucleation, all the crystals started growing at the same time).
3. The growth of primary nuclei can be mono-, bi-, or three-dimensional.
4. The growth of each single unit is completely free, that is, independent of the previously crystallized polymer and of the development of other centers.

The Avrami equation in its logarithm form, which is the most commonly used, reads: $\log[-\ln(1 - X)] = \log k + n \log t$, where X is the crystallinity degree developed at time t; k contains the rate constants of the stages of nucleation, that is, primary nucleation and crystal growth, and the exponent n is the sum of two contribution, that is, $n = m + a$, where $a = 1$ if nucleation is thermal and $a = 0$ if nucleation is athermal; and m reflects the habit of the growing nuclei and is assumed to have the values of 1, 2, or 3 for linear, circular, or spherical growth, respectively. The general problem of interpretation of the experimental data using the Avrami equation is that fractional exponents from data-fitting are more usual than exceptional. Moreover, the experi-

---

[a]Address for communication: +39-971-474227 (voice); +39-971-474223 (fax); Pucciariello @unibas.it (e-mail).

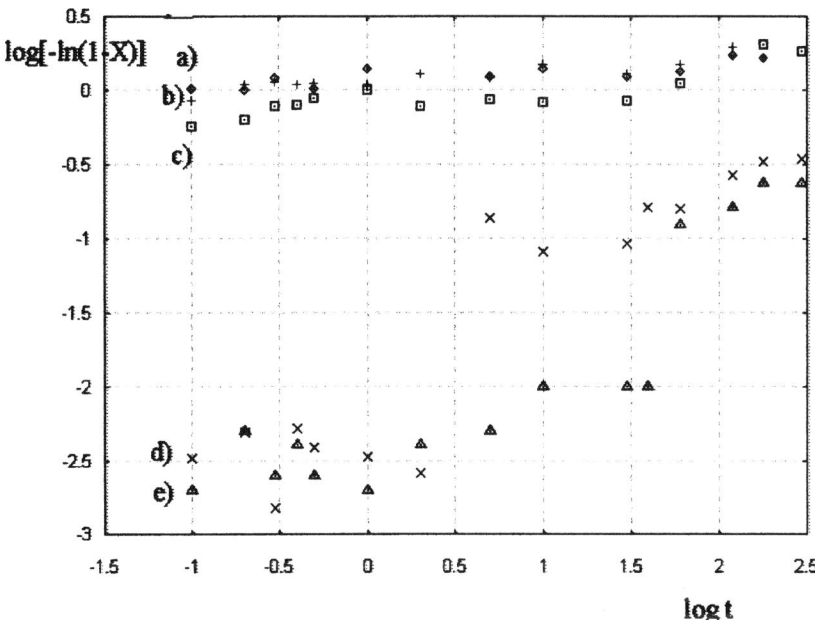

FIGURE 1. Development of crystallinity as $\log[-\ln(1-X)]$ versus $\log t$ for isothermal crystallizations from the melt of PTFE. Actual crystallizations: **(d)** $T_c = 319°C$, **(e)** $320°C$. Secondary crystallizations: **(a)** $T_c = 280°C$, **(b)** $285°C$, **(c)** $318°C$.

mental data frequently deviate at long crystallization times, showing slower increases in the crystallinity than expected from the initial data. In this work we report the study of the crystallization kinetics of polytetrafluoroethylene (PTFE) through differential scanning calorimetry (DSC). The study is further complicated by the difficulty of melt-crystallizing PTFE. In fact, it was reported that crystallization from the melt for this polymer is so rapid that isothermal experiments are prevented: the sample crystallizes during the cooling from the melt to the selected crystallization temperature.[3]

EXPERIMENTAL

Thermal analysis was performed by a differential scanning calorimeter (DSC7, Perkin Elmer) and the data analyzed by the Pyris series thermal analysis system running under Microsoft Windows NT 4.0 on a Pentium Compaq Prolinea 5133 computer. Runs were performed on 10 ± 0.5 mg samples in a nitrogen atmosphere. Before each run the baseline was recorded and optimized, then subtracted from the corresponding DSC curve. The heats of fusion were calculated from the peak areas, and their uncertainty is at most ± 0.3 J/g. Isothermal crystallizations were performed in the DSC apparatus in a nitrogen atmosphere for times ranging from 0.1 to 300 minutes. Each sample was taken to 400°C at 10°C/min, held at this temperature for

TABLE 1. Avrami constants for isothermal crystallizations (n_1 and k_1) and for secondary crystallization (n_2 and k_2) calculated for PTFE as a function of the crystallization temperature T_c

T_c (°C)	n_1	k_1 (min^{-n1})	n_2	k_2(min^{-n2})
280			0.057	0.99
285			0.078	0.99
318			0.12	0.97
319	0.62	16.4		
320	0.58	20.5		

5 minutes (a procedure that was reported to produce negligible thermal degradation) in order to be assured of its complete melting, then cooled to the selected crystallization temperature T_c at the maximum rate allowed by the instrument and taken at T_c for a predetermined time t. Each experiment was repeated until no crystallization during the rapid cooling was observed. Then the sample was reheated at 10°C/min to 400°C. The area under the resulting melting peak is taken as a measure of the crystalline fraction obtained at the selected T_c at the time t. This procedure, although it involves more drudgery than the usually adopted one, reduces errors in the determination of crystallinity originating from the very small DSC signals corresponding to isotherms and allows the evaluation of crystallinity using a very reliable reference value, i.e. the enthalpy of fusion of a perfect crystal of PTFE.[4]

RESULTS

1. Stochastic behavior was observed: sometimes the sample crystallized during the rapid cooling from the melt to T_c, sometimes it did not. Therefore, we repeated each experiment until no crystallization on cooling was observed.
2. For $T_c < 319°C$, crystallization is complete even for the lowest crystallization time allowed by the instrument (6 sec): in this case only secondary crystallization kinetics can be observed.
3. In FIGURE 1 the data of $\log[-\ln(1 - X)]$ versus $\log t$ are reported for actual crystallizations (**d** and **e**) and for secondary crystallizations (**a**, **b**, and **c**). For the first cases a linear trend of the data is observed without the usual breakdown of crystallinity versus time. The value of the kinetic parameters, namely n_1 and k_1, obtained by fitting the data through the Avrami equation, are reported in TABLE 1. The values of n_1 are very low, indicating a monodimensional, highly restricted crystal growth. For the second case the linear trend of the data is also observed, corresponding to the second stage of crystallization, the first one not being detectable at these crystallization temperatures. The kinetic parameters, n_2 and k_2 are also reported in TABLE 1.

DISCUSSION

The values of n_1 are similar to those obtained for liquid crystalline polymers, according to the rod-like and stiff chain character of their molecules.[5,6] At this point let us remember the mesophase character of PTFE in its high temperature phase (polymorph I), recognized as a *condis* (*con*formationally *dis*ordered) crystal.[7] Moreover, morphological studies on melt-crystallized PTFE have shown that its microstructure is constituted of extended chain lamellae, where sufficiently long polymer chains may still be folded at the lamellar surface.[8] Following another viewpoint, melt-crystallized PTFE can be regarded as a chain-folded polymer whose lamellar thickness is much greater than nearly all other crystalline polymers.[9] This kind of microstructure would anyway, correspond to a monodimensional crystal growth, that is, $n \leq 1$, as obtained by us. However, it should be noted that the obtained values of n_1 are not integers, as often observed for most polymers. This has been considered in the past as proof that the parameters in the Avrami equation do not have a physical meaning. By contrast, as one of us has proposed, the noninteger values of n_1 may be related to the *fractal dimension* of the crystallites, taking into account that the functional form of Avrami equation is the same as that defining the dimensionality of an object.[10] At this point let us come back to FIGURE 1 curves a, b, and c, corresponding only to a secondary crystallization process. In fact, as previously mentioned, in these cases crystallization is so fast that the first stage cannot be detected, at least through the method used here. The slight increase in crystallinity versus time may only be due to crystallization of initially amorphous portions of macromolecules and/or to crystal perfection. The very low values observed for n_2 are those expected.

REFERENCES

1. WUNDERLICH, B. 1976. Macromolecular Physics: Crystal Nucleation, Growth, Annealing. Vol. 2. Academic Press. New York.
2. AVRAMI, M. 1941. J. Chem. Phys. **9:** 177.
3. OZAWA, T. 1984. Bull. Chem. Soc. Jpn. **57:** 952.
4. LAU, S.F., H. SUZUKI & B. WUNDERLICH. 1984. J. Polym. Sci. Part B: Polym. Phys. **22:** 379.
5. HEDMARK, P.G., P.E. WERNER, M. WESTDAHL & U.W. GEDDE. 1989. Polymer **30:** 2068.
6. PUCCIARIELLO, R. & C. CARFAGNA. 1994. Colloid Polym. Sci. **272:** 1501.
7. GREBOWICZ, J., S.Z.D. CHENG & B. WUNDERLICH. 1986. J. Polym. Sci. Part B: Polym. Phys. **24:** 675.
8. MELILLO, L., B. WUNDERLICH & Z.U.Z. KOLLOID. 1972. Polymere **250:** 417.
9. BASSETT, D.C. & R. DAVITT. 1974. Polymer **15:** 721.
10. HARRISON, A. 1995. Fractals in Chemistry. Oxford University Press. Oxford, England. 1995.

Conformational Chaos of an Elastin-Related Peptide in Aqueous Solution

VINCENZO VILLANI[a] AND A.M. TAMBURRO

Universitá della Basilicata Dipartimento di Chimica, Via N. Sauro, 85, 85100 Potenza, Italy

INTRODUCTION

Which is the flexibility of a peptide in dilute aqueous solution? In other words, *Which is its dynamical conformational behavior?* Of course, a simple answer is not possible: it will depend upon the *particular* primary structure and on the experimental conditions. Nevertheless, *universal* behaviors controlled by the nonlinear forces at stake are possible.

We are involved in aqueous solutions of essentially flexible and amphiphilic peptides, and because of their ambiguous nature we expect to observe a complex dynamics.

We will deal with a glycine-rich peptide, the tetrapeptide Ac-Gly-Leu-Gly-Gly-NMe, whose sequence frequently recurs in the tropoelastin chain and may characterize the elastic performance of the elastin protein.[1] In a series of previous papers,[2–5] we have characterized the relative stability of conformers and the dynamic behavior of the molecule, either isolated or in solution, detailing the experimental avalaible NMR and CD data.[6]

Amazing, nonlinear dynamic behavior with conformational solitons has been evidenced in the *in vacuo* molecule and a chaotic one for the molecule in aqueous solution. This has been related to the entropic mechanism of the elasticity of elastin as a *transition chaos-soliton* from the relaxed to the stretched form in aqueous medium.[5] In general, we have hypothesized that a transition to chaos could be the universal behavior at the basis of the rubber elasticity.[5] Now, due to a molecular dynamics (MD) simulation in aqueous solution and an original data analysis, we have investigated the complexity and instability of the motion, showing the space–time chaos of the system that could be the basis of the role played by the sequence in biological systems.

MODEL

The MD simulation at standard temperature and pressure by means of the Berendsen method[7] was performed using AMBER 4.1 software.[8] The periodic boundary conditions have been applied, and the cutoff distance criterion of 9Å for nonbonded interactions used.

---

[a]Address for correspondence: +39-971-474227 (voice); +39-971-474223 (fax); Villani@uni-bas.it (e-mail).

In our MD simulation the integration time step δt was 2 fsec, and energies and coordinates were stored every 40 steps ($\Delta t = 0.08$ psec). The time period of 1 nsec was simulated.

A solution box of volume $32.2337 \times 30.4357 \times 27.3159$ Å^3 and density 0.9733 g cm^{-3} with one tetrapeptide Ac-Gly-Leu-Gly-Gly-NMe molecule solvated by 852 waters is the starting point of our simulation. This is the state obtained by MD simulated annealing reported in a previous work[9]: in this way the initial transient state is absent and the whole MD-simulation develops close to the most probable state at thermal equilibrium.

The diffusive properties of some important global variables of the peptide have been considered. In particular, the trajectory of the end-to-end vector $\mathbf{R}_{ee}$ has been analyzed. The end-to-end vector $\mathbf{R}_{ee}$ is defined by the difference of position components of the Cartesian coordinate of the end carbon atoms of acetyl and N-methyl-amide groups.

METHODS

Mean Squared Displacement and Hurst Exponent

The mean squared displacement $\langle \mathbf{R}^2(\tau) \rangle$ is defined as the time-dependent difference correlation function[10]:

$$\langle \mathbf{R}^2(\tau) \rangle := \langle (\mathbf{R}(t_0) - \mathbf{R}(t_0 + \tau))^2 \rangle_{t_0}$$

where τ is the correlation time. The time average denotes the averaging over different time origins t_0.

The time scaling law[10] of the mean squared displacement of the diffusing variable is:

$$\langle \mathbf{R}^2(\tau) \rangle \sim \tau^{2H}$$

where H is the critical exponent of Hurst.[10] We obtained the diffusive exponent H from the slope of the corresponding bilogarithm scale plot:

$$\ln \langle \mathbf{R}^2(\tau) \rangle \sim 2H \ln(\tau)$$

Fractal Dimension

How long is a dynamic path? The length of a trajectory in its space as a function of the time-resolution step T was measured, obtaining a dimension, measure of the jaggedness of the trajectory.[10] In this way the length of the trajectory of the vector $\mathbf{R}$ has been measured and from the log–log plot:

$$\ln(L(n)) \sim d \ln(n)$$

(where $n = T/\Delta t$) the corresponding fractal dimension[10] $D = 1 - d$ estimated.

Intrinsic Divergence Exponents

Similarly to the Lyapunov exponent,[11] we define the intrinsic divergence exponent (IDE) as a measure of the instability of the corresponding trajectory. The intrinsic divergence displacement D_X of a certain trajectory is calculated as vectorial displacement of delayed series with respect to the stationary one. The equations read:

$$D_\mathbf{R}(\tau) := |\mathbf{R}(t_0) - \mathbf{R}(t_0 + \tau)| \sim e^{l\tau}$$

and

$$\ln D_\mathbf{R}(\tau) \sim l\tau$$

where τ is the delay time with respect to the stationary series at time t_0 and l the intrinsic divergence exponent.

Dynamical Fourier Spectra

This method, introduced for the first time in the study of the isolated tetrapeptide, is a very powerful tool for characterizing the nonlinear dynamic behavior.[4] We have taken into account the moduli and the components of the selected global variables. A family of delayed and bounded time series called traveling trajectory pockets (TTP) was generated. In particular, TTP of 500-psec length starting at $t_n = nP$ (n is an integer number and P = 50 psec) from the initial point of MD simulation, have been considered. This family of TTP was used to calculate the corresponding Fourier spectra. In this way a family of power spectra $F_n(\omega, t_n)$ called dynamical fourier spectrum (DFS) was constructed.

RESULTS AND DISCUSSION

1. The peptide in solution is characterized by high intramolecular flexibility: large amplitude motions of molecular backbone and dynamical pattern of H-bonds.
2. The perfectly irregular trajectory $\mathbf{R}_{ee}(t)$ suggest a chaotic intramolecular dynamics according to the expected Brownian walk in solution.
3. The fractal dimension D = 1.6 of $\mathbf{R}_{ee}(t)$ is typical of fractional Brownian motion. The departure from the ideal (D = 1.5) is due to the solute–solvent interactions. The scaling exponent d = 0.6 is in agreement with the Flory's chain size exponent in good solvents[12] ν = 3/5, testifying to the role of water and the equivalence of scaling laws on the chain length or time basis.
4. The observed anomalous diffusion (H = 0.4 instead of 0.5) is in agreement with the fractional Brownian motion of peptides in the complex medium. The delay of diffusion is due to the H-bonds solute-solvent network.
5. The asymptotic positive value of the l_{ee} exponent testifies to the chaotic motion of intramolecular peptide dynamics.
6. The intramolecular vectors (and in particular the end-to-end one) show nonstationary DFS characterized by low-frequency modes damping to the high frequencies, typical of fractional Brownian spectra.
7. The modulus of intramolecular vectors shows stationary, ordinary Brownian DFS, testifying to the independent motions of the corresponding vectorial components and the expected disorder at thermal equilibrium. In contrast, the corresponding spectrum for the *in vacuo* peptide showed the typical soliton mode-sharing of essentially quasiperiodic motion.

8. The dynamics of intramolecular vectors is space-time self-similar, testifying to a disordered behavior of the peptide chain on large observation scales.

CONCLUSIONS

a. The expected stochastic molecular behavior is the observed solution of the dynamical deterministic-nonlinear problem.

b. Water behaves as a good solvent for the peptide and, destroying the intramolecular H-bond network (typical for the nonhydrated peptide), induces the large amplitude nonlinear motions responsible of peptide conformational chaos.

c. The relaxed peptide in solution is in a dynamic, high-entropy state according to the proposed mechanism of the transition to chaos for the elasticity of elastin of which our tetrapeptide is a representative component.

REFERENCES

1. TAMBURRO, A.M. 1990. In Elastin: Chemical and Biological Aspects. A.M. Tamburro & J.M. Davidson, Eds.: 126. Congedo Publ. Italy.
2. VILLANI, V. & A.M. TAMBURRO. 1951. J. Chem. Soc. Perkin Trans. **2**: 1951.
3. VILLANI, V. & A.M. TAMBURRO. 1995. J. Biomol. Struct. Dyn. **12**: 1173.
4. VILLANI, V., L. D'ALESSIO & A.M. TAMBURRO. 1997. J. Chem. Soc., Perkin Trans. **2**: 2375.
5. VILLANI, V., L. D'ALESSIO & A.M. TAMBURRO. 1997. In Elastin and Elastic Tissue. A.M. Tamburro, Ed.: 31. Mario Armento & C. Publ. Potenza, Italy.
6. TAMBURRO, A.M., V. GUANTIERI, L. PANDOLFO & A. SCOPA. 1990. Biopolymers **29**: 855.
7. BERENDSEN, H.J.C., J.P.M. POSTMA, W.F. VAN GUNSTEREN, A DI NOLA & J.R. HAAK. 1984. J. Chem. Phys. **81**: 3684.
8. PEARLMAN, D.A., D.A. CASE, J.C. CALDWELL, G.L. SEIBEL, U. CHANDRA SINGH, P. WEINER & P.A. KOLLMAN. 1991. AMBER 4.0. University of California, San Francisco, CA.
9. VILLANI, V. & A.M. TAMBURRO. 1998. J. Mol. Struct. (Theochem) **431**: 205.
10. BUNDE, A. & S. HAVLIN. 1992. Fractals and Disordered Systems. Springer-Verlag. Berlin.
11. SCHUSTER, H.G. 1995. Deterministic Chaos. VCH. Weinheim, Germany.
12. DE GENNES, P.G. 1979. Scaling Concepts in Polymer Physics. Cornell University Press. Ithaca, NY [fourth printing 1993].

Vanadium Uptake by Yeasts

MARIA ANTONIETTA ZORODDU[a]

Dipartimento di Chimica, Via Vienna 2, 07100 Sassari, Italy

Saccharomyces cerevisiae cells reduce vanadate V(V) to vanadyl V(IV) as a detoxification mechanism. Cells resume growth as V(V) concentration in the medium decreases as a consequence of its reduction to V(IV); the final number of cells is proportional to initial vanadate concentrations.

The results of vanadyl production experiments demonstrate that vanadate reduction occurred during the exponential phase of growth. EPR spectroscopy was used to determine the concentration of vanadyl and its speciation inside the cells and in the supernatants.

Toxicology of vanadium has become an area of great interest because of the increasing amounts of vanadium in the environment as a result of its use in industrial processes. The study of the essentiality and toxicology of vanadium is currently an area of great interest. The mechanisms for both toxic and beneficial effects are not well understood. When vanadium compounds in the (IV) or (V) oxidation states are given to animals, the vanadium is found in the vanadyl VO^{2+} form.[1,3]

In order to understand how inorganic processes can be involved in cellular regulation, we have been studying the alteration of metabolism caused by different concentrations of vanadium in the vanadate form. It is known that *Saccharomyces cerevisiae* reduces V(V) to V(IV)[4–8]; here, we report that the detoxification mechanism is related to this reduction and that cells can gain energy for growth by this reduction.

MATERIAL AND METHODS

Saccharomyces cerevisiae cells, strains S288c and ε1278b, were grown in YEPD medium (1% yeast extract, 2% peptone, 2% glucose, pH 5–6) or in GYNB (2% glucose, 0.7% yeast nitrogen base w/o a.a. Difco), as previously described.[5,7]

Appropriate volumes of the sodium orthovanadate (Sigma) filtered–sterilized stock solution were added at the start of each experiment to obtain the desired concentrations (from 5 to 20 mM) of monovanadium units in the growth medium. When buffer solutions were added (Tris, Hepes), the concentrations used were in the range of 50–200mM.

Vanadium in the (V) oxidation state in the media was determined as reported in the literature.[8] Cell growth was monitored by carrying out a cell count using a hemocytometer slide under a light microscope as well as in terms of absorbance at 600 nm using a Jasco-Uvidec 610 spectrophotometer, and cell viability was measured by the methylene blue staining procedure. Small differences were found using different strains.

---

[a]Address for correspondence: 079-229559 (fax); zoroddu@ssmain.uniss.it (e-mail).

lutions were added to the culture media, the period of stasis was proportional to the concentration of the added buffers. These results suggest that the detoxification mechanism strictly depends on the reduction of vanadate to vanadyl.

In fact, the viability of yeast cells is accompanied by an efflux of H+ in the medium[10] that can in our case heighten concomitantly the following reaction where vanadate and vanadyl are joint by the redox couple:

$$H_2VO_4^- + 4H^+ + 1e^- \Rightarrow VO^{2+} + 3H_2O \ (E_0 = 1.31 \text{ V}, E_{pH} = 7 = -0.34 \text{ V})$$

and imply the resumed growth; vanadate appears to be more toxic than vanadyl species and the vanadate reduction to vanadyl hence contributes to its detoxification.

In conclusion, our results show that the detoxification mechanism is related to the vanadate V(V) reduction to vanadyl V(IV), which is accompanied by cellular activity. Besides, the final number of cells reached was higher in comparison to the greater concentration of vanadate species in the medium, although the toxicity of vanadate militates against such an effect. Microorganisms that reduce metal ions as a detoxification mechanism have an important role in the remediation of environments contaminated with metal ions and certain organic compounds.

REFERENCES

1. SIGEL, H., Ed. 1995. Metal Ions in Biological Systems. M. Dekker. New York.
2. REHEDER, D. 1991. Structure and function of vanadium compounds in living organisms. Biometals **5:** 3–12.
3. WILLSKY, G.R. 1990. *In* Vanadium in Biological Systems. N.D. Chasteen, Ed.: 1–24. Kluer Academic Press. Amsterdam.
4. ZORODDU, M.A., R.P. BONOMO, A.J. DI BILIO, E. BERARDI & M.G. MELONI. 1991. EPR study on vanadyl and vanadate ion retention by a thermotolerant yest. J. Inorg. Biochem. **43:** 731–738.
5. ZORODDU, M.A., M. FRUIANU, R. DALLOCCHIO & A. MASIA. 1996. Electron paramagnetic resonance studies and effects of vanadium in *Saccharomyces cervisiae*. Biometals **9:** 91–97.
6. ZORODDU, M.A. & A. MASIA. 1997. A novel dimeric oxovanadium species identified in *Saccharomyces cervisiae* cells. Biochem. Biophys. Acta **1358:** 249–254.
7. MANNAZZU, I., E. GUERRA, R. STRABBIOLI, A. MASIA, G.B. MAESTRALE, M.A. ZORODDU & F. FATICHENTI. 1997. Vanadium affects vacuolation and phosphate metabolism in *Hansenula polymorpha*. FEMS Microbiol. Lett. **147:** 23–28.
8. PRIYADARSHINI, U. & S.G. TANDON. 1961. Spectrophotometric determination of vanadium (V) with *N*-benzoyl-*N*-phenylhydroxylamine. Anal. Chem. **33:** 435–438.
9. FITZGERALD, J.J. & N.D. CHASTEEN. 1974. Determination of vanadium content of protein solutions by electron paramagnetic resonance spectroscopy. Anal. Biochem. **60:** 170–180.
10. MASIA, A., M.A. ZORODDU, S.V. IVORY & G.M. GADD. 1998. Enrichment with a polyunsatured fatty acid enhances the survival of *Saccharomyces cervisiae* in the presence of tributyltin. FEMS Microbiol. Lett. **167:** 321–326.

A Modeling Approach to the Understanding of Very Complex Dynamics in a Planktonic Community

GRACIELA ANA CANZIANI[a]

Núcleo Consolidado de Matemática Pura y Aplicada, Facultad de Ciencias Exactas, Universidad Nacional del Centro de la Provincia de Buenos Aires, 7000 Tandil, Argentina

INTRODUCTION TO THE PROBLEM

Phaeocystis spp. are unicellular algae, found in temperate oceans and polar seas, where they produce massive blooms. They are also found in tropical waters, where blooms are rare. They occur in at least two different forms, single and colonial. Solitary cells are free-living, 3 to 10 µm in diameter, and can be either flagellated or nonmotile. Colonies are formed from nonmotile cells and consist of cells embedded in a mucilaginous matrix, where they grow and divide. Colonies can break into smaller colonies as the result of shear, but two colonies can not aggregate to form a larger one. Colony diameter varies from 10 µm to 3 mm; but, under particular environmental conditions, much larger ones have been found. Because of the wide range of sizes (up to five orders of magnitude) *Phaeocystis spp.* have the potential of serving as a resource for numerous types of grazers, mainly microzooplankton—such as ciliates, tintinnids, or dinoflagellates—and also larger zooplankton—such as copepods and euphausiids. Massive blooms of *Phaeocystis spp.* have received considerable attention because of their broad impact on the environment. They can have a negative effect on fisheries, with the mucilage clogging the gills of fish, and on tourism, because of the accumulation of sludges and thick foam layers on the beaches, causing serious environmental and economic problems. Also *Phaeocystis spp.* are one of the main phytoplankton communities producing dimethylsulfo-propionate (DMSP), a precursor of dimethylsulfide (DMS), which is known to have an impact on the chemical quality of the atmosphere and on global climate regulation. Because *Phaeocystis spp.* can be large contributors to primary production both in polar regions and in temperate seas, the *Phaeocystis*-dominated planktonic system could play a significant role in global climate, particularly in the processes of removal of CO_2 from the atmosphere and sedimentation of carbon in the ocean.

Much of the research conducted in order to understand the complexity of *Phaeocystis* life history and its role as a dominant primary producer in different environments has been accomplished in the last fifteen years,[1] basically because of the impact of the increasing intensity of the blooms. Despite constant research efforts, many important questions remain unanswered. The many factors involved in the

---

[a]Address for correspondence: ++54-2293-447104 (voice/fax); canziani@exa.unicen.edu.ar (e-mail).

control of the vernal blooms, together with the complexity of *Phaeocystis* life cycle, make any predictions on the development or fate of blooms difficult. Using these conditions, we constructed a mathematical model as a tool to organize data, investigate the relative effect of different factors considered to be important in the development and termination of a bloom, and test some hypotheses on aspects that have not yet been fully studied in the field.

MATHEMATICAL MODELS

Initially, mathematical models were developed as tools to study the importance of different factors in maintaining the stability of marine ecosystems. For example, Steele[2] developed a trophic dynamic model including nitrate, phytoplankton, herbivorous zooplankton, and predator and used it to prove that the rates of energy flow in an ecosystem are the most significant parameters. Later, Steele and Frost[3] reformulated the problem, arguing that by defining a food web only in terms of biomass or energy, the species composition and the age structure of the population are left aside, thus it is *"describing an ecology without species."* This new approach reveals the importance of size structure in the effects of predation and in community dynamics. A few years ago, Steele and Henderson[4] analyzed several recent N-P-Z models, in order to stress the importance of predation on zooplankton in the behavior of the system, as a way to find more general explanations for the difference in patterns observed in the oceans, which are independent of physical phenomena. These approaches to the exploration of the behavior of planktonic ecosystems are a few examples of the usefulness of mathematical models as tools for unveiling counterintuitive mechanisms and revealing patterns, as well as for orienting theoretical speculation and suggesting directions of field research.

Usually, models used in the study of planktonic ecosystems have aggregated populations into compartments representing different trophic levels: phytoplankton, zooplankton, predators. Even more detailed models group species into populations of identical, average individuals. It is true that for practical modeling purposes it is inconvenient and usually just impossible to work with large sets of observables. We have to reduce their numbers, either by ignoring part of them or by grouping them into functional units, and focus our attention on a proper subset. The view we have of the system using this subset is necessarily a partial view. The very essence of an abstraction is to reduce the description of a system to a simpler and hopefully more tractable form. Nevertheless, too much aggregation could be misleading. Some phenomena require a finer degree of resolution in their treatment, particularly when coupled processes occur in different time or spatial scales.

In our case, because of the broad range of sizes *Phaeocystis* covers, the diversity of its grazers, and the dependency on size of the impact of grazing, the organization and structure of the community is of great importance. Here we mean not only the size structure of each population and the size relation between phytoplankton, grazers, and predators, but also the timing of events relating them within the community. We have to consider the fact that the bloom lasts for about 60 days, so that the sequence of events is rather rapid.

CONSTRUCTION OF THE MODEL

The proposed model, consisting of both structured and unstructured components, is constructed in modular form, with each module representing different relationships between components, in order to analyze the role and the magnitude of each potential controlling factor in bloom dynamics. It includes nutrients (nitrogen and phosphorus), grazers (microzooplankton and macrozooplankton), and *Phaeocystis* (single cells and colonial forms) compartments, as well as components representing other phytoplankton, bacteria, effects of shear, and the process of sedimentation (FIG. 1). The model is formulated using ordinary and partial differential equations as well as integral equations.

The *Phaeocystis* module discriminates between single cells and colonies. Single cells grow until they double in volume and then divide, their growth rate depending on the surface of each cell and on the concentration of nutrients in the water. Nutrient uptake follows an additive model that is a variation of the Monod function for the simultaneous limitation of growth by two nutrients.[5] The growth rate is proportional to the balance of uptake and losses. Losses are assumed to be proportional to cell volume. An initial cell size distribution is assumed to be a truncated normal distribution covering the range between a minimum and a maximum cell size for the species. Colonial cells follow similar equations, but rates differ. When a colonial cell divides, the two daughter cells remain within the mucilaginous envelope. Because a colony originates from one cell and all cells in a colony are subject to the same environment, we assume synchronous divisions within the colony.[6,7] This results in the doubling

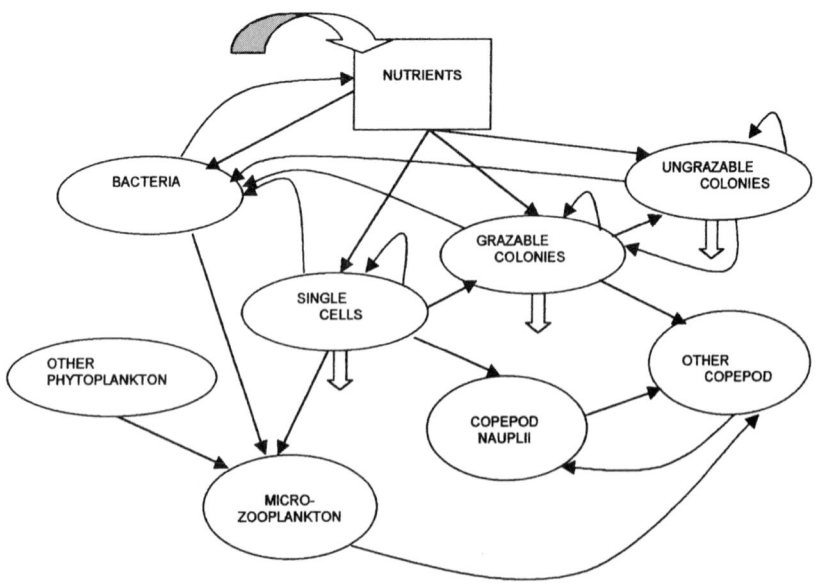

FIGURE 1. Model of a *Phaeocystis*-dominated plankton community.

of colonial cell number. A system of colony sizes is constructed in order to track colonies individually. For functional purposes, we also distinguish two compartments for colonies: one containing colonies of grazable sizes, and the other for ungrazable sizes.

As the algal population evolves, nitrogen and phosphorus concentrations vary proportionally to total cell surface and to the respective additive functions for the simultaneous limitation of growth by two nutrients. The ratios C:N:P within cells are assumed to be constant 106:16:1. Although ratios change as the bloom progresses, this is a reasonable approximation in view of numerical simplicity.

The two types of grazers are modeled differently. Microzooplankton (ciliates) feed only on single cells. In this case population models need not be structured because their individual size is within the same order of magnitude of that of single cells. The ciliate population is represented in an aggregated form, assuming a type II functional response to algal abundance and a mortality that is density dependant, thus reflecting a crowding effect. This simple approach seems sufficient to obtain a satisfactory qualitative behavior.

On the other hand, larger grazers such as copepods need to be modeled in a structured manner because individuals in different stages of development graze on colonies of different sizes. We represent their population in five groups: nauplii, early copepodites (CI–CIII), late copepodites (CIV–CVI), female adults, and male adults. We use individual growth equations to follow each cohort through its development into consecutive stages. We assume that growth and maintenance costs are proportional to individual mass, parameters depending on the stage of development of each individual.[8] If we call $f_2(m)$ the individual growth function resulting from the growth equation and $G_2(m, t)$ the density of copepods of mass m at time t, then the copepod population follows the size structured McKendrick-von Foerster equation

$$\frac{\partial G_2}{\partial t} + \frac{\partial}{\partial m}(f_2 G_2) = -\mu_2(j(m))G_2$$

with boundary condition

$$G_2(m_0, t) = \int_{m_4}^{m_M} \beta(m, t) G_2(m, t) dm$$

where m_0 is the mass at hatching, m_4 and m_M the minimal and maximal masses respectively, for adult females, and $\beta(m,t)$ the birth function.

Also the *Phaeocystis* population follows a size structured McKendrick-von Foerster model. For each cell type (ℓ) we denote by $\rho_0(V^{(\ell)}, t)$ the density of single cells of this type that have the volume $V^{(\ell)}$ at time t, and $\rho_{2i}(V^{(\ell)}, t)$ the density of colonies having 2^i cells of volume $V^{(\ell)}$ at time t, for $i = 1, \ldots, n_{max}/2$. Then the *Phaeocystis* population is described by the $(1 + n_{max}/2)$ equations

$$\frac{\partial \rho_i}{\partial t} + \frac{\partial}{\partial V}(f_* \rho_i) = -\text{losses}$$

for $i = 0, 1, \ldots, n_{max}/2$, where f_* denotes the individual growth function f_s or f_c, depending on the case. The boundary conditions are

$$\rho_0(V_0^{(\ell)}, t) = (2 - \chi)\rho_0(2V_0^{(\ell)}, t)$$

$$\rho_1(V_0^{(\ell)}, t) = \chi\rho_0(2V_0^{(\ell)}, t)$$

where χ is a number such that $0 < \chi < 1$ and represents the probability of a solitary cell forming a colony; for $i > 1$,

$$\rho_i(V_0^{(\ell)}, t) = 2\rho_{i-1}(2V_0^{(\ell)}, t)$$

The term labeled losses on the right hand side of the population equation includes terms representing sinking out of the upper layer of the water column, grazing by ciliates, and/or grazing by copepods. It is assumed that flagellated cells do not sink, but all colonies obey Stoke's Law for small spherical particles. Ciliates graze only on single cells, at a rate proportional to their density following a type II functional response to algal abundance. The effect of copepods grazing on the size-structured colonial population is realized through two different controls: a *preferred size* function $p(i, j(m))$ that depicts the size distribution of particles ingested at each stage j of development of the individual,[9] and a stage dependent *diet* function $D(j(m))$ that determines the portion of the individual's daily total demand, in terms of carbon, met by consuming *Phaeocystis* cells.

Copepods also predate on microzooplankton, so we define $D_c(j(m))$ the stage-dependent diet function that determines the portion of the individual copepod's daily carbon demand met by consumption of ciliates. We assume that adult copepods tend to include a larger proportion of microzooplankton in their diet while earlier stages tend to consume a larger proportion of algae.

Other modules considered include the presence of other types of phytoplankton available to ciliates and a source of nutrients that varies with time. Also, the effects of turbulence in the water column under the form of shear that breaks the colonies, transforming the size distribution was investigated. There is very little information available in the literature on this point, consisting basically of references relating periods of calm or mild winds to the abundance of larger colonial sizes. We propose several expressions for the probability of breaking a colony of a given size depending on the intensity of shear and investigate potential effects of these processes on the dynamics of the community. Colonies often break into two equal portions, but also can break into two asymmetric pieces. In order to cover this cases, the probabilities of breaking into even and uneven portions are defined, and a set of intermediate size classes (indexed by odd numbers) containing $3*2^i$ cells per colony is considered.

Finally, modules that include the bacterial loop in the system and the effects of senescence are developed to investigate the role of bacteria in the development and termination of the bloom. In natural systems a delay in the peak densities of bacteria relative to *Phaeocystis* is observed,[10] but no single satisfactory explanation has been proposed yet. The bacteria model takes into account the abundance of monomers S exuded by *Phaeocystis* cells as well as two types of polymers H_1 and H_2 released into the water column by colonial lysis and/or the breaking of colonies induced by shear. We also consider the possible effects of the production of acrylic acid, an antibiotic, during the exponential growth phase of the bloom on the bacterial population. Details on most of the models developed can be found in References 11 and 12.

HOW ARE MODELS USED?

Simulations are performed in a modular manner, first investigating interactions in a simple setting and gradually adding more complexity. Initially, a minimal set of components is used to evaluate the model's behavior compared to a collection of benchmarks and explore patterns produced. This rudimentary model, including only the nutrient, algae, and microzooplankton compartments, reveals the underlying feedback mechanisms in the system and points to the most significant parameters. Then the effects of the inclusion of additional components and links are analyzed, bringing the sequence of scenarios closer to perceived reality. This process not only helps understand the relative importance of interactions, controls, and feedback, but also validates the model itself. When simulations reach a stage where data or information on certain processes are not available, the model is used to investigate the impact of assumptions on community-level phenomena, such as structure and dynamics, and evaluate whether they are plausible. This information may contribute to the design of experiments that may validate the assumptions.

It is pertinent to note at this stage of the discussion what any modeler should keep in mind, and this is that the model can only reflect what has been taken into account in its construction. When building a sequence of models in a modular manner as proposed here, each step forward into a more complex structure reveals a richer pattern of interactions, but this does not imply that the information obtained from a simpler model is less valuable. With this process of successive steps, it is easier to detect feedback mechanisms that may remain hidden under a more global approach, due basically to the nonlinear nature of the interactions.

The sequence of models permits us to detect interesting interactions and sensitive responses to some factors. For example, the basic model is on the one hand sensitive to a physiological factor, such as the half-saturation constant that describes the character of the microzooplankton functional response to algal abundance, and on the other hand it is also sensitive to a temporal phenomenon such as the timing of copepod grazing. In other words, what species of microzooplankton are present and when the macrozooplankton start grazing are important questions. In most cases grazing does not terminate a bloom, but by depressing algal densities, it enhances nutrient consumption for the remaining population, leading to the formation of larger colonies that sink faster.

The behavior of the system strongly depends on the balance between growth rates and removal rates in both *Phaeocystis* and microzooplankton populations during the first days of the bloom. When copepods prey on microzooplankton, the variety of responses of the system to changing predatory pressure and initial microzooplankton densities increases, in some cases yielding oscillations. The responses are not necessarily the expected ones because of the existence of colonial sizes that escape consumption and the underlying feedback mechanisms. For example, there are cases where an increase in the predatory pressure of copepods on ciliates results in an increase of ciliate densities: an initial depression of ciliate density causes the algal abundance to increase faster, which in turn results in a higher growth rate for ciliates. Under this scenario the *Phaeocystis* single cell population is not necessarily driven to extinction. An interesting pattern emerges in which the system is not bottom or top controlled, but rather the controls shift temporarily from nutrients to grazers and back as growth rates vary with time, since they depend dynamically on resource abundance.

The presence of diatoms or an alternate resource for microzooplankton available early in the bloom tends to homogenize the results, making them less dependent on initial microzooplankton densities. The inflow of nutrients during the bloom, even at high concentrations, does not have an observable effect on the *Phaeocystis* population. An accumulation of nutrients is observable only if the inflow starts in the early stages of the bloom; if it starts later, nutrients are consumed instantly.

The inclusion of shear in the system totally transforms the size distribution of colonies in ways that depend on the intensity of the shearing stress as well as on the timing of it. Because of the effects of shear, more colonies are available to copepods for grazing. But even though removal by grazing is sufficiently intense to surpass removal by sinking from the mixed layer, it is not capable of terminating a bloom. The impact of size-dependent grazing is clearly observed in cases when shear occurs later in the bloom, transforming large, ungrazable colonies into small-sized ones at a time when nutrients have been depleted and colonies have reached peak sizes.

One interesting result is that the effect of shear is intensified either by a very low predatory pressure of copepods on microzooplankton, in which case colonies have more nutrients available to them and can grow larger, or by a relatively high predatory pressure, when colonies, though smaller, are more abundant. There is evidence that an important portion of the colony is lost when colony cleavage occurs, hence shear contributes to the increase of loose organic matter (polymers) in the water column and to particle aggregation. An increase in the intensity of shear causes a decrease in the sedimentation flux.

The module for bacterial dynamics is under study so that we don't yet have a complete set of results to comment on. Nevertheless, we can mention that we are investigating the possibility that the very rapid process of hydrolysis–polymer uptake by bacteria may cause a rapid bloom termination by solubilization of mucilage. Also, an increase in bacterial density coupled to a decrease in chlorophyll in the water column and a manganese gradient could be an indicator of algal senescence.[13]

CONCLUSIONS

The models generate abrupt changes in plankton densities as are often observed in the oceans. Varying a parameter over a reasonable range of values shows that there exist thresholds where a small variation can produce a smooth but significant change in the dynamic outcome, transforming the system from one type of dynamic behavior to another. The results of simulations indicate that the system has a strong dependence on the balance between changing growth rates and removal rates in both *Phaeocystis* and protozoan populations during the first days of the bloom, the nutrients and the copepods playing alternate roles as indirect controls as the bloom progresses. Because of the importance of the timing of events, the structure of the colonial population, and the complexity of underlying feedback mechanisms, the system does not exhibit the classical pattern of trophic chain oscillations. The inclusion of components representing bacterial activity or effects of shear accentuates the importance both of the timing of event occurrence and duration and the size structure of the colonial population that when coupled to a multitude of feedback mechanisms, lead to a wide range of feasible outcomes that reproduce a variety of behaviors observed in nature. Because the community appears to be driven by delicately balanced relationships, the multitude of apparently distinct field observations in the literature seem feasible and noncontradictory.

This modeling approach based on the development of a sequence of individual-based, size structured models, appears to be a useful tool for the study and understanding of the sensitive dynamics of a very complex system. During field research it often becomes difficult to monitor a large set of variables under natural conditions, and the risk of ignoring some important factor is high. James Clerk Maxwell wisely observes that "the success of any physical investigation depends upon the judicious selection of what is to be observed as of primary importance." Moreover, "the success of any modeling venture depends upon a judicious selection of observables and means for encapsulating these observables within the framework of suitable formal mathematical systems." (From *Reality Rules* by John Casti, John Wiley, New York.) One advantage of this type of model resides in the fact that the individual models are based on rather detailed knowledge of local low-level processes that can be studied under experimental conditions. Multimodular models permit to integrate the parts and study the outcome of simultaneous interactions within the system, including or excluding factors in order to investigate their relative importance, and particularly they allow the detection of counterintuitive feedback mechanisms.

The whole purpose of making models is to be able to bring a measure of order to our experiences and observations, and to understand the mechanisms involved, as well as to make specific predictions about certain aspects of the world as we experience it. Models are necessarily imperfect, but they help us think about what we observe. In a short story entitled "El Memorioso Funes," Jorge Luis Borges makes an account of his contacts with a very unusual man who had an amazing memory and could remember every leaf of every tree of every forest, every cloud he had ever seen pass by. In fact, remembering took him exactly the same amount of time that the original action had taken. All this represents a scientist's worst nightmare:

> "Locke, en el siglo XVII, postuló (y reprobó) un idioma imposible en el que cada cosa individual, cada piedra, cada pájaro y cada rama tuviera un nombre propio; Funes proyectó alguna vez un idioma análogo, pero lo desechó por parecerle demasiado general, demasiado ambiguo. En efecto, Funes no sólo recordaba cada hoja de cada árbol de cada monte, sino cada una de las veces que la había percibido o imaginado. [...] Sospecho, sin embargo, que no era muy capaz de pensar. Pensar es olvidar diferencias, es generalizar, abstraer. En el abarrotado mundo de Funes no había sino detalles, casi inmediatos."[b] [Borges[14]]

But, as Borges well suspects, Funes was incapable of thinking, for to think is, among other things, to forget differences, to generalize, to synthesize, to abstract. In Funes' overcrowded world, there were only immediate details. As modelers well

---

[b]"In the seventeenth century Locke postulated (and rejected) an impossible language where every individual object, every stone, every bird, and every branch had its own name; Funes planned some time ago an analogous language, but rejected it because it seemed too ambiguous to him. In fact, Funes not only remembered every leaf of every tree of every woodland, but also every time he had perceived or imagined it. [...] Nevertheless, I suspect he was not very capable of thinking. To think is to forget differences, to generalize, to abstract. In the overcrowded world of Funes there were only details, almost immediate details." [personal translation]

know, a good model is a compromise between the need for abstraction and the need for details. In order to build a good model, the modeler needs to have clearly in mind the questions that need to be answered. The process of building models in a modular manner helps reformulate those questions as the complexity of the model increases, because it changes at each step the perspective from which the real world is observed. In other words, the process helps one to think.

SUMMARY

Mathematical models have been used in the last 25 years in order to study different aspects of planktonic community structure in aquatic ecosystems rather than to provide predictive tools. Those studies reveal the importance of size structure in the effects of predation and in the complex dynamics of the community. Here we propose a multimodular mathematical model as a tool to better understand the behavior of a very complex plankton ecosystem and evaluate the relative importance of different components in its dynamics. In this case, we focus on a *Phaeocystis*-dominated plankton community. *Phaeocystis spp.* are unicellular algae, found in temperate oceans and polar seas, where they produce massive blooms, but can also be found in tropical waters. They occur in at least two different forms, single and colonial, and are characterized by a complex life cycle. The many factors involved in the control of the vernal blooms in polar and temperate waters, as well as the complexity of *Phaeocystis* life cycle, the intricacies of the timing and densities associated with its numerous grazers, the role of bacteria in lysis and degradation, the effects of shear, and the process of sedimentation, make it difficult to predict the development as well as the fate of a bloom. Because of the broad range of sizes *Phaeocystis* cover, the diversity of its grazers, and the dependency on size of the impact of grazing, the organization and structure of the community are of great importance. The proposed model, consisting of both structured and unstructured components, is constructed in modular form, with each module representing different relationships between components, in order to analyze the role and the magnitude of each potential controlling factor in bloom dynamics.

REFERENCES

1. LANCELOT, C. & P. WASSMANN, Eds. 1994. Special issue: ecology of *Phaeocystis*-dominated ecosystems. J. Mar. Syst. **5**(1).
2. STEELE, J.H. 1974. The Structure of Marine Ecosystems. Harvard University Press. Cambridge, MA.
3. STEELE, J.H. & B.W. FROST. 1977. The structure of plankton communities. Phil. Trans. R. Soc. London **280**: 485–534.
4. STEELE, J.H. & E.M. HENDERSON. 1992. The role of predation in plankton models. J. Plankton Res. **14**: 157–172.
5. O'NEILL, R.V., D.L. DEANGELIS, J.J. PASTOR, B.J. JACKSON & W.M. POST. 1989. Multiple nutrient limitations in ecological models. Ecol. Modelling **46**: 147–163.
6. KORNMANN, V.P. 1955. Beobachtungen an *Phaeocystis* Kulturen. Helg. Wiss Meers. **5**: 218–233.
7. ROUSSEAU, V., D. VAULOT, R. CASOTTI, V. CARIOU, J. LENZ, J. GUNKEL & M. BAUMANN. 1994. The life cycle of *Phaeocystis* (Prymnesiophyceae): evidence and hypothesis. J. Mar. Syst. **5**: 23–39.

8. HARRIS, R.P. & G.-A. PAFFENHOFER. 1976. Feeding, growth, and reproduction of the marine copepod *Temora longicornis* Muller. J. Mar. Biol. Assoc. UK. **56:** 675–690.
9. BERGREEN, U., B. HANSEN & T. KIORBOE. 1988. Food size spectra, ingestion and growth of the copepod *Acartia tonsa* during development: implications for determination of copepod production. Mar. Biol. **99:** 341–352.
10. BILLEN, G. & S. BECQUEVORT. 1991. Phytoplankton-bacteria relationship in the Antarctic marine ecosystem. Polar Res. **10:** 245–253.
11. CANZIANI, G.A. & T.G. HALLAM. 1996. A mathematical model for *Phaeocystis spp.* dominated plankton community dynamics. Part I: The basic model. Nonlinear World, Special Issue on Structured Communities. **3:** 19–76.
12. CANZIANI, G.A. & T.G. HALLAM. 1995. El Problema de la Depredación y de la Turbulencia en el Modelo Matemático de una Comunidad Planctónica. Actas de la VI Reunión de Trabajo en Procesamiento de la Información y Control. Bahia Blanca. pp. 437–444.
13. DAVIDSON, A.T. & H.J. MARCHAND. 1987. Binding of manganese by Antarctic *Phaeocystis* and the role of bacteria in its release. Mar. Biol. **95:** 481–487.
14. BORGES, J.L. 1944. Ficciones. *In* Obras Completas de Jorge Luis Borges. Emecé Editores, Buenos Aires, 1974.

Complexity and Emergence in Models of Chemically Stressed Populations

THOMAS G. HALLAM[a,b] AND ERIC T. FUNASAKI[c]

[a]*Department of Ecology and Evolutionary Biology, University of Tennessee, Knoxville, Tennessee 37996, USA*
[c]*Department of Mathematics and Computer Science, Georgia Southern University, Statesboro, Georgia 30460, USA*

INTRODUCTION

The population ecology of stress presented here assumes that a proper level of investigation is the individual organism and its associated response to stress. The motivation for this principle is that stressor impact occurs at the level of the individual, not at the population level. Even though the target site of a stressor, such as a toxic chemical, may be a specific cellular receptor, the exposed, affected individual is the appropriate reference point for extrapolation to the population level. Consideration of individual variability is necessary to properly develop an appropriate theoretical basis for ecotoxicology. Variation in the distribution of physiological characteristics of individuals in a population together with the biogeochemical environment of the population determines the characteristics of the effects resulting from chemical exposure.[1] A fundamental feature of nature is that these characteristics change temporally, and often change as the result of disruptive perturbations of the system. We first discuss change in populations due to chemical stress perturbation and delineate some propositions that provide emergent phenomena. In addition, for the class of physiologically structured models under consideration, we indicate some of the complexity issues associated with population dynamics and structure that can arise because of the perturbing actions of a chemical stressor.

The concept of population susceptibility intrinsically relates to the existence of physiological variation because the type of population model we consider is focused on the physiology of individuals. Susceptibility that occurs due to physiological differences is temporally related to the growth dynamics of the individual.

We look at theoretical aspects of the stressor effects, from lethal to sublethal, of a reversible, lipophilic narcotic on a dynamic physiologically structured population, as they relate to complex dynamic behavior and emergent phenomena. Chemical stresses are perturbations that may completely change system behavior. We refer to *nominal* behavior as a system proposition preceding a perturbation. A proposition that is a persistent deviation from a characteristic associated with the nominal population behavior following a perturbation is called *a population emergent phenomenon*.

---

[b]Address for correspondence: 423-974-3065 (voice); 423-974-3067 (fax); thallam@utk.edu (e-mail).

Physiologically structured models yield two interesting propositions relevant to emergent phenomena associated with chemical stressor perturbations. First, a nominal behavior for unstressed populations is "survival of the fittest." A population model composed of multiple ecotypes has the "survival of the fittest" dominance property: when the simulation is performed for a sufficiently long period of time, the population evolves to a single ecotype, which is the fastest growing organism among all of the ecotypes. Gause's principle of competitive exclusion predicts this dominance (see also Henson and Hallam[2] for analogues in structured populations). Second, for static populations, an acute exposure to a nonpolar narcotic chemical results in "survival of the fattest".[3] In survival of the fattest, lipid content of the individual provides protection from chemical stress wherein a fatter individual can withstand an acute nonpolar narcotic stress better than a lean individual.

When a system deviation occurs, emergent properties of a population occur as a consequence of physiological attributes of the ecotypes that compose the population and the strength of the perturbation. Hence, life history traits are important for the study of emergent phenomena. Life history traits that are attributes of the species, such as the length of the juvenile period, J, or a periodic reproduction interval, P, are characteristics that determine the asymptotic dynamics of a population that reproduces clonally and is composed of a single ecotype.[4] Life history traits can be modified by chemical or physical stressors, and the subsequent changes produced in the asymptotic attractor of the population dynamics sometimes can be predicted. When a population contains multiple ecotypes, the dynamics are much less predictable, and more complex dynamics can result.

Even though the models are interesting, and do not deserve such a fate, they have been relegated to the APPENDIX. The models include a system of ordinary differential equations that describe the individual dynamics, an exposure model that governs the transfer of chemical from the environment into the organism and determines the magnitude of the chemical stressor in the organism, an effects model that relates the chemical in the organism to the effect of the perturbation on the individual, and a partial differential equation that generates population dynamics. The individual and population models were parameterized for *Daphnia magna* for use in the simulations.

The population model focuses on the effects of chemicals on a single trophic level. The dynamics of this tropic level are determined by physiological processes of individuals, reproduction and mortality. Even though variability in resources is fundamental to population dynamics, these models provide a resource, and a quality of resource as measured by the lipid content, to an ecotype at a fixed rate, but the rate can vary among ecotypes.

LETHAL EFFECTS AND EMERGENCE

In this section, the stressor perturbation effect assessed is the mortality of exposed organisms, provided that lethal concentrations are attained in the individual. The assumption that no sublethal effects are assumed to occur is unnecessary, but it yields a setting sufficient to illustrate that emergent phenomena occur for physiologically structured populations.

TABLE 1. Ecotypes of individuals used

Ecotype Number	A_1	x_L	Resource x	Age at First Blood	Structure Mass (mg) after Last Reproduction	Lipid Mass (mg) after Last Reproduction	Number of Eggs in Last Brood	Population Extinction Threshold
1	0.68×10^{-6}	0.2125	0.325×10^{-6}	8.05	0.1320	0.01401	9	0.40×10^{-4}
2	0.68×10^{-6}	0.25	0.325×10^{-6}	8.15	0.1150	0.01506	10	0.40×10^{-4}
3	0.68×10^{-6}	0.2875	0.325×10^{-6}	8.30	0.1159	0.02030	11	0.40×10^{-4}
4	0.68×10^{-6}	0.2125	0.45×10^{-6}	6.70	0.2816	0.02896	21	0.53×10^{-4}
5	0.68×10^{-6}	0.25	0.45×10^{-6}	6.80	0.2457	0.03141	24	0.53×10^{-4}
6	0.68×10^{-6}	0.2875	0.45×10^{-6}	6.95	0.2581	0.04572	25	0.53×10^{-4}
7	0.68×10^{-6}	0.2125	0.575×10^{-6}	6.15	0.4874	0.04848	39	0.63×10^{-4}
8	0.68×10^{-6}	0.25	0.575×10^{-6}	6.20	0.4320	0.05473	44	0.63×10^{-4}
9	0.68×10^{-6}	0.2875	0.575×10^{-6}	6.30	0.4489	0.07919	46	0.62×10^{-4}
10	0.80×10^{-6}	0.2125	0.325×10^{-6}	9.20	0.08444	0.008685	6	0.33×10^{-4}
11	0.80×10^{-6}	0.25	0.325×10^{-6}	9.30	0.06826	0.008388	7	0.33×10^{-4}
12	0.80×10^{-6}	0.2875	0.325×10^{-6}	9.45	0.08115	0.01437	7	0.33×10^{-4}
13	0.80×10^{-6}	0.2125	0.45×10^{-6}	7.30	0.1883	0.01933	14	0.47×10^{-4}
14	0.80×10^{-6}	0.25	0.45×10^{-6}	7.40	0.1653	0.02104	16	0.47×10^{-4}
15	0.80×10^{-6}	0.2875	0.45×10^{-6}	7.50	0.1673	0.02911	17	0.46×10^{-4}
16	0.80×10^{-6}	0.2125	0.575×10^{-6}	6.50	0.3363	0.03393	26	0.57×10^{-4}
17	0.80×10^{-6}	0.25	0.575×10^{-6}	6.60	0.2912	0.03641	30	0.56×10^{-4}
18	0.80×10^{-6}	0.2875	0.575×10^{-6}	6.70	0.3079	0.05422	31	0.56×10^{-4}
19	0.92×10^{-6}	0.2125	0.325×10^{-6}	10.75	0.06117	0.006566	4	0.27×10^{-4}
20	0.92×10^{-6}	0.25	0.325×10^{-6}	10.80	0.05380	0.006758	5	0.28×10^{-4}
21	0.92×10^{-6}	0.2875	0.325×10^{-6}	10.95	0.05112	0.008822	5	0.28×10^{-4}
22	0.92×10^{-6}	0.2125	0.45×10^{-6}	7.95	0.1295	0.01276	10	0.41×10^{-4}
23	0.92×10^{-6}	0.25	0.45×10^{-6}	8.05	0.1198	0.01553	11	0.41×10^{-4}
24	0.92×10^{-6}	0.2875	0.45×10^{-6}	8.20	0.1172	0.02013	12	0.41×10^{-4}
25	0.92×10^{-6}	0.2125	0.575×10^{-6}	6.90	0.2460	0.02530	18	0.51×10^{-4}
26	0.92×10^{-6}	0.25	0.575×10^{-6}	7.00	0.2108	0.02665	21	0.51×10^{-4}
27	0.92×10^{-6}	0.2875	0.575×10^{-6}	7.10	0.2204	0.03863	22	0.51×10^{-4}

NOTE: Life history data for each of the ecotypes in the population simulated here. See text for explanation and definition of parameters.

Dynamic Populations and Chronic Exposures

Persistence of a dynamic population after chronic exposure is determined not only by the lipid distribution, which is fundamental to response to acute exposures, but also by the growth of the individuals in the population and population-level processes such as birth and mortality rates. The twenty-seven ecotypes described in TABLE 1 were coupled with a fixed set of mortality parameters to delineate characteristics of the *Daphnia* population. The population with these 27 ecotypes displays the "survival of the fittest" dominance property in that the climax population is dominated by, and ultimately consists of, individuals of ecotype 9 of TABLE 1 for any initial distribution of ecotypes that includes ecotype 9.

To demonstrate that the perturbation of chemical stress can affect population structure and lead to emergent phenomena in a model environment, the model population was exposed to several different toxic chemicals, and the time evolution of the stressed population was studied.[5] The two behavioral propositions indicated above, survival of the fattest and survival of the fittest, can be destroyed under chemical perturbation, and emergent phenomena occur.

Chemical Stress Case 1

A 7-day exposure to a 16-parts-per-million (ppm) aqueous concentration, initiated on day 24, to a chemical having an octanol–water partition coefficient, K_{OW}, of 10^4 decreases the population to two cohorts of ecotype 19, which filters at the lowest rate and feeds at the lowest resource level with the lowest lipid content. These emergent, dominant individuals are the slowest growers and the leanest of any of the ecotypes. The toxic stress has completely reversed the ecotypic succession proposition (where the fastest growers formed the climax population) and also violates the survival of fattest to the greatest opposite extreme proposition (the leanest individuals survive the chronic stress).

Chemical Stess Case 2

Model simulations showed that exposure to a different chemical can result in a completely different population structure. A 7-day exposure initiated at day 24 to a chemical with an octanol--water partition coefficient of 4.5×10^6 at a concentration of 4×10^{-6} (4 ppm) results in a population composed of a mixture of ecotypes 13, 14, and 15. These emergent individuals are relatively close in size, and lipid content and reproductive capability and are intermediate growth ecotypes. They are neither the faster nor slower growing organisms in our population.

Chemical Stress Case 3

Another interesting situation results from an 8-day exposure, initiated at day 24, to a chemical having an octanol–water partition coefficient of 2×10^4 at a concentration of 8×10^{-6} (8 ppm). This leads to a population consisting of many ecotypes. When the simulation is continued after termination of toxicant release, the population is ultimately dominated by the fastest growing ecotype of the survivors. This climax ecotype dominates only after a long time period because of the pressure of another similarly fast-growing ecotype. At the end of the exposure period, the two ecotypes that ultimately dominate the population were a very minor part of the population. When the time of exposure is extended, an additional 0.5 day, the two cohorts of the dominant ecotypes are eliminated so that they are no longer a component of the population.

COMPLEXITY ARISING FROM SUBLETHAL EFFECTS

Sublethal effects can affect physiological processes of individuals by reducing growth and reproductive processes. This, in turn, produces changes in population structure. Although populations can go to extinction as a result of a sublethal exposure, this occurs, not via death toxic stress, but rather by growth reduction and repro-

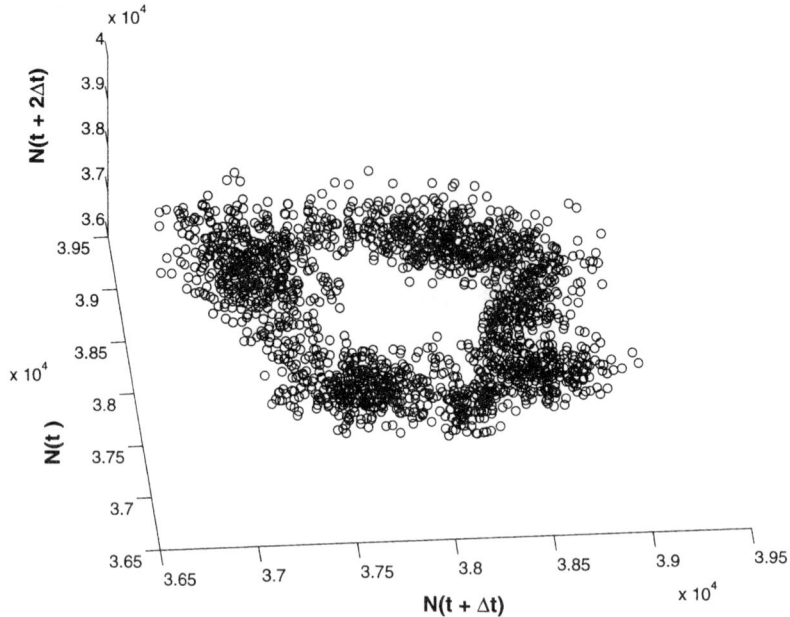

FIGURE 1. The "fuzzy torus" phase portrait of the trajectory of the population model. $N(t)$ is the total number in the population at time t. Δt is 9/20 day.

ductive limitation of individuals.[6] We turn now to the dynamics that result from system perturbation caused by exposure to sublethal levels of chemical and discuss the asymptotic nature of the behavior of a population composed of individuals of a single ecotype to develop the basis for complexity, as determined by multiple ecotypes, in structured models.

The numerical asymptotic attractors of the family of population models, as obtained from simulations, can be characterized by the population-level life history parameters J, the length of the juvenile period, and P, the periodic reproduction time[4]; in addition, the computational time step plays an important role in the numerical character of the asymptotic attractor because of discrete events such as determination of the intervals in which the births occur. To indicate the basis of the asymptotic dynamics of the population are grounded in the life history parameters, Funasaki and Hallam[4] have considered a population in a constant environment where resource input to each individual is maintained at a constant level. The simulations were studied on the time scale of 10,000 days, which is 200 generations for the model organisms.

A perspective used to analyze the total population size time-series data from the *Daphnia* population model (Equation (A-1) of the APPENDIX) was phase portrait reconstruction.[7,8] Lagging one-dimensional time-series data from a dynamical system with itself an appropriate number of times results in the reconstruction of the higher dimensional phase portrait of that system. The best way to choose the lag to be used has not been agreed upon; here, we use the lag that corresponds to the first time that the autocorrelation function of the time-series data becomes zero.[9] Total population

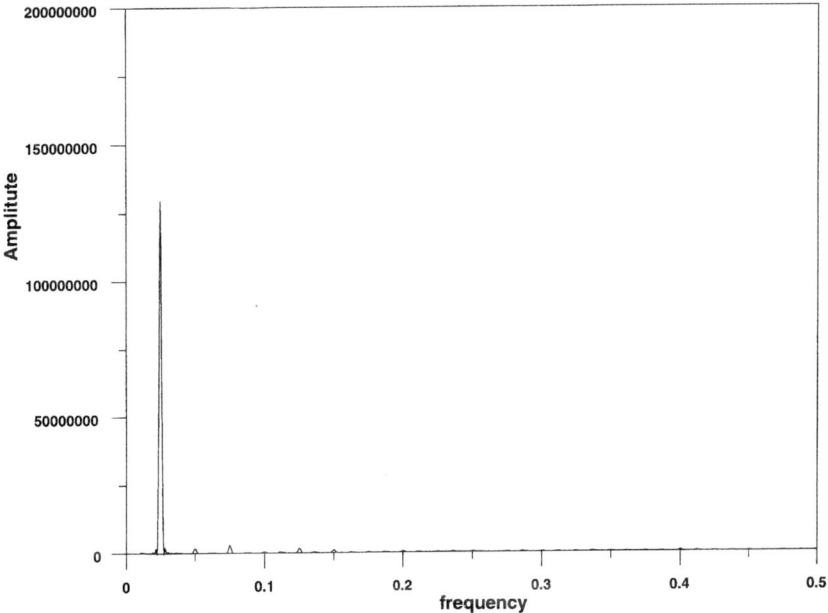

FIGURE 2. The frequency spectrum of the time series associated with the phase portrait of FIGURE 1. The dominant frequency and the largest subdominant frequency correspond to the 20-point and the 1-point attractor predicted by the birth interval model.

size time-series data was lagged with itself by the lag obtained from the autocorrelation function and twice that lag to produce a phase portrait in three dimensions.

FIGURE 1 shows the phase portrait for ecotype 12 for a value of the perturbation parameter r (see APPENDIX). The phase portraits present the behavior for days 9900 to 10,000. Each point in the time-series corresponds to 1/20 of a day. For FIGURE 1, the time lags used were 9/20 and 18/20 of a day. For each value of $r = C_W$, the phase portraits are strikingly different[4] and in the case of FIGURE 1, it appears as a fuzzy torus.

In FIGURE 2 we show the time-series power spectrum for a 10-day interval, days 9990 to 10,000, that correspond to the phase portrait in FIGURE 1. The plot shows the competing long and short periods that produce the torus-shaped attractor.

We have been able to predict the characteristics of attractors associated with sublethal stress employing a birth interval concept with some success by using the juvenile period length J, the periodic reproduction time P, and the computational time step.[4] The basic idea is to delineate the number of possible intervals in which births can occur for the given reproductive period and a computational time step size. For the *Daphnia* population model (A-1), births occur as discrete events, and so it is possible to have births in a proper subset of the birth intervals. In this case, a phase locking occurs and an attractor is predicted from this birth frequency with the number of filled birth intervals determining the characteristics of the phase portrait. Taking the total number of possible birth intervals and dividing by the actual number of birth intervals filled yields the number of points associated with the phase portrait. In the fuzzy torus case, even though the population is composed of a single ecotype, there are two juvenile period lengths associated with the individuals because of the variable allocation of lipid to eggs. Because of this variability, newborns with a slightly larger initial amount of lipid reach the reproductive threshold one time step before newborns with a slightly lower initial amount of lipid. One cohort has a juvenile period length of 9.95, and another has a juvenile period length of 10.0. With a time iteration of 1/20 day, these lead to the conjunction of the dynamics of a 1-point cycle and a 20-point cycle, respectively. This situation is indicative of the behavior of a population composed of multiple ecotypes and demonstrates that the model population with multiple ecotypes can lead to complex behavior in which predictability wanes with additional composition of attractors.

DISCUSSION

Emergent phenomena that occur at the population level of ecological organization exist in our models because of chemical impact on the physiology of individuals. For lethal effects, the emergent phenomena discussed here represent a change in dominant ecotype that results as consequences of ecotypic susceptibility to stress. It is difficult to determine which ecotype will be dominant following the stress perturbation because of the complexity of interactions between chemical, organism physiology, and population ecology of births and deaths.

Complexity in the dynamics of populations subjected to sublethal chemical stresses is illustrated by considering an example of a population consisting of a single ecotype. Because of a threshold, this ecotype produces two cohorts that have very different dynamics. One component cohort and its associated offspring approach a single point attractor. The other component of the population approaches a 20-point attractor. The total population attractor is a conjunction of these two attractors and produces a complicated phase portrait. It is a interesting problem to unravel the components from these composite dynamics.

SUMMARY

To explore the ecology of stress, we use the concept that the individual is the fundamental ecological unit, and we represent the dynamics of growth and reproduction of individuals in a modeling framework. Impacts of stress, as measured through exposure and effects, are incorporated at the individual level; and individuals, along with the response to toxic stress, are accumulated to describe stressed population dynamics. This review of some of our research on stress ecology emphasizes that stress, by changing physiological characteristics of individuals, can modify the structure and dynamics of the population. A physiological stress perturbation can produce emergent phenomena uncharacteristic of nominal system behavior. Complexity in population ecology is explored through an example of a population composed of a single ecotype.

REFERENCES

1. KOH, H.L., T.G. HALLAM & H.L. LEE. 1997. Combined effects of environmental and chemical stressors on a model *Daphnia* population. Ecol. Modell. **103**: 19–32.
2. HENSON, S.M. & T.G. HALLAM. 1994. Survival of the fittest: Asymptotic coompetitve exclusion in structured population and community models. Nonlinear World **1**: 385-402.
3. LASSITER, R.R. & T.G. HALLAM. 1990. Survival of the fattest: Implications for acute effects of lipohiolic chemcials on aquatic populations. Environ. Toxicol. Chem **9**: 585–595.
4. FUNASAKI, E.T. & T.G. HALLAM. 1999. Prediction of dynamics in structured populations. In preparation.
5. HALLAM, T.G., et al. 1990. Toxicant induced mortality in *Daphnia* populations. Environ. Toxicol. Chem. **9**: 597–621.
6. HALLAM, T.G., G.A. CANZIANI & R.R. LASSITER 1993. Sublethal narcosis and population persistence: a modeling study on growth effects. Environ. Toxicol. Chem. **12**: 947–954.
7. OLSEN, L.F. & W.M. SCHAFFER. 1990. Chaos versus noisy periodicity: alternative hypothesis for childhood epidemics. Science **249**: 499–504.
8. TAKENS, F. 1981. Detecting strange attractors in turbulence. *In* Dynamical Systems and Turbulence, Warick. D.A. Rand & L.S. Young, Eds. Springer-Verlag. New York.
9. OTT, E., T. SAUER & J.A. YORKE. 1994. Coping with Chaos. John Wiley and Sons, Inc. New York.

Appendix

THE MODELS

We consider a family of physiologically based populations, determined by a perturbation stress representation function, X, that affects the growth rate of individuals and is determined by the level of a chemical or physical stressor in the organism. Each population is represented by a collection of an individual ecotype model that mimics a (generic) life history, a physiologically structured population model that incorporates the individual dynamics, and a perturbation model to generate different ecotypic characteristics of individuals that compose the family of populations.

The model organism has a life history consisting of an embryonic period, a juvenile period and an adult stage. The end of the embryonic stage is marked by the onset of feeding, whereas the end of the juvenile period, is determined by attainment of a threshold size, which occurs at the time, J, when adulthood is reached and first reproduction occurs. Reproduction, assumed to be clonal where daughters have a physiology and environment identical to that of their mother, is a discrete process that occurs periodically with period P after the organism reaches adulthood.

For these simulation studies, a mathematical model of a *Daphnia magna* population, a McKendrick-von Foerster hyperbolic partial differential equation,[5] is employed:

$$\frac{\partial \rho}{\partial t} + \frac{\partial \rho}{\partial a} + \frac{\partial(\rho g_L)}{\partial t} + \frac{\partial(\rho g_S)}{\partial m_S} = -\mu\rho - \nu(B)\rho \quad \text{(A-1)}$$

where $\rho = \rho(X; t, a, m_L, m_S)$ is the density of the model *Daphnia* population indexed by stressor functional response X, at time t of age a with mass of lipid and mass m_L of structure m_S; g_L is the accumulation rate of lipid and g_S is the growth rate of structure in the individual daphnid; μ is the density-independent mortality rate; is a density-dependent mortality function; and $B = B_X(t)$ is the total population biomass given by

$$B_X(t) = \int_0^\infty \int_0^\infty \int_0^\infty (m_L + m_S)\rho(X; t, a, m_L, m_S) da\, dm_L dm_S \quad \text{(A-2)}$$

In the *Daphnia* population model (A-1), a uniform density-dependent mortality that harvests equal numbers of individuals, independent of size or age, is used to control the biomass size of the *Daphnia* population.

Initial and boundary conditions are needed for the model to be well posed mathematically. The initial distribution of individuals, $\rho(X; 0, a, m_L, m_S)$, describes the *Daphnia* population at time $t = 0$. The boundary condition, $\rho(X; t, 0, m_{L_0}, m_{S_0})$, describes the birth process for the *Daphnia* population where m_{L_0} and m_{S_0} represent the mass of lipid and mass of structure of individuals of age $a = 0$, respectively.

Each *Daphnia* population model (A-1) is solved using the method of characteristics that transforms the hyperbolic partial differential equation model into a system of ordinary differential equations.[5] This system of ordinary differential equations is then solved numerically.

The individual model, represented in (A-1) by g_L and g_S, consists of a coupled system of nonlinear ordinary differential equations that describes the accumulation of lipids and structure (proteins and carbohydrates) in an individual daphnid:

$$\frac{dm_L}{da} = g_L(X; a, m_L, m_S) = \frac{X A_{0L} x_L m_S}{A_1 m_S^{1/3} + A_2 x} - f_L(m_L, m_S) \quad \text{(A-3a)}$$

$$\frac{dm_S}{da} = g_S(X; a, m_L, m_S) = \frac{X A_{0S} x_S m_S}{A_1 m_S^{1/3} + A_2 x} - f_S(m_L, m_S) \quad \text{(A-3b)}$$

where m_L is the mass of lipid, m_S is the mass of structure, a is age, g_L is the accumulation rate of lipid, g_S is the growth rate of structure, A_{0L} is the assimilation efficiency for lipid, A_{0S} is the assimilation efficiency for structure, x_S is the density of structure in the resource, x_L is the density of lipid in the resource, A_1 is the reciprocal of the constant allometrically relating filtering rate to organism length squared, A_2 is the reciprocal of the constant allometrically relating maximal ingestion rate to organism length squared, x is the total resource density, and f_L and f_S are the loss functions of the individual model.[1,6] Included in these loss functions are terms that represent allocations of lipids and structure to the birth process for an individual daphnid. Physiological processes are modeled by (A-3) on a continuous time scale; reproduction is a discrete process. After a daphnid reproduces for the first time, it continues to reproduce every 4 days, the periodic period P as assumed above. An important component of the daphnid birth process is that only daphnids of length 2.5 mm or greater can reproduce; here the time for attaining length 2.5 is $J = J(X)$.

X is the perturbation function applied to the growth rates that creates different ecotypes and the family of populations. X, as incorporated into the individual daphnid model (A-3), has the same formulation as a sublethal effects model for nonpolar narcotic chemicals.[6] The model reduces the rate of accumulation of lipid and the rate of growth of structure for an individual daphnid by the factor

$$X = \begin{cases} 1 & r < A \\ 1 - \dfrac{\alpha(r-A)}{\beta + r} & A \leq r \leq B \\ 0 & r > B \end{cases} \quad \text{(A-4)}$$

where is the perturbation basis parameter and is taken to be the chemical concentration in the water, A is the experimental no effect threshold; and B is a no-growth limitation. The parameters α and β allow data fitting to effect prescribed perturbations.

The use of different constant r values yields different individual models with correspondingly distinct J parameters; but it is assumed in the simulations that the constant periodic reproduction parameter P remains unaffected by the perturbation.

The Influence of Classic Sciences on Ecology and Evolution of Ecological Studies

VINCENT HULL[a] AND MARGHERITA FALCUCCI

Laboratorio Centrale di Idrobiologia di Roma, Ministero per le Politiche Agricole, Via del Caravaggio, 107 - 00147 Rome, Italy

> *Si todos los rìos son dulces*
> *de dònde saca sal el mar?*
> *Libro de las preguntas* [*P. Neruda*]

ECOLOGY, ECOSYSTEMS, ENVIRONMENTALISM, AND CULTURAL EVOLUTION

Contemporary citizens of the Western world increasingly are found to flee city life as often as possible, in the supposition that it is desirable to escape from crowds and seek open space, return to nature, and escape pollution.

Consider the amount of automobile traffic or the many airplane flights from any major airport as indicators of the demand to travel large distances. Beyond the extraordinary abilities current transport systems give to average citizens to cover large distances in short times, modern transportation also transforms the notion of place and space in modern societies. In turn, the evolution of Western societies contributes to the changes found in inhabitants and their desires.

We cannot ignore the materialism of the eighties, but we also cannot deny that over the last 30 years we have witnessed an increased awareness of the environment and the development of environmental movements. Furthermore, many political ideologies that began with the industrial era have waned over time, and the large social changes of the last century have moved people into a new sensitivity toward the environment.

Words like environment, ecology, and ecosystems are commonplace in today's vocabulary, a simple sign of the widespread gain of environmentalism. Of course, often these words are used incorrectly: a classic example is the use of the term "ecological" to describe a commercial product and guarantee its quality or safety.

Conflict abounds in Western societies where economic policies are based on a constant increase in consumption. Yet, there is an awareness that these policies beget degradation of the biosphere.

Where is the unitary science to inform our understanding of society and nature? Of development? Of nonanthropocentric earth studies? It does not exist. The science closest to meeting these requirements is ecology, which has only belatedly and begrudgingly been recognized in the academic sphere. Ecology—or any science that attempts to bridge the complexity of nature, economics, and human society—must break the molds of standard science, which inevitably makes it a heretical science.

---

[a]Address for correspondence: +39 6 5140296 (voice/fax); hull@www.inea.it (e-mail).

FOR AN ECOLOGICAL ECOLOGY

Over the last two decades, an array of neologisms across all sciences developed in an attempt to merge ecology with the traditional sciences. Classical sciences developed a new face, more appropriate to the study of nature. Many theories, old and new, emerged about the natural world, sometimes with no clear or direct connection with nature, that nevertheless satisfied the desire to make the scientific disciplines relevant to ecology. Some have remained unfilled promises to explicate nature. Mandelbrot's fractal theory suggests a geometric interpretation of natural structures. Chaos theory, an outgrowth of meteorological physics, has been applied to ecological studies with a view to dynamics. Complexity theory is deeply imbued with ecological principles. From general complexity studies emerge new disciplines: economic ecology, urban ecology, ecological engineering, and systems ecology. Absent from all these new fields of study is an ecological ecology. Let us examine the guidelines for an ecological ecology.

Ecologists study the natural world through an examination of individual species and populations, ecosystems, plant and animal communities, and the physical environment. It has become evident to these scientists that natural systems require special tools to analyze the field measurements. Although ecology employs analytic methods from classical physics and chemistry, the fact that it is a historical science (that is, history is part of the science), renders traditional scientific analysis inadequate. Consider, for example, the frequency with which ecologists gather precise data and many physical measurements. Consider the example of temperature measurements. Why are they crucial? The justification is that living organisms react to temperature. Still, is it really necessary in most situations to know the temperature to a 10th or 100th of a degree? In fact, low-cost precise temperature measurements justify the precision of the data, not the need for the data. Transition from winter to spring brings warmer temperatures, which trigger many biological responses. Can we say that all, or even most, organisms can detect the difference between a rise to 14.1 rather than 14.3°C? With the spring, many changes take place in nature, so we know temperature is important to biological organisms and represents a major focus in ecological studies. Yet rather than emphasize the understanding of temperature, we focus on developing more precise thermometers.

In another case, consider mathematical models of the predator–prey interaction type; these represent the abstract concept of trophic relations, and they have successfully provided coherent theoretical interpretation. Supporting the theory was an examination of the data, specifically the parameters that determine the outcome of the models. But aquatic ecologist has directly measured natural mortality in plankton, especially as defined by the Lotka-Volterra equations? Moreover, the mortality parameter is not constant, so which time and place and circumstance is appropriate to consider? Yet natural mortality is a fundamental parameter in the equations of Lotka and Volterra, and it is always considered to be constant. There is no lack of examples in ecological modelling that show a difference between the parameter requirements and the actual availability of data, resulting in considerable literature debating the type of mathematical model needed for ecological studies. Leaving this aside for the moment, let us first examine other aspects of ecological studies.

Natural systems are remarkably self-regulating, through feedback systems. Feedback-loop studies, despite their recognized importance in regulation of equilibrium and stability, are a neglected area of ecological research. For example, when the chicks in a nest peep, the mother responds with food: the obvious effect is to both satisfy the hunger and reinforces the chicks' behavior of producing auditory signals when hungery. The mother's ability to find food is dependent on the link between the bird population dynamics and food availability. Prey and predator relations have similar dynamics, a classic example of a negative-feedback system. Positive-feedback systems are more difficult to resolve because of their inherent dynamics toward destabilization; yet these may be key to ecological adaptation and evolution.

Evolution and adaptation of biological and ecological systems characterize the distinction between the classical physical sciences and ecological sciences. History determines ecological conditions, unlike mechanistic physics and chemistry. Biotic communities cannot be reconstructed *de novo* irrespective of the past, evolution is irreversible, and ecosystems cannot be separated into parts.

Ecological science has shown the inadequacy of classical scientific methods, but has failed to create many alternative analytical tools and research paradigms.

Ecologists must examine the underlying processes of interactions and feedback; the co-evolution of organisms; transformations between organisms and their physical environment; the relationship of scale, both time and space, with dynamics; and those structural relationships capable of evolving. Of course the ecological literature is replete with these arguments, but in the main ecologists continue to rely on analytic methods borrowed directly from the other sciences or modified only slightly. This means that we do not study ecology in an ecological fashion, and this begs the question of whether the underlying assumptions made by classical science are valid in ecology.

An unusual characteristic of natural systems is the functional role of shape, color, size, sound, and texture in dynamics. Unlike physical systems in which few, if any, of these factors matter to the underlying theory or dynamics, in ecological systems these "information" packets may be key to the dynamics of ecosystems. In summary, ecological models have had a greater propensity to emphasize the math over the ecology, formal analysis over the biotic interactions; and simplification to enhance solvability and computability over approaches that reflect the complexity of ecological systems.

WHERE IS AN ECOSYSTEM?

Without doubt, systems analysis owes much to ecological studies. Natural systems showed the error in the practice of separating single variables, since the entire process required understanding of both the whole and the interconnections. Many will recall a classic example: a pile of bricks becomes a house only when ordered. The ordered bricks form a structure; the structure becomes a house only when it acquires a social context. A house possesses a high level of information that is not imparted by the sum of the number of bricks. The inanimate nature of bricks enables them to exist either as a pile or as a house. In natural systems we find the components exist to make the whole, and the whole exists to make the components. What organism survives in a vacuum? Natural systems take the notion of the parts making up the whole and reverse it to make the whole contribute to the existence of the parts.

We will return to the matter of history, information, and memory in ecological systems.

Ecologists have contributed to leading ecological studies down the garden path through dichotomies that are misleading. For instance, classical ecology is the study of the relationships between living beings and their physical environment.[1] Furthermore, ecological studies are further subdivided. One subdivision is autoecology—the study of an individual species and its physical environment. For example, an autoecological study would determine the distribution of specific trees on a mountain slope as a result of the impact of the physical condition of altitude on plant physiology. The second kind of ecology is synecology, which suggests the study of relationships among organisms. A stereotypic example is trophic studies of the interactions between prey and predator. During the 1970s ecological thinking marched toward an integration of both approaches in recognition that living organisms, their physical environment, time, space, organizational structure, regulatory feedback, and various auto-organizational tendencies, must be analyzed as an integrated structure.

Network representation gained acceptance in ecology. To be comprehensive, ecological nets must be multidimensional, to include structural and temporal relationships. As a case in point, the mycorrhizal fungi are symbionts, special fungi that attach to the roots of higher level plants; the fungal hyphae penetrate the interior radical cells of plants. More than 90% of plants have a mycorrhizal structure, which is both a physiological and ecological relation. The fungi obtain organic carbon from the plant host, while providing an efficient transfer of nitrogen and phosphorous nutrients to the plant. Clearly this is a structural dependency that forms a network between specific fungi and their hosts. What becomes less intuitive and more amazing is recent evidence that there is a network among the mycorrhizal fungi, spreading worldwide. If true, then the entire world's vegetation should be interconnected in a huge plant internet.

Like the preceding discussion of ecology as a discipline, so too the concept of ecosystem needs to be revisited. In broad terms, an ecosystem is a community of organisms and their physical environment interacting as a single ecological unity.[2] In fact, in practical terms identifying the borders of an ecosystem is difficult, maybe impossible.

Consider a lake ecosystem. It would appear that there is no doubt about the borders of this ecosystem, but is this true? Remember, the definition of an ecosystem is the community interacting with its physical environment. The lake community also depends on the shoreline vegetation and animals. Nearby reeds or fallen trees release nutrients that enter the lake; beavers obstruct water flow; insects that live on the shore lay larvae in the lake. All of these organisms form part of the network vital to the lake, yet would be outside the traditional vision of the lake ecosystem. The dilemma lies in avoiding definitions that entrap the research, yet also recognize the need to accept ecological research that will itself be an adaptive process of study.

On the other hand, just because there is a biological entity present, it does not establish the structure of the ecosystem. Following the lake example, suppose we were to imagine that all the fish were removed by overfishing. It seems that this removal would destroy the lake ecosystem, or at least create instability because an important ecological component is missing, thus creating an imbalance in the food web network. Yet it would be a mistake to say the lake ecosystem has disappeared. Continuing in this vein, suppose now that all the plankton and benthic species disappear.

Again, we imagine this depauperate lake as something abhorrent to our ideal, but still the lake ecosystem remains because the biotic bacterial community exists. Finally, if every trace of biotic organism disappears, we have finally arrived at a non-ecological lake system, just purified bath water. At this point we can try a thought-reversal experiment. To this pure water we return the fish. Not surprising, these fish would not survive, given that there is no support for their survival. On the other hand, return the bacteria, and these simple creatures would survive. Now the plankton and benthic communities can also flourish. In short, the history of the lake development would make a difference. To reach a state of being, an ecosystem requires a sequence and process, just as it took place over the millennia. A structured organization can follow many paths from a single point of development.

From this imaginary experiment two conclusions can be anticipated and a third can be reached that comes as a surprise. First, natural systems evolve with an irreversible history; time provides a unique ownership to the world around us. Seasonal events, night and day sequences, climates, and geological processes interpenetrate the evolution of natural organisms that likewise alter the physical environment to create new circumstances. In our lake experiment, the fish do not create conditions for the bacteria, but the sequence of bacteria alters the lake physical conditions to provide a precursor for the plankton and fish.

The second thing we learn is that evolution bears with it information in the form of hierarchical structures. Clearly, a lake with a fish population contains more ecological information than a lake with just bacteria. If the imagined experiment is correct, the surprise conclusion is that the elements of an ecosystem carrying the greatest amount of information still require that the information exist as an ordered hierarchy. In contrast, the biotic elements with the least information do not require a hierarchy of information, even if they eventually adapt into a more complex system.

FROM VIRTUAL EXPERIMENTS TO REAL SITUATIONS

Although the preceding abstract examples provide a pedagogical and theoretical foundation, examples from data provide another perspective. Even better are cases in which an ecological system is under stress. Take the case of the many Italian coastal lagoons that become anoxic in early morning during the summer. These events have huge economic repercussion on tourism, fishing, bathing, and other activities that rely on a healthy aquatic environment. Our laboratory has monitored anoxic lagoons over several years. During this time we have observed that the organisms that suffer the largest loss are surprisingly the higher-order aerobic species: fish being the first. From a strictly biological point of view, setting aside for the moment the implications for people and economics, the loss of the fish population is not catastrophic to the lagoon. Certainly, the aerobic population perishes but the whole microbial pool of the sulfur cycle survives. If we could avoid attaching human value, then we might cynically conclude anaerobic lagoons have life. As humans, we might feel that a lagoon with only lowly bacteria isn't worth much, even if the bacteria might "think" differently. However, it is not mere anthropocentrism to say the lagoon is not the same when the fish and other species disappear. With them goes ecological information. Economic considerations aside, the vertebrates and other organisms contain the heritage of the past and the prospect for the future in their gene pools. Fortunately, lagoons are resilient, and the lower organizational level of the sulfate-reducing bacterial community eventually yields as the dystrophic conditions abate. More remarkable, the lagoon has a "memory" in the sense that with the recurrence of conditions leading to anoxia, usually a change in the climate, it triggers another dystrophic phase. Each new occurrence brings a faster loss of the fish and other higher level aerobic species.

Eutrophication along the Adriatic coast produces biological responses similar to those of the lagoons. Trophic instability in the Adriatic was noticed as early as the 18th century.[3] In the nearly 200 years since, the periods between recurring eutrophic episodes have grown shorter, until now eutrophic conditions arise nearly every two to three years. Prevailing thought is that the Adriatic episodes have increased because of ever-increasing nutrient runoff. Nutrient runoffs, along with other factors, are most likely the cause of the memory response of the Adriatic ecosystem, causing intervals of large eutrophic episodes to become shorter. We might say that natural ecosystems write and recall their own history, while human input reminds the ecosystems of past occurrences.

MATHEMATICAL ECOLOGY

If a poll were taken among ecologists, most would place the birth of mathematical ecology sometime during the 1920s with the work by Volterra and Lotka. The same poll might also find many ecologists agreeing that a mathematical analysis appropriate to ecological systems needs further work. Part of the reason would be that Lotka and Volterra established the use of time-invariant parameters, which have outlived their usefulness. Questions that were originally thought important during the predator–prey era of mathematical ecology have been replaced by the need to understand seasonality, variability, succession, and transient events. Furthermore, unlike the original search for answers about uniformly reproducing prey pursued by uniformly foraging predators across a uniform habitat, modern ecologists focus on differences, changes over time and space, and interactions that change.

Defenders of mathematical ecology would correctly point out that models are only models; models contribute to the broader perspective by means of abstraction. Modelers would claim any attempt to impose criteria for detail and precision misconstrues the use of models. Of course this defense is correct, yet insufficient. Even most modelers would not be inspired by an analysis that concludes big fish eat little fish. The ecological system of interest is not just the big fish and little fish, but the entire set of dynamics and relations among them, the interweaving of material exchange and information exchange. Explanations cannot be superficial if we intend to understand the environment.

Lotka and Volterra do have a message for ecologists. Fundamentally, it is to accept as necessary formal, mathematical tools. Then change the tools as the original questions become irrelevant and new questions arise. Whatever the original Lotka-Volterra models provided that was of value, they did not examine structural relations and time-varying events. Their determinism and reductionism will not suffice, but without having studied the case we would not have known this.

ECOLOGY AND ECOLOGICAL MODELS

Numerical simulation models rest on parameter identification and measurement. The more parameters identified and measured, the greater the belief that the model represents an ecological system. No effort is spared to identify all the parameters. First, it is hoped that every parameter can be gotten from a direct field measurement. If this fails, then the hope is to perform a laboratory experiment. Modelers know this will not always be possible. So a little intuition, extrapolation from the literature, and guesses become surrogate means to gain parametric values. The fact is that any sophisticated simulation model uses values from diverse sources and not necessarily from actual field data. As ecological studies expand, they include variables that are more than species numbers and population abundance. Inevitably, this increases the number of immeasurable variables and parameters. Inevitably, the models or parameter measurements lead to ambiguity, and this is not a strength of the numerical simulation approach.

Faced with ecological models that poorly represent an ecosystem, we can hardly be surprised that using parameters that define a specific relationship for one ecosystem hardly suits even a similar one in another location. We cannot simply get around the fact that nutrient-loading rates, vegetation nutrient uptake rates, predation rates, and all the other parameters of one environment are not equal to the values of another, although spacially close. Even in the same location, values can differ: Each ecosystem's spatial scale, micro-climate, and nonlinearity prevents a generalized simulation model becoming transferable.

In 1958, Redfield[4] measured carbon, nitrogen, and phosphorous in the western north Atlantic and found these occurred in a ratio of 106:16:1. Nearly all phytoplankton growth models use this ratio in the stoichiometric equations of photosynthesis, yet few ecologists would argue that this relationship and these values are constant, invariant, and unchangeable, like some ecological speed of light constant. Ecologists do not challenge the tenet that adaptation, evolution, and ecological change take place in every ecological community. Current ecological science would accept that phytoplankton do alter their assimilation rates of nutrients based on environmental conditions and the physiological characteristics of individual species respond to external conditions. Furthermore, ecologists would not be surprised to find that different species have evolved to use different ratios of ambient nutrients. Few would believe the same phytoplankton species always behave in the same way, despite different environmental conditions. Physiological constraints are inherent to species and are important, as suggested by Liebig's law. Yet, ecologists know that a physiologic response to field conditions negates the notion of a fixed and unchanging stoichiometry.

Healthy skepticism of ecological models would be prudent. Not every parameter under every possible condition can be measured. Rather, the models must capture the complexity of natural systems without being simplistic or overly deterministic.

Seeking a common language among scientists interested in ecology has become a priority for the future. Specifically, a means must be found to avoid equivocation because of a plethora of terms and definitions to cover every possible circumstance. Intuition among field ecologists needs an explication; observing nature is not the same as understanding it. Definitions of a few words have been established, but gather all those researchers interested in an effort toward the "scientification" of ecology: the study of inter- and intradependent systems in which life signifies its essentiality and the environment its receptiveness. We can anticipate that the next development to move ecology forward will come when the culture of ecologists evolves.

ACKNOWLEDGMENTS

The authors would like to thank Charles J. Puccia for his comments and suggestions which served to improve the presentation and for his review of the final manuscript.

REFERENCES

1. CLARKE, G.L. 1967. Elements of Ecology. John Wiley & Sons, Inc. London/New York/Sydney.
2. LINCOLN, R.J., G.A. BOXSHALL & P.F. CLARK. 1982. A Dictionary of Ecology. Cambridge University Press. New York.
3. FONDA UMANI, S., E. GHIRARDELLI & M. SPECCHI. 1989. Gli episodi di "mare sporco" nell'Adriatico dal 1729 ai giorni nostri. Regione Autonoma Friuli Venezia Giulia—Direzione Regionale Ambiente, Tirieste.
4. REDFIELD, A.C. 1958. The biological control of chemical factors in the environment. Am. Sci. **46:** 205–221.

A Tentative Fourth Law of Thermodynamics, Applied to Description of Ecosystem Development

SVEN E. JØRGENSEN[a]

DFH, Institute A, Miljøkemi, Universitetsparken 2, 2100 Copenhagen Ø, Denmark

INTRODUCTION

The three laws of thermodynamics can be considered to be constraints on the development of ecosystems: only processes that follow the conservation principle (first law) and that consume exergy (produce entropy or dissipate energy) (second law) are possible. A flow of energy (exergy) through the system, which means that the system must be open or at least nonisolated, is absolutely necessary for its existence (partly deduced from the third law). A flow of exergy through the system is also *sufficient* to form an ordered structure (also called a dissipative structure[45]). Morowitz[36] calls this latter formulation the fourth law of thermodynamics, but it would be more appropriate to expand this law to encompass a statement about *which* ordered structure among the possible ones will be selected, or which factors determine how an ecosystem will grow and develop? This expanded version was formulated as a tentative fourth law of thermodynamics by Jørgensen,[16,25] but the content was already expressed in Jørgensen and Mejer,[20] Mejer and Jørgensen,[34] and in Jørgensen.[13] This paper focuses on this tentative and expanded version of the fourth law of thermodynamics and its implications for ecosystem properties and development. It is to a high extent based on a recent presentation of the tentative fourth law of thermodynamic; see Jørgensen[25] and Jørgensen et al.[24]

Exergy is defined as the work the system can perform when it is brought into equilibrium with the environment or another well-defined reference state. If we presume a reference environment that represents the system (ecosystem) at thermodynamic equilibrium, which means that all the components are inorganic at the highest possible oxidation state (all the free energy has been previously used to do work) and homogeneously distributed in the system (no gradients), the situation illustrated in FIGURE 1 is valid. The chemical energy embodied in the organic components and the biological structure contributes by far the most to the exergy content of the system, so there seems to be no reason to assume a (minor) temperature and pressure difference between the system and the reference environment.

Under these circumstances we can calculate the exergy content of the system as if it comes entirely from the chemical energy: $\Sigma(\mu_c - \mu_{co})N_i$. Only what Szargut[56] calls chemical exergy is included in the computation of exergy The physical exergy[54,55] is omitted in these calculations because there is no temperature and pres-

---

[a]Address for correspondence: 45-35375744 (fax); sej@mail.dfh.dk (e-mail).

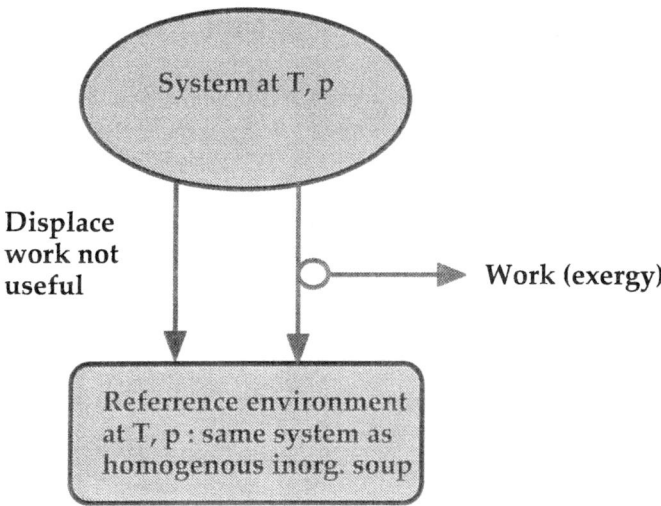

FIGURE 1. The exergy calculated in this paper is using the same system pictured here, but as a homogeneous, inorganic soup without any gradients and free-energy-containing compounds.

sure difference between the system and the reference system. By means of these calculations, we find the exergy of the system compared with the same system at the same temperature and pressure but in the form of an inorganic soup without any life, biological structure, information or organic molecules. Because $(\mu_c - \mu_{co})$ can be found from the definition of the chemical potential replacing activities by concentrations, we get the following expressions for the exergy

$$Ex = RT \sum_{i=0}^{i=n} c_i \ln c_i / c_{ieq} \tag{1}$$

where R is the gas constant, T is the temperature of the environment (and the system; see FIG. 1), while c_i is the concentration of the ith component expressed in a suitable unit, for example, for phytoplankton in a lake c_i could be expressed as mg/l or as mg/l of a focal nutrient. c_{ieq} is the concentration of the ith component at thermodynamic equilibrium and n is the number of components. $c_{i,eq}$ is of course a very small concentration (except for $i = 0$, which is considered to cover the inorganic compounds), but is not zero, corresponding to a very low probability of forming complex organic compounds spontaneously in an inorganic soup at thermodynamic equilibrium.

In thermodynamic terms, a growing system is moving away from thermodynamic equilibrium. At thermodynamic equilibrium, the system cannot do any work, the components are inorganic and have zero free energy, and all gradients are eliminated. Everywhere in the universe you will, however, find structure and gradients resulting from growth and developmental processes. Dissipation attempts to tear down the structures and eliminate the gradients, but dissipation cannot operate unless the gradients were present in the first place. An obvious question is therefore, What determines the buildup of gradients?

Biological systems have a particularly large number of possibilities for moving away from thermodynamic equilibrium, that is, to build up gradients. It is therefore crucial in ecology to know which pathways among the possible ones an ecosystem. will select for development. That would be the key to describing the processes characteristic for ecosystem development. Considering the ontic openness presented in Jørgensen et al.[23] focusing on the third law of thermodynamics, it would, however, be more appropriate not to discuss the selection of components and processes for development of ecosystems but rather to discuss the propensity of direction for development.[57] The possible solution to the problem discussed here may be formulated as a hypothesis:

> If a system receives a through-flow of exergy (a) the system will utilize this exergy flow to move away from thermodynamic equilibrium. (b) If t more than one pathway to move away is offered from thermodynamic equilibrium, the one yielding most *stored* exergy (measured in J/m^2 or J/m^3) by the prevailing conditions, that is, with the most ordered structure and the longest distance to thermodynamic equilibrium, will have a propensity to be selected.

The tentative fourth law of thermodynamics may also be considered as an extended version of "Le Chatelier's principle." Formation of biomass may be described as:

Energy + nutrients = molecules with more free energy (exergy) and organization

If we pump energy into a system, the equilibrium will, according to Le Chatelier's principle shift toward a utilization of the energy. It means that molecules with more free energy and organization are formed. If more pathways are offered, the pathways that give most relief, that is, use the most energy and thereby form molecules with the most embodied free energy (exergy) will win according to the proposed tentative law of thermodynamics.

Because it is not possible to prove the first through third laws of thermodynamics by deductive methods, the tentative fourth law can only at the best be proved by inductive methods. It implies that the tentative fourth law should be investigated (a falsification is attempted) in as many concrete cases as possible.

The next section is devoted to a presentation of these concrete cases in which the hypothesis has been tested, in the sense that falsifications have been attempted. A method for calculating important contributions to exergy is presented and discussed in the third section. The two following sections examine the consistency of the hypothesis with other theories describing ecosystem development. The sixth section focuses on typical ecosystem properties and examines whether they are consistent with the proposed tentative fourth law of thermodynamics. The seventh section presents structural dynamic modeling in which applying exergy is the goal function. Recent developments in ecological modeling strongly support the hypothesis formulated as a tentative fourth law of thermodynamics, as observations reconcile with the structural changes predicted by use of exergy as the goal function. The following two sections give other examples of structural dynamic modeling by application of exergy as this modeling approach is considered strongly supportive of the

TABLE 1. Exergy (kJ/equiv) available to build ATP for various oxidation processes of organic matter at pH = 7.0 and 25°C

Reaction	Available kJ/equiv
$CH_2O + O_2 \to CO_2 + H_2O$	125
$CH_2O + 0.8NO_3^- + 0.8H^+ \to CO_2 + 0.4N_2 + 1.4H_2O$	119
$CH_2O + 2MnO_2 + H^+ \to CO_2 + 2Mn^{2+} + 3H_2O$	85
$CH_2O + 4FeOOH + 8H^+ \to CO_2 + 7H_2O + Fe^{2+}$	27
$CH_2O + 0.5SO_4^{2-} + 0.5H^+ \to CO_2 + 0.5HS^- + H_2O$	26
$CH_2O + 0.5CO_2 \to CO_2 + 0.5CH_4$	23

tentative fourth law of thermodynamics. The last section will summarize the important results and conclusions of the paper.

OBSERVATIONS SUPPORTING THE HYPOTHESIS EXPRESSED IN THE TENTATIVE FOURTH LAW OF THERMODYNAMICS

It has been possible to find ecologically relevant case studies in which several pathways are available to utilize the flow of exergy and in which the exergy gained by the system can be compared.[25] These (few) case studies support the tentative fourth law of thermodynamics, just as we unsuccessfully attempted to falsify the hypothesis.

The sequence of oxidation of organic matter (see, for instance, Schlesinger[50]) is as follows: by oxygen, by nitrate, by manganese dioxide, by iron (III), by sulfate, and by carbon dioxide. It means that oxygen, if present, will always outcompete nitrate which will out-compete manganese dioxide, and so on. The amount of exergy stored as a result of an oxidation process is measured by the number of ATP molecules formed. ATP represents the storage of 42 kJ exergy per mole. The available exergy in form of ATP molecules decreases in the same sequence as indicated above, as it should be expected if the tentative fourth law of thermodynamics is valid (TABLE 1).

Numerous experiments have been performed to imitate the formation of organic matter in the primeval atmosphere on earth 4×10^9 years ago (see, for instance, Jørgensen[25]). Various sources of energy have been sent through a gas mixture of carbon dioxide, ammonia, and methane. Analyses have shown that a wide spectrum of various compounds including several amino acids is formed under these circumstances, but generally only compounds with rather large amounts of free energy (i.e., high exergy storage) will form an appreciable part of the mixture.[36]

Evolution has always been toward organisms with an increasing number of information genes (genes actually utilized) and more types of cells, that is, toward storage of more energy due to the increased information content.

At the biochemical level, we find that different plants operate three different biochemical pathways for the process of photosynthesis: (a) the C3 or Calvin Benson cycle, (b) the C4 pathway, and (c) the crassulacean acid metabolism (CAM) pathway. The latter pathway is less efficient than the two other possible pathways, measured as grams of plant biomass formed per unit of energy received. Plants using the

CAM pathway can, however, survive in a harsh, arid environment, which plants following C3 and C4 pathways cannot. The photosynthesis will, however, switch to C3 as sufficient water is available (see Shugart[52]). The CAM pathways give, in other words, the highest exergy storage under harsh, arid conditions, while the two other pathways give the highest exergy storage under other conditions. These observations are completely in accordance with the tentative fourth law of thermodynamics.

Givnish and Vermeli[9] made the assumption that leaves optimize their size by balancing the payoff of having leaves of a given size against that of maintaining leaves of a given size. By means of this assumption, which corresponds to optimization of exergy storage, they can explain how the size of leaves depends on the solar radiation and humidity in a given environment.

It should also be mentioned that the general relationship between animal body size, W, and population density, D, is $D = A/W$,[44] where A is a constant. The highest possible packing of biomass is therefore independent of the size of the organisms, which supports the idea that an ecosystem attempts to optimize exergy storage. If, for instance, the exergy dissipation should be maximized which is not the case, then the packing would be in accordance with the following relationship: $D = $ constant/$W^{0.65-0.75}$ (see Peters[44]), because the specific respiration is proportional to W in the exponent 0.65–0.75.

If a resource (for instance, one of the limiting elements for plant growth) is abundant, it will recycle faster. This is a little strange, because recycling is not needed when a resource is nonlimiting. Nevertheless, it has been shown that the exergy stored[25] increases when an abundant resource recycles faster. Ulanowicz[57] claims that ascendency also increases with faster recycling of an abundant resource, and modeling studies (also see below) have shown that ascendency and storage of exergy are closely correlated.

Brown et al.[3] and Marquet and Taper[31] have examined the patterns of animal body size. They can explain the frequency distributions for the number of species as a function of body size by optimization of fitness. Fitness may be defined as the rate that resources,[4] in excess of those required for maintenance of the individual, can be used for reproduction. This definition of fitness is close to "ability to increase the stored exergy." Their results may therefore be considered to support the presented hypothesis.

Several modeling cases, mostly of aquatic ecosystems, have been examined, and they have all supported the tentative law by calculating exergy indexes to be used for a description of the structural dynamic changes in the focal ecosystem. Models with dynamic structure are models based on nonstationary, time varying differential equations. Structural dynamic models have also been developed through knowledge about which species with which properties (used as parameters in the model) would be dominant under which circumstances (see for instance Recknagel et al.,[46] Reynolds,[48] and Patten.[42] The structural dynamic models refered to here in support of the tentative fourth law of thermodynamics have, however, applied optimization of an exergy index to describe the current changes of the parameters (see the various case studies in Jørgensen,[14–18,25] Jørgensen and Padisák,[21] Jørgensen and de Bernardi,[19] Nielsen,[37,38] and Caffaro and Bocci,[6] Caffaro et al.[7]

The exergy index is calculated as shown in Jørgensen et al.[22,24] and in Jørgensen.[25] In these references it is shown that a good approximation of the exergy storage of an ecosystem can the concentrations of the various components, c_i, multiplied by weighting factors, β_i, reflecting the exergy that the various components possess because of their chemical energy and the information embodied in the genes:

$$Ex = \sum_{i=n}^{i=0} \beta_i c_i \qquad (2)$$

The index 0 covers all the inorganic components that of course in principle should be included, but in most cases can be neglected. This is because the contributions from detritus, and even to a greater extent from the biological components, are much higher because of an extremely low concentration (probability) of these components in the reference system (the ecosystem converted into an inorganic dead system). The inorganic components, including oxygen, may have different concentrations in the observed system and the reference system, but becuse the ratios between the concentrations of inorganic components in the two systems are much lower than the ratios of the concentrations of biological components, the contribution from inorganic components to exergy is negligible compared with the contribution of the biological components.

The calculation of the exergy index accounts for the chemical energy in the organic matter as well as for the information embodied in the living organisms. It is measured by the extremely small probability of forming the living components, for instance algae, zooplankton, fish, mammals, and so on spontaneously from inorganic matter. The weighting factors are shown in TABLE 2. They presume that the exergy is expressed in detritus exergy equivalent, inasmuch as β for detritus is 1. They may also be considered quality factors reflecting how developed the various groups are and to which extent they contribute to the exergy by means of their information content, which is reflected in the computation. This is entirely according to Boltzmann,[1] who gave the following relationship for the work, W, that is embodied in the thermodynamic information:

$$W = RT \ln N \qquad (ML^2T^{-2}) \qquad (3)$$

where N is the number of possible states among which the information has been selected. N is as seen the inverse of the probability to obtain the valid amino acid sequence for a species spontaneously.

Exergy calculated by Equation (2) has some shortcomings,. We therefore propose to consider the exergy found by these calculations be considered a *relative exergy* index:

1. We account only for the contributions from the organisms' biomass and information in the genes. Although these contributions are probably the most important ones, the possibility that other important contributions are omitted cannot be completely excluded.

2. We don't account for the information embodied in the network—that is, the relations between organisms. The information in the model network that we use to describe ecosystems is negligible compared with the information in the genes, but we cannot exclude the possibility that the real, much more complex network may contribute considerably to the total exergy of a natural ecosystem.

TABLE 2. Approximate number of nonrepetitive genes[a]

Organisms	Number of Nonrepetitive Genes	Conversion Factor[b]
Detritus	0	1
Minimal cell[c]	470	2.3
Bacteria	600	2.7
Algae	850	3.4
Yeast	2000	5.8
Fungus	3000	9.5
Sponges	9000	26.7
Moulds	9,500	28.0
Plants, trees	10,000–30,000	29.6-86.8
Worms	10,500	30.0
Insects,	10,000–15,000	29.6–43.9
Jellyfish	10,000	29,6
Zooplankton	10,000–15,000	29.6–43.9
Fish	100,000–120,000	287–344
Birds	120,000	344
Amphibians	120,000	344
Reptiles	130,000	370
Mammals	140,000	402
Human	250,000	716

[a]Sources: Cavalier-Smith,[5] Li and Grauer,[29] and Lewin.[28]
[b]Based on energy contained in detritus; 1 g detritus has on average 18.7 kJ energy.
[c]Morowitz.[36]

3. We have made approximations in our thermodynamic calculations. They are all indicated in the calculations and are in most cases negligible.

4. We can never know all the components in a natural (complex) ecosystem. Therefore, we will only be able to utilize these calculations to determine exergy indices of our simplified images of ecosystems, for instance of models.

Exergy indices are, however, useful, as they have been successfully used as goal functions (orientors) to develop structural dynamic models. The *difference* in exergy by *comparison* of two different possible structures (species composition) is decisive here. Moreover, exergy computations always give only relative values, because the exergy is calculated relative to the reference system.

EXERGY AND DARWIN'S THEORY

Darwinism and neodarwinism have given the answer, when one species is considered, to the question raised in the Introduction: which pathways among the many possible ones are selected by an ecosystem? Darwin's answer is: the best fitted, that is, with the properties best coordinated to the prevailing conditions, will survive. Survival in this context means that the biomass of the species will be maintained or maybe even increased. An organism or a population is exposed to many constraints, determined by the forcing functions on the ecosystem and the other organisms living in it. The question is, Who is winning the competition for the resources? The winner's award is survival and even growth. The exergy or chemical energy that can be used to do work in mineral oil is about 42 kJ/g. For biomass with an average composition of proteins, carbohydrates, and fat, the chemical energy (free energy, exergy can be calculated to be approximately 18.7 kJ/g (the details of this calculation are given in Jørgensen et al.[22]). If the exergy found in accordance with Equation (2) and the β-values in TABLE 1 is multiplied by 18.7, the exergy will be expressed in kilojoules.

Brown[4] defines fitness as the rate at which resources in excess of those required for maintenance can be utilized for reproduction. He uses dW/dt, reproductive power, to find the optimal body mass, W. So, he is asking the question, Which size is best fitted? The answer is found by a determination of the size with the highest growth potential, that is, the size yielding the biggest increase of the biomass corresponding to the biggest increase of the exergy in the system.

An ecosystem encompasses, however, many species. They cannot all obtain the biggest biomass independently of the other species—the species are interdependent. Darwin considered this complication, as the expression "prevailing conditions" is anticipated to include all the abiological and biological constraints imposed on the species that is, including the constraints originating from other species. Evolution and coevolution over a very long period have, however, implied that the species have adapted to each other. They have been able to find how they can move further away from thermodynamic equilibrium (get more growth) if they cooperate by adjusting their properties to each other and to the prevailing external factors. The effect of this cooperation is consistent with Patten.[41] He shows that the indirect effect often exceeds the direct one. For instance, a predator–prey relationship may also be beneficial for the prey as the result of a number of factors, including faster circulation of the nutrients.

Darwin's theory presumes that populations consist of individuals that have the following characteristics:

1. On average the individuals should produce more offspring than needed to replace themselves upon death; this is the property of high reproduction. Translated into thermodynamics, more possible pathways for utilization of the energy flow are developed than the system and its energy flow can sustain. It implies that a competition among the pathways, even among pathways that are only slightly different, will be established.

2. The individuals that have offspring that resemble their parents more than they resemble randomly chosen individuals in the same population; this is the property of inheritance. Thermodynamically, it means that the properties that have shown a better ability to use the energy flow to move as far away from thermodynamic equilibrium as possible by construction of more biomass will, to a high extent, be preserved. Genetics can explain how this is possible.

3. The individuals vary in heritable traits influencing reproduction and survival due to differences in fitness to the prevailing conditions; this is the property of variation. The modernized neodarwinism is able to give a long list of mechanisms that can create new pathways. It implies that new possibilities are steadily created to meet the challenge of utilizing the energy flow to move away from thermodynamic equi-

librium. These possibilities are tested under the prevailing conditions, and the successful ones are preserved according to the second listed characteristic.

Evolution can therefore continue on the shoulders of the already found successful solutions and steadily find new and better solutions, that is, select the best genes from among all the present genes including the ones that are continuously emerging through mutations and sexual recombination.

This theory implies that properties will continue to be changed by selection processes to give the best possible survival under the prevailing conditions, which for plants include grazing and for grazers the availability of food. One species cannot change the properties of the other species directly, but all species must consider all the other species in their effort to find a feasible combination of properties that offers a higher probability of survival. This explains how species become adapted to each other (coevolve) and can cooperate on the joint goal of moving as much as possible away from thermodynamic equilibrium. In principle, each of the species is striving toward its own goal: to achieve the highest possible growth for its own species. These goals cannot be reached if the species don't adapt to the other species, because they are all a part of the living conditions; therefore, the result will be that the species together move as much as possible away from thermodynamic equilibrium, that is, give the system the highest possible exergy. It will in many cases coincide with the highest or close to the highest biomass for most of the species, at least on a long-term basis.

The conclusion from these considerations is that exergy, because it measures the distance from thermodynamic equilibrium of the entire ecosystem, seems to be a good candidate for quantifying survival and growth in the Darwinian sense for the entire ecosystem. Calculations of the exergy of ecosystems make it possible to unite the chemical energy of the organic matter and the information (in the sense of Boltzmann) embodied in the species.

OTHER ORIENTORS DESCRIBING GROWTH OF AN ORGANISM OR ECOSYSTEM DEVELOPMENT

Boltzmann[1] proposed that "life is a struggle for the ability to perform work," which is exergy. The ability of a single species to perform work is proportional to its biomass. Margalef,[58] Straskraba,[59,60] and Brown[61] have proposed using biomass as goal function.

Lotka[30] proposed using maximum power as a goal function to describe ecosystem development. Maximum power is defined as the transformation of energy to perform work per unit of time.[40] The transformation of energy to perform work is correlated with the amount of energy available (stored) in the system. The more energy stored, the more energy can be transformed to perform work. This correlation was demonstrated by Salomonsen.[62] He showed that the ratio of energy and maximum power of two lakes at significantly different levels of eutrophication was approximately the same. Is it a "what was first the chicken or the egg" problem? Or is Mauersberger[32,33] "minimum entropy principle" right and is this principle consistent with the maximum power principle or with the principle of maximum exergy storage? Are there situations where ecosystems attempt to minimize the energy dissipation,[12] and maximize the storage of exergy in spite of the general relationship that more structure (more exergy stored) will require more exergy (energy spent on maintenance of already stored exergy? This question will be further discussed in the next section.

Odum[40] has introduced another goal function named emergy. The idea is that it is not the actual content of energy that counts but the embodied energy. It is the energy (originating from the ultimate source of energy, solar energy) that it costs to construct the considered component. If it costs 10 units of phytoplankton, for instance, to construct one unit of zooplankton, the energy of zooplankton should be multiplied by 10 to obtain the embodied energy in zooplankton relative to the energy of phytoplankton. It has been shown that there is a good correlation between exergy index and emergy when the calculations are based on realistic ecosystem models derived from data and observations.[18] Different organisms have, however, developed different strategies for the prevailing conditions in different ecological niches. In accordance with the emergy calculations, a tree adapted to the shade should have lower emergy than a tree adapted to full sunlight, if we presume that they have the same biomass and growth rate. Because we know that they may have an equal role in ecosystem development and they have approximately equal exergy (which is dependent on the embodied information), this postulated difference in emergy cannot be correct. Emergy calculates how much solar energy it costs to build the structure while exergy expresses the actual work capacity. If emergy is calculated for an entire ecosystem, the differences between exergy and emergy will level out, but emergy for individual species/components may sometimes provide an erroneous measure of the potential for developing the ecosystem further away from thermodynamic equilibrium.

Kay and Schneider[27] propose that the development of an ecosystem is best described as follows: the system will use all possible avenues to capture per unit of time as much of the incoming exergy as possible, and the captured exergy will be degraded (dissipated) to cover the exergy needed for maintenance (see also Chapter 11 in Ref. 27). Because the degradation of exergy is a focal point in Kay and Schneider's theory, they call their formulation an extended version of the second law of thermodynamics. The relationship between the use of exergy per unit of time can be described by use of the following equation (see Jørgensen et al.[23]):

$$\Delta Ex_{cap} = \Delta Ex_{bio} + \Delta E_{resp+eva} \qquad (ML^2T^{-3}) \qquad (4)$$

where ΔEx_{cap} is the exergy captured by the system per unit of time, ΔEx_{bio} is the exergy stored (accumulated) by the structure per unit of time, $\Delta E_{resp+eva}$ is the exergy degraded by respiration and evapotranspiration processes into heat energy per unit of time. Kay and Schneider mean that the system will optimize $\Delta E_{resp+eva}$, whereas the previously presented formulation of the fourth law of thermodynamics claims that $F(t) = \int \Delta Ex_{bio}(t)\,dt$ (measures the integrated growth over time from 0 to t) is optimized. Very little of the exergy captured per unit of time is used for building new structure (except for a system at the early stage), so the exergy degradation per unit of time = energy converted to heat per unit of time = $\Delta E_{resp+eva}$ is for most (at least the most developed) ecosystems $\gg \Delta Ex_{bio}$. This implies that $\Delta E_{resp+eva} \approx \Delta Ex_{cap}$. $\Delta Ex_{bio}(t)$, resulting from an integration of ΔEx_{bio}, determines the size of the structure and how well organized the ecosystem is. It determines how much exergy the system has on stock for later consumption (including for dissipation). It also determines how much exergy the system can capture per unit of time in the future and how much exergy the system dissipates in maintenance, because there is a close relationship between the size of the structure on the one hand and the exergy captured per unit of time and the exergy used for maintenance per unit of time on the other hand.

Note, however, that Equation (4) presents a relationship between rates that can hardly be used to optimize the development of an ecological system, because:

1. Rates cannot continuously increase, because they have upper limits. It is, for instance, impossible to capture more than 100% of the solar radiation (or rather 85–90% because of physical constraints). Longer term development will therefore need a function derived from integration of rates over time to be able to describe continuous development under prevailing conditions.

2. Rates in ecological systems are continuously changed, because the environment determining the rates varys over time. It is therefore significant to account for

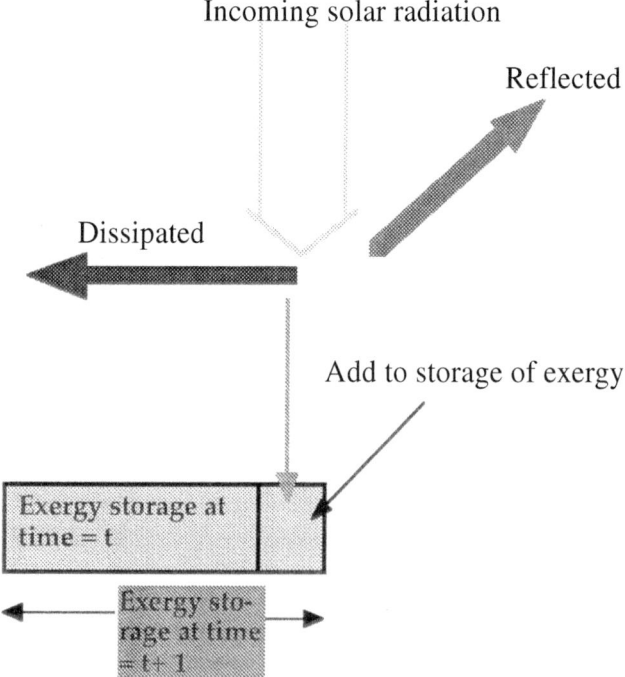

Extended second law of thermodynamics: optimization of dissipated exergy or captured exergy = dissipated + add to storage of exergy—all per unit of time.

Tentative fourth law of thermodynamics: optimization of exergy storage as f(time) corresponding to integration over time of "add to storage of exergy per unit of time."

FIGURE 2. The difference between the extended second law of thermodynamics and the tentative fourth law of thermodynamics is illustrated.

the results (integration) of rates over time, because the system cannot be determined by an occasional high rate for a short, insignificant time.

3. In their formulation, the thermodynamic laws apply variables derived from integration over time, for instance, "energy is conserved" and "entropy will increase for all real processes."

The two hypotheses, "to optimize exergy dissipation or exergy captured" and "to optimize the exergy of the system under the prevailing conditions" are just two sides of the same coin, when we describe the development of an ecosystem from the early stage to the mature stage. (See also the next section of the paper.) Increased exergy stored in the structure of the system under development will also enable the system to capture more exergy as already mentioned, but the discrepancy between the two theories occurs when the system has attained the maximum rate of capturing exergy (as mentioned above, 85–90% of the incoming solar radiation).

An illustration of the concepts presented and their relation to the conservation and dissipation principles and to the hypotheses of Kay and Schneider on the one side and Jørgensen on the other is shown in FIGURE 2.

DEVELOPMENT OF ECOSYSTEMS

Ulanowicz[56] uses growth and development as the extensive and intensive aspects of the same process. Growth implies increase or expansion, whereas development focuses on the increase in organization that is considered independent of the size of the system. Ulanowicz considers growth and development to be aspects of a unitary process, and he applies the concept ascendency (which is strongly correlated with exergy as mentioned above) to cover both the changes in size and organization.

The successional development of ecosystems from an early to a mature stage (see for instance Odum[39]), illustrates that the two concepts, exergy storage and exergy utilization are parallel. An ecosystem at an early stage of development, for instance an agricultural field, has minimal exergy storage and utilization The biomass per square meter is small compared with the mature system, that is, the exergy storage is small. The structure is simple and only a small amount of energy (exergy) is needed for respiration or growth, as they are both to a certain extent proportional to the biomass. The total surface area of the plants is furthermore small, which implies that they are not able to catch and use as much solar radiation.

As the system develops the structure becomes more complicated: animals with more information per unit of biomass, that is, with more genes, populate the ecosystem, and the total biomass per square meter increases. It implies that both the exergy storage and the exergy needed for maintenance increase (see FIG. 3). A very mature ecosystem, for instance a natural forest, has a very complex structure and well-organized food webs. It contains a high concentration of biomass per square meter and contains much information in a wide variety of organisms. The entire structure tries to use solar radiation either directly or indirectly, resulting in a high utilization of the solar energy flux.

The catabolic energy demand is related to the total biomass and the overall organization. It represents the exergy needed for maintaining the ecosystem far from thermodynamic equilibrium. This is parallel to what is experienced by man-made systems: A large town with many buildings of different types (skyscrapers, cathedrals, museums, scientific institutes, etc.) obviously needs much more maintenance than a small village consisting of a few almost identical farmhouses.

The development of ecosystems may also be described (Kay and Schneider[26,51]) as a steady growth of a gradient between the ecosystem and thermodynamic equilibrium. The force to break down the gradient will increase with increasing gradient. This tendency to break down the gradient is represented by respiration and evapotranspiration, which spend exergy and produce entropy. As long as the exergy received from solar radiation can compensate for this need of exergy to maintain the gradient, it is possible for the system to stay far from thermodynamic equilibrium. **If even more exergy can be captured than is needed for maintenance of the gradient, the surplus exergy increases the stored exergy, which means that the system moves further away from thermodynamic equilibrium and thereby increases the gradient even more.**

The amount of information stored in biomass may still increase in a mature ecosystem as a result of: (1) immigration of (slightly) better fitted species and (2) emergence of new genes or genetic combinations. This latter possibility is covered by the concept of "evolution."

The system stops growing in biomass when the most limiting inorganic component has been fully utilized for biomass construction. Then the mature stage of the ecosystem has been attained. Nutrients and water are often the limiting factors in the growth of plants. These resources cycle, which provides possibilities for formation of new biomass with perhaps more information, but the *total* biomass is not changed by this reallocation of resources. This constraint by the law of conservation is essential for the development of more and more complex living structures. As living organisms compete for limited food supplies, they invent and develop thousands of new and ingenious strategies.[47] Some species invest in movement; speed can be a valuable asset both for capturing prey and for avoiding predators. Others use protective armour or chemical poisons. Each species thus defines the terms under which it engages in the harsh business of life. Better feedback mechanisms to assure maintenance of a high biomass level under changed circumstances, better buffer capacities, better specialization to populate all possible ecological niches, and better adapted organisms to meet the variability of forcing functions are all developed. Thus, the biomass is maintained at the highest level over a longer time, and the information level will increase because of the steady development of better feedback mechanisms and more self-organization. Both contributions are reflected in a higher exergy The exergy of the mature system can therefore still grow further, namely by increasing the information. In other words, the system uses its resources better and becomes more fitted to the prevailing conditions. Adaptation and specialization require information, which implies that a better fitness to prevailing conditions is more probable on the part of a system with more stored information in the genes.

Neither energy nor information is conserved. Exergy is lost by all transfers of energy, but energy and information are also lost by death of organism as β in Equation (2) decreases from a value $\gg 1$ to 1 (detritus exergy equivalent is applied as a unit). Exergy and information may, however, be gained when phytoplankton is converted to zooplankton by grazing. **Exergy and information are therefore not cycling in the same manner as mass and energy. Their distribution in ecosystems as ener-**

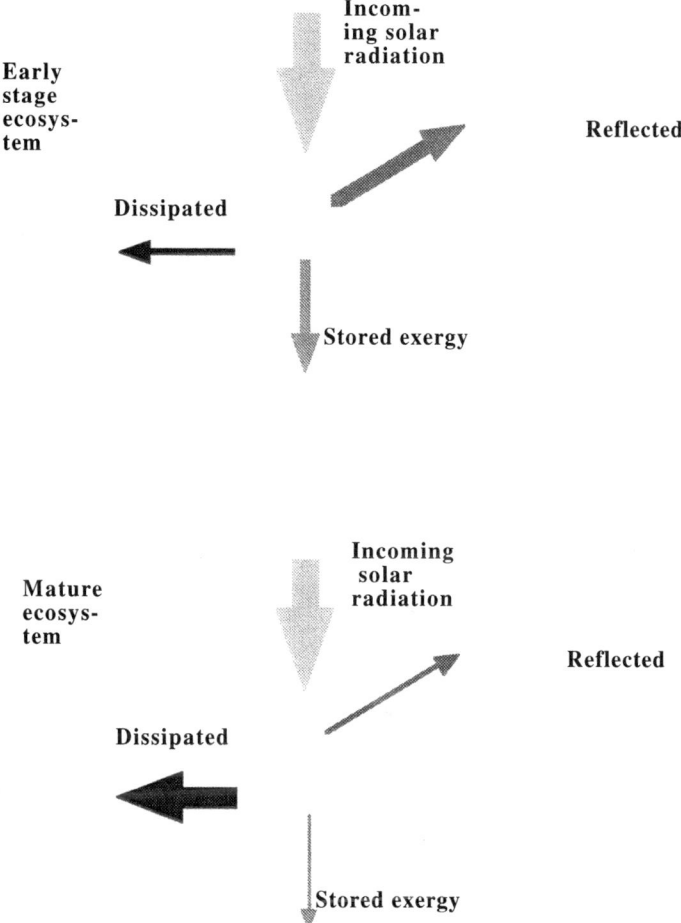

FIGURE 3. The exergy allocation for an early stage and a mature stage is compared.

gy and matter will in any case follow a complicated pattern determined by the life processes.

The two hypotheses of maximization of exergy storage and of exergy capture are, as already pointed out, not consistent when we have to describe the further development of a mature system; in this domain Mauersberger's minimum principle may be valid (see Mauersberger[32,33]):

Ecosystems locally decrease entropy production by transporting energy-matter from more probable to less probable spatial locations. Thereby, less exergy is lost and more exergy stored.

A possible hypothetical formulation of the tentative fourth law of thermodynamics trying to unite the hypotheses of Kay and Schneider with the hypotheses of Mauersberger may be:

If a system is moved away from thermodynamic equilibrium by application of a flow of exergy it will use all avenues available to build up as much dissipative structure (store as much energy as possible). An ecosystem at an early stage will try to store more exergy by increasing the amount captured (decrease the reflection, whereas a mature ecosystem will gain more exergy by decreasing the exergy lost by the maintenance processes) mainly achieved by increase of the information content of the system. These two situations are illustrated in FIGURE 3.

The overall description of the development of ecosystems is illustrated in FIGURE 4. In the first phase, the structure and its biomass increase rapidly, mainly due to the rapid growth of r-strategists. The gradients, the amount of exergy captured by the system, and the exergy required for maintenance are all increasing. A transition phase between the first and third phase is shown. In the third phase, the limiting elements are used up, which implies that a further increase in the physical structure measured by the biomass is not possible. A better use of the resources and more storage of information can, however, continue, and the two processes will work hand in hand. More exergy cannot be captured or dissipated, but the resources can be reallocated to store more exergy and more development toward relatively less dissipation of exergy (Mauersberger's principle). A development toward K-strategists is therefore favored, inasmuchas they will have bigger size and therefore less specific energy needs. Microorganisms also shift from r-strategists with quick exploitation of the resources to K-strategists with slow exploitation of resources.[8] Late-successional K-strategy plants produce litter that is poor in nutrients and simple sugar but high in lignine,[10] which changes the physico-chemical environment of the top soils. These processes cause modifications in the composition as well as in the activity of micro-

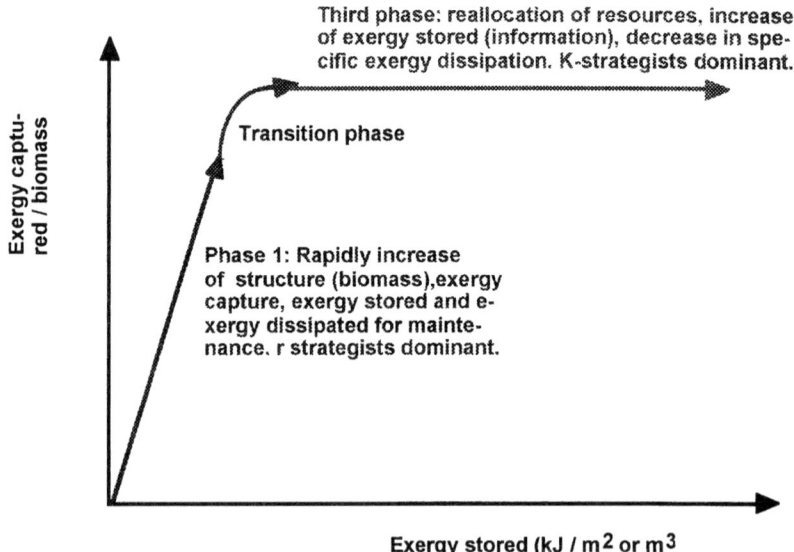

FIGURE 4. The exergy captured or the biomass is plotted versus exergy stored.

TABLE 3. Exergy utilization and exergy storage

Ecosystem	Percent Exergy Utilization	Exergy Storage (kJ/m^2)
Quarry	6	0
Desert	2	73
Clear cut	49	594
Grassland	59	940
Fir plantation	70	12700
Natural forest	71	26000
Old deciduous forest	72	38000
Tropical rain forest	70	64000

bial communities, favoring the K-selected organisms. The shift from r-strategists to K-strategists may therefore be considered a process with synergistic effects.

Figures from satellite measurements support thsi description of ecosystem development. A forest captures much more exergy than a desert or a grassland, but a 50-year-old forest or a 200-year-old forest or even an old rain forest all capture approximately 80% of the incoming solar radiation.[27] It can, however, be shown,[25] that the stored exergy increases when a forest gets older, and a rain forest has more exergy stored than a temperate forest.

TABLE 3 shows the exergy utilization for different types of systems.[27] In the same table, the exergy storage is shown for some typical "average" systems. **As can be seen, there is a steep linear relationship between exergy storage and exergy capture when the system is under development from the early to a mature stage, but a mature system can still develop its exergy storage, although the exergy captured has attained the practical maximum of about 70–80% of the total solar energy received by radiation.** This points toward Ex_{bio} rather than ΔEx_{cap} or degradation of exergy as a general optimizer, although the optimization is parallel when the ecosystem is under development from early stage to the mature stage.

A parallel with economic systems may be used to illustrate the difference between Ex_{bio} and ΔEx_{cap}. When an enterprise or a country is under development, it is important to increase the turnover of the unit, which is parallel with ΔEx_{cap}, at the maximum rate. The turnover is, of course, dependent on the investment already made. In the long run it is more important for the firm (or country) to increase the *active* investment in infrastructure, production facilities, sales network, and so on. The enterprise or country making the most useful investments will be in the best position for competition. At a particular point in this development the investment in education and information becomes crucial—a clear parallel with the development of ecosystems, where investment in information seems particularly beneficial for the mature ecosystem.

A description of the development shown in FIGURE 4 has been attempted by application of a model. The model accounts for the exergy in kilojoules per m^2. It has the following stated variables: nutrients, plants (r-strategists), plants (K-strategists), herbivores, carnivores, and detritus. The photosynthesis (uptake of carbon dioxide from the air) is regulated by the exergy flow of solar radiation in the sense that a maximum (80–90%) of the incoming solar radiation can be captured by the plants to cover maintenance (respiration) and growth and replace the grazing of the herbi-

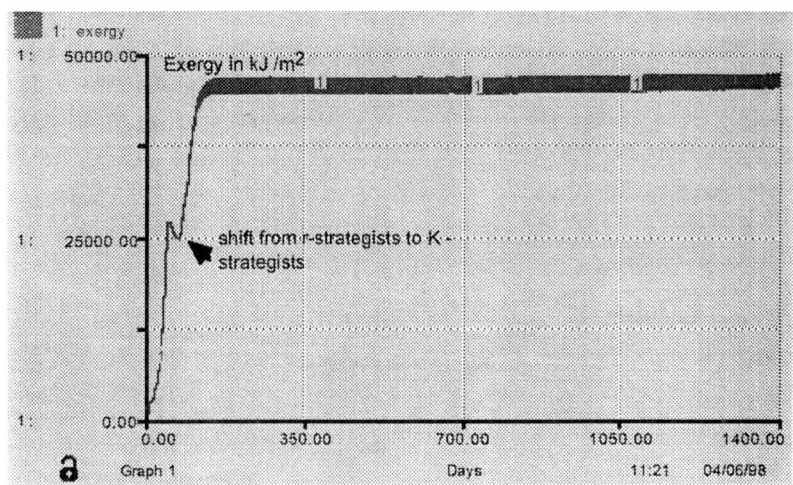

FIGURE 5. Results of a realistic model of a competition between r-strategists and K-strategists. For further detail about the model see the text.

vores. The model results (see FIG. 5) show that the r-strategists grow rapidly in the first phase, but in the long run are replaced by the K-strategists. The model also shows rapid growth in the first phase and very slow growth of the stored energy at a later stage, when the incoming flow of exergy in form of solar radiation becomes limiting. These results are consistent with Johnson.[12] He has found that when ecosystems are relatively isolated, competitive exclusion results in a relatively homogeneous system configuration that exhibits low dissipation.

In contrast, physical dissipation is much higher in ecosystems that are less fully developed. Johnson[12] concludes that "ecosystem structure is a function of two antagonistic trends: one toward a symmetrical state resulting in least dissipation and the other toward a state of maximum attainable dissipation." The latter is obtained by a rapid growth of biomass and structure, implying a more effective capture of the exergy contained in solar radiation. The first trend is obtained by a reallocation of the biochemical elements, which because of mass conservation limit the amount of biomass and structure. A development toward more effective organisms (less dissipation relative to the biomass, which is the case for bigger organisms; see also Straskraba et al.[53]), requiring less exergy for maintenance, will imply that although (almost) the same amount of energy is captured, the stored exergy can still increase. Inevitably, the configurations of the interactions become more mutualistic, self-reinforcing, and self-entailing.

This is consistent with Salthe,[49] where three phenological rules of thermodynamically open systems are proposed. As the system develops from immaturity through maturity to senescence:

1. There is an average monotonic decrease in the intensity of energy flow (flow per unit mass) through the system. The gross energy flow increases monotonically against a limit.

2. There is a continual, hyperbolic increase in complexity (= size + number of types of components + number of organizational constraints) or, generally, an ever-diminishing rate of increase in stored information.

3. There is an increase the system's internal stability (its rate of development slows down). Originally, this was stated as Minot's Law in developmental physiology.

Patten and Fath[43] have shown that increased cycling implies increased exergy storage under steady-state conditions. This was demonstrated as a general mathematical consequence of steady-state network theory. FIGURE 6 illustrates the results

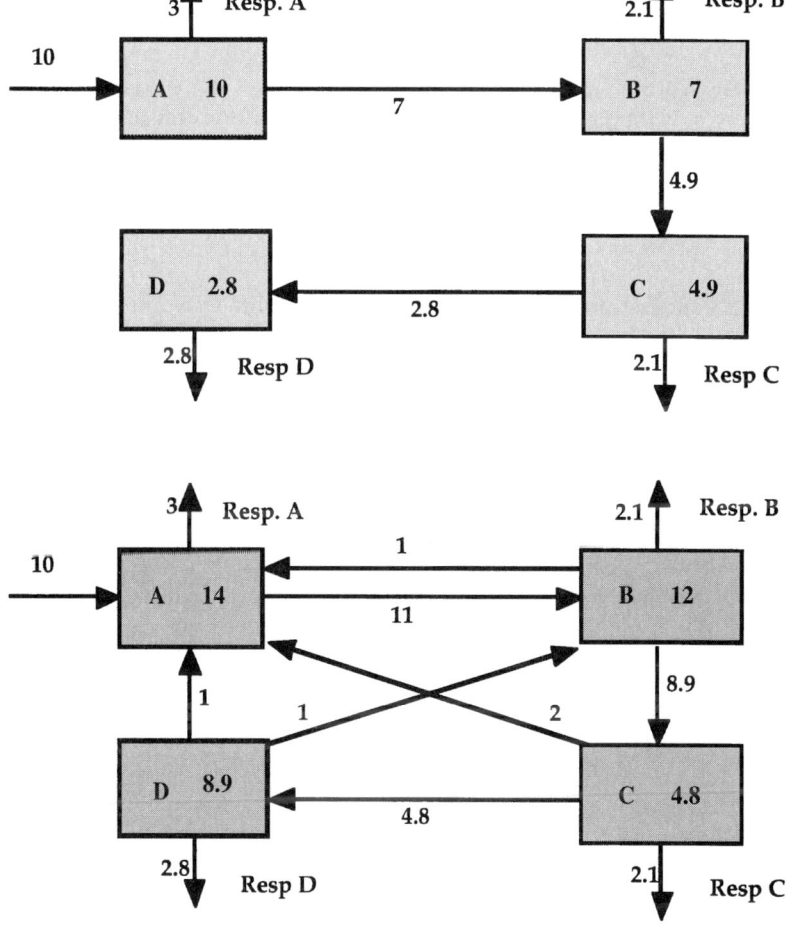

FIGURE 6. *Above*: an ecosystem with no recirculation and an input of 10 units. *Below*: the same system, but now with recirculation. The steady-state values of the four stated variables (compartments) are shown, if it is presumed that all flows are first order, donor determined.

by comparison of two networks at steady state with and without cycling. It is presumed that the process rates are first-order, donor-determined reactions. The cycling inevitably implies that the compartments (meaning the exergy stored), increase in size. The input of exergy (exergy captured) is the same in the two situations and corresponds to the dissipation of exergy while the two networks are in steady state. The specific dissipation of exergy understood as the exergy dissipated per unit of biomass, will of course decrease as the cycling increases, while the flow of exergy through the system will increase due to the cycling and resultant better utilization of the first path of exergy.

Ecosystem are dynamic. It is therefore a simplification to assume a steady state. On the other hand, a steady state may be interpreted as a freezing of the system in a situation assuming that ΔEx_{bio} is 0 (see Eq. (4)), which gives a realistic comparison of two situations: with and without cycling. It seems therefore appropriate (according to Patten and Fath[43]) to set up the following pertinent hypothesis:

Increased cycling implies that the exergy storage and the exergy through flow increase simultaneously with a decrease in specific exergy dissipation. This is in accordance with the interpretation of the tentative fourth law of thermodynamics presented here.

The role of natural disturbances such as fire or storms should be discussed in this context. When a forest is burned (for details see Botkin and Keller[2]), complex organic compounds are converted into inorganic compounds. Some of the inorganic compounds from the wood are lost as particles of ash that are blown away or as vapors that escape into the atmosphere and are distributed widely. Other compounds are deposited on the soil surface. These are highly soluble in water and are readily available for vegetation uptake. Therefore, immediately after a fire there is an increase in the availability of chemical elements, which are taken up rapidly, especially if there is a moderate amount of rainfall.

The pulse of inorganic nutrients can then lead to a pulse in the growth of vegetation. This in turn provides an increase in nutritious food for herbivores. The pulse in chemical inorganic elements can therefore have effects that extend through the food chain. Challenges to find new opportunities to move even further away from thermodynamic equilibrium are therefore created, which may explain how natural disturbances can have a long-term positive effect on the growth of ecosystems in the broadest sense of this concept.

This description is according to Holling's cycle.[11] FIGURE 7 is a modified version of this cycle presented by Ulanowicz,[58] in which Holling's cycle is modified in almost the same way as FIGURE 7, although the x-axis is not specific exergy = exergy/total biomass in Ulanowicz's presentation of Holling's cycle, but represents mutual information resulting from the flow structure. The basic idea is, however, the same. The renewal phase corresponds to rapidly increased biomass, the exploitation phase to a rapid increase in the level of information, and conservation to a very slow increase in both biomass and information. The destruction phase will, because of an external impact (forcing function), reduce both the amount of biomass and the information stored in this biomass, but new possibilities are thereby created for the utilization of emergent mutations and sexual recombinations. After each round in Holling's cycle, the biomass can probably hardly be higher as it is limited by the

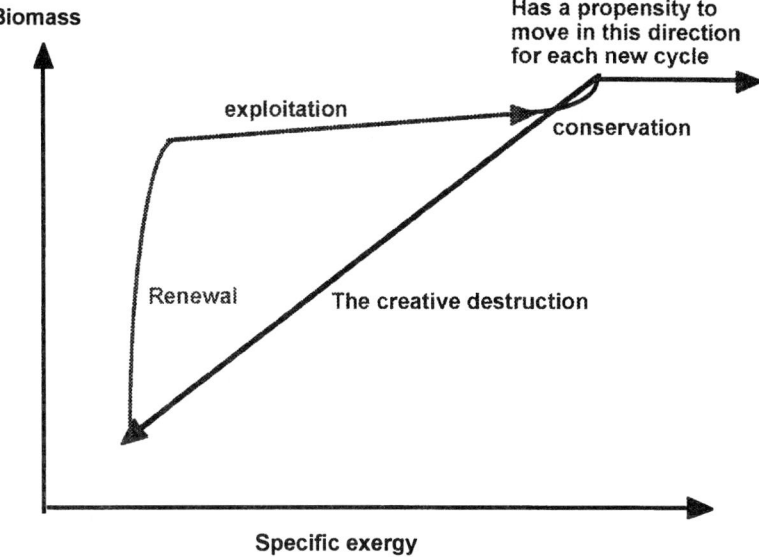

FIGURE 7. Holling's cycle is shown by use of biomass versus specific exergy = exergy/biomass.

presence of essential elements, but as testing of new mutations and sexual recombinations takes place, new and perhaps better combinations of properties will emerge. Consequently, there is a propensity for the exergy and the specific exergy to increase for every new round in Holling's cycle.

It is interesting in this context that scientists and visitors to Yellowstone National Park find the rapid recovery of the park remarkable. The park flourishes a decade after the fires that took place in 1988. In 1988, across Yellowstone National Park, tens of millions of trees have sprung from the ashes of what United States assumed was an ecological disaster. The fires of 1988 opened the forest canopy to abundant sunlight and enriched the soil with nutrients from dead trees. Now, 10 years after the fires, the branches of robust saplings that were not supposed to reproduce for another decade or two bristle with seed cones. Park officials even are studying whether or not to set controlled burns to clear out more deadwood and other volatile "fuel" in areas missed by the 1988 blazes.

SUMMARY AND CONCLUSIONS

A fourth law of thermodynamics has been proposed as a hypothesis. The tentative hypothetical law may be formulated as follows: **If a system receives a through-flow of exergy, the system will utilize this exergy flow to move away from thermodynamic equilibrium. If more than one pathway to move away from thermodynamic equilibrium is offered the one yielding the most stored exergy under the prevailing conditions, that is, with the most ordered structure and the longest**

distance to thermodynamic equilibrium, will have a propensity to be selected.
This hypothesis may also be formulated as an extended version of Le Chatelier's principle.

The tentative fourth law of thermodynamics has been shown to be in accordance with several observations in which more pathways are competing. It provides strong support for the tentative hypothesis we have named the fourth law of thermodynamics, as the pathways selected by all the referred observations are the ones showing the highest amount of stored exergy

The hypothesis is also in accordance with the general description of ecosystem development by Odum,[39] including Holling's cycle.

The tentative fourth law of thermodynamics makes it possible to unite Kay and Schneider's maximum dissipation hypothesis (the extended second law of thermodynamics), valid for the development of an ecosystem from an early to a mature stage, and Prigogine's and Mauersberger's minimum entropy production hypothesis, valid for a mature ecosystem. The two stages of development, the early stage and the mature stage, can, however, both be explained by the tentative fourth law of thermodynamics.

It can be concluded that the tentative fourth law of thermodynamics can explain the development of ecosystems in all phases. Several case studies (see also Jørgensen[25] and Jørgensen et al.[24]), including many structural dynamic models, support the tentative law of thermodynamics, but because of its general character it is absolutely necessary to test it with many more case studies that allow falsification before its more general use can recommended.

REFERENCES

1. BOLTZMANN, L. 1905. The Second Law of Thermodynamics. Populare Schriften, Essay No. 3 (address to Imperial Academy of Science in 1886). Reprinted in English in: Theoretical Physics and Philosophical Problems, Selected Writings of L. Boltzmann. D. Reidel. Dordrecht, The Netherlands.
2. BOTKIN, D. & E. KELLER. 1995. Environmental Science, Earth as a living Planet. John Wiley and Sons, New York. 630 pp.
3. BROWN, J.H., P.A. MARQUET & M.L. TAPER. 1993. Evolution of body size: consequences of an energetic definition of fitness. Am. Natur. **142:** 573–584.
4. BROWN, J.H. 1995. Macroecology. The University of Chicago Press. Chicago, IL.
5. CAVALIER-SMITH, T. 1985. The Evolution of Genome Size. Wiley. Chichester, England. 480 pp.
6. COFFARO, G. & M. BOCCI. 1997. How *Ulva rigida* and *Zostera marina* compete for resources: a quantitative approach with application to the Lagoon of Venice. Ecol. Model. **102:** 81–96.
7. COFFARO, G., M. BOCCI & G. BENDORICCHIO. 1997. Structural dynamic application to space variability of primary producers in shallow marine water. Ecol. Model. **102:** 97–115.
8. GERSON, U. & I. CHET. 1981. Are allochthonous and autochtonous soil microorganisms r- and K-selected? Rev. Ecol. Biol. Sol. **18:** 285–289.
9. GIVNISH, T.J. & G.J. VERMELJ. 1976. Sizes and shapes of liana leaves. Am. Natur. **110:** 743–778.
10. HEAL, O.W. & J. DIGHTON. 1986. Nutrient cycling and decomposition in natural terrestrial ecosystems. *In* M.J. Mitchell & J.P. Nakas, Eds.: 14–73. Microfloral and faunal interactions in natural and agro ecosystems. Nijhoff & Junk. Dordrecht.

11. HOLLING, C.S. 1986. The resilience of terrestrial ecosystems: Local surprise and global change. *In* Sustainable Development of the Biosphere. W.C. Clark & R.E. Munn, Eds.: 292–317. Cambridge University Press. Cambridge, England.
12. JOHNSON, L. 1990. The thermodynamics of ecosystem. *In* The Handbook of Environmental Chemistry, Vol. 1. The Natural Environmental and The Biogeochernical Cycles. O. Hutzinger, Ed.: 2–46. Springer Verlag. Heidelberg.
13. JØRGENSEN, S.E. 1982. A holistic approach to ecological modelling by application of thermodynamics. *In* Systems and Energy. W. Mitsch *et al.*, Eds. Ann Arbor, MI.
14. JØRGENSEN, S.E. 1986. Structural dynamic model. Ecol. Model. **31:** 1–9.
15. JØRGENSEN, S.E. 1988. Use of models as an experimental tool to show that structural changes are accompanied by increased energy. Ecol. Model. **41:** 117–126.
16. JØRGENSEN, S.E. 1992. Development of models able to account for changes in species composition. Ecol. Model. **62:** 195–208.
17. JØRGENSEN, S.E. 1992. Parameters, ecological constraints and energy Ecol. Model. **62:** 163–170.
18. JØRGENSEN, S.E. 1994. Fundamentals of Ecological Modelling. 2nd edit. (Developments in Environmental Modelling, Vol. 19). Elsevier. Amsterdam. 628 pp.
19. JØRGENSEN, S.E. & R. DE BERNARDI. 1997. The application of a model with dynamic structure to simulate the effect of mass fish mortality on zooplankton structure in Lago di Annone. Hydrobiologia **356:** 87–96.
20. JØRGENSEN, S.E. & MEJER, H.F. 1977. Ecological buffer capacity. Ecol. Model. **3:** 39–61.
21. JØRGENSEN, S.E. & J. PADISDK. 1996. Does the intermediate disturbance hypothesis comply with thermodynamics? Hydrobiologia **323:** 9–21.
22. JØRGENSEN, S.E., S.N. NIELSEN & H. MEJER. 1995. Emergy, environ, energy and ecological modelling. Ecol. Model.**77:** 99–109.
23. JØRGENSEN, S.E., B.C. PATTEN & M. STRASKRABA. 1999. Ecosystem emerging: 3. Openness. Ecol. Model. In press.
24. JØRGENSEN, S.E., B.C. PATTEN & M. STRASKRABA. 1999. Ecosystem emerging: 4. Growth. Ecol. Model. Submitted.
25. JØRGENSEN, S.E. 1997. Integration of Ecosystem Theories: A Pattern. 2. edition. Kluwer Academic Publ. Dordrecht/Boston/London. 400 pp. (1st edition, 1992).
26. KAY, J. & E.D. SCHNEIDER. 1990. On the applicability of non-equilibrium thermodynamics to living systems [internal paper]. Waterloo University. Ontario, Canada.
27. KAY, J. & E.D. SCHNEIDER. 1992. Thermodynamics and measures of ecological integrity. *In* Proceedings of "Ecological Indicators.": 159–182. Elsevier. Amsterdam.
28. LEWIN, B. 1994. GENES V. Oxford University Press. Oxford, England. 620 pp.
29. LI, W.-H. & D. GRAUER. 1991. Fundamentals of Molecular Evolution. Sinauer. Sunderland, MA. 430 pp.
30. LOTKA, A.J. 1922. Contribution to the energetics of evolution. Proc. Natl. Acad. Sci. USA **8:** 147–150.
31. MARQUET, P.A. & M.L. TAPER. 1998. On size and area: Patterns of mammalian body size extremes across landmasses. Evol. Ecol. **12:** 127–139.
32. MAUERSBERGER, P. 1983. General principles in deterministic water quality modeling. *In* Mathematical Modeling of Water Quality: Streams, Lakes and Reservoirs (International Series on Applied Systems Analysis, 12). G.T. Orlob, Ed.: 42–115. Wiley. New York.
33. MAUERSBERGER, P. 1995. Entropy control of complex ecological processes. *In* Complex Ecology: The Part-Whole Relation in Ecosystems. B.C. Patten & S.E. Jørgensen, Eds.: 130–165. Prentice-Hall. Englewood Cliffs, NJ.

34. MEJER, H.F. & S.E. JØRGENSEN. 1979. Energy and ecological buffer capacity. *In* State-of-the-Art of Ecological Modelling. S.E. Jørgensen, Ed.: 829–846. (Environmental Sciences and Applications, Vol. 7). Proceedings of a Conference on Ecological Modelling, 28 August–2 September 1978, Copenhagen. International Society for Ecological Modelling. Copenhagen.
35. MOROWITZ, H.J. 1968. Energy Flow in Biology. Academic Press. New York.
36. MOROWITZ, H.J. 1992. Beginnings of Cellular Life. Yale University Press. New Haven, CT and London.
37. NIELSEN, S.N. 1992. Application of Maximum Exergy in Structural Synamic Models Ph.D. Thesis. National Environmental Research Institute. Copenhagen, Denmark.
38. NIELSEN, S.N. 1992. Strategies for structural-dynamical modelling. Ecol. Model. **63:** 91–102.
39. ODUM, E.P. 1969. The strategy of ecosystem development. Science **164:** 262–270.
40. ODUM, H.T. 1983. System Ecology. Wiley Interscience. New York. 510 pp.
41. PATTEN, B.C. 1991. Network ecology: indirect determination of the lifeenvironment relationship in ecosystems. *In* Theoretical Studies of Ecosystems: The Network Perspective. M. Higashi & T.P. Burns, Eds.: 288–351. Cambridge University Press. Cambridge, England.
42. PATTEN, B.C. 1997. Synthesis of chaos and sustainability in a nonstationary linear dynamic model of the American black bear (*Ursus americanus Pallas*) in the Adirondack Mountains of New York. Ecol. Model. **100:** 11–42.
43. PATTEN, B.C. & B.C. FATH. 1999. Environmental theory and analysis. Submitted.
44. PETERS, R.H. 1986. The Ecological Implications of Body Size. Cambridge University Press. Cambridge, England.
45. PRIGOGINE, I. 1980. From Being to Becoming: Time and Complexity in the Physical Sciences. Freeman. San Fransisco, CA. 260 pp.
46. RECKNAGEL, F., T. PETZHOLD, O. HAEKE & F. KRUSCHE. 1994. Hybrid expert system DELAQUA—toolkit for water quality control of lakes and reservoirs. Ecol. Model. **71:** 17.
47. REEVES, H. 1991. The Hour of Our Delight. Cosmic, Evolution, Order and Complexity. Freeman. New York. 246 pp.
48. REYNOLDS, C.S. 1996. The plant life of the pelagic. Verh. Internat. Verein. Limnol. Stuttgart, December. **26:** 97–113.
49. SALTHE, S.N. 1993. Development and Evolution: Complexity and Change in Biology. MIT Press. Cambridge, MA. 257 pp.
50. SCHLESINGER, W.H. 1997. Biogeochemistry. An Analysis of Global Change. 2nd edition. Academic Press. San Diego/London/Boston/New York/Sydney/Tokyo/Toronto. pp 680.
51. SCHNEIDER, E.D. & J.J KAY. 1994. Life as a Manifestation of the Second Law of Thermodynamics. Mathl. Comput. Model. **19**(6–8)**:** 25–48.
52. SHUGART, H.H. 1998. Terrestrial ecosystems in changing environments. Cambridge University Press. Cambridge, MA. 534 pp.
53. STRASKRABA, M., S.E. JØRGENSEN & B.C. PATTEN. 1999. Ecosystem emerging: 3. Dissip. Ecol. Model. In press.
54. SZARGUT, J., D.R. MORRIS & F.R. STEWARD. 1988. Energy Analysis of Thermal, Chemical and Metallurgical Processes. Hemisphere Publishing. New York/Washington/Philadelphia/London. Springer-Verlag. Berlin/Heidelberg/New York/London/Paris/Tokyo. 312 pp.
55. SZARGUT, J. 1998. Energy Analysis of Thermal Processes: Ecological Cost. Presented at a workshop in Porto Venere, May 1998.
56. ULANOWICZ, R.E. 1986. Growth and Development. Ecosystems Phenomenology. Springer-Verlag. New York/Berlin/Heidelberg/Tokyo. 204 pp.

57. ULANOWICZ, R.E. 1997. Ecology, the Ascendent Perspective. Columbia University Press. New York. 201 pp.
58. MARGALEF, R. 1968. Perspectives in Ecological Theory. University of Chicago Press. Chicago, IL.
59. STRASKRABA, M. 1979. Natural control mechanisms in models of aquatic systems. Ecol. Model. **6:** 305–322.
60. STRASKRABA, M. 1980. The effects of physical variables on freshwater production: analyses based on models. *In* The Functioning of Freshwater Ecosystems. E.D. Le Cren & R.H. McConnell, Eds.: 13–31. Cambridge University Press. Cambridge, England.
61. BROWN, J.H. 1995. Macroecology. University of Chicago Press. Chicago, IL.
62. SALOMONSEN, J. 1992. Properties of exergy. Power and ascendency along a eutrophication gradient. Ecol. Model. **62:** 171–182.

Energy Analyses as a Tool for Sustainability: Lessons from Complex System Theory

MARIO GIAMPIETRO,[a] KOZO MAYUMI,[b,c] AND GIANNI PASTORE[a]

[a]*Istituto Nazionale della Nutrizione, via Ardiatina, 546-00178, Rome, Italy*
[b]*Faculty of Integrated Arts and Sciences, University of Tokushima, 1-1 Minamijosanjima, Tokushima City 770-8502, Japan*

INTRODUCTION

Part 1 of this paper presents an innovative approach for study of the evolution and stability of socioeconomic systems. The approach is based on (1) several distinct views of socioeconomic systems obtained by non-equivalent descriptions of those system on different hierarchical levels and (2) equations of congruence of flows of matter, energy, human time and money across different hierarchical levels to link non-equivalent views. Because a socioeconomic system may be described as a nested dissipative adaptive system (holarchy), a few related concepts in complex system theory are discussed. Particular focus is on the crucial analysis of unavoidable conflict between short-term goals and long-term goals that affect every holarchy. Part 1 also presents a method for describing evolution of socioeconomic systems in parallel, on different hierarchical levels, an approach allowing study of the exergy budget of various nested elements of a holarchy.

Part 2 first describes the procedure used to set up a database of 107 countries and comprising more than 90% of world's population. Four applications of the approach described in Part 1 are presented: (1) BEP is an indicator of development obtained by combining only biophysical variables. BEP is better than GNP in correlating with a set of more than 20 traditional indicators of development used by the World Bank. (2) A common trajectory of development for the 107 countries and their evolution is described in an appropriate state space. (3) Equations of congruence across levels can link demographic variables, level of development, existing technology, and availability of natural resources. (4) "demographic transition" based on the dataset and approach used can be studied in terms of a shift from one metastable equilibrium of the dynamic societal energy budget to another.

PART 1: THEORETICAL MODEL

Rationale of the Model

The triadic reading of dissipative hierarchical systems proposed by Salthe[32] allows definition of three levels of interest: (1) the socioeconomic system as the focal level; (2) the ecosystem within which the socioeconomic system operates as the

---

[c]Address for correspondence: +81-886-56-7175 (voice/fax); mayumi@ias.tokushima-u.ac.jp (e-mail).

higher level; and (3) the set of individual households and economic sectors operating within the socioeconomic system as the lower level.

The present methodological approach is discussed in detail in Giampietro,[8,10] Giampietro et al.,[15] and Giampietro and Mayumi.[16] Practical applications and validation are in Giampietro,[9,11–13] Giampietro et al.,[18] Giampietro and Pastore,[17] and Pastore et al.[27] The current approach has two main characteristics: (1) several distinct views of the same socioeconomic system are obtained on different space–time scales and (2) a set of equations of congruence of flows of matter, energy, human time, and money across levels can link these different nonequivalent views. Each distinct, view-dependent description (e.g., what is perceived as "good" or "bad" by individual households, by national economies, by natural ecosystems) defines a set of indicators of "good" and "bad." Clearly, different sets of indicators of performance depend on both space-time scale and "encoding," select description of the interaction between human and ecological systems over a defined space–time scale.

Choice of a method of encoding is unavoidably arbitrary. However, biophysical constraints allow a check of whether or not different scenarios are feasible and those constraints can examine reciprocal effect of parallel changes on different hierarchical levels.

Main Theoretical Concepts

Socioeconomic Systems as Nested Dissipative Hierarchical Systems (Holarchies)

A dissipative system is hierarchical when it operates on multiple space–time scales with different process rates.[26] Such a system can be analyzed through division into successive sets of subsystems (see Simon,[34] p. 468). Alternative nonequivalent methods of description (encoding) exist for the same system.[38]

Each component of a dissipative nested hierarchical system may be called a "holon," a term introduced by Koestler[21] to stress that a holon has a double nature. A holon is a whole made of smaller parts that is simultaneously part of a larger whole (Allen and Starr,[2] pp. 8–16). Holons have implicit duality and composite structure at the focal level. Because of their interaction with the rest of the hierarchy, however, holons perform functions that contribute to "emergent properties" observable only from higher levels of analysis. A nested hierarchy of dissipative systems can be termed a "holarchy."[21]

Because of peculiar means of functioning in cascade on parallel scales, the behavior of a holarchy requires examination of both structural stability and relational functions. In fact, analyzing a holarchy only in terms of structures (ceteris paribus or steady-state) implicitly assumes (1) initial conditions reflecting the history of the holarchy and (2) a stable higher level holon for which structures of the holarchy perform functions. Similarly, functions in a certain holon require structural stability of other holons at the lower level.[34] Description of the dynamics of a focal-level holon such as society as a whole must face both the issue of structural constraints (how or what occurs at lower level holons) and the issue of functional constraints (why or what occurs at higher level holons). The complex behavior of holarchy needs complementary descriptions.

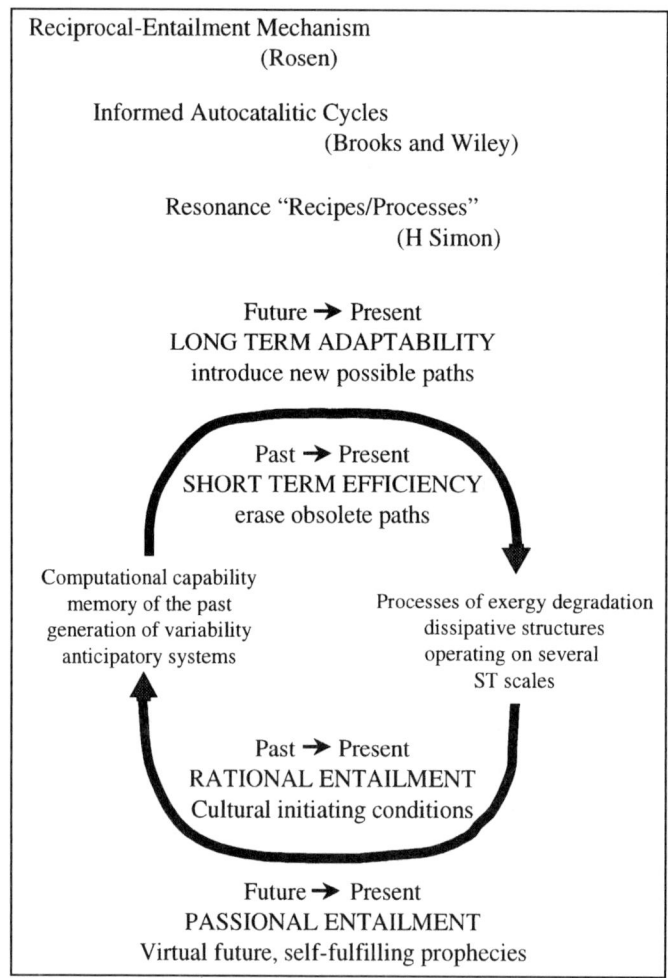

FIGURE 1. Resonance between controls and processes.

Resonance between Useful Energy and Useful Information

According to Simon (see pp. 477–482),[34] dissipative systems may be described in terms of a resonance, "recipes inducing processes which in turn makes better recipes." Prigogine[29] observes that living systems can establish resonance between coded information (e.g., DNA) that induces physical process (e.g., metabolism) and physical processes that generate coded information.

The triadic model in FIGURE 1 shows resonance between the "computational capability of society" and the "processes of exergy degradation," resonance of the sort described by Simon and Prigogine. Both human societies and ecosystems may be seen as nested dissipative hierarchical systems stabilized by "informed autocatalytic cycles"[3,4] or by a "mechanism of reciprocal entailment between two systems of entailments operating across hierarchical levels."[30,31]

The proxy that assesses investments of "computational capability" defined at the hierarchical level of society is the profile of allocation of human time on different activities. A given level of technology can be assessed according to the magnitude of the ratio of exosomatic to endosomatic energy metabolized by society, and it is important to calculate how much the biological metabolism of humans is amplified by the exosomatic energy metabolism of machines. This exosomatic metabolism based on the level of technology can be seen as a measure of the amplification of human activity that boosts process of self-organization by adding new levels of organization to human biological metabolism.

The proxy that assesses investments of "useful energy" (controlled processes of exergy degradation) is energy input consumed in various economic sectors. Assessments of this energy can be used as indicators of the investment in terms of exergy degradation allocated to stabilizing structures and functions.

Brooks et al.[4] epitomize resonance between energy flows stabilizing information flows and information flows stabilizing energy flows: "biophysical systems are complex, thermodynamic systems stabilized far from thermodynamic equilibrium by a process of self-organization induced by informed autocatalytic cycles."

Efficiency versus Adaptability

H. T. Odum[24,25] indicates that in economic and ecological systems based on a dynamic equilibrium of energy flows, cost of energetic investment must be repaid in order for those systems to be stable. According to Lotka[22] and Morowitz,[23] the autocatalytic process that sustains dissipation of economic and ecological systems must be able to increase its rate of exergy degradation through larger and faster matter cycling.

The autocatalytic process is related to two functions in the evolution of dissipative systems,[33] and dissipative systems must be able to stabilize a specific rate of exergy degradation related to two goals. The first goal of dissipative systems is to guarantee short-term stability of current dissipative structures that maintain existing metabolism of matter flows by using existing favorable gradients. Short-term stability relates to efficiency according to boundary conditions, a set of aims and information about boundary conditions stored in the systems. The second goal of dissipative systems is to guarantee long-term stability of the process of dissipation through maintaining high compatibility of patterns of self-organization in the face of a changing environment. Long-term stability relates to adaptability,[6] the ability to be efficient according to unknown future aims and unknown future boundary conditions of the systems. Adaptability can only be obtained by developing and maintaining a repertoire of diverse possible behaviors within dissipative systems. In other words, adaptability can be obtained by expanding state space, which in turn depends on expanding the computational capability of the dissipative systems of controls determining possible behaviors of those systems.

Description of Socioeconomic Systems through Encoding

Investments in Adaptability and Efficiency

Ulanowicz[36] divides the network of matter and energy flows in an ecosystem into two parts, a hypercyclic part and a purely dissipative part. A hypercycle is a net energy producer for the rest of the ecosystem, comprising activities that use free energy outside the ecosystem (e.g., solar energy, stocks of energy inputs). The hypercycle generates positive feedback into an ecosystem by introducing degradable exergy at a higher rate than exergy is consumed. The hypercycle drives the whole ecosystem and keeps it away from thermodynamic equilibrium. The purely dissipative part of an ecosystem comprises activities that are net energy degraders, but this dissipative part controls the entire process of energy degradation and stabilizes the whole system. A purely hypercyclic ecosystem cannot remain stable, for without the dissipative part positive feedback "will be reflected upon itself without attenuation, and eventually the upward spiral will exceed any conceivable bounds " (see Ulanowicz,[36] p. 57).

Following Ulanowicz's idea, it is possible to assume that (1) processes of exergy degradation and the fraction of computational capability (human time) invested in productive economic sectors aim at improving "efficiency" and (2) processes of exergy degradation and computational capability invested in household and service sectors aim at improving "adaptability." FIGURE 2 shows the total amount of energy consumed by society (ET) divided into CI (energy for activities related to efficiency) and FI (HH + SS) (energy for activities related to adaptability).

1. HH is energy investment in household sector activities, purely dissipative activities consuming net energy in the short term. These activities are not strictly in the form of defined roles or protocols and include sleeping, per-

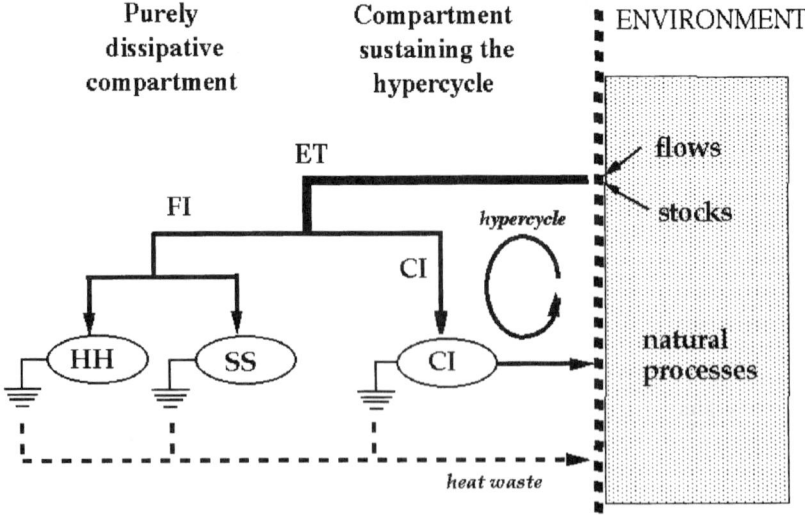

FIGURE 2. Structure of exosomatic energy flows in society.

sonal care, leisure time, and activities performed by the economically inactive population.

2. SS is energy investment in service sector activities, which are also dissipative, but take the form of defined social roles such as job positions and service activities like police, army, health care, education, and insurance.

3. CI is energy investment in activities in productive economic sectors having positive return in terms of energy flows. These activities take the form of defined social roles (e.g., job positions), in the energy and mining sector, the manufacturing sector in modern economies, the food security sector, and the environmental security sector. These activities generate hypercycle.

Regarding investment of human time (selected proxy for computational capability), the total time available to a society is:

- THT = total human time = number of individuals (population) × hours in a year (hy)
- hy = 8760 (hours in a year)
- WS = work supply = B + C = amount of time (hours) that the economically active population allocates to work annually as opposed to sleeping, leisure, and so forth
- A = THT − WS = Non working time, including sleeping time, leisure time, and all the time of the non-working population;
- C = hours of work delivered in productive sectors of the economy
- B = hours of work delivered in the service sector of the economy

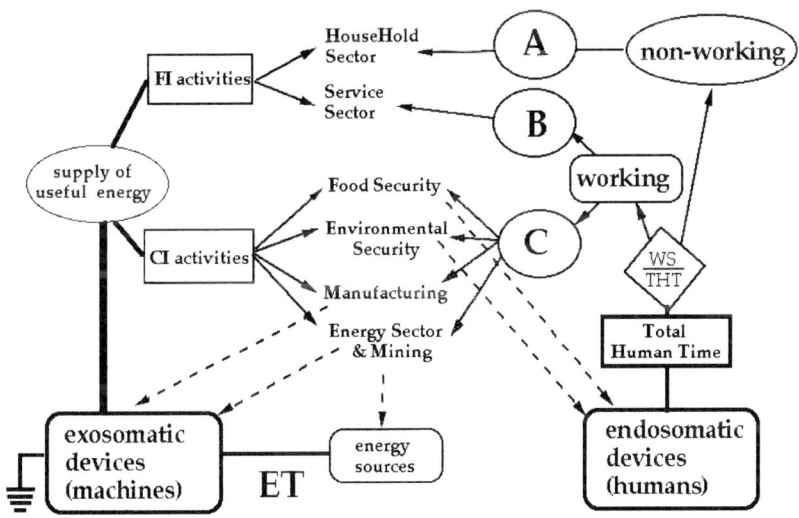

FIGURE 3. Parallel allocation of exosomatic energy and human time in society.

FIGURE 3 shows parallel allocation of energy and human time on various activities of the economy, balancing efficiency and adaptability.

Dynamic Exergy Budget

The energy throughput at which society's exergy budget can be stabilized (demand = supply) is defined by (1) socioeconomic characteristics of society generating demand and (2) characteristics of the exosomatic autocatalytic loop generating supply. Several variables can characterize the socioeconomic organization of society (demand side) and the nature of the exosomatic autocatalytic loop of energy (supply side). The dynamic equilibrium between demand and supply can then be studied by using existing relationships between groups of parameters determining demand and supply.

Demand side. The flow of exosomatic energy consumed by society (ET) can be expressed as:

$$ET = (MF \times ABM) \times (Exo/Endo) \times THT \qquad (1)$$

where ET = energy throughput or flow of exosomatic energy (joules/year); MF = metabolic flow = flow of metabolic energy per kg of body mass of humans (joules/kg hr); ABM = average body mass = total mass (kilogram) of population divided by population; and Exo/Endo = ratio between exosomatic and endosomatic energy flows.

The ratio of working time to total human time can be expressed as:

$$WS/THT = (B + C)/(\text{population size} \times \text{hours in a year}) \qquad (2)$$

Combining relations (1) and (2) definies bioeconomic pressure (BEP), which measures exosomatic energy throughput consumed at the level of society per hour of labor time in productive sectors of the economy:

$$BEP = ET/C = (ABM \times MF) \times (Exo/Endo \times (THT/C) \qquad (3)$$

Supply side. CI/C = exosomatic energy throughput per hour of labor in productive sectors (MJ/hr). The value of this parameter is defined by technical coefficients (inputs/outputs of productive sectors) which in turn are affected by (i) existing technology and (ii) quality of accessible natural resources.

C/THT = fraction of total human time (THT) allocated to activities in productive sectors. ET/CI = return for the socioeconomic system of energetic investment in productive sectors. The value of this parameter is also defined by a set of technical coefficients for productive sectors.

$$SEH = ET/C = (ET/CI) \times (CI/C) \qquad (4)$$

where strength of the exosomatic hypercycle (SEH) is defined as the exosomatic energy throughput (societal power) generated at the level of society per unit of work delivered in productive sectors. SEH on the supply side is the analog of BEP on the demand side, and SEH depends on two characteristics of the exosomatic compartment:

1. ET/CI measures how much exosomatic energy throughput of society is "eaten" by the hypercycle. ET/CI relates to the output-to-input energy ratio of processes that

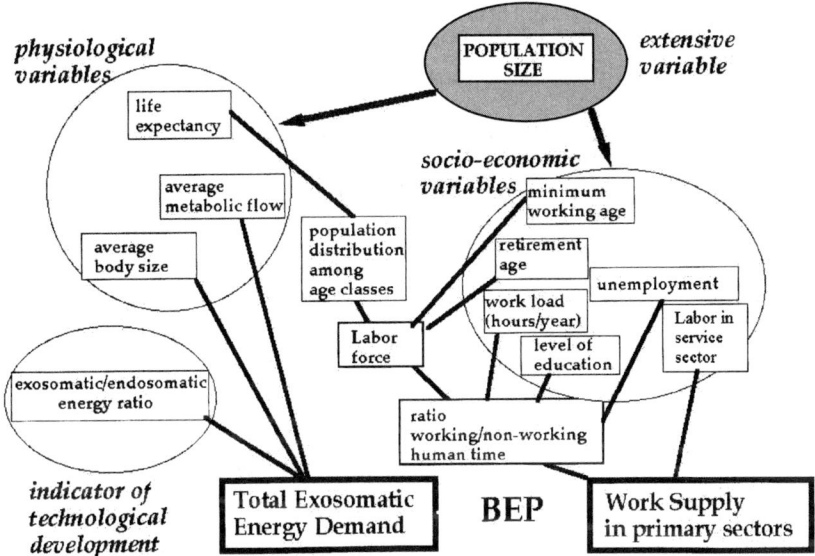

FIGURE 4. Variables defining the bioeconomic pressure (BEP).

make resources available to the economy. ET/CI can be expressed as a combination of technical coefficients (input/output ratios) for each of the productive sectors.

2. CI/C measures power level per worker in productive sectors. WS/THT decreases in developed countries because of an aging population, longer education, and smaller work load. In developed countries B/(B + C) increases because the service sector absorbs a large fraction of available work supply. Continuous decrease of C/THT is feasible only if there is concomitant increase in CI/C.

Identities and Congruence of Biophysical Flows across Levels

Relations between BEP and SHE lead to dynamic exergy budget:

$$(ABM \times MF) \times (Exo/Endo) \times (THT/C) = (ET/CI) \times (CI/C) \qquad (5)$$

A dynamic exergy budget links the physiological and socioeconomic variables in FIGURE 4 with and technological variables in FIGURE 5. Technical coefficients (input/output values) in FIGURE 5 include labor as input and the household sector as an economic sector. Dynamic equilibrium between demand and supply can be studied in terms of relationships linking groups of parameters that determine demand and supply.

The model in this paper defines quantity (ET/C) by using three different encodings of relevant qualities of socioeconomic systems:

1. ET/C is determined by parameters reflecting socioeconomic characteristics such as demographic structure, income, retirement age, and work load at the hierarchical level of the whole society. For example, ET/C = (ABM × MF) × (Exo/Endo) × (THT/C).

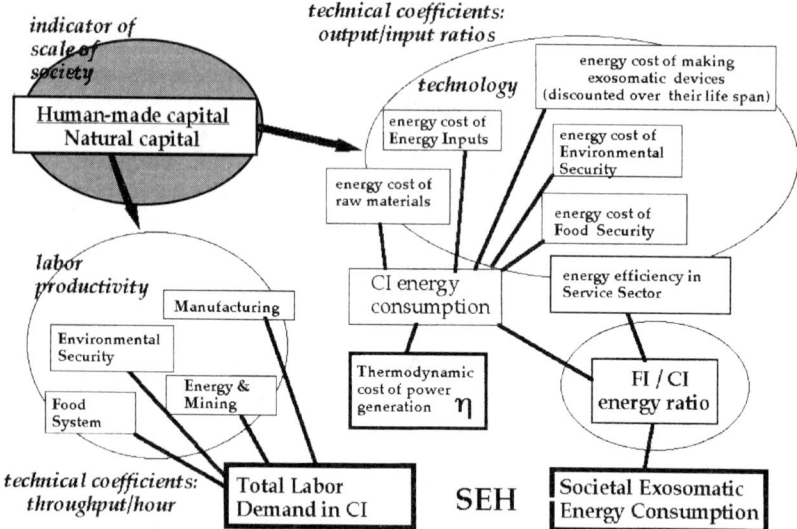

FIGURE 5. Variables defining the strength of the hypercycle (SEH).

2. ET/C is also determined by parameters reflecting technology, such as technical coefficients or input/output of different economic activities in different economic sectors. For example, in ET/C = (ET/CI) × (CI/C), the value of ET/CI and CI/C can be expressed as the sum of values of corresponding parameters describing performance of individual productive sectors such as agriculture, energy, and mining.

3. ET/C results from characteristics of the existing set of household types (existing range of life-styles) and distribution of individual households over that set. For example, in ET/C = $\Sigma ET_i / \Sigma C_i$, ET_i is the metabolism of household type i, and C_i is the amount of working time that household i invests in productive sectors.

The result of three different encodings may appear trivial (ET = ET), but each formulation of this identity links different characteristics of society at different hierarchical levels. This new model defines the same flows in redundant ways, but each time the model uses a different combination of nonequivalent descriptions. For example, the new model considers overall data assessed at the national level, data referring to different economic sectors reflecting technical coefficients, demographic variables, data characterizing metabolism of household types, and curves of distribution of households over the set of possible types. Parameters values obtained by adopting these non-equivalent descriptions are affected by internal constraints determined through forcing congruence of biophysical flows across different hierarchical levels.

Checking Sustainability across Hierarchical Levels

Interface focal/lower level (based on intensive variables). This new model can deal with the question of whether or not the current material standard of living is techni-

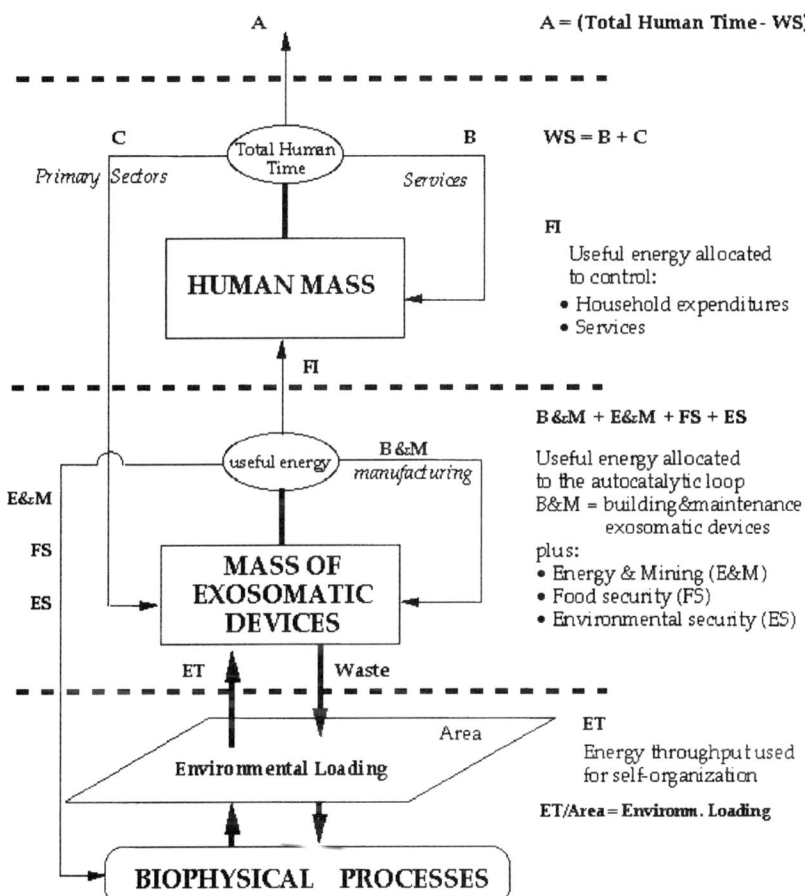

FIGURE 6. Intensive and extensive variables describing the economic process within its biophysical environment.

cally feasible and culturally acceptable, because the model considers congruence between three parameters:

1. BEP is a good indicator of development.

2. BEP* is the minimum acceptable level of material standard of living, below which lower level components (individual households) lose their sense of belonging to the holarchy. Aspiration for improvement in standard of living expressed by individual households generates "internal pressure," pushing the socioeconomic system to increase SEH continuously in order to make energy and matter throughputs faster and faster. Aspiration for improvement is particularly important when there are big gradients of BEP values among countries. Rapid increase in BEP is necessary for developing countries to approach BEP values of developed countries.

3. SEH is determined by technical coefficients dependent on technology and quality of natural resources.

There are two questions regarding the compatibility focal/lower level: (1) Is it possible to have a BEP value desired by people greater than the SEH value achievable by technology? (2) Can BEP be greater than BEP* by Λ? (Λ is the difference that people are willing to accept to preserve societal identity).

Interface focal/higher level (using intensive and extensive variables). It is necessary to consider whether amounts of inputs taken from ecosystems and amounts of wastes dumped into ecosystems are compatible with stability of processes of self-organization occurring in ecosystems with which a society interacts. To consider this topic, two concepts are necessary: (1) environmental loading (EL) and (2) critical environmental loading (CEL) (H.T. Odum, 1983, 1986).[24,25] EL is defined as human interference in the activity of natural systems, and can be obtained by comparing (i) assessments of the scale of human activity (input demand and waste production) with (ii) assessments of the scale of ecosystem activity (regenerative capacity and absorbing capacity). CEL may be defined as the maximum level of EL compatible with stability of the process of self-organization of ecosystems with which society interacts. Consideration of the compatibility focal/higher level leads to the question of whether or not current EL is less than CEL.

FIGURE 6 shows that the hierarchical nature of the model permits linking information related to the internal organization of the system (BEP and parameters determining SEH) to information about environmental loadings (size and nature of societal metabolism in relation to size and nature of ecological processes providing the life-support system for the economy).

PART 2. VALIDATION OF THE MODEL USING HISTORICAL DATA

Database

Database in this paper is based on 107 countries comprising more than 90% of the world's population, using official data of the UN, FAO, and World Bank.

Parameters for Calculating BEP

(1) "ABM × MF"

- ABM is calculated by considering average weights (by age and sex classes) and population structure as reported in James and Schofield,[20] based on the total population of 1992 reported in *World Tables*, published by the World Bank.[40]
- MF is computed separately for age and sex classes of each country following database and protocols in James and Schofield[20] and merged into national averages.

(2) "Exo/Endo"

- The annual flow of exosomatic energy is evaluated according to UN energy statistics[37] for commercial and traditional biomass consumption in 1992, using a conversion factor of 29.3076 terajoules per thousand metric tons of coal. A minimum value of 5/1 is adopted for countries, with a resulting value of exo/endo <5 since official statistics tend to underestimate the contribution of animal power, and biomass for cooking and for building shelters in rural communities.[14]

- Computation of the annual flow of endosomatic energy starts with "ABM × MF" value and population size of 1992 as reported in *World Tables*.[40]

(3) "THT/C"

- The fraction of economically active population and distribution of labor force in different economic sectors derive from UN statistics, including 1990–1993 data.
- For each country, transport is divided between productive sectors and service sectors, according to working time spent in productive sectors and service sectors.
- Workload is a "flat" value of 1800 hours/year, including vacations, absences, and strikes.

Material Standard of Living and Socioeconomic Development

Basically, conventional indicators are from *World Tables*,[40] with 24 indicators divided into three groups. Data used in calculating these 24 indicators come from the FAO (*FAO Yearbook*[7]), the UN (*Statistical Yearbook*[37]), and the World Bank (*Social Indicators of Development*[39]). Data refer to the latest available year between 1991 and 1993. Data on prevalence of malnutrition in children come from ACC/SCN.[1]

(i) *Nutritional status and physiological well-being (8 indicators)*: (1) life expectancy, (2) energy intake as food, (3) fat intake, (4) protein intake, (5) average BMI adult, (6) prevalence of child malnutrition (Wt/Ht < 2 z-score of NCHS reference growth curve), (7) infant mortality, and (8) percent low birth weight.

(ii) *Economic and technological development (7 indicators)*: (9) GNP per capita, (10) percent GDP from agriculture, (11) $ARL_\$$ (average return of labor in terms of added value = GNP/WS), (12) percent of labor force in agriculture, (13) percent of labor force in services, (14) energy consumption per capita, (15) percent of GDP expended for food.

(iii) *Social development (9 indicators)*: (16) television/1000 people, (17) cars/1000 people, (18) newspaper/1000 people, (19) phones/100 people, (20) population/physician ratio, (21) population/hospital bed ratio, (22) pupil/teacher ratio, (23) illiteracy rate, (24) access to safe water (percent of population).

BEP as an Indicator of Development for Socioeconomic Systems

The more developed a society is, the smaller is the fraction of human time used to run productive economic sectors. However, energy throughputs within productive economic sectors dramatically increase as asociety develops. These two trends can be explained in terms of balancing adaptability and efficiency. With a high rate of energy dissipation (faster consumption of natural resources), it is highly probable that the society will eventually face changes in boundary conditions. Systems consuming more must invest more in developing adaptability.

The database in this paper can be used to check whether or not BEP can be an indicator of development within the present model of analysis. The database shows that BEP correlates strongly with classic economic indicators of development (TABLE 1). Therefore, BEP could replace GNP as an indicator of development according to a socioeconomic perspective. BEP is a good indicator of material standard of living, according to the conventional economic perspective and also directly links different perspectives (readings) of the process of development, reflecting various nonequivalent descriptions on different scales.

BEP reflects three views of the material standard of living in a society, from three different hierarchical levels of analysis:

1. "ABM × MF" (endosomatic metabolism per capita (MJ/hour) refers to physiological hierarchical level. The higher this value is, the better human physiological conditions are in that society. The present database shows that the feasibility domain of ABM × MF is within a minimum of 0.33 and a maximum value of 0.43.

2. "Exo/Endo energy ratio" (exosomatic metabolism per capita) refers to socioeconomic hierarchical level and short-term efficiency. According to our database and previous studies on preindustrial societies, the feasibility domain of the Exo/Endo energy ratio is within a minimum value of 5 and a maximum value of 90.

3. "THT/C" (total human time available in the society/working time allocated in productive economic sectors) refers to socioeconomic hierarchical level and long-term adaptability. The database shows that the feasibility domain of THT/C is within a minimum value of 10 and a maximum value of 45. THT/C reflects the social implications of development, assessing allocation of human controls on long-term returns rather than on short-term returns (adaptability versus efficiency).

The database shows that the feasibility domain of BEP is within a minimum value of 18 MJ/hr, and a maximum value of 1500 MJ/hr. Increased value of this parameter reflects ability to increase the fraction of resources that a socioeconomic system actually invests in adaptability

Internal Constraints on the Evolutionary Pattern of Socioeconomic Systems

The model should show similarities in trajectories of development of various societies. The need for congruence of flows of energy and human time across hierarchical levels imposes constraints on the shape of possible paths.

FIGURE 7 graphs the path of development of six indicators against BEP and shows that 107 countries cluster around a given trajectory. Even more striking is the analysis of the same trajectory if one of the three factors determining BEP appears on the x-axis. In FIGURE 8, values taken by six indicators of development graph against the value of Exo/Endo and it is easy to identify a threshold value for Exo/Endo of 25/2, above which trajectory of development seems to reach a plateau. Similar observations apply to other two factors making up BEP: for ABM× MF, the plateau is about 9 MJ/day or 0.4 MJ/hour; for THT/C, the threshold value is 30/1.

TABLE 1. Correlation between BEP and some major indicators of development

Economic Indicators of Development	log (BEP), r
log (Gross National Product)	0.89
% of GNP from agriculture	0.85
US$ of added value per hour of paid labor	0.90
% of work force in agriculture	0.93
% of work force in services	0.88
log (energy consumption per capita)	0.98
% of income spent on food	0.89
Physiological Indicators of Development	**log (BEP), r**
Life expectancy	0.88
Energy intake (in the diet)	0.83
Fat intake (in the diet)	0.80
Protein intake (in the diet)	0.79
Children malnutrition	0.83
Infant mortality	0.86
Low birth weight	0.62
Social Indicators of Development	**log (BEP), r**
Log (TV sets/inhabitants)	0.90
Log (cars/inhabitants)	0.90
Log (newspapers/inhabitants)	0.89
Log (phones/inhabitants)	0.89
Log (population/physician)	0.87
Log (population/hospital bed)	0.76
Pupils/teacher	0.74
Illiteracy rate	0.67
Primary school enrollment	0.58
Access to safe water	0.81

Links across Levels and Feasibility of Future Scenarios

Analysis of socioeconomic systems interacting in biophysical terms with their environment applies to discussion of feasibility of future scenarios. A set of characteristics describing a given socioeconomic system's demographic structure and material standard of living may be defined. Then it is possible to calculate: (1) technical coefficients required in specific economic sectors to match SEH demand generated by the "envisioned society" and (2) correspondent environmental loading, technological achievement required to keep environmental loading of a society below a critical value.

Two applications deserve consideration:

1. It is possible to analyze performance of farming systems using several sets of indicators reflecting different perceptions of "improvements" on different hierarchical levels. It is also possible to link various effects generated by changes within the

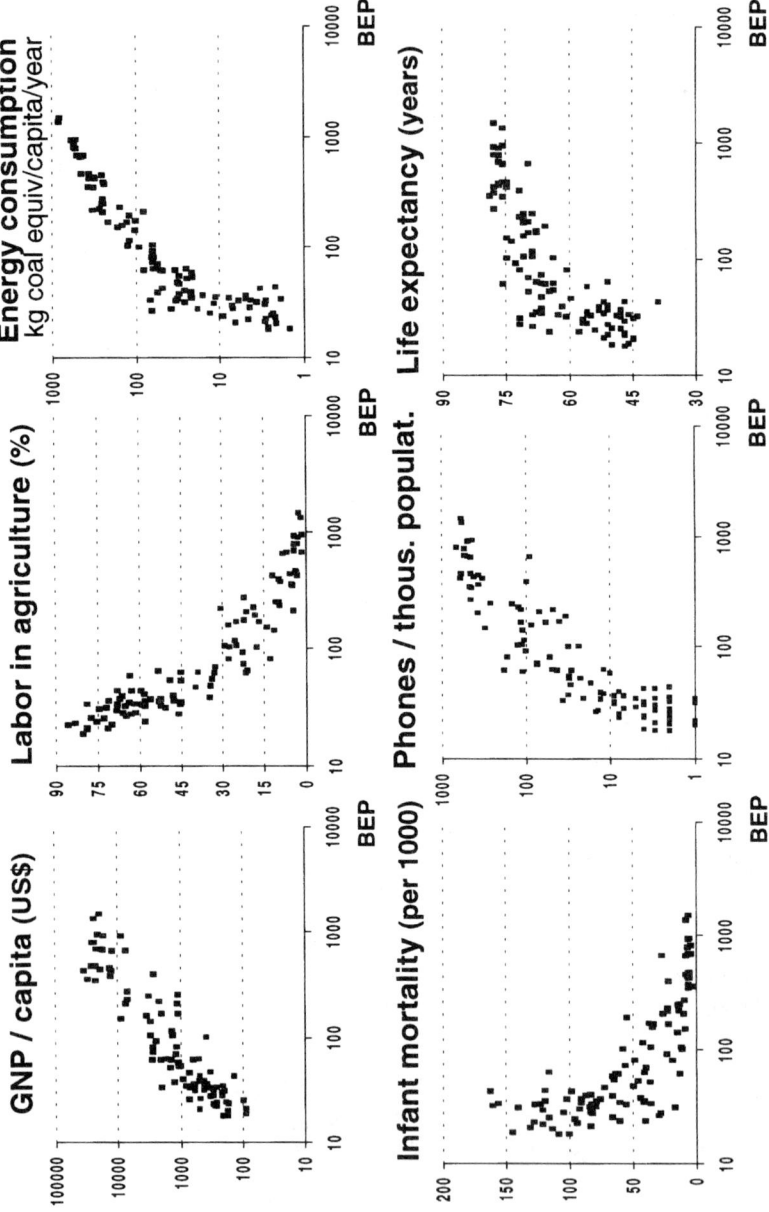

FIGURE 7. Correlation between BEP and major indicators of development.

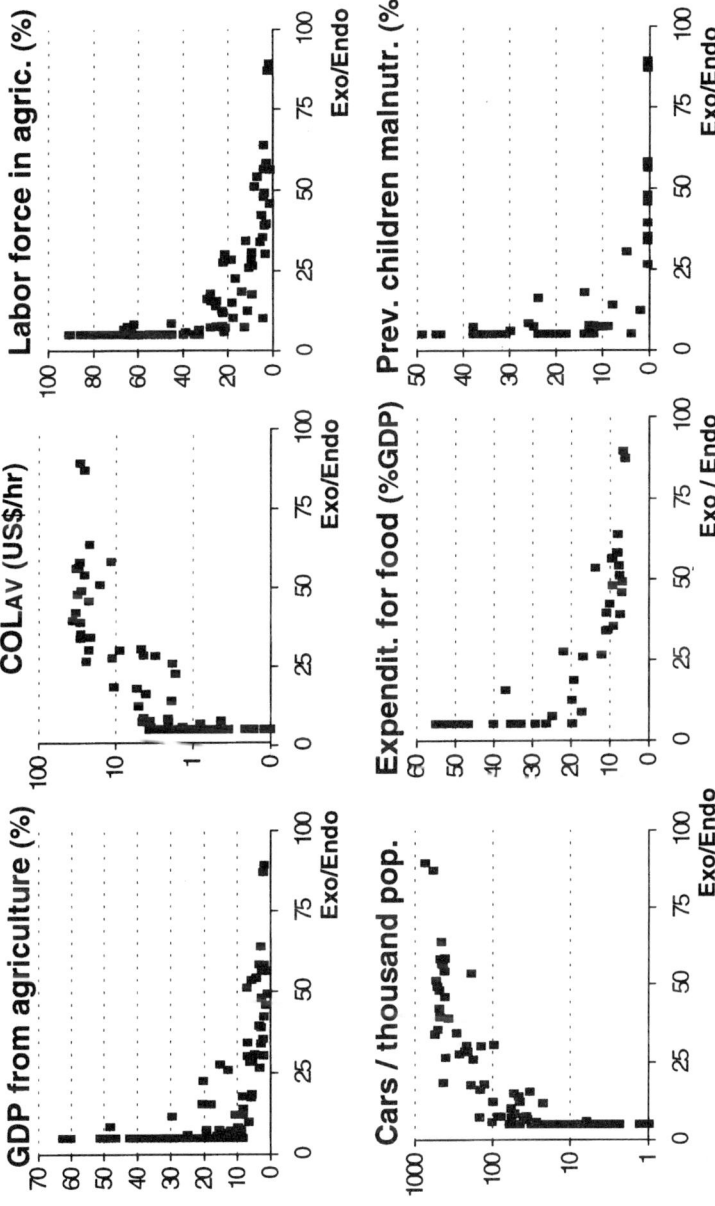

FIGURE 8. Correlation between Exo/Endo ration and major indicators of development.

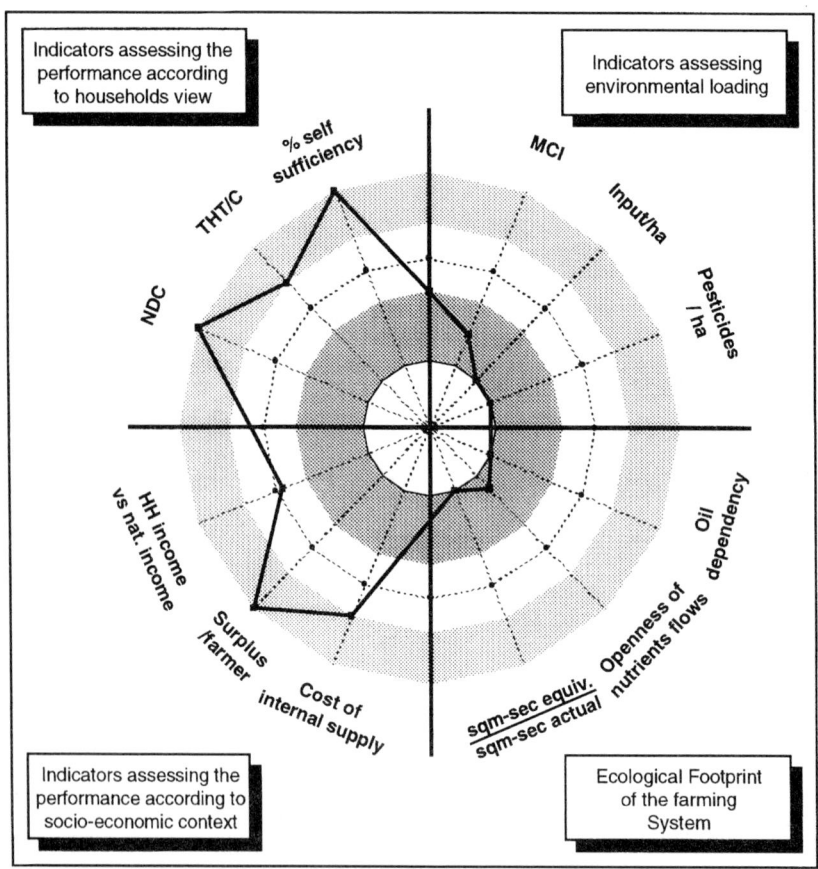

FIGURE 9. Indicators from four different perspectives.

hierarchical system. The application in FIGURE 9 has been developed in a four-year field project in China to study the mechanism of intensification of rural areas in relation to the issue of sustainability.[27] In this radar-type graph, several sets of indicators are divided into four quadrants, each of which contain a set of indicators related to the perspective of a particular holon (e.g., farmer household, national economy, agroecosystem). Each axis has a "viability domain," the range of values within which the holon can be considered stable. Equations of congruence link changes in values of indicators within a quadrant and changes induced in values of indicators within another quadrant. For example, lowering the price of rice by imposing a "fixed price" policy results in a worsening of the situation for "holon" farmers (reduction of income) and improvement of the situation for the "holon" government of China (reduction of food cost for the cities). Considering the distance of the indicators from the edges of the viability domain (minimum acceptable values) and blending socioeconomic and biophysical readings allow evaluation of correspondent trade-offs between economy and ecology.

2. It is possible to discuss whether or not large-scale biofuel production is a feasible option for a fuel-developed society heavily dependent on fossil energy stocks.[18] An energy sector running on biofuels must be compatible with the socioeconomic characteristics of society providing a SEH compatible with existing BEP. The energy sector should also be compatible with ecological constraints and have a demand for natural resources such as arable land and fresh water that is compatible with current supply. Available data on modern biofuel systems can be used to estimate biophysical requirements per unit of net energy supply. Depending on production system, requirements per gigajoule (1 GJ = 10^9 joules) of net energy are 0.015–0.100 ha of arable land, 200–400 tons of fresh water, and 0.6–5.5 hours of labor.[15] In a developed society (BEP > 400 MJ/hour) work supply in the energy sector is only a small fraction (generally less than 5%) of the work force in productive sectors. To achieve a small fraction of the work force in productive economic sectors, throughput per hour of labor in the energy sector must be on the order of 10,000 MJ/hour of labor.

In Italy, with a population of 57 million, only 7.3% of 499 billion hours of human time available in 1991 were spent in paid work. Of this labor supply, 60% was absorbed by the service sector, 30% by the industrial sector, and 9% by agriculture, fishery and forestry, leaving only 1% (360 million labor hours) for the entire energy sector.[19] Total energy consumption in Italy in 1991 was 6,500,000 terajoules, implying that the Italian energy sector delivered almost 18,000 MJ of energy throughput per hour of labor. This throughput was achieved through using mainly fossil energy (about 90%), suggesting that a developed society requires energy throughput per hour of labor in the energy sector in the range of 10,000 to 20,000 MJ/hr. Such levels are well beyond the 250–1,600 MJ/hr range of values achievable with biofuels.[15] To use biofuels as energy sources with a much smaller throughput per hour of labor and high levels of energy consumption per capita, an energy sector would have to absorb between 20 and 40% of the labor force. Such a scenario is incompatible with the current profile of labor allocation in various economic sectors.

Demographic Transition and Dynamic Exergy Budget

BEP is the product of three terms, two of which are affected by demographic changes. SEH depends on two parameters linked to technological development and availability of natural resources, parameters both scale-dependent and affected by population size. Constraint of balance (BEP = SEH) is in reality affected by demographic changes, making it necessary to add extensive variables. It is possible to impose congruence between ET supply (as a function of SEH and population size) and total ET demand (as a function of BEP and population size). Some useful relations are:

$$(THT/C) \times (C/CI) \times CI = pop \times hy \tag{6}$$

$$ET\ supply = SEH \times C = (ET/CI) \times (CI/C) \times C \tag{7}$$

$$ET\ demand = BEP \times C = BEP \times (C/THT) \times pop \times hy \tag{8}$$

ET at a particular point i in time can be expressed as:

$$ET_i = (Exo/Endo)_i \times pop_i \times (ABM \times MF)_i \times hy \tag{9}$$

When SEH > BEP, socioeconomic systems adjust parameters to increase ability to use surplus useful energy and to expand activity assessed by larger ET. At a given point in time $i + 1$:

$$ET_{i+1} > ET_i \tag{10}$$

which can be formulated as:

$$(Exo/Endo)_{i+1} \times pop_{i+1} \times (ABM \times MF)_{i+1} > (Exo/Endo)_i \times pop_i \times (ABM \times MF)_i \tag{11}$$

Relation (11) can also be written as:

$$(ET/C)_{i+1} \times (C/THT)_{i+1} \times pop_{i+1} > (ET/C)_i \times (C/THT)_i \times pop_i \tag{12}$$

Relation (12) gives a different view of possible ways of expanding ET. When surplus is absorbed by an increase in the Exo/Endo ratio rather than by an increase in population, the solution implies an increase in ET/C linked to a reduction of C/THT. Fraction of working time in productive sectors is also an indicator of development. Put another way, depending on path of expansion followed by the socioeconomic system, increases in ET can transform either into either (1) increased population size or (2) improved material standard of living (combined change of ET/C and C/THT).

Changes in ET/C and THT/C are subject to (1) lag time in building technical infrastructures needed to expand Exo/Endo (job positions available in economic sectors), (2) speed of demographic changes determining dependency ratio within existing households, and (3) cultural constraints to change the profile of allocation of human time on different sets of activities (moving to a different job or different housing type). These three factors create an "intrinsic" lag-time in the adjustment of values of ET/C and THT/C to new conditions, a lag-time that determines how much of the increase in ET results in population growth rather than improved BEP.

The Path of Expansion of ET

Starting from Relation (12), it is possible to see paths of expansion in ET by examining relative changes among the three parameters. An increase in size of the system (ET) can imply changes in (1) ET/C, (2) ET per capita (Exo/Endo), and (3) C/THT. Population becomes stable when ET/C increase on the supply side matches a combination of changes in C/THT and Exo/Endo on the supply side. Due to the limited range of possible values of the ABM × MF, where maximum increase amounts to about 20% of initial value, increase in ET per capita translates almost directly into increase of Exo/Endo ratio.

It may be helpful to imagine a process of expansion of ET, starting from a hypothetical socioeconomic system characterized by stable population due to high fertility and mortality. It is assumed that parameter population is subject to fluctuations and initial population size is small. Also, it is assumed that improvements in exosomatic autocatalytic loop (e.g., new technologies, use of higher-quality resources, better knowledge and management of socioeconomic activities) make surplus of energy (SEH > BEP) available to the socioeconomic system. Surplus energy can be absorbed by increase in population and/or by increase in energy consumption per capita, increases that will be reflected in a changing pattern of human activity (changes in THT/C). In the short term, population growth increases THT/C because of the increased number of children in the system. In the long term, increase in size of the socioeconomic system due to population growth further increases THT/C because of increasing B/(B+C). With the expanding size of the system, fraction of total work supply allocated to administration and other services gradually increases. However, the possibility of increasing ratio THT/C is limited in pre industrial societies, due to lack of devices to amplify the power of workers in productive sectors. CI/C in preindustrial societies is small; and historically, animal, wind, and water power were decisive factors in determining local development. "Power devices" are by their very nature location-specific, do not provide a continuous, reliable flow of power, and prevent preindustrial socioeconomic systems from reaching type 2 equilibrium on a large scale. Large-scale complex societies require division of population into widely different social classes to allow effective taxation of farmers by a central administration. However, the process of economic development could not avoid a high degree of instability in preindustrial societies.[15,35]

In modern industrialized society, population growth does not imply reduction in SEH because (1) supply of energy input no longer relates to land availability and (2) surplus of ET is easily absorbed by an increase in exo/endo ratio.

FIGURE 10 shows two possible points of equilibrium. All variables are intensive to avoid complicated three-dimensional representation, but an extensive variable such as population size can easily be introduced on a third axis perpendicular to the plane that indicates changes in the size of the system (ET or population). Clearly,

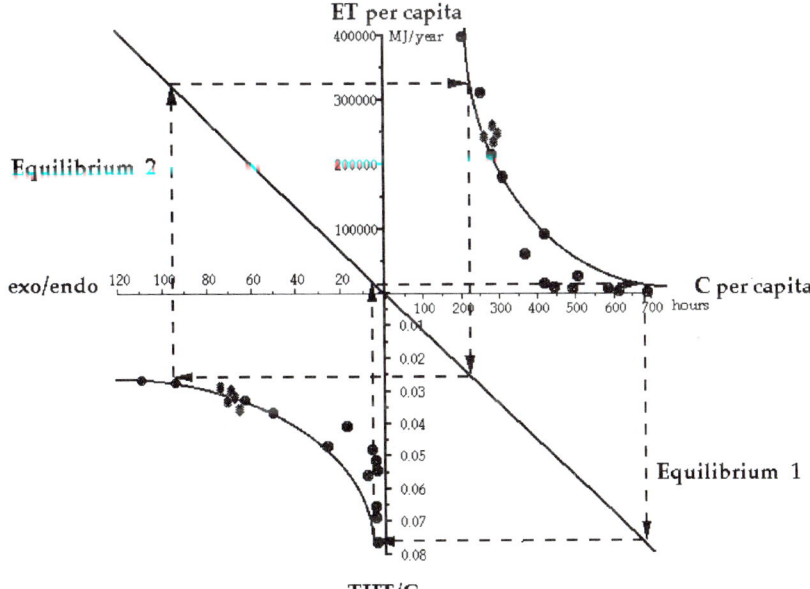

FIGURE 10. The two equilbria of the demographic transition.

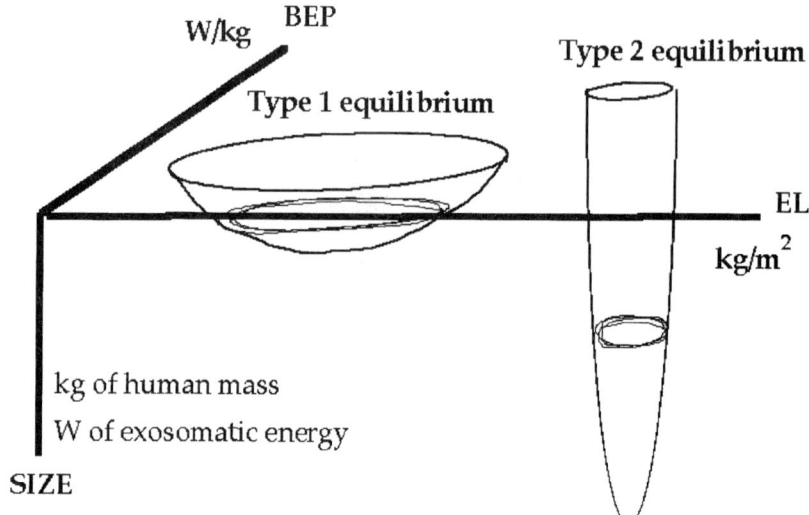

FIGURE 11. Two basin attractors for socioeconomic systems in the space BEP-EL-SIZE.

type 2 equilibrium is achieved only in a system of much larger size than a system of type 1 equilibrium. The figure shows synchronic analysis of 107 countries (*white dots*) grouped in clusters and diachronic analysis of average values for OECD countries from 1970 to 1990 (*black dots*) calculated over historic series of OECD countries.[28] Note that values resulting from both synchronic and diachronic analysis are on the same curve.

The reciprocal relation among a set of variables generates two basins of attraction for the energy budget of a socioeconomic system. Trajectory of development of socioeconomic systems can be described in a three-dimensional phase space with

Axis 1, energy dissipation per unit of human control (ET per capita); Axis 2, size of the system (population size); Axis 3, environmental loading.

Axis 3 indicates the intensity of self-organization socioeconomic activity relative to the intensity of the natural processes that guarantee stability of boundary conditions. This ratio affects ET/C, and therefore THT/C and is itself affected by population size. Type 1 and type 2 equilibria in FIGURE 10 can be imagined as the three-dimensional phase space shown in FIGURE 11.

Reconsidering the Classic Representation of Demographic Transition

Traditional graphic representation of demographic transition is a sigmoid curve connecting two values of stabilized population sizes, and the horizontal axis generally represents time. However, data for societies that have completed demographic transition (e.g., France and Sweden) and societies in different stages of transition (e.g., Burundi and Singapore) suggest that speed of transition differs dramatically from country to country.[5] So, the variable "time" is not directly linked to changes during transition from one metastable equilibrium to another.

Exo/Endo ratio may be a better variable for the horizontal axis, providing a better explanation for the transition process. At the beginning of the transition (stage 1), socioeconomic systems cannot assume larger ET values and can replicate themselves only through "seedlings," expanding into other ecosystems. Growth generates redundancy; there is more of the same thing (replication), but no real development (no qualitative changes).

Socioeconomic systems that can expand their autocatalytic loop of exosomatic energy enter a transitional phase in which they can maintain SEH > BEP. In this phase, socioeconomic systems can expand size (ET) by increasing both population and Exo/Endo ratio. These two forms of expansion have different intrinsic lag times, and therefore the two processes proceed at different speeds.

Changes in structure of the exosomatic autocatalytic loop (industrialization) enable socioeconomic systems to absorb the entire surplus of ET by increasing Exo/Endo ratio while maintaining fixed population size. After these changes, socioeconomic systems have completed demographic transition. The final stage is coupled to increased material standard of living and changes in profile of human time allocation over different activities (increased THT/C). This transformation is linked to dramatic change in social patterns of organization.

Current description of demographic transition in terms of indicators of fertility and mortality is just one possible description of the demographic transition. As discussed in Part 2, a plot of 24 indicators of development against Exo/Endo ratio for different countries (FIG. 8) shows another description of the trajectory that countries follow in their transition between two metastable equilibria of a dynamic energy budget. The process of demographic transition based on our model can be a good alternative to the traditional description of the same process.

SUMMARY

Part 1 presents an energy analysis model for the study of the evolution and stability of socioeconomic systems. The model describes socioeconomic system as a nested dissipative adaptive system (holarchy), allowing study of the exergy budget of various nested elements of a holarchy. Part 1 also deals with the unavoidable conflict between short-term goals and long-term goals affecting the holarchy.

Part 2 first sets up a database referring to 107 countries and presents four applications of the model presented in Part 1: (1) BEP (an indicator of development) is better than GNP in correlating with traditional indicators of development; (2) a common trajectory of development for the 107 countries and their evolution in state space; (3) equations of congruence across levels linking demographic variables, level of development, technology, and natural resources; (4) description of "demographic transition" in terms of a shift from one metastable equilibirum of the dynamic exergy budget to another.

REFERENCES

1. ACC/SCN. 1993. Second Report on the World Nutritional Situation. FAO. Rome.

2. ALLEN, T.F.H. & T.B. STARR. 1982. Hierarchy. The University of Chicago Press. Chicago.
3. BROOKS, D.R. & E. O. WILEY. 1988. Evolution as Entropy. University of Chicago Press. Chicago.
4. BROOKS, D.R., J. COLLIER, B.A. MAURER, J.D.H. SMITH & E.O. WILEY. 1989. Entropy and Information in evolving biological systems. Biol. Philos. **4:** 407–432.
5. CHESNAIS, J.C. 1992. The Demographic Transition. Clarendon Press. New York.
6. CONRAD, M. 1983. Adaptability: The Significance of Variability from Molecules to Ecosystems. Plenum Press. New York.
7. FAO. 1995. Production Yearbook 1994. FAO Statistic Series No. 125. FAO. Rome, Italy.
8. GIAMPIETRO, M. 1994. Sustainability and technological development in agriculture: a critical appraisal of genetic engineering. BioScience **44**(10): 677–689.
9. GIAMPIETRO, M. 1994. Using hierarchy theory to explore the concept of sustainable development. Futures. **26**(6): 616–625.
10. GIAMPIETRO, M. 1997. The link between resources, technology and standard of living: a theoretical model. In Advances in Human Ecology. Vol. 6. L. Freese, Ed.: 73–128. JAI Press. Greenwich, CT.
11. GIAMPIETRO, M. 1997. Socioeconomic pressure, demographic pressure, environmental loading and technological changes in agriculture. Agric. Ecosyst. Environ. **65:** 201–229.
12. GIAMPIETRO, M. 1997. Socioeconomic constraints to farming with biodiversity. Agric. Ecosyst. Environ. **62:** 145–167.
13. GIAMPIETRO, M. 1998. Energy budget and demographic changes in socioeconomic systems. In Ecology, Society, Economy: Life Sciences Dimensions. U. Ganslasser & M. O'Connor, Eds.: 327–354. Filander Press. Germany.
14. GIAMPIETRO, M., S.G.F. BUKKENS & D. PIMENTEL. 1993. Labor productivity: a biophysical definition and assessment. Hum. Ecol. **21**(3): 229–260.
15. GIAMPIETRO, M., S.G.F. BUKKENS & D. PIMENTEL. 1997. The link between resources, technology and standard of living: examples and applications. In Advances in Human Ecology, Vol. 6. L. Freese, Ed.: 129–199. JAI Press. Greenwich, CT.
16. GIAMPIETRO M. & K. MAYUMI. 1997. A dynamic model of socioeconomic systems based on hierarchy theory and its application to sustainability. Struct. Change Econ. Dynam. **8**(4): 453–470.
17. GIAMPIETRO, M. & G. PASTORE. 1999. A model of analysis to study the dynamics of rural intensification in China. In Special Issue of Critical Reviews in Plant Sciences. Paoletti *et al.*, Eds. CRC Press. Boca Raton, Fl. In press. (May)
18. GIAMPIETRO, M., S. ULGIATI & D. PIMENTEL. 1997. Feasibility of large-scale biofuel production: Does an enlargement of scale change the picture? BioScience **47**(9): 587–600.
19. ISTAT. 1992. Annuario Statistico Italiano. Istituto Centrale di Statistica. Rome.
20. JAMES, W.P.T. & E.C. SCHOFIELD. 1990. Human Energy Requirement. Oxford University Press. Oxford, England.
21. KOESTLER, A. 1969. Beyond atomism and holism — The concept of the Holon. In Beyond Reductionism. A. Koestler & J. R. Smythies, Eds.: 192–232. Hutchinson. London.
22. LOTKA, A.J. 1956. Elements of Mathematical Biology. Dover Publications. New York.
23. MOROWITZ, H.J. 1979. Energy Flow in Biology. Ox Bow Press. Woodbridge, CT.
24. ODUM, H.T. 1983. System Ecology. John Wiley. New York.
25. ODUM, H.T. 1996. Environental Accounting: EMergy and decision making. John Wiley. New York.

26. O'Neill, R.V. 1989. Perspective in hierarchy and Scale. *In* Perspectives in Ecological Theory. J. Rougharden, R.M. May & S. Levin, Eds.: 140–156. Princeton University Press. Princeton, NJ.
27. PASTORE, G., M. GIAMPIETRO & LI JI. 1998. Understanding the dynamics of rural intensification in China: land–time crossed budget of five villages in Hubei province. *In* Special Issue of Critical Reviews in Plant Sciences. Paoletti *et al.*, Eds. CRC Press. Boca Raton. Fl. In press.
28. PASTORE, G., M. GIAMPIETRO & K. MAYUMI. 1996. Bio-economic pressure as indicator of material standard of living. Presented at the Fourth Biennial Meeting of the International Society for Ecological Economics: Designing Sustainability. Boston University, Boston, MA. August 4–7, 1996.
29. PRIGOGINE, I. 1978. From Being to Becoming. W.H. Freeman. San Francisco.
30. ROSEN, R. 1985. Anticipatory Systems: Philosophical, Mathematical and Methodological Foundations. Pergamon Press. New York.
31. ROSEN, R. 1991. Life Itself: A Comprehensive Inquiry into Nature, Origin, and Fabrication of Life. Columbia University Press. New York.
32. SALTHE, S.N. 1985. Evolving Hierarchical Systems: Their Structure and Representation. Columbia University Press. New York.
33. SCHNEIDER, E.D. & J.J. KAY. 1994. Life as a manifestation of the second law of thermodynamics. Math. Comp. Model. **19:** 25–48.
34. SIMON, H.A. 1962. The Architecture of Complexity. Proc. Am. Philos. Soc. **106:** 467–482.
35. TAINTER, J.A. 1988. The Collapse of Complex Societies. Cambridge University Press. Cambridge, UK.
36. ULANOWICZ, R.E. 1986. Growth and Development: Ecosystem Phenomenology. Springer-Verlag. New York.
37. UNITED NATIONS. 1995. Statistical Yearbook 1993. U.N. Department for Economic and Social Information and Policy Analysis. Statistical Division. New York.
38. WHYTE, L.L., A.G. WILSON & D. WILSON, Eds. 1969. Hierarchical Structures. American Elsevier Publishing Company. New York.
39. WORLD BANK. 1995. Social Indicators of Development 1995. The Johns Hopkins University Press. Baltimore, MD.
40. WORLD BANK. 1995. World Tables 1995. The Johns Hopkins University Press. Baltimore, MD.

Virtual Biospheres: Complexity versus Simplicity

YURI M. SVIREZHEV[a]

Potsdam Institute for Climate Impact Research, P.O. Box 60, 12 03 D-14412, Potsdam, Germany

> *When we study the History of Science we discover two mutual contrary phenomena: either behind an apparent complexity a simplicity is hidden or, on the contrary, an evident simplicity conceals within itself an extraordinary complexity.*
>
> —H. Poincaré, 1894

PART A. DYNAMICS OF THE BIOSPHERE

Introduction and Virtual Biospheres Concept

Two very important problems exist in modern globalistics.[1]

- how the *biosphera machina* operates, and
- whether our Earth biosphere is unique or whether any other virtual biospheres exist.

A different view of these questions would be to ask how the complexity of the biosphere connects with these problems. This complexity may connect with a sufficiently complex arrangement of the *biosphera machina* including a lot of different positive and negative feedback systems, a tangled network of different causal loops, and so forth. But the complexity may also be defined by different multiple equilibria when either one equilibrium or others can be attained as a result of multiple bifurcations when the evolutionary tree has a very complex topology. In order to answer these questions, we consider the system "biosphere + climate" as a nonlinear system with multiple equilibria. Note that in considering this problem we shall remain within a framework of simple, zero-dimensional models.

It is necessary to say a few words about the history of this problem. It seems to us that it was first formulated in 1926 (at a qualitative level, of course) by V. Vernadsky[2] in the form of an idea about the interdependence between vegetation and climate. Then Kostitzin[3] realized this idea in the form of the first mathematical model for co-evolution of the atmosphere (climate) and biota. It is interesting that he obtained the *époques glaciers* as self-oscillations of this system. Recently, Watson and Lovelock[4] further developed Vernadsky's idea. They considered the causal loop between surface temperature and two types of vegetation (by means of albedo). The

---

[a]Address for correspondence: +49-331 288 2971 (voice); +49-331 288 2600 (fax); juri@pik-potsdam.de (e-mail).

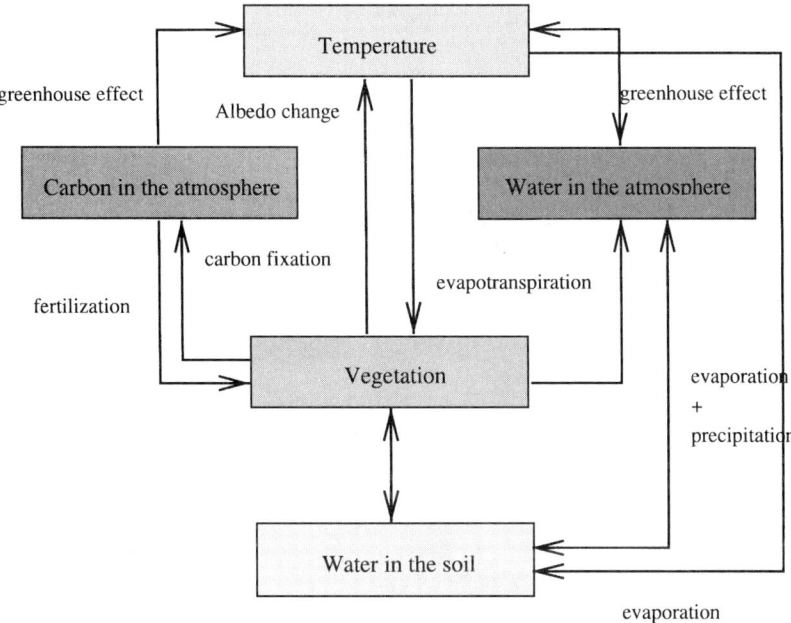

FIGURE 1. Variables and causal loops for Svirezhev–von Bloh's model.

competition between them for a "temperature ecological niche" generates different spatial vegetation and temperature patterns ("Daisy World"). In 1997, Schellnhuber, Block, and von Bloh realized this model using cellular automata in the form of a two-dimensional structure.[5] This approach highlighted the important role of fluctuations in the formation of spatial patterns.

Recently, the so-called "virtual biospheres" concept was formulated.[6] In accordance with this, the contemporary Earth's biosphere is one of many possible (virtual) biospheres corresponding to multiple equilibria of some strongly nonlinear dynamic system "climate + biosphere."[b] In the course of planetary history and our own planet's evolution, this system passed through several bifurcation points, at which random factors (small perturbations) determined which branch of the solution the system would take. A moving force of this evolution could be the evolution of Earth's green cover," which has, in turn, several bifurcation points: for instance, the appearance of terrestrial vegetation and, later on, deciduous trees.

Zero-Dimensional Model for the "Biosphere + Climate" System (Svirezhev–Von Bloh's Model)

The structure of the model and its causal loops is shown in FIGURE 1, and both the model and some results of its analysis are described in detail in References 7 and

---

[b]Note that this concept contradicts Vernadsky's "ergodicity axiom," according to which the contemporary Earth's biosphere is unique, beyond dependence on initial and previous states.

8. Therefore, in this work I shall deal more with the interpretation of these results than with the results themselves.

It is obvious that vegetation dynamics depends on the temperature, precipitation (more correctly, on soil moisture), and concentration of carbon in the atmosphere. On the other hand, temperature dynamics depends on the concentrations of carbon and water vapor in the atmosphere and on the albedo of the planetary surface. For instance, the albedo of "white sands" desert is equal to 0.4; for coniferous forest it is about 0.1. The model includes the following submodels: the global carbon cycle, the global hydrologic cycle, global vegetation, and the equation for annual global temperature. It is obvious that the system solutions depend on all the system parameters, but only three of them can be considered *bifurcation* parameters, for which changing the type of solution constitutes a principal changing. These parameters are:

- total amount of carbon in the system, A, including carbon in the atmosphere (in the form of carbon dioxide) and carbon of the biosphere contained in organic matter;
- total amount of water, B, including water vapor in the atmosphere, fresh water, and water contained in the so-called "active layer" of the ocean (surface layer involved in the fast global hydrological cycle),
- the product $\beta = P_m \cdot \tau$, where P_m is so-called a "potential" productivity of global vegetation, that is, a maximum of productivity that can be reached under optimal environmental conditions by a particular sort of vegetation, and $\tau = 1/m$ is the residence time of carbon in biota.

What kind of bifurcation is possible in this system? We can observe the change of the planet's "status" from a "cold (ice) desert" either to a "cold green" planet or to a "hot green" planet (first bifurcation). Then, either a "wet hot" planet, covered by tropical rain forest, or a "dry hot" planet (savannah) develops from hot green planet as a result of a second bifurcation. Analogously, either a "wet cold" (temperate forest) or a "dry cold" (steppe) planet arises from green cold planet (see FIG. 2). In this FIGURE 2, the bifurcation diagram is presented as a branching structure depending on one general value. In fact, it depends on all three bifurcation parameters.

As a first step we shall study global mechanisms that are formed by two causal loops: (1) *vegetation* $\Rightarrow$ *albedo* $\Rightarrow$ *temperature* $\Rightarrow$ *vegetation* and (2) *vegetation* $\Leftrightarrow$

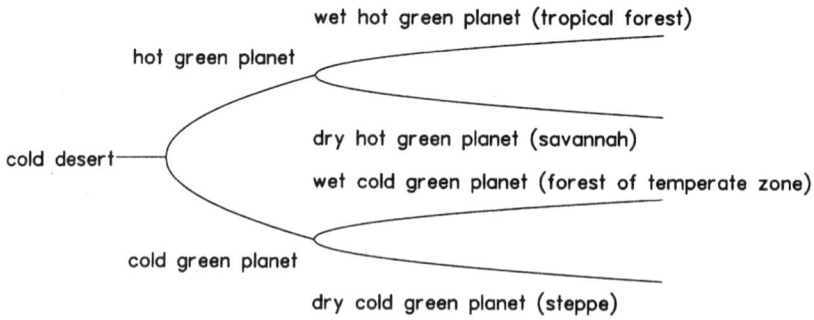

FIGURE 2. Bifurcation diagram for the system "climate + biosphere."

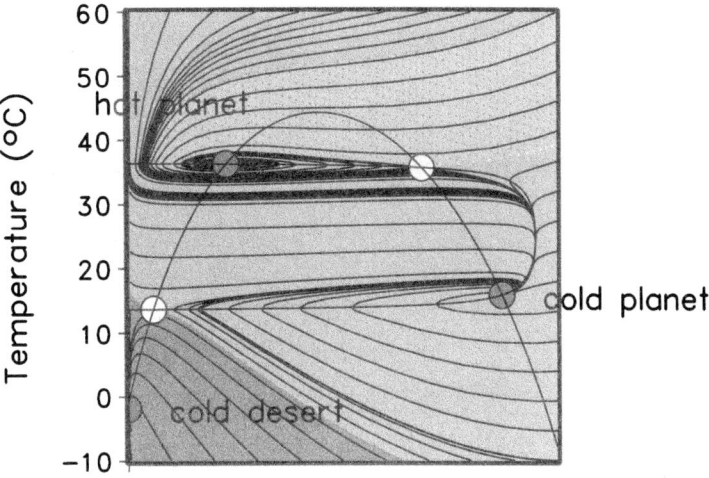

FIGURE 3. Phase portrait projected onto the temperature–vegetation subspace (T, N) when there are fife different equilibria, three of them are stable. T is the annual planetary temperature and N is the biomass of vegetation (or its density).

atmospheric carbon $\Rightarrow$ *temperature* $\Rightarrow$ *vegetation*. In order to get a visual impression of the system's behavior, the concept of phase portraits has been used. Because the system exhibits a hierarchy of time scales for water t_w, temperature t_T, and vegetation t_N with $t_w \ll t_T < t_N$, it is possible to use a projection onto the two-dimensional (T, N) subspace for the phase portraits (see FIGS. 3 and 4).

Climate, Vegetation, and the Global Carbon Cycle

The system can have up to five different equilibria: three of them are stable and the two other are unstable. The projection of its phase portrait onto the plane (T, N), where T is the annual planetary temperature and N is the biomass of vegetation (or, its density) is shown in FIGURE 3.

You can see there are three attractive domains corresponding to three equilibria (the values of parameters correspond approximately to ones for our planet):

1. "cold desert" when the planet has no vegetation and the planetary temperature is about $-3°C$;
2. "cold green" planet with rich vegetation and relatively low temperature, approximately equal to the Earth's current temperature, $15°C$;
3. "hot" planet with poor vegetation and relatively high temperature, $36°C$, that is, typical for Earth's hot deserts.

It is interesting that the domain corresponding to the second equilibrium has a maximal area (see FIG. 3); and, for instance, the trajectories, which start from the state "hot planet practically without any vegetation" (all the points that are located

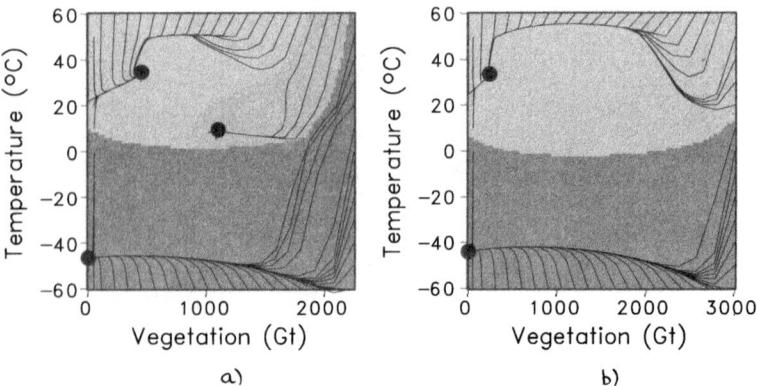

FIGURE 4. Phase portrait projected onto the temperature–vegetation subspace (T, N) for two different values of total amount of carbon, A, in the system (in gigatons, $1Gt = 10^9$ tons): **(a)** $A = 1500$ Gt; **(b)** $A = 2250$ Gt.

within upper-left corner of FIG. 3), go to the second equilibrium, interpreted as a contemporary state of Earth. Such behavior illustrates the stabilizing role of the biosphere in the formation of planetary climate, that is, a special feedback loop from the biosphere to climate.

Climate, Vegetation, the Global Carbon and Water Cycles

The total amount of carbon and water in the system together with the product of maximum ("potential") productivity of the vegetation and its residence-time of carbon were identified as the bifurcation parameters of the system, that is, they determine the number of possible stable states of the system. For increasing carbon in the system and fixed water content, and product $\beta = P_m \cdot \tau$ different regimes of system behavior exist and are discussed in the following.

1. For low values of carbon A in the system, only one equilibrium without any vegetation ("cold" desert) is stable. Independent of the initial conditions in the phase space, any vegetation dies out exponentially, and the cold desert is reached, ice finally covering the planet. The concentration of greenhouse gases is too low to prevent the planet from becoming glaciated. For higher values of A, the behavior slightly changes: one equilibrium is stable as before, but a finite interval in time of occurrence of vegetation covering the planet exists. Because the vegetation extracts carbon out of the atmosphere, the planet cools down and glaciates completely at the final state.

2. Above a critical value of A, three stable nodes appear (FIG. 4a): one equilibrium without any vegetation corresponds to the cold desert, while the other two are stable nodes with vegetation. One of them is characterized by a large amount of vegetation and low temperature ("cold" planet which is interpreted as the contemporary state), the other by a low amount of vegetation and high temperatures ("hot" planet). Because of the topology of the

basins of attraction a decrease of temperature in the cold planet state forces a glaciation of the planet and the cold desert state is reached. Look at FIGURE 4a, one sees that the contemporary climatic equilibrium is disposed sufficiently close to a dangerous boundary (the boundary of the stability region). Indeed, a "nuclear winter" would be such a result of a specific anthropogenic impact (see, for instance, Pittock et al.[9]). But the phenomenon of nuclear winter is a temporary climatic one: after a relaxation time the climate system would return to its predisturbed state. But now we are just beginning to understand that the current climatic equilibrium is situated very close to a stability border and a sufficiently strong perturbation could force the system to cross the border. As a result the climate would evolve toward a new equilibrium, which can be interpreted as a "ice desert."[8] A "nuclear winter" would be permanent! On the other hand, a slight decrease in vegetation is followed by a drastic change in vegetation, and the hot planet scenario is reached.

3. By increasing A the configuration is again changed. The stable node corresponding to the "cold" planet disappears, and two stable states are obtained. If the carbon in the system is further increased, vegetation becomes almost extinct, and the state of a "hot" desert is reached. For instance, the emission into the atmosphere of 600 Gt of carbon (as a result of use during the 100 years of the "business as usual" current economic scenario) leads to a loss of the stability of contemporary equilibrium. After that, the system "biosphere + climate" evolves to a state that could be characterized as "hot desert" (with temperature about 36°C, in which vegetation biomass would be reduced 8–10 times).

Certainly, all these estimations are approximate.

Complexity versus Simplicity: The Cost of Complexity

Let us look at FIGURE 5 where the number of possible stable states of the system as a function of the bifurcation parameters A (total amount of carbon) and B (total amount of water) is plotted.

Assuming that the probability of different numbers (one, two, three) of eqililibria is proportional to corresponding area in FIGURE 5, then the most probable state is the state with one equilibrium, the state with two equilibria is less probable, and the state with three equilibria can be realized with the least probability. An impression arises that nature pays for the increase of its own diversity (which can be measured by the number of stable equilibria) by the corresponding decrease of the probability of realization of such a state.

Let us introduce a simple measure of complexity, W, which is equal to the number of stable equilibria of the system. We can see that the system's complexity is changed when the system walks in the space of parameters A, B, and β. If we compare the domain areas corresponding to different complexity we can say that in order to increase its complexity the system must donate an opportunity to realize this complexity.

How should we define the probability for the system to realize a different number of equilibria? The problem is to measure the areas of corresponding domain. We can

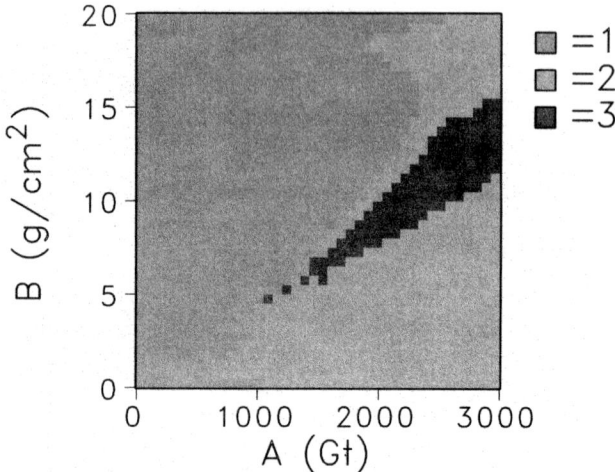

FIGURE 5. Number of stable states as a function of the total amount of carbon, A, and water, B, in the system (expressed in a mass of water column *per* square unit, grams per cm^2). The maximum productivity of the vegetation, P_m, and the residence-time of carbon, τ, are fixed: $P_m = 200$ Gt/year, $\tau = 20$ years.

do this using the so-called Halmos' measures. The idea of this approach is the following. In the plane (T, N) we define the sequences of squares, imbedded into each other with the origin as a common vertex. Let their areas be equal to $S_1 < S_2 < \ldots < S_n < \ldots$. For each nth square, we can calculate the domain areas corresponding to one, two, three equilibria, s_n^1, s_n^2, s_n^3, and then the fractions $p_n^1 = s_n^1/S_n, p_n^2 = s_n^2/S_n, p_n^3 = s_n^3/S_n$. The unknown probabilities are defined as $p_i = \lim_{n \to \infty} p_i^n$ for $i = 1, 2, 3$.

Using classic Boltzmann's idea we can introduce an entropy measure of the system complexity, H, in the form:

$$H = \sum_{i=1}^{3} p_i \ln W_i = \sum_{i=1}^{3} p_i \ln i \approx 0.69 p_2 + 1.1 p_3 \quad (1)$$

It is obvious that $\min H = 0$ and $\max H = \ln 3 \approx 1.1$.

The factors, which stand before values p_2 and p_3, can be interpreted as some of the costs of complexity. For instance, nature pays almost one and a half times higher for maintenance of the system with three stable equilibria than it pays for maintenance of the two equilibria system.

And finally, I would like to formulate one hypothesis concerning an increase of the system complexity in the course of the system evolution.

> *If the entropy of complexity, H, increases in the course of the system's evolution (as a result of change of the system's structure) then the probability of realization of the state with a maximal number of stable equilibria decreases.*

Apparently, there is a compromise between these two processes. It wil be interesting to test the hypothesis. Note that this entropy measure is not directly applied to systems with chaotic behavior.

PART B. THERMODYNAMICS OF THE BIOSPHERE

Is the Biosphere in a Steady State? Some Thermodynamic Calculations

It is clear that the hypothesis about quasi-stationary state of the contemporary biosphere plays one of the most important roles in globalistics.[1] Should this statement be tested in by some way? Various estimations show that the energy balance of the biosphere is fulfilled with sufficient accuracy, and this speaks to an advantage for the stationary hypothesis, but this is insufficient. From the viewpoint of the thermodynamics of open systems, the balance between the internal entropy production and its export into environment must be also fulfilled. In order to estimate the balance we suggest the use of one thermodynamic criterion based on the "zero-dimension thermodynamic model of the biosphere." This model contains the following compartment: atmosphere (A), biota (B), pedosphere (P), and hydrosphere (H). The model's characteristic time is equal to 10^3 years.

Let dS_{iE} be the annual production of entropy by the open system, Earth; let dS_{eE} be the annual export of entropy into the Cosmos. If the system "Earth + Cosmos" is in a dynamic equilibrium, then $d\, S_{iE} = -dS_{eE}$ where

$$dS_{eE} = 4/3 * dE *(1/T_s - 1/T_E) \sim -1.8 * 10^{22} \text{ J/K* Year,}$$

$$T_s \cong 5800 \text{ K}; \quad T_E \cong 260 \text{ K.}$$

Let dSj be the entropy production by the jth compartment, and $dSkj$ be the export of entropy from the jth compartment to the kth one. Calculating a compartment's entropy and entropy fluxes between compartments, we use the ideas and methods from the brilliant Morowitz book.[10]

For the model compartments we get the following:

Atmosphere. We use the polytropic model of the "static" atmosphere, which is the mixture of ideal gases N_2, O_2, CO_2, argon, and H_2O vapors. Then, the total entropy contents in the atmosphere is equal to $S_A = 3.49*10^{22}$ J/K. Because carbon dioxide is one of the "life gases," we then calculate separately the corresponding entropy $S_A^{CO_2} = 1,026*10^{19}$ J/K.

Biota (Biosphere). We assume that "Biota" is submitted into a thermostat with the temperature $T = 14°C$, that is, with the temperature equaling to the annual average temperature of our planet. If we consider the following standard (averaging) composition of biomass:

- liquid H_2O — 44%
- fixed H_2O — 6%
- cellulose — 37.5%
- proteins — 8.4%
- carbonhydrates, lipids, etc. — 4.1%,

then $S_B = 9.05*10^{18}$ J/K. Excluding the water entropy, we immediately get the entropy of dry biomass equaling $S_B = 2.9*10^{18}$ J/K.

The main uncertainty is regarding how to estimate the fraction of H_2O (from 40% up to 90% for different plant species). We developed the special method in order to minimize any influence of uncertainty, so that the corrected value $S_B = 1*10^{19}$ J/K (if taking into account the marine biota).

Calculation of entropy balance for biota gives (data were taken from Costanza and Neil[11]) the following:

- Solar radiation utilized by vegetation (directly): $dS_{eB} = 2.4*10^{20}$ J/K*Year.
- Water balance for vegetation: $dS_{HB} = 2.59*10^{20}$ J/K*Year.
- Atmosphere–biota interaction. The value, dS_{AB}, is added of the entropy fluxes accompanying the processes of carbon dioxide and oxygen diffusion through stomata and evapotranspiration

$$dS_{AB} = dS_{AB}^{CO_2} + dS_{AB}^{O_2} + dS_{AB}^{H_2O}$$

$$= 1.1*10^{18} + (-1.05*10^{18}) + (-4.96*10^{20}), \text{ in J/K*Year}.$$

- Soil–biota interaction: $dS_{PB} = -8.25*10^{17}$ J/K*Year.

Because all these values possess different exponents, it is senseless to attempt to sum them with accuracy up to the level of characteristics, and we shall sum only their exponents. As a result we get:

Exponent 20: $dS_1 = dS_{eB} + dS_{HB} + dS_{AB}^{H_2O} = 3.85*10^{18}$ J/K*Year.

Exponent 18: $dS_1 = dS_{AB}^{CO_2} + dS_{AB}^{O_2} = 4.4*10^{18}$ J/K*Year.

The value of dS_1 will be less if we take into account an "entropy jump" caused by the two phase transitions: fixed water $\Rightarrow$ liquid water $\Rightarrow$ vapor. In this case $dS_1 = 1 \div 2 *10^{18}$ J/K*Year.

Thus, with accuracy up to the level of two exponents, the entropy balance for the biosphere (or, more correctly, for biota) in 1970 was equal to zero; that is, the biosphere was in dynamic equilibrium. This can be seen in detail in FIGURE 6.

Diversity of the Biosphere

Life on Earth is present in different and various forms, and it is necessary to support all this diversity. This is a major thermodynamic role of solar energy. Otherwise, in accordance with the second law, life would be fully uniform, and what is more, it would be not exist. How can we estimate what sort and what quantity of work is performed by sunlight? For this, the following simple thermodynamic model is used (see also Svirezhev and Svirejeva-Hopkins[12]).

Let the biota contain of n different classes. These classes could be represented by different taxonomic units: biomes, ecosystems, communities, and species. Each class has the own mass N_i, so the total mass of biota is equal to $N = \Sigma_{i=1}^{n} N_i$. (For instance, if these classes are biological species then $n \sim 3*10^6$). We assume that at some initial moment of time they were mixed up some "prebiosphere" substance and this "prebiosphere" system had not any structure. What is the manner in which such

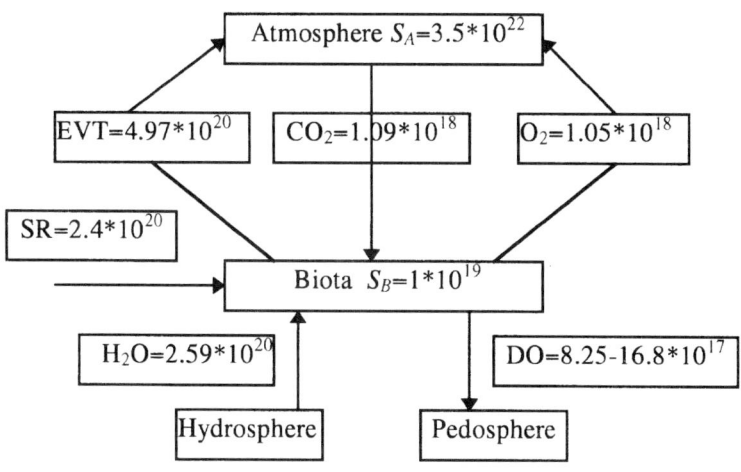

$$dS_B \cong 0$$

FIGURE 6. Annual entropy balance for the biosphere (biota) in 70s. Flux: SR, solar radiation; CO_2, carbon dioxide (net); H_2O, liquid phase of water; DO, dead organic matter. All storage values are in J/K; all fluxes are in J/K*year.

an ordered structure as the biosphere had been appeared? We think that it is a result of work of one super-being, the *"biosphere demon"* (let us remember Maxwell's demon). He feeds on negative entropy (*negentropy*) of sunlight, and distributes n different types of biosphere "particles" into n boxes, removing them from one "prebiosphere" pool.

As a result, each ith box contains N_i particles of the ith type with unit mass, and a new structure appears which we may call the *"biota."* Certainly in general, the individual masses of different particles must differ from each other, but, as a first approximation, we assume they are equal. The transition from a fully mixed system ("pre-biosphere chaos") to the "structured" biosphere is accompanied by an entropy reduction according to the formula[13]

$$S = -\kappa N \sum_{i=1}^{n} p_i \ln p_i \qquad (2)$$

where $p_i = N_i/N$ and κ is the specific entropy of the mass unit of some prebiosphere substance. In our case this substance is a mixture of chemical elements, from which living matter can be constructed: 106 molecules of CO_2 + 90 molecules of H_2O + a few molecules of some other substances. Note that when we speak about a "mixture," we do not have in mind a real physical mixture; most likely this would be a sum of already existing elements, but—and this is very important—they are not still reacting, and "living matter" is not still being formed. If we remember that the specific molar entropies for CO_2 and H_2O (vapor) are equal to 50 cal/mol*K and 16 cal/mol*K, correspondingly (at 15°C), then $k \approx 1.07$ cal/gram*K.

Let us now consider the energetic properties of this process. In order to produce N' units of a new biomass, the same amount of old biomass must be eliminated, that is, a corresponding amount of dead organic matter is decomposed (since the biosphere is in a steady state). Therefore, the value of N' must be equal to the net annual production of the stationary biosphere, Pr. As a result of decomposing processes, a quantity of heat that is equal to the caloric equivalent of carbon contained in N' must be given off. If we assume that the entropy production, S', accompanying these processes is approximately equal to their thermal effect, then $S' = Pr_c/T$ where Pr_c (in calories) is the carbon equivalent of the net annual production and $T = 287°K$ is the annual average planetary temperature.

In accordance with the principle of dynamic equilibrium, the entropy production S' must be balanced by the entropy production caused by exchange processes between the biosphere and its environment. We assume (and this is our basic hypothesis) that this entropy production is equal to S, that is, the decrease of entropy which is described by Equation (2) and corresponds to the process of new biomass formation. In other words, the value Pr_c is work performed by the "biosphere demon" in one year.

If we remember that the value $H = -\sum_{i=1}^{n} p_i \ln p_i$ is the information entropy or diversity of the system then

$$S = \kappa NH = S' = Pr_c/T \quad \text{and} \quad H = (Pr_c/N) \cdot (1/\kappa T) \qquad (3)$$

Because N is the annual biosphere production, Pr, expressed in grams of dry biomass, then the ratio $Pr_c/N = Pr_c/Pr$ is the energy (carbon) content of one gram of dry biomass. (In speaking about dry biomass, we take into account not only carbon, nitrogen, and phosphorus, but also fixed water.) It is known that one gram of biomass contains 0.5 gram of carbon, which corresponds to 4500 cal. Then $Pr_c/N = 4500$ cal/gram and

$$H = (4500 \text{ cal/gram})/((1.07 \text{ cal/gram*K})*287°K) = 14.65 \approx 15 \qquad (4)$$

If we multiply this value by 1.44, we get the result in bits: $H_b = 21$.

Let us estimate the probability of spontaneous creation of the biosphere. It is equal to

$$W = e^{-H} = e^{-15} = 3*10^{-7} \qquad (5)$$

Therefore it is very small. But if the contemporary biosphere is a result of a sufficiently large number of attempts L, in accordance with the simple probabilistic model,[14] the probability of its creation will be equal to:

$$W_L = LW/(1+LW) \qquad (6)$$

How do we evaluate the number of attempts? The photosynthetic biosphere with vegetation has existed for approximately 10^9 years.[15] The average time for biosphere renovation is equal to $\tau = B/Pr_c$ where $Pr = 5.4*10^{20}$ cal/year is the annual net production in caloric units, and B is the total biomass of the biosphere in the same units, equal to $8.3*10^{21}$ cal.[16] Then $\tau = 15$ years, and if we assume that one attempt is nothing else than one cycle of the biosphere renovation, then $K = 10^9/15 = 0.66*10^8$ and

$$W_L = (0.66*10^8*3*10^{-7})/(1 + 0.66*10^8*3*10^{-7}) = 0.95 \qquad (7)$$

Thus, the probability of the biosphere creation is close to one. In other words, *there is nothing surprising in the existence of the contemporary biosphere.*

It is also very interesting that the probability W depends neither on the mass of the biosphere, nor on its productivity [you can see it from Eq. (3) for H]. The probability W depends on only two factors: (1) the work of the climatic machine that determines the earth's temperature and (2) the type of main energetic reaction which determines the κ-value.

The photosynthesis reaction uses two gases as the basic substances for formation of living matter: carbon dioxide and water vapor. Certainly, we can imagine other hypothetical reactions that would use other elements and substances (for instance, silicon instead of carbon), but this would give other values of K and, as a consequence, other values of diversity. As a result, the probabilities of existence of such virtual biospheres would differ from a similar probability for the actually existing one. Note, however, that the probability W_L depends implicitly on the total mass of the biosphere and its productivity, because the number of attempts (L) is defined by these values.

If the value of H were known, the number of components (elements, elementary units, etc.) of the biosphere could be estimated. If these elements are relatively independent and they occur with almost the same frequencies, then $H \approx \ln n$, and $n \approx e^H = e^{15} = 3.3*10^6$.

It is interesting that this number coincides with the number of biological species on our planet. An impression develops that our "demon" uses species as boxes. On the other hand, if elementary units inside the biosphere are organized in hard structures (like trophic chains and trophic levels) with the exponential distributions of frequencies p_i, then $n \approx H$, that is, a number of these elementary units would be relatively small.

Exergy of the Biosphere

Reading the book *Chemical Evolution of Earth* by Vinogradov,[17] I turned my attention to the table, containing the comparison of the chemical compositions for living and nonliving matter (biota and Earth's crust, repectively). Part of this table (the original table contains the information for almost all chemical elements) is reproduced here as TABLE 1.

If we assume that Earth's crust (nonliving matter) is a system in thermodynamic equilibrium, then we can calculate the exergy of living matter (that is, the exergy of the biosphere, where the nonliving matter of Earth's crust is considered as a reference state (see Jørgensen[18]). In other words, we consider the biosphere to be a chemical system (for instance, an "active membrane") that either concentrates or disperses chemical elements depending on their basic concentrations in Earth's crust.

TABLE 1. Chemical composition of living (biota) and nonliving (crust) matter (in percent of total weight)

	Element						
	H	C	Si	Al	O	N	P
Biota	$1.05*10^1$	$1.8*10^1$	$2.0*10^{-1}$	$5.0*10^{-3}$	$7.0*10^1$	$3.0*10^{-1}$	$7.0*10^{-2}$
Crust	1.00	$3.56*10^{-1}$	$2.6*10^1$	7.45	$4.9*10^1$	$4*10^{-2}$	$1.2*10^{-1}$

Let s_i be the concentration of ith element in the biota and s_i^* be the same in the crust. In accordance to S.-E. Jørgensen,[18] the exergy of the biosphere will be

$$Ex = B \cdot RT \sum_{i=1}^{n} \left[\frac{s_i}{\mu_i} \ln\left(\frac{s_i}{s_i^*} \cdot \frac{B}{B^*}\right) - \frac{s_i - s_i^*(B^*/B)}{\mu_i} \right] \quad (8)$$

where T is the annual average temperature of the biosphere 288°K, R is the gas constant (1.987 cal/K*mol), μ_i is the atomic weight of ith element, and B and B^* are the total amount of matter in the biosphere and Earth's crust, respectively. It is obvious that we can introduce the new value $e_x = Ex/B$. This value can be called specific exergy and it is equal to the amount of exergy contained in one gram of living matter.

It is very important that specific exergy depends on not only the relative elements of composition of both living matter and crust (i.e., the percentage of different chemical elements in these systems in the biosphere and crust), but, to the significant degree, its value depends on the ratio $r = B/B^*$, that is, on the relative value of crust matter involved into the global biogeochemical cycles. It will be clearer if to rewrite Equation (8) in the form

$$e_x = RT \left\{ \sum_{i=1}^{n} \frac{s_i}{\mu_i} \ln\left(\frac{s_i}{s_i^*}\right) + \ln\left(\frac{B}{B^*} - 1\right) \cdot \sum_{i=1}^{n} \frac{s_i}{\mu_i} + \frac{B^*}{B} \sum_{i=1}^{n} \frac{s_i^*}{\mu_i} \right\}$$

or

$$e_x = K + \sigma(\ln r - 1) + \frac{\sigma^*}{r} \quad (9)$$

where

$$K = RT \sum_{i=1}^{n} \frac{s_i}{\mu_i} \ln\left(\frac{s_i}{s_i^*}\right), \quad \sigma = RT \sum_{i=1}^{n} \frac{s_i}{\mu_i}, \quad \sigma^* = RT \sum_{i=1}^{n} \frac{s_i^*}{\mu_i}$$

Taking into account that all concentrations in Vinogradov's table are evaluated in relative weight units (%) using the Equation (9), we can calculate the values of K, σ, and σ*: K = 185, σ = 94, σ* = 31. All the values are measured in cal/gram.

TABLE 2. Partial specific exergy for different chemical elements (in cal/gram of living matter)

	Element							
	H	C	Si	Al	O	Na	Mg	Fe
e_x^i	86.9	25.3	5.07	1.57	1.4	0.57	0.49	0.42

Let us calculate the value of specific exergy for different values of r. The simplest hypothesis is if we assume that the total amount of matter has not been changed in the course of the transition from a nonliving state to the living one, that is, the peculiar conservation law of matter is realized and $B = B^*$. Thus, $r = 1$. In other words, there is dynamic equilibrium between the biota and crust, between the living and nonliving matter of the biosphere (that is understood in Vernadsky's broad sense), where all the components of crust has been involved into the "big living cycle." Let us remember that in according to Vernadsky,[2] Earth's crust is a result of biosphere activity, and is the trace of the past biospheres. By setting $r = 1$ in Equation (9), we immediately get the value of specific exergy for one gram of living matter. It is equal to 122 cal/gram. Because $e_x = \Sigma_{i=1}^{n} e_x^i$, where is the contribution of partial exergy, corresponding to the ith element, into the total exergy, it is interesting to compare these contributions (see TABLE 2).

In fact, the main contribution belongs to hydrogen (~71%). This includes water, carbon-hydrates, and so forth. The second place is occupied by a carbon (~21%). Then silicon, aluminium and oxygen are disposed of (4.15%, 1.3%, and 1.12%). The contribution of others is negligible.

Because the summary contribution of hydrogen, carbon, and silicon is equal to 96.2%, we call our biosphere the hydrogenous–carbonate–silicon biosphere. If we compare the "carbon exergy" of one gram of living matter (25.3 cal) and the so-called "carbon equivalent" (~2 kcal per one gram of raw biomass), we can conclude that part of the "structural," "creative" exergy is equal to 1.25%—this is very small in comparison with "heat" exergy. The latter is equal to the number of calories obtained in the process of biomass burning.

However, Jørgensen has recently suggested a new measure of exergy based on the genetic complexity of different organisms with respect to detritus.[19] In accordance with this, if the "exergy cost" of detritus is equal to 1 then the "exergy cost" of most plants and trees will be situated in the vicinity of 30–70. Note that global vegetation is a leading actor in our biosphere. If the free energy of 1 gram of detritus is equal to 4.4 kcal/gram, then the specific exergy of living matter in the biosphere must be equal to 140–300 kcal/gram. is this not a contradiction? How can we resolve this?

Let us remember Vinogradov's estimations[17] for the total amount of crust matter, 10^{25} grams, and for the total biomass of the biosphere, 10^{21} grams, so that $r = 10^{-4}$. Then the specific exergy will be equal to ≈309 kcal/gram. Comparing this value with Jørgensen's specific exergy, we see these values are close. It seems to me that this coincidence is very interesting. Note that the main contribution to specific exergy is represented by term σ^*/r, that is, the term corresponding to processes that are working against the entropy and that separate a thin film of living matter from an immense mass of crust. The latter, in turn, is the entropy storage of the past biospheres. It is clear that the film is thinner, and the ability of one unit of living matter to perform such work must be higher. In other words, the value of r is less, the specific exergy must be bigger. But this situation is typical for an old biosphere when it has been in equilibrium for a long time.

Let us imagine the young biosphere hat has developed on the thin, young crust. This young biosphere is very aggressive, and all the crust matter is involved in processes of chemical interaction and exchange with the biosphere. Then we should consider the case with $r = 1$, which was discussed previously. We see that the main role in the formation of comparatively low exergy (K) is determined by the chemical composition of living matter. In other words, at the first stages of the biosphere formation the exergy of living matter is determined, mainly, by its chemical composition and, as a consequence, chemical composition also determines what sort of chemical processes are used by life in the formation of own matter.

REFERENCES

1. SVIREZHEV, YU. M. 1998. Globalistics: a new synthesis. Philosophy of global modelling. Ecol. Model. **108:** 53–65.
2. VERNADSKY, V.I. 1926. The Biosphere. Gostekhizdat. Leningrad.
3. KOSTITZIN, V.A. 1935. Evolution de l'atmosphere: circulation organique, epoques glaciares. Hermann. Paris.
4. WATSON, A.J. & J.E. LOVELOCK. 1983. Biological homeostasis of the global environment: the parable of Daisyworld. Tellus **35B:** 286–289.
5. VON BLOH, W., A. BLOCK & H.-J. SCHELLNHUBER. 1997. Self-stabilisation of the biosphere under global change: a tutorial geophysical approach. Tellus **49B:** 249–262.
6. SVIREZHEV, YU.M. 1994. Simple model of interaction between climate and vegetation: virtual biospheres. IIASA Seminar. Laxenburg. Austria.
7. SVIREZHEV, YU.M. & W. VON BLOH. 1997. Climate, vegetation and global carbon cycle: the simplest zero-dimensional model. Ecol. Model. **101:** 79–92.
8. SVIREZHEV, YU.M. & W. VON BLOH. 1998. A zero-dimensional climate-vegetation model containing global carbon and hydrological cycle. Ecol. Model. **106:** 119–127.
9. PITTOCK, A.B. et al. 1986. SCOPE 28: Environmental Consequences of Nuclear War. Vol.1. Physical and Atmospheric Effects. Wiley. Chichester/New York.
10. MOROVITZ, H.J. 1968. Energy Flow in Biology. Academic Press. New York.
11. COSTANZA, R. & C. NEIL. 1982. The energy embodied in the products of the biosphere. In Energy and Ecological Modelling. W. Mitsch, R. Bosserman & A. Klopathek, Eds.: 743–755. Elsevier. Amsterdam /London/New York.
12. SVIREZHEV, YU.M. & A. SVIREJEVA-HOPKINS. 1997. Diversity of the biosphere. Ecol. Model. **97:** 145–146.
13. LANDAU, L.D. & E.M. LIFSHITZ. 1964. Statistical physics. Nauka. Moscow.
14. CHERNAVSKY, D.S. & N.S. CHERNAVSKAYA. 1984. Problem of the new information in evolution. In Thermodynamics and Control of Biological Processes. A. Zotin, Ed.: 247–254. Nauka. Moscow.
15. RUTTEN, M.G. 1971. The origin of life by natural causes. Elsevier. Amsterdam/London/New York.
16. SVIREZHEV, YU.M., V.F. KRAPIVIN & A.M. TARKO. 1985. Modelling of the main biosphere cycles. In Global Change. T.F. Malone & J.G. Roederer, Eds.: 298–313. Cambridge University Press. Cambridge, England.
17. VINOGRADOV, A.P. 1959. Chemical Evolution of the Earth. The USSR Academy Scientific Publisher. Moscow.
18. JØRGENSEN, S.-E. 1992. Integration of Ecosystem Theories: A Pattern. Kluwer Academic Publishers. Dordrecht/Boston/London.
19. JØRGENSEN, S.-E. 1995. Exergy and ecological buffer capacities as measures of ecosystem health. Ecosyst. Health **1:** 150–160.

A Comprehensive Atmospheric Chemistry Model for the Description of Dynamics of Reactive Pollutants

G. BARONE,[a,b] P. D'AMBRA,[c] D. DI SERAFINO,[c,d] G. GIUNTA,[c,e] AND A. RICCIO[a]

[a]*Department of Chemistry, University "Federico II" of Naples, via Mezzocannone 4, 80134 Naples, Italy*
[c]*Center for Research on Parallel Computing and Supercomputers (CPS)-CNR, Naples, Italy*
[d]*Department of Mathematics, The Second University of Naples, Naples, Italy*
[e]*Institute of Mathematics, Naval University, Naples, Italy*

INTRODUCTION

An air quality model consists of the mathematical description of atmospheric transport and chemical transformation of reactive pollutants, that is, in the prediction of how species concentrations change in response to changes in emission over urban (~10^4 km^2, ~1 day) to regional (~10^6 km^2, ~3 day) to global (~10^8 km^2, ~30 days) scales.

These models can be used for various purposes, for example, to assess the responsibility for actual emission levels, to determine the most effective emission control strategies needed to comply with the air quality standards, or for planning purposes.

The basic approach is often Eulerian: atmospheric dynamics are simulated by a three-dimensional, coordinated system fixed with the surface terrain. In this case the general dynamics equation is the so-called continuity equation:

$$\frac{\partial c_i}{\partial t} = -\nabla \cdot (\mathbf{u} c_i) + \nabla \cdot (\mathbf{K} \cdot \nabla c_i) + R_i(z, t) + D_i(\mathbf{x}, t) + E_i(\mathbf{x}, t) \quad i = 1, \ldots, N \quad (1)$$

where $c_i(\mathbf{x}, t)$ are the concentration of the N gas-phase chemical species, $u = (u, v, w)$ is the flow field, $\mathbf{K}$ the eddy diffusion tensor, R_i the kinetics term due to chemical reactivity, T the temperature, and E_i and D_i the emission and deposition terms, respectively. The solution of the above system of coupled nonlinear partial differential equations requires a large amount of input data, that is, spatial and temporal emission inventory, wind, diffusion tensor, temperature, humidity, mixing height, land use, and surface roughness fields. Moreover, the high degree of stiffness, associated with the very different chemical reactivity, requires the use of expensive implicit methods for its numerical solution and of powerful computational facilities, that is, parallel computers.

Recently, we started the numerical simulation of the dynamics of atmospheric pollutants over the Campania region (in Southern Italy),[1–3] which is one of the most

---

[b]Address for correspondence: +39 081 5476502 (voice); +39 081 5527771 (fax); barone@chemna.dichi.unina.it (e-mail).

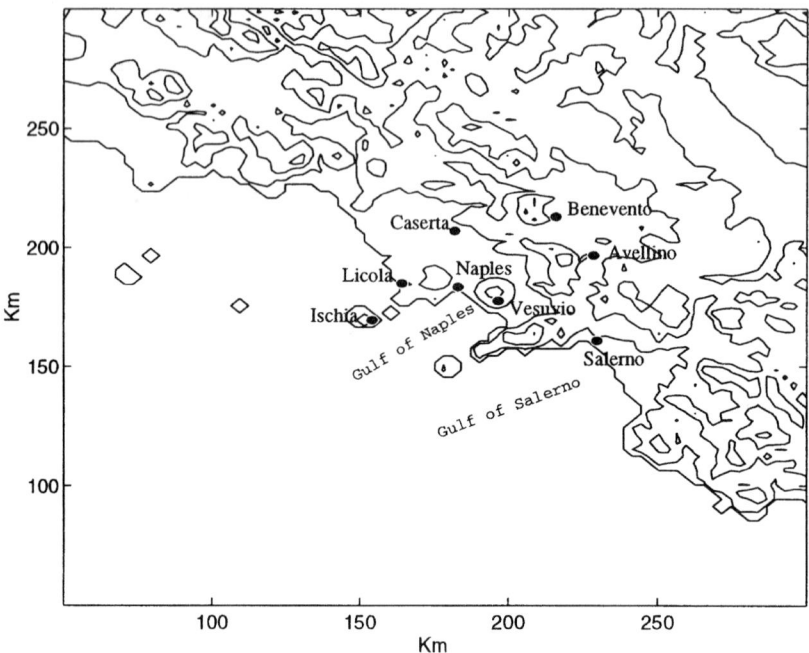

FIGURE 1. Campania region topography. Contour lines are at 0, 100, 500, 1500, and 2000 m.

densely populated European areas. As a consequence, the main urban areas of the Campania region are characterized by the constant presence of photochemical smog during the summer months, due to the frequently occurring atmospheric stagnant conditions and high levels of anthropogenic emissions; whereas winter months are characterized by high concentrations of primary pollutants (particularly VOC and NO_x).

To gain insight into the physical and chemical mechanisms that give rise to high air pollutant concentrations over this area, a research activity is being carried out at the Center for Research for Parallel Computing and Supercomputers (CPS). The Department of Chemistry of the University of Naples "Federico II" and the Faculty of Environmental Science of the Naval University of Naples are also involved in this interdisciplinary project. This last year of activity has led to the development of parallel software, which we named PNAM (Parallel Naples Airshed Model),[4] for the numerical simulation of air pollution episodes on the mesoscale for MIMD distributed-memory computers. The final goal is to develop a system that can be used for forecasting purposes.

In the following section we present the results of the application of this software to an air pollution episode that occurred in the Naples urban area on July 26, 1995.

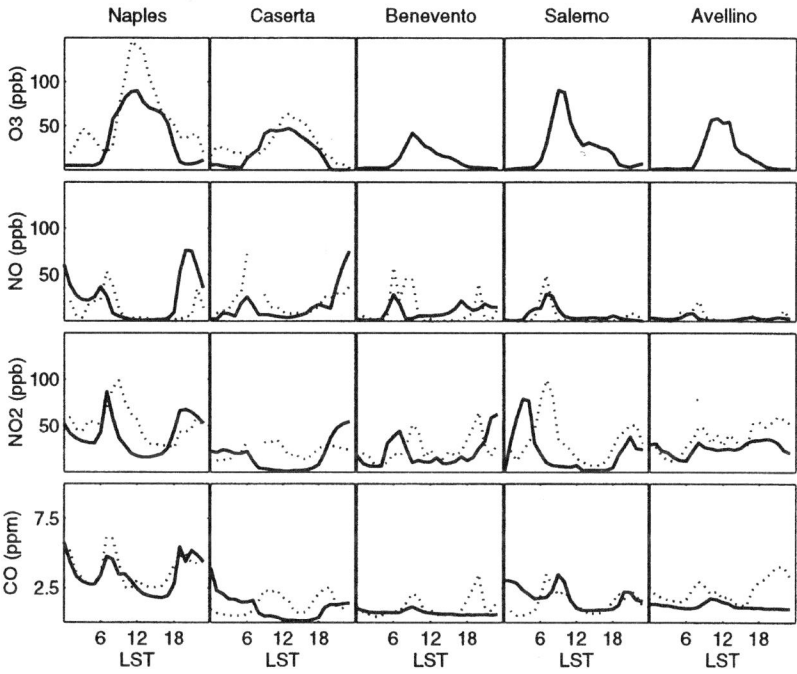

FIGURE 2. Computed (*solid line*) and measured (*dotted line*) profiles of pollutant concentrations in the Campania region on July 26, 1995.

RESULTS

The Campania regional topography, along with the localization of the main urban areas, is shown in FIGURE 1. Measured air quality data are available from the monitoring network of the Campania region (MARC). A comparison of these data with simulated data is shown in FIGURE 2. Unfortunately, measured O_3 data are available only from the Naples and Caserta monitoring stations. The computed results appear promising, because they show the PNAM's capability for reproducing the temporal and spatial patterns of measured air quality data, although some discrepancies are evident. Let us analyze, for example, the situation in Naples. During the night, the simulated O_3 concentration is lower than that observed; PNAM predicts an almost zero concentration, below the background concentration of about 30 ppb. This underestimation is probably in conjunction with the overestimation of NO, which consumes O_3 during the night. There could be various explanations for this discrepancy. During the night concentrations become very sensitive to meteorological parameters, such as wind speed and mixing layer height (which are difficult to estimate because of their very low values); moreover, our emission inventory is affected by large uncertainties (as in any air quality model) and a small difference in the temporal and spatial allocation of sources can cause significant differences. The peak in O_3 concentration is also missed at the Naples monitoring station: the observed concentration reaches about 150 ppb, while the computed concentration peak is about 100 ppb. The NO experimental concentration shows a sharp peak at the Caserta station in the early morning, which is probably due to local effects.

ACKNOWLEDGMENTS

This work was supported financially by the Ministry of University and Scientific Research (MURST), Rome (Cofin.MURST 97 CFSIB).

REFERENCES

1. RICCIO, A. 1997. Reti di reazioni chimiche e/o fotochimiche interagenti con processi diffusionali e/o convettivi. Ph.D. thesis, University of Naples, Naples, Italy. [In Italian.]
2. BARONE, G., P. D'AMBRA, D. DI SERAFINO, G. GIUNTA & A. RICCIO. 1998. Numerical simulation of air pollution phenomena in the Neapolitan urban area (Southern Italy): first experiences. *In* Large-Scale Computations of Engineering and Environmental Problems. M. Griebel, P. Iliev, P. O. Margenov & P. S. Vassilevski, Eds. Vol. 62: 128–135. Kluwer.
3. BARONE, G., P. D'AMBRA, D. DI SERAFINO, G. GIUNTA, A. MURLI & A. RICCIO. 1998. Parallel numerical simulation of air pollution in Southern Italy. Presented at the NATO Advanced Research Workshop on Large Scale Computations in Air Pollution Modelling. Sofia, Bulgaria, July 6–10, 1998.
4. BARONE, G., P. D'AMBRA, D. DI SERAFINO, G. GIUNTA, A. MURLI & A. RICCIO. 1998. Application of a parallel photochemical Air Quality Model to the Campania region (Southern Italy). Presented at the APMS'98 Conference. Paris, France, October 26–28, 1998.

Modeling the Complex Behavior of Microorganisms

S. BASTIANONI,[a] S. MARTINI, M. PORCELLI, AND C. ROSSI

Department of Chemical and Biosystem Sciences, University of Siena, Pian dei Mantellini 44, 53100 Siena, Italy

INTRODUCTION

The cell is the fundamental unit of higher organisms, and even in unicellular species it still contains subcellular structures with specialized functions. Different approaches have been used to investigate cell organization and function. These include study of cell cycles, uptake processes, energy metabolism, communications with other cells, storage of genetic information, and so forth. Molecular studies of cell organization have revealed numerous important biological roles and functions. However, methods of studying a second level of cell biomolecular organization need to be developed. The biological significance of a cell consituent can only be completely understood by considering the complex network of interactions that take place in the system. Innovative points of view and methods are needed to investigate the organization of complex systems and their responses to external stimuli.

In vivo NMR techniques together with selective ^{13}C sugar substrate enrichments have been used to study metabolic processes (see, e.g., Refs. 1, 2, and 3). Because of its "noninvasive" nature, NMR spectroscopy can be useful in kinetic studies in which several samplings are required during metabolic activity.[4–6] This technique also reveals the metabolic route of enriched carbon-13 nuclei from substrate to the endproduct.

Experimental results from organized complex systems like intracellular organisms require an appropriate approach for their interpretation. In the past, theoretical models have been proposed and used to elucidate metabolic steps and calculate kinetic parameters. Such models must deal with a large number of interactions and must be flexible enough to adapt to the existing approaches and experimental results.

In this study, our interest is the metabolism of the yeast *Saccharomyces cerevisiae*, chosen because of its efficiency in converting glucose to ethanol. Ethanol from sugar fermentation is currently a biofuel. Biological processes for biofuel production from agricultural residues, such as sugar, starch, cellulose, and hemicellulose, are of considerable importance. The replacement of fossil fuels is an urgent priority in the effort for the reduction of atmospheric carbon dioxide (CO_2) concentrations and global temperature.

In previous papers,[7,8] we analyzed the glucose metabolism of *Saccharomyces cerevisiae* by *in vivo* ^{13}C-NMR spectroscopy, using the energy flow approach and the compartmental models introduced by Odum.[9] The model we proposed then con-

---

[a]Address for correspondence: 39-0577-232088(voice); 39-0577-232004 (fax); simo@spectrum.chim.unisi.it (e-mail).

siders the cellular metabolic reactions of the yeast cells in terms of activation, inhibition, and feedback activities of the yeast cells and the substrate, with a range of initial glucose concentration between 85 and 200 g/l. The aim of this paper is to examine the possibility of extending the validity of this model to the degradation of an initial sugar concentration of 20 g/l by *Saccharomyces cerevisiae*, where the scarcity of substrate from the very beginning of the process induces different performances in the metabolism.[8]

EXPERIMENTAL

Saccharomyces cerevisiae strain KL-144A was grown at 31°C in a liquid medium containing 6 g yeast extract, 0.5 g L-cysteine HCl, 5.6 g KH_2PO_4, 7 g K_2HPO_4, 1.0 g/l $NaHCO_3$, 1.5 g $(NH_4)_2SO_4$, 0.15 g $MgCl_2 \cdot 6H_2O$, 0.01 g $FeSO_4 \cdot 6H_2O$, and 3 g/l sodium citrate $\cdot 2H_2O$. Samples with a cell density of 2.5×10^9 cell/ml were used. For all determinations pH was kept constant during the fermentation process. The initial glucose concentration was 20 g/l. The number of transients necessary for a spectrum with a high signal-to-noise ratio was reduced by adding 5 g/l of [1-^{13}C] 90% enriched glucose (from Stohler Isotopic Chemicals) to each sample. ^{13}C-selective enrichment of substrate also allows the transfer process of ^{13}C isotopes to end products to be followed.

^{13}C-NMR spectra were recorded with a Varian XL-200 spectrometer operating at 200.085 MHz and 50.288 MHz for proton and carbon, respectively. Carbon spectra were recorded under broad-band proton decoupling conditions using a low-power MLEV-16 pulse in order to avoid sample temperature effects. Ten-millimeter coaxial tubes, containing 99.75% D_2O in the outer part, were used for the NMR measurements. The ^{13}C-NMR spectra were recorded in 10-min blocks until the end of the fermentation process.

RESULTS AND DISCUSSION

As in the general model presented in our previous papers, glucose consumption is described as the overlap of two kinetic processes: an autocatalytic one and one depending only on active cell concentration. Yeast activity and fermentation are modeled as flows from interactions between glucose and active cells, meaning that part of the glucose is used by the yeast for the production of ethanol and the remainder to feed the cells. Inhibition of glucose conversion, due to ethanol build-up, was modeled as a flow from active to inhibited cells. This approach has been used with success for systems in which the initial concentrations of glucose were 85 and 200 g/l. The equations of this model are:

$$\frac{dG}{dt} = -k_{1d} \cdot G \cdot C - k_{2d} \cdot C$$

$$\frac{dC}{dt} = -k_{1a} \cdot G \cdot C + k_{2a} \cdot C - k_i \cdot C \cdot E$$

$$\frac{dE}{dt} = k_{1p} \cdot G \cdot C + k_{2p} \cdot C$$

where G and E are glucose and ethanol concentrations measured in g/l; C is an index related to the quantity of active cells; and $k_{j\alpha}$ are the parameters of the model.

The application of this model in the case of the degradation of an initial sugar concentration of 20 g/l by *Saccharomyces cerevisiae* was performed with the program MLAB.[10] The fitting procedure was done by maintaining constant the value of the k_i parameter, since the effect of a unit quantity of ethanol does not depend on the initial concentration of glucose in the solution. The values of the other parameters were allowed to vary, to consider the effect of a lower initial concentration with respect to the range 85–200 g/l. The description of the trend of the degradation was good ($R^2 = 0.996$). The values of the paramenters show that globally the microrganism transformed glucose into ethanol less efficiently than at 85 and 200 g/l. In particular the balance of the degradation is shifted more toward the autocatalytic pathway, meaning that the degradation process becomes relatively slower as it proceeds. It is interesting to note that the parameter k_{2a} remains zero as in the previous cases: the pathway depending only on the active cells' concentration is still one of "guaranteed minimum," even though the cells in this case seem to need less substrate for a minimum activity level.

A regular difference, although not very significant, was found between the predicted trend and the experimental data. The plot of the residues (experimental values − predicted values) is shown in FIGURE 1 and reveals oscillations in the dynamics of glucose, oscillations that seem to have a decreasing amplitude. The residues for

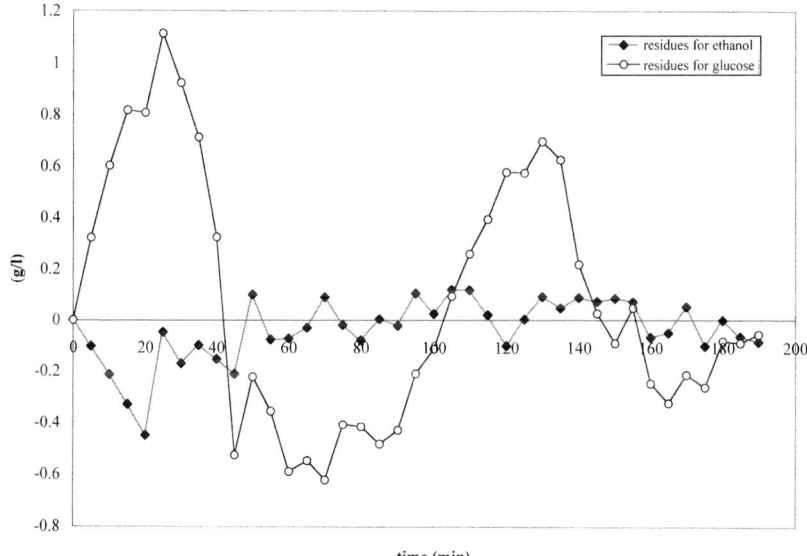

FIGURE 1. Plot of the difference between experimental and predicted data (residues) of the glucose and ethanol dynamics.

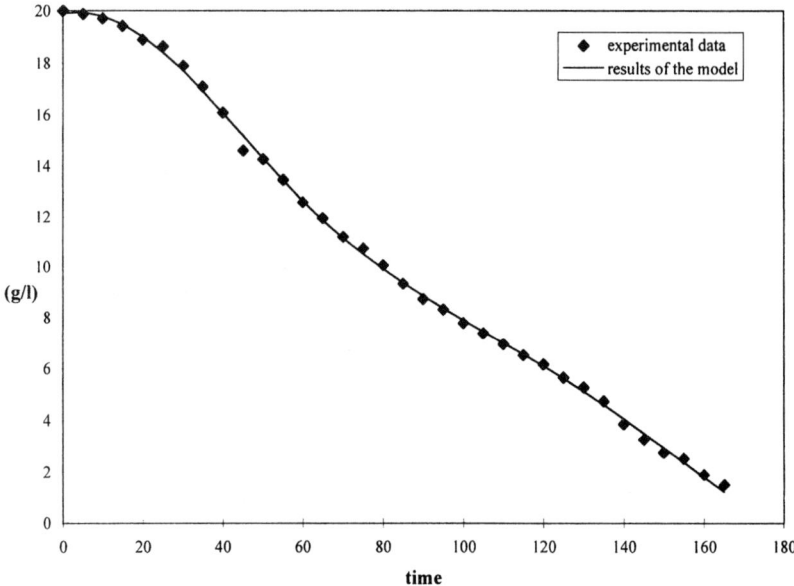

FIGURE 2. Fit of the function $g(t) = (mt + q) + A[\sin(2\pi ft + \varphi)]e^{-kt}$ to the experimental data of glucose.

the ethanol dynamics are much less relevant and show no regularity. The Fourier transform of these data gave a frequency of about 1.45×10^{-4} Hz, or a period of around 115 minutes in the degradation of glucose, with a coefficient of variation of about 10%.

Starting from these results, we used a mathematical model to have a better understanding and a more correct measure of the oscillating behavior of the degradation of glucose by *Saccharomices cerevisiae*. The model used is the following:

$$g(t) = (mt + q) + A[\sin(2\pi ft + \varphi)]e^{-kt}$$

where $g(t)$ is glucose concentration as a function of time and f the frequency of the oscillations. This function was chosen because: (a) a straight line did not fit the glucose degradation trend; (b) the sum of a straight line with a sine curve is the simplest way of representing a function that decreases and oscillates at the same time; (c) the negative exponential function was introduced because of the decreasing value of the amplitude of the oscillations (see FIG. 1).

The function fits the experimental data almost perfectly ($R^2 = 0.999$), if the last four points are ignored (FIG. 2). They correspond to the last 15 minutes of fermentation, when glucose concentrations dropped below 1 g/l.

The period of the oscillations was also determined by a fitting procedure (performed with MLAB) and was found to be of 123.3 minutes, in accordance with the results of the Fourier transform.

CONCLUSIONS

We have shown that, even with different values for the parameters, the model we have proposed for the description of the metabolism of *Saccharomices cerevisiae* is also valid for a lower concentration of 20 g/l. The plot of the residues shows that only the general trend of the glucose degradation is followed and that oscillating behavior is present; the ethanol production is, instead, well described.

We have used a sinusoidal function to fit the data of the glucose degradation, and a period of around 120 minutes results from the fitting procedure. The biochemical reason for such oscillating behavior is still under discussion.

ACKNOWLEDGMENT

This project was carried out with the support of MURST 97 CFSIB (Italian Ministry of University and Scientific Research).

REFERENCES

1. SHULMAN, R.G., T.R. BROWN, K. UGURBIL, S. OGAWA, S.N. COHEN & J.A. DEN HOLLANDER. 1979. Science **205**: 160–176.
2. JACQUEZ, J.A. 1996. Compartmental Analysis in Biology and Medicine. BioMedware. Ann Arbor, MI.
3. BEECHEM, J.M. 1992. *In* Methods in Enzimology. Vol. 210. L. Brand & M.J. Johnson, Eds.: 37–54. Academic Press. San Diego.
4. WÜTHRICH, K. 1989. Science **243**: 45–47.
5. ULIANOV, N.B., U. SCHMITZ, A. KUMAR & T.L. JAMES. 1995. Biophys. J. **68**: 13–21.
6. ROSSI, C., A. DONATI, D. MEDAGLINI, M. VALASSINA, S. BASTIANONI & E. CRESTA. 1995. Biomass Bioenergy **8**: 197–202.
7. BASTIANONI, S., A. GASTALDELLI, C. BONECHI, C. MOCENNI & C. ROSSI. 1996. Biochem. Biophys. Res. Commun. **227**: 41–46.
8. BASTIANONI, S., A. DONATI, A. GASTALDELLI, N. MARCHETTINI, D. RENZONI & C. ROSSI. 1996. Biochem. Biophys. Res. Commun. **227**: 53–58.
9. ODUM, H.T. 1972. *In* Systems Analysis and Simulation in Ecology. Vol. II. B.C. Patten, Ed.: 139–211. Academic Press. New York.
10. BUNOW, B. & G. KNOTT. 1992. MLAB, a mathematical modeling laboratory. Civilized Software Inc. Bethesda, MD.

The Effects of Vertical Mixing Parameterization on 3-D Models of a Pelagic Ecosystem

A. BELLUCCI, A. CRISE, G. CRISPI, AND C. SOLIDORO

Osservatorio Geofisico Sperimentale, P.O. Box 2011, 34016 Trieste, Italy

One of the major physical forces acting on the biological processes of the upper ocean is turbulent mixing, mainly produced by wind-stirring and internal wave breaking.[1] Turbulent eddies, and the associated overturning, displace phytoplankton and other biotracers along the water column. Conversely, density stratification inhibits vertical transport by reducing the length-scale of energy-containing eddies near the pycnocline. Wind stress and density fields are characterized by strong space and time variability, so that sophisticated turbulence closure schemes are needed in order to correctly describe subgrid-scale processes. Several attempts in this direction have been proposed in the literature, although most of them concern one-dimensional domains. On the other hand, 1-D models are forced to artificially increase vertical diffusion to compensate for the absence of vertical advection, and therefore a three-dimensional approach is preferable when the vertical dynamics have to be fully resolved.

The aim of this paper is to compare the effects of different vertical mixing parameters on the variability of the lower trophic levels in the Mediterranean pelagic ecosystem, by means of a three-dimensional coupled ecological–hydrodynamical model.[2] The hydrodynamic forces are obtained by means of a primitive equation model forced with climatological monthly mean winds, while ecological dynamics are described by use of an aggregated model, based on inorganic nitrogen, phytoplankton, and detritus.

We have compared a simple A-physics closure scheme (hereafter AP) with a more complex parameterization such as the one proposed by Pacanowski and Philander (hereafter PP[3]). Whereas in the AP scheme, constant eddy coefficients are used, in the PP parameterization momentum and tracer diffusivities vary in space and time depending upon vertical density gradient and velocity shear, through the dimensionless Richardson number.

A well-developed surface mixed layer, together with a sharper vertical density gradient, is the most striking feature evidenced by PP when compared to AP results (FIGS. 1 and 2). The modified structure of the mixed layer can lead to detectable differences between the velocity patterns in the two simulations.

The geostrophic adjustment of the density field under the effect of a sheared-wind force, produces a horizontal buoyancy gradient due to the pycnocline slope, which is stronger in PP with respect to the AP case. The increased horizontal density gradients are in turn responsible for an enhanced vertical shear in the horizontal velocity components, according to the thermal wind equations[4]:

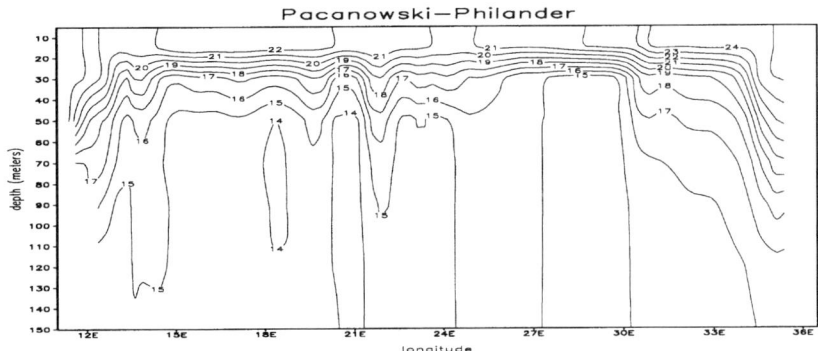

FIGURE 1. Temperature as a function of depth and longitude along a zonal section in the levantine basin as simulated by PP parameterization.

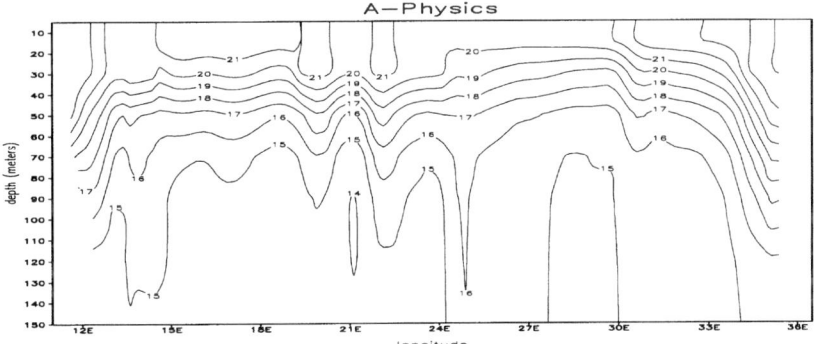

FIGURE 2. Temperature as a function of depth and longitude along a zonal section in the levantine basin as simulated by AP parameterization.

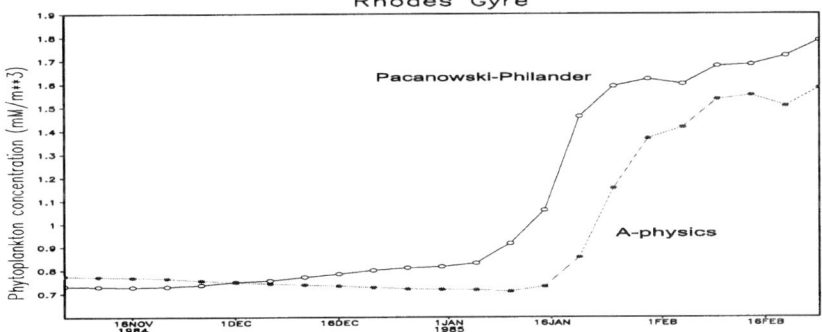

FIGURE 3. Time evolution of phytoplankton average concentration (upper 200 meters in the Rhodes gyre core).

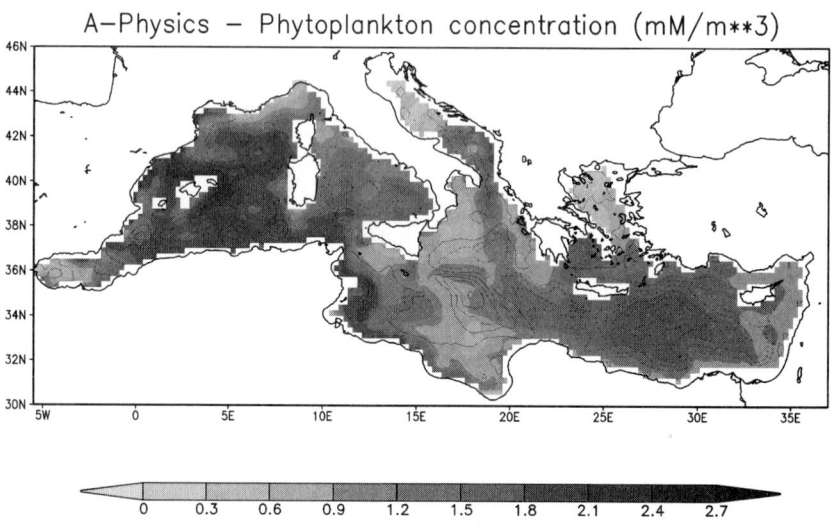

FIGURE 4. Spatial distribution of phytoplankton average concentration (upper 200 meters) for A-physics parameterization. Streamfunction is superimposed as a contour line.

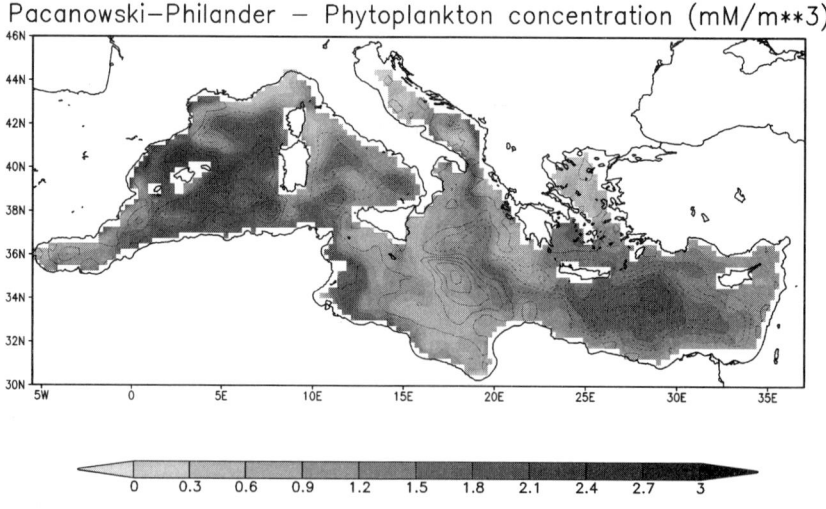

FIGURE 5. Spatial distribution of phytoplankton average concentration (upper 200 meters) for PP parameterization. Streamfunction is superimposed as a contour line.

$$-f\frac{\partial v}{\partial z} = \frac{g}{\rho_0}\frac{\partial \rho}{\partial x} \qquad -f\frac{\partial u}{\partial z} = \frac{g}{\rho_0}\frac{\partial \rho}{\partial y}$$

where u and v are the horizontal components of mean velocity, f is the Coriolis parameter, g the gravitational acceleration, ρ the density, and ρ_0 a constant reference density.

This affects the geostrophic vorticity field, which is intensified in both cyclonic and anticyclonic gyres. A direct effect on tracer transport is therefore observed: the intensified large-scale flow makes the advective transport prevailing on the horizontal diffusive transport inside the gyres, and the tracers are advected along streamlines rather than across them. In this way the gyres behave as a "tracer trap," modifying space and time tracer distribution. Looking at the space variability of phytoplankton concentration in the upper layer, the PP physics produces more pronounced gradients, giving rise to a sort of patchy distribution on a synoptic scale, particularly evident in the eastern basin (FIGS. 4 and 5). Also, in the gyre area, tracers have a higher residence time. Because FIGURES 4 and 5 refer to the surface layer, these effects are more easily recognizable in the cyclonic gyres, characterized by upwelling phenomena, than in the anticyclonic one. Ekman suction pumps nutrients (inorganic nitrogen) in the cyclone's core, making them available to phytoplankton uptake. In PP simulation nutrients are trapped in the gyre for a longer time, while in the AP case the horizontal turbulent diffusion is more effective in spreading them outside. As a consequence, when we adopt the PP parameterization, phytoplankton growth inside a cyclone is more rapid and effective, as shown in FIGURE 3 for the Rhodes gyre. We can finally argue that different choices of turbulence closure schemes result in significatively different space-time distributions of modeled biological variables.

REFERENCES

1. DENMAN, K.L., & A.E. GARGETT. 1983. J. Limn. Oceanogr. **28:** 801–815.
2. CRISE, A., G. CRISPI & E. MAURI. 1998. A seasonal three-dimensional study on the nitrogen cycle in the Mediterranean Sea. Part I. Model implementation and numerical results. J. Mar. Sys. **18**(1–3): 287–312.
3. PACANOWSKI, R.C. & S.G.H. PHILANDER. 1981. J. Phys. Oceanogr. **11:** 1443–1451.
4. CUSHMAN-ROISIN, B. 1994. Introduction to Geophysical Fluid Dynamics. Prentice Hall. Englewood Cliffs, NJ.

Identification, Characterization, and Remediation of Contaminated Sites: A Case Study

ELENA COLLINA, MARINA LASAGNI, AND DEMETRIO PITEA[a]

Dipartimento di Chimica Fisica ed Elettrochimica, Università degli Studi di Milano, via C. Golgi, 19-20133 Milano, Italy

This study is part of a project funded by the Italian National Research Council for the development of a general strategy for the management and remediation of contaminated soils.

A 38000 m^2 contaminated area in Crespiatica (Lodi) was chosen for experimental activity. From 1961 to 1986 five different industries have operated at the site. Their activities included production of phtalic anhydride, recovery of solvents, and handling and storage of exhaust oils. Possible contaminants deriving from these activities include: the chemicals themselves, residues from the production cycle, residues from solvent distillation, and oil residues contained in the six storage tanks still present. Moreover, after cesation of the industrial activities, the site was used for illegal disposal of plastic bags containing tar residues (so-called "big-bags") and burial of waste of unknown origin. Therefore, the pattern of contamination is very complex, due to organic and inorganic sources of contamination. Among the former, naphthalene is present with the highest concentrations, being the starting material for phtalic anhydride production.

In the past years, a large number of chemical analyses were performed on soil and groundwater samples and on samples collected in the oil tanks and from the big-bags. Thus, the situation is further complicated by the different laboratory practices and methods used for analyzing contaminants, which led, in some cases, to very different results.

Various activities were carried out in Crespiatica regarding site assessment, waste disposal, and soil remediation.

Two different soil quality standards (the Dutch standard and that of the Lombardia region) were applied, and the results compared, to verify if a univocal classification was achieved and to assess to what extent the choice of statistical data descriptors influences classification.

Even if the number of samples satisfied the guidelines' sampling requirements, data distribution was verified before any calculation of statistical descriptors. After this analysis, we decided to use statistical descriptors that are not derived on the basis of a given distribution (i.e., maximum value, 0.95th percentile, and upper quartile) instead of commonly used descriptors such as mean, mean ± standard deviation,

---

[a]To whom correspondence may be addressed: +39 02 26603 253 (voice); +39 02 70638129 (fax); pitea@sg2.csrsrc.mi.cnr.it (e-mail).

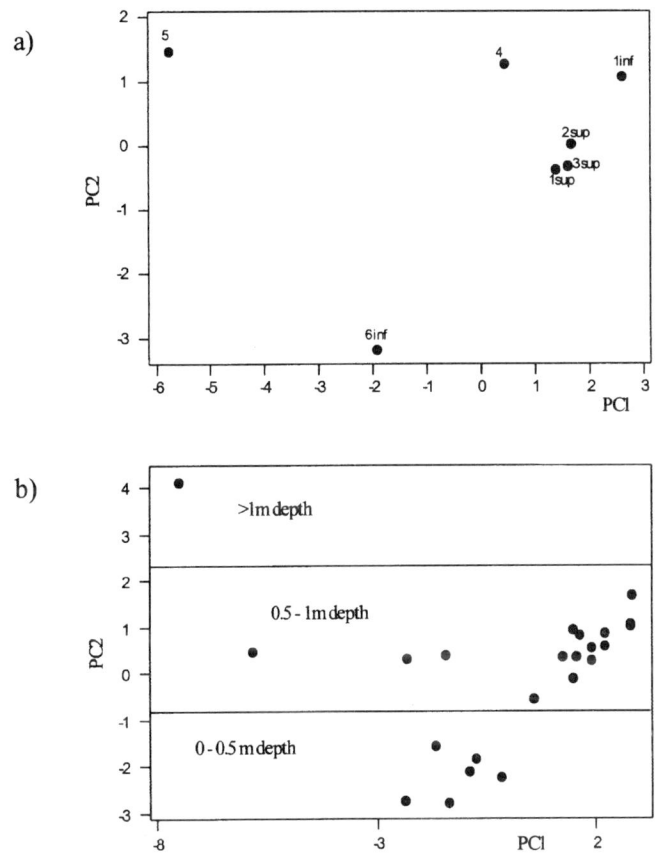

FIGURE 1. Score projection in the PC1–PC2 plane for (**a**) oil and (**b**) soil samples.

upper confidence limit. In any case, the conclusion was that the site is contaminated and needs remediation.

To investigate the homogeneity of soil, tanks, and big-bag samples before final disposal, principal component analysis (PCA), was performed. PCA is a multivariate eigenvalue/eigenvector method designed to extract information in a data set. New variables, called PC or principal components, are obtained as independent linear combinations of the original ones. For each component, the absolute value of the "loading" indicates how much the variable contributes to that PC. The "score" projection in the PC plane indicates the presence of grouping among objects.

PCA was performed on three data sets of autoscaled analytical results, where the samples represent the objects and the analytical parameters the variables: (1) Oil samples from the tanks, 7 objects × 15 variables; (2) Samples from big-bags, 5 objects × 20 variables; and (3) Soil samples, 23 objects × 16 variables.

For data set 1, PC1 and PC2 sum up 73.9% of the information content of the data set. From the score projection in the PC1–PC2 plane (FIG. 1a), we may notice a

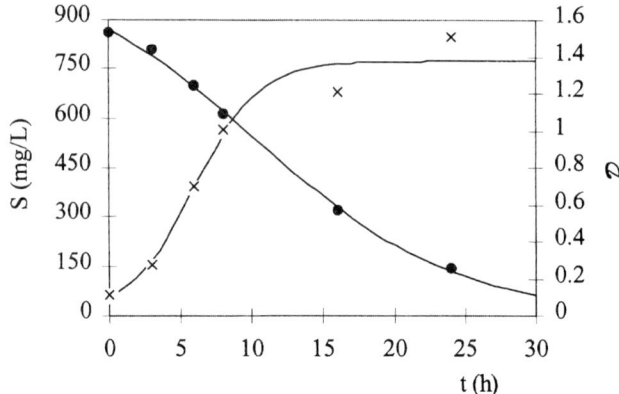

FIGURE 2. Experimental data at 30°C and fitting curves for naphtalene disappearance (•) and bacterial growth (×).

group of three objects, namely, the samples collected from the superior part of tanks 1, 2, and 3; the other objects are quite scattered. In fact, tanks 1 and 2 stored carter and hydraulic oil, tank 3 stored carter oil; tank 6 stored industrial oil; the content of tanks 4 and 5 is unknown. For data set 2, the first four PC sum up 100% of the original information. Sample homogeneity is established. For data set 3, PC1 and PC2 sum up 62.0% of the information content of the data set. From the score projection in the PC1-PC2 plane (Fig. 1b) we may notice the grouping of samples collected at the same depth. This result is related to the illegal burial of waste which determined the presence, at different depths, of materials of different density.

Labor tests were performed to study the microbial degradation kinetics of naphthalene and to determine the environmental factors that enhance or deplete degradation, in order to evaluate the applicability of an *in situ* bioremediation process.

From soil samples collected on the site, microorganisms capable of degrading naphthalene have been selected. Experiments have been performed in flasks at different temperatures, measuring at different times the population density as the absorbance $\mathcal{D}$ of 540 nm wavelength radiation and the residual napthalene concentration, S (mg/l).

A logistic kinetic model was used to model the disappearance with time of naphtalene concentration. Three sigmoidal functions (logistic and Gompertz with three parameters, Richards with four parameters) were used to describe the bacterial growth curve, with microbiologically relevant parameters.

The "best" parameters were obtained by an iterative procedure minimizing the residual sum of squares starting from initial guess values. At each experimental temperature (6, 23, 30, 37°C), the fitting was very good. As an example, the fitting at 30°C is reported in FIGURE 2.

The activation and thermodynamic parameters determined for both processes from the kinetic constant values are reported in TABLE 1 ($\Delta G^{\#}$ calculated at 293 K). The values are practically the same, as is expected, when the processes studied are coincident.

TABLE 1. Activation and thermodynamic parameters

Parameters	Naphtalene Disappearance	Bacterial Growth
Activation		
ln A	8.3 ± 0.8	10 ± 3
E_a (kJ mol^{-1})	26 ± 2	31 ± 6
Thermodynamic		
$\Delta H^{\#}$ (kJ mol^{-1})	23 ± 2	28 ± 7
$\Delta S^{\#}$ (kJ K^{-1} mol^{-1})	−(0.252 ± 0.007)	−(0.24 ± 0.02)
$\Delta G^{\#}$ (kJ mol^{-1})	97 ± 2	98 ± 7

ACKNOWLEDGMENTS

We thank the Italian National Research Council for financial support (Grants CT95.04515.ST74.115.30850, CT96.05317.ST74.115.30850, and CT97.05219. ST74.115.30850).

Classical Thermodynamic–Statistical Approach of Open Systems Deviating from Maximum Entropy

C. DEJAK,[a] R. PASTRES, G. PECENIK, AND I. POLENGHI

Departimento di Chimica Fisica, Università di Venezia, Dorsoduro 2137, 30123 Venezia, Italy

Nowadays, several models of aquatic ecosystems describe the pluriannual evolution of the system reasonably well, after having been validated against time series of experimental data. From the analysis of the trends in time of this data, the problem of tending to an exergy maximum, emerged. This problem can be analyzed not only in an evolutionary frame, but also in the context of classical thermodynamics, as the characteristic functions of isolated and open systems tend toward some extremes at equilibrium. In this regard, one must keep in mind that characteristic functions of closed or open systems, which depend on different sets of independent state variables, tend toward opposite extremes. An example is given by the entropy, which is maximum at equilibrium in isolated systems, and the Massieu potential, which represents the internal entropy for systems connected to a thermostat and which is minimum with respect to the reciprocal of the temperature. Similarly, in an energetic representation, the internal energy of an isolated system tends to a minimum value with respect to entropy, which is one of its characteristic variables, whereas the Helmholtz free energy of a closed system is maximum at equilibrium with respect to the temperature. This unusual aspect is also a characteristic of state variables both mechanical and chemical; extensive, if isolated (volume and mole number); or intensive, if connected to a manostat (pressure) or a chemostat (chemical potential). The mathematical demonstration of this, by means of the Legendre transormation, is simple,[1] but its intuitive explanation is not so easy, as is usually the case in phenomenological thermodynamics, without the support of statistical explanations.

One should resort to the statistical thermodynamics, for example, in order to describe the evolutive thermodynamics of steady states or even dissipative systems. Only in this way was Prigogine able to introduce the concept of local equilibrium, an essential concept that gives real meaning to the utilization of thermodynamic functions, otherwise nonexistent outside the equilibrium. But in statistical thermodynamics there are at least three possible ways to approach the problem. First, the simple original theory of Boltzmann, which is the most intuitive way to approach the problem; second, the classical approach, which in this field is represented by the complete phase space theory within the ergodic condition; and third, the quantistic treatment, which is often based on well-established postulates, but it is difficult to relate to equally intuitive macroscopic analogies, as is the first approach.

---

[a]Address for correspondence: 0039-41-2578528 (voice); pastres@unive.it (e-mail).

According to the ultrasimplified treatment of Boltzmann, in a molecular phase space, an isolated system (microcanonic) could be exemplified for a very intuitive view by particles with only one degree of freedom (at least the utilized one). In this example all particles move with the same absolute velocity, along parallel trajectories but in two opposite senses. They hit two perpendicular mirrors, placed one in front of the other. The collisions between the particles and these two exactly parallel walls are assumed to be elastic ones. Therefore, the resulting distribution of the velocities is completely random. Usually, this example is used to show that simple systems with the same kinetic energy could also follow different, nonintercommunicating trajectories and, therefore, they could be nonergodic. It is demonstrated that, in a microcanonic reference of any single particle in these nonergodic conditions, the system arrives at an equilibrium: it is at its highest probable state when half of its particles move in the opposite direction of the other half. Also, the entropy S (and the Shannon index, $s \cong S/Nk$, with N = number of particles and k = Boltzmann constant) reaches its maximum value under these nonergodic conditions for the whole system composed by these isolated particles.

But it is well known that just a small deviation of only one particle from the parallel trajectory is enough to introduce a chaotic collapse within the whole system such that, consequently, it is no longer decoupled and can then reach the ergodic condition and a new, completely different equilibrium state. It is, however, a difficult task to develop an intuitive approach for this behavior like, for example, the ultrasimplified Boltzmann scheme in which velocity vectors are only directed along the three Cartesian axes. For this purpose it is possible to utilize a model created by Paul and Tatiana Ehrenfest,[2] which contributed to the resolution of fervent debates among thermodynamic scientists at the end of the 19th century. They consider not only one kind of particles, but also point molecules moving only along orthogonal directions and colliding against larger squared molecules, randomly placed in fixed positions of the motion plane. The system looks like a large billiard table, where square ashtrays, much larger than the billiard balls, are randomly placed. These fixed molecules have their sides directed diagonally with respect to the motion direction, so that they can reflect the point molecules only perpendicularly. We added to this model a zigzag boundary which, in analogy, is made up of reflecting surfaces diagonally placed. A global path is reached after a sort of Poincarè period, that is, a period after which the particles (the billiard balls) come out of this sort of black body, all with the same parallel velocity they had when they entered it. It is possible to count the number of configurations that the velocities of the four point particles assume, step by step, along their long and entangled path. The highest order of degree, characterized by parallel velocities is reached only before and after such a Poincarè period, and it is very rare. It is surprising that the opposite configuration with the highest degree of disorder, corresponding to all four velocities being different, also has a very low frequency.

These frequencies can be calculated (FIG. 1) in good agreement with the above analyzed experimental data, throwing four tetrahedron dice with labeled faces. The model could also become three dimensional considering the motion of point-like molecules along the three orthogonal directions, as in Boltzmann's ultrasimplified approach. In this system, the large fixed molecules become octahedrons, with multifaced walls similar to the beautiful shapes of the mihrab of a mosque. The analogy with the frequencies of dice throw s in the 3-D case can be extended to considering the throwing of hexahedrons (cubes). This is similar to the famous game of five-dice poker, with the difference that one more die is used. In FIGURE 2 it is shown that the case of highest disorder (maximum entropy or maximum Shannon index as its Stirling approximation $s = S/Nk = -\ln W/N = (-1/N)\ln(N!/\Pi_i n_i!) = (-1/N)\ln(N^N/\Pi_i n_i^{ni} = \ln\Pi_i(ni/N)ni/N = \Sigma_i(ni/N)\ln(ni/N))$ is very rare, and the frequency maximum is positioned well away from it.

These results can be intuitively explained using as a "model" the five-die poker. The Shannon index, that is, the informational entropy, in this case depends only on the number of dice and therefore on the number of different pictures that characterize a dice throw, but does not depend on the number of faces, that is on the number of pictures. The Shannon index is the same for cubic dice with six faces and tetrahedral dice with four faces. For example, $s = 5!/(1!)^5$ when the pictures are all different, $s = 5!/2!(1!)^3$ for a couple, $s = 5!/1!(2!)^2$ for a double couple, and so on. But the observed frequencies (which tend to the probabilities) do not have a simple relation with s, because, for example, with six cubic dice a couple is the most likely outcome, while with tetrahedral die a double couple is more likely than a single one.

This is due to the fact that analyzing the $6^5 = 7776$ events of the first case and over the $4^5 = 1024$ events of the second case one must take into account not only the number of repeated appearance of pictures on the dice (2,2+2, 3,2+3,4,5), but also the combinations of the nature of the figure that occupies specific places. The same reasoning would apply if we substitute dice with ecological niches and pictures with biological species. Therefore, the composite probability is obtained by multiplying the probability of the positions by that of the pictures. When the pictures are all different, the latter probability is respectively $\binom{6}{5} = 6$ and $\binom{4}{5} = 0$. Analogously, a more complex calculation leads to the probability of a couple, when the correction factor

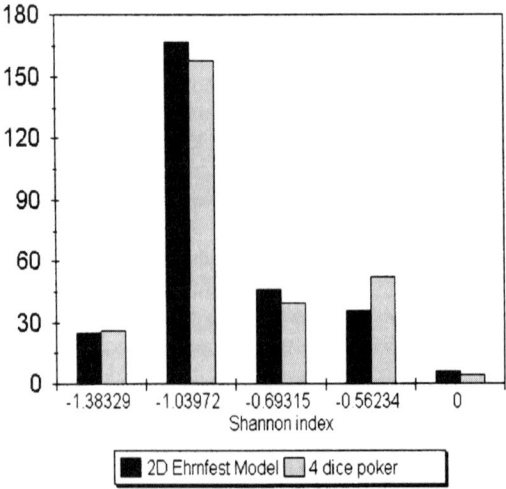

FIGURE 1. Distribution of the directions of the four particles moving on 2-D Ehrenfest model at each time step, in comparison with the distribution obtained by throwing four tetrahedral dice, as a function of the Shannon index (informational entropy).

is $\binom{6}{3}\binom{6-3}{1}$ = 60 and $\binom{4}{3}\binom{4-3}{1}$ = 4, thus obtaining 3600/7776 = 46.3% and 240/1024 = 23.4%, respectively.

For a double couple the correction factor is $\binom{6}{1}\binom{6-1}{2}$ = 60 and $\binom{4}{1}\binom{4-1}{2}$ = 12, thus obtaining probabilities of 1800/7776 = 23.1% and 360/1024 = 35.2%, respectively. The probabilities for the other four possible configurations can be calculated in the same way. The total probability is now maximal for a couple in the first case and for the double couple in the second case, in accordance with the experience. The same result is also obtained for four tetrahedric dice (see FIG. 1) and for 6 cubic, dice (see FIG. 2). What is remarkable is that the maximization of the simple Shannon index does not suffice for describing more complex systems, in which there could be even more species than niches: this is evident from the example of the poker game with six different pictures and only five dice, which is the configuration with the maximum Shannon index but with more pictures (or species) than dice (or niches). For the different configurations, this coincides with the necessity for exchanges of species, different from the ones of the first configuration, with a reservoir. In the case of more and more complex clusters (couple, triples, etc. and their repetition), the species needed by the system for a single configuration are fewer and fewer than those contained in the reservoir, whereas the complexity of the configurations increases.

This explains the results presented in FIGURE 2, where the frequencies of a full-hand (3+2+1, with s = 1.011) exceeds by far that of a triple couple (2+2+2, s = 1.098), as a consequence of the necessity of introducing more statistical parameters for dealing with the cases of repeated clusters—identical, in the second case, or not, in the first one. In less simple cases, these complications are multiplied to infinity,

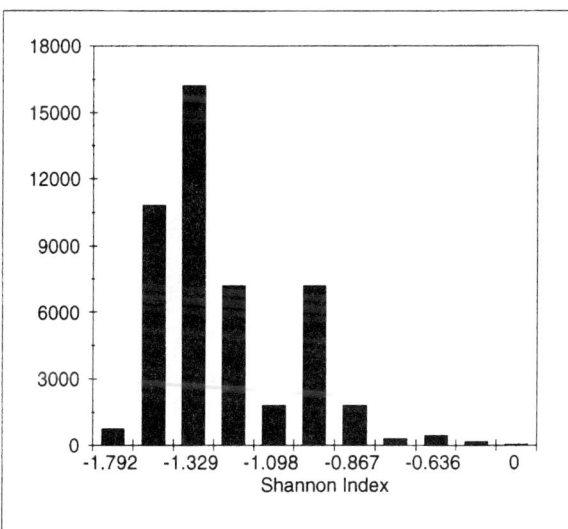

FIGURE 2. Distribution obtained by throwing six cubic dice, as a function of Shannon index (informational entropy). This distribution simulates the one that would be obtained using a 3-D version of the Ehrenfest model, with large octahedral particles placed in a 3-D cartesian system.

as long as more and more complex clusters are obtained. The maximum frequency shifts towards values of the Shannon index closer and closer to 0. In open systems, this leads to a tendency towards a mimimum internal entropy, even though such a minimum is never reached: for Massieu potential it would be reached only when the absolute temperature vanishes, but this contrasts with the third principle.

It follows that for nonisolated but interacting molecules (also if in the strange but simple Ehrenfest exemplification) the perfect disorder is not the most frequent case: if the system is connected with a tank, new configurations are found, and the ensemble of complexions changes quantitatively and qualitatively.

Many of the complex patterns obtained by throwing six dice can also be reproduced following a different classical Boltzman scheme. This scheme considers six different levels of energy with the constraint that the total energy must be constant. This is the canonic system in which even a single molecule interacts with the system of other molecules that can be considered to be like a thermostat, isolated from the environment together with the molecule itself. This third classical representation of the same phenomenon (an increase in the number of complex patterns for the interaction between system and thermostat with respect to an isolated system) allows the computation of a partition function and, as is well known, the determination of all the thermodynamic functions, following the Boltzman approach. The free energy, the internal energy, the entropy, and the Massieu potential can also be calculated as a function of their characteristic independent variables. The same calculation can be done in a thermodynamic-statistical phase space, first for an isolated molecule and an isolated thermostat of more molecules of perfect gas and then for the system composed by these subsystems, when they are brought into contact. The entropy difference can be easily calculated and corresponds to a contribution of the logarithm of the total energy.

In conclusion, using these models characterized by a gradually increasing degree of complexity but also gradually decreasing intuitability, it is possible to describe the detachment from tending to a maximum disorder through a mechanicistic–probabilistic analogy. If on the contrary, a system interacts with a tank of an extensive property (like the heat) with which it has in common only the correspondent intensive variable (i.e., the temperature), the Massieu potential (internal entropy) does not tend to a maximum but to a minimum. This aspect can easily be extended to other different mechanical and chemical variables with respect to their characteristic potentials: the Plank potential, the enthalpy, the Gibbs free enthalpy, the grand-potential and the exergy: it is also possible to make a further extension into evolutive states, steady states, or dissipative structures.

It can be noted that for simple equilibria this fact leads to extremes, such as maximum free energy with respect to the temperature and maximum enthalpy with respect to the pressure, but the boundaries of their interval of definition is therefore only of theorctical interest. On the contrary, for chemical reactions, which are more complex as they involve the presence of nonlinear terms with respect to the concentrations of the reactants, these equilibria assume concrete importance, as they are reached for finite values of gran-potential or exergy. Even more complex, but not substantially different, is the case of enzymatic kinetics in molecular biology.

It is very important to be able to figure out intuitively the way in which a nonisolated system, in contact with environmental tanks whose intensive variables (temperature, pressure, and chemical potential) fluctuate, tries to reach the highest internal order, while the whole isolate system (system + tanks) tends toward maximum disorder. This is possible if some degree of disorder is somehow "discharged" in the tanks. On a very large scale, this happens for our planet, with the discharge of radiant infrared energy toward outer space; whereas, on a smaller scale, the same phenomenon occurs with every living cell, which takes up nutrients and discharges its excrements. In both cases there is an input of valuable energy, that is, the radiant (solar) energy in the first case and valuable chemicals in the second one; whereas in both cases the output is highly degraded energy, which is necessary for the whole system to keep working. These cases are both in accordance with the second law, likewise with the Carnot machine, which cannot work without two reservoirs (hot and cold).

ACKNOWLEDGMENTS

This research has been partially funded by the National Reseach Council, as part of the national project Cofin.MURST 97 CFSIB.

REFERENCES

1. DEJAK, C. 1998. Oikos **3:** 105.
2. EHRENFEST, P. & T. EHRENFEST. 1909. Z. Phys. Chem. **10:** 19–20.

Use of Exergy and Structural Exergy as Ecological Indicators for the Development State of Homogeneous Lake Ecosystems

A. LUDOVISI[a] AND A. POLETTI

Department of Chemistry, University of Perugia, Via Elce di Sotto, 8-06123 Perugia, Italy

INTRODUCTION

For the last few decades a number of goal functions for the evaluation of the developmental state of ecosystems have been proposed: Jørgensen[1] gave a review and a comparison of these functions and showed that exergy has a good theoretical basis in thermodynamics, a close relation to information theory, and good correlation with other goal functions. Jørgensen[2] introduced exergy, structural exergy, and ecological buffer capacities as measures of ecosystem health and showed that these three concepts can combine the three components in Costanza's index[3]: system vigor, system organization, and system resilience.

The goodness of the exergetic approach is tested here comparing three IBP sites (Lake Batorin, Lake Myastro, Lake Naroch) that possess very similar morphological, hydrological, chemical and biological features, but show different trophic states.

METHODOLOGY

Calculation of Exergy and Structural Exergy

Exergy is defined as the maximum amount of work that a system can perform when it is brought to thermodynamic equilibrium with its environment. It is possible[1] to calculate the relative exergy contribution of an ecosystem as:

$$Ex = R \cdot T \cdot \sum_{i=1}^{n} W_i \cdot C_i \qquad (1)$$

where R is the gas constant, T is the absolute temperature, C_i is the concentration of the ith organic component in the ecosystem, and W_i is the information stored in non-repetitive genes. Exergy expresses the information that the biomass is carrying.

Structural exergy is exergy calculated relative to the total biomass:

$$Ex_{st} = \sum_{i=0}^{n} \left(\frac{C_i}{C_t}\right) \cdot W_i \qquad (2)$$

---

[a]Address for correspondence: 39-07505855584 (phone); 39-075-5855599 (fax); aludo@hotmail.com (e-mail).

where C_t is the total biomass concentration, which is the sum of all C_i including inorganic matter available for growth of biomass ($i = 0$). The Ex_{st} becomes independent of the nutrient level and measure the ability of the ecosystem to use the available resources.

According to Jørgensen,[2] exergy increases with increased eutrophication of lake ecosystems, because eutrophication implies more biomass mainly in the form of phytoplankton. The structural exergy will also increase with eutrophication at a low level of eutrophication: when the lake becomes eutrophic or highly eutrophic, the structural exergy will decrease, because of the reducing abundance of more complex organisms (e.g., top-carnivorous fish).

Data

Lake Batorin, Lake Myastro, and Lake Naroch are located in northwestern Byelorussia among glacial morains and are part of the basin of the River Neman; the lakes are adjacent and connected to each other and their outflows are slow.

Biological and physic-chemical data of the three ecosystems are reviewed in a very detailed study of biological productivity made by Winberg et al.[4] Environmental data for the lakes examined are shown in TABLE 1.

Despite the similarity of the nutrient concentrations, the lakes strongly differ in terms of community structure. Winberg et al.[4] classify Lake Naroch as mesotrophic, Lake Myastro as eutrophic and Lake Batorin as highly eutrophic.

The comparison of the biocenosis of the three lakes shows that phytoplankton, bacterioplankton, zooplankton, and fish biomass concentrations increase with eutrophication degree, while macrophytes, biomass concentration is decreasing, probably because of the reduction of water transparency.

Concerning the zooplankton community, the relative role of Cladocera becomes more important with increasing eutrophication, the role of Copepoda less important: for example, in Lake Naroch, Myastro, and Batorin, Cladocera averaged 31, 46, and 67%, respectively, while Copepoda constituted 63, 47, and 30%. The increasing of relative abundance of cladocerans with increasing eutrophication has been pointed out by Canale and De Bernardi.[5]

RESULTS AND DISCUSSION

Exergy and structural exergy values calculated on the basis of average biomass concentrations of fish, macrobenthos, macrophytes, zooplankton, bacterioplankton, and phytoplankton are shown in TABLE 1.

While exergy doesn't show a clear relation with trophic state, structural exergy decreases from mesotrophic lake Naroch to highly eutrophic Batorin.

We found that structural exergy has a good linear correlation with two of the most important indicators of trophic state as phytoplankton biomass concentration (FIG. 1a), and Secchi depth ($R = 0.9747$). Structural exergy also shows a very good linear correlation with average depth of the lakes (FIG. 1b).

No correlation has been found between structural exergy and phosphorus and nitrogen concentrations.

TABLE 1. Environmental data and thermodynamic values for the lakes examined[a]

Lake Features	Naroch	Myastro	Batorin
Morphometrical data			
Surface area (km^2)	80.09	13.1	6.25
Average depth (m)	9	5.4	3
Maximum depth (m)	24.8	11.3	5.5
Volume ($m^3 \times 10^6$)	739	70	18.7
Water Retention time	High	High	High
Area of watershed (km^2)	279	133	27
Biological data			
Maximum depth of macrophyte distribution (m)	9.5	6	3.5
Percent lake area occupied by macrophyte	30	17	23
Seasonal maximum macrophyte biomass (g/l)	1.1E-02	4.0 E-03	3.5 E-03
Phytoplankton biomass (g/l)	8.2 E-04	3.8 E-03	9.0 E-03
Bacterioplankton biomass (g/l)	6.0 E-5	2.1 E-4	6.0 E-4
Zooplankton biomass (g/l)	8.2 E-05	3.4 E-04	6.1 E-04
Benthos biomass (g/l)	2.2 E-04	4.2 E-05	2.7 E-04
Fish biomass (g/l)	1.8 E-04	4.6 E-04	1.1 E-03
Chemical data			
D.O. (% saturation)	100		110–120
pH	8.0–8.6	8.0–8.6	8.0-8.6
Cl^- (mg/l)	2.9–3.5	2.9–3.5	2.9–3.5
P/PO_4^{3-} (mg/l)	0.01	0.01	0.01
N/NH_3 (mg/l)	0.06–0.1	0.06–0.1	0.06–0.10
N/NO_3^- (mg/l)	0.07	0.07	0.07
Transparency (Secchi depth, m)	5.0–6.0	0.9–2	0.3–0.6
Trophic classification	Mesotrophic	Eutrophic	Highly eutrophic
Exergy (J/l)	1.27 E9	8.2 E8	1.25 E9
Structural exergy (J/g)	1.06 E11	9.20 E10	8.25 E10

[a]Standing crops and chemical data are averaged for the vegetative season (May–October) based on values reported in Tables I–IX in Winberg *et al.*[4]

The inconsistent result found by applying the exergy as ecological indicator of the trophic state is a consequence of the high primary production of submerged macrophytes in Lake Naroch, which balances, in terms of exergy, the relatively low concentrations of fish, phytoplankton, bacterioplankton, and zooplankton biomasses. The relevance of macrophytes as ecological indicators for the trophic state has been emphasized by several investigators,[6,7] who include macrophytes in a multivariate approach to trophic classification.

The eutrophication process involves strong modifications of the relative abundance of species and of the trophic structure of an ecosystem, besides increasing the total biomass concentration. Moreover, eutrophication is accompanied by reduction of species diversity of algae[9] and, in general, of every trophic level.

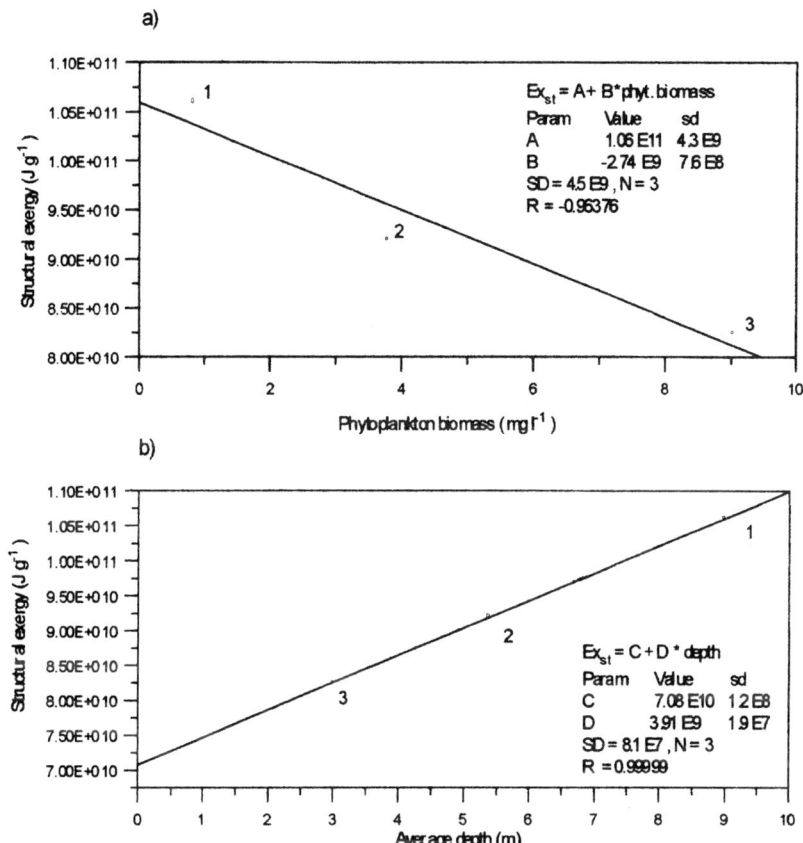

FIGURE 1. Linear regression between structural exergy and (**a**) phytoplankton concentration, (**b**) average depth of Lake Naroch (1), Lake Myastro (2), and Lake Batorin (3).

The primary producers, such as phytoplankton and submerged macrophytes, are competitors for limiting factors such as nutrients and, particularly, underwater irradiance which is the most important factor in determining the amount of lake area covered by submerged plants as well as the amount of plant biomass.[8] As a consequence of the increase in phytoplankton biomass, the eutrophication process is often accompanied by decreasing macrophytes. The amount of exergy or information embedded in the unity of phytoplankton biomass is 10 times lower than that carried by macrophytes.

This considered, exergy doesn't have an obvious relationship with the trophic state.

In a comparative study of 15 lakes, Christensen and Pauli[10] show that exergy has a positive correlation with the total biomass of phytoplankton and macrophytes. This result is confirmed by comparing Lakes Naroch, Myastro and Batorin: a good linear correlation ($R=0.982$) has been found between total exergy and total plant biomass.

These results are also consistent with the observations made by Xu,[11] who found that exergy is more sensitive to the system biomass than to trophic state, while structural exergy has a positive correlation with trophic state and biodiversity.

It should be pointed out that these conclusions come from a comparison of ecosystems homogeneous with reference to hydrological, climatic, geological, and hydrochemical conditions. In this context the lake morphology, with particular reference to water depth, seems to be one of the most important driving forces in determining the trophic structure of the lakes. This is not surprising because it is known that the long-term development and aging of lake ecosystems are accompanied by depth decrease as consequence of inflows of matter into the system.

CONCLUSIONS

The results of the exergetic analysis applied to a small but homogeneous set of lake ecosystems reveal that structural exergy may be considered a good ecological indicator of the developmental state of lake ecosystems; the decreasing Ex_{st} indicates the development of the lake towards a trophic structure in which the primary production is mostly due to less complex organisms such as phytoplankton rather than to macrophytes.

On the other hand, exergy doesn't show a clear relationship with trophic state, but it has positive correlation with the total primary production.

The study emphasises the role of lake morphology, with particular reference to water depth, as the driving force of lake development and aging and points out that the evaluation of trophic state by commonly measured water quality parameters as nutrient concentrations could be, in some cases, inadequate.

ACKNOWLEDGMENTS

The authors are very grateful to Professor S. E. Jørgensen and Dr. S. N. Nielsen of the Royal Danish School of Pharmacy for their valuable suggestions and support. Thanks also to Professor M. I. Taticchi and Dr. P. Pandolfi of the University of Perugia for their useful discussion.

REFERENCES

1. JORGENSEN, S.E. 1994. Review and comparison of goal functions in system ecology. VIE MILIEU **44**(1): 11–20.
2. JØRGENSEN, S.E. 1995. The application of ecological indicators to assess the ecological condition of a lake. Lake Reservoir Res. Manage. **1:** 177–182.
3. COSTANZA, R. 1992. Ecosystem Health. Columbia University Press. New York.
4. WINBERG, G., V.A. BABITSKY, S.I. GAVRILOV, G.V. GLADKY, I.S. ZAKHARENKOV, R.Z. KOVALEVSKAYA, T.M. MIKHEEVA, P.S. NEVYADOMSKAYA, A.P. OSTAPENYA, P.G. PETROVICH, J.S. POTAENKO & O. YAKUSHKO. 1972. Biological productivity of different types of lakes. Productivity problems of freshwater. Proceedings of the IBP-UNESCO. UNESCO. New York.
5. CANALE, C. & R. DE BERNARDI. 1997. Variazioni strutturali nel popolamento zooplanctonico del Lago Maggiore nel periodo 1948–1992 a seguito della sua evoluzione

trofica. Atti 12° Congresso A.I.O.L., Isola di Vulcano, September 18–21, 1996 (I vol.). (In Italian with English summary.)
6. CARLSON, R.E. 1977. A trophic state index for lakes. Limnol. Oceanogr. **22**(2): 361–369.
7. PORCELLA, D.B., S.A. PETERSON & D.P. LARSEN. 1980. Index to evaluate lakes restoration. J. Environ. Eng. Div. ASCE **106**(EE6): 1151–1169.
8. DUARTE, C.M. 1986. Littoral slope as a predictor of the maximum biomass of submerged macrophyte communities. Limnol. Oceanogr. **31**: 1072–1080.
9. WEIDERHOLM, T. 1980. Use of benthos in lake monitoring. J. Water Poll. Control. Fed. **52**: 537–557.
10. CHRISTENSEN, V. & D. PAULI. 1993. Trophic models of aquatic ecosystems. ICLAN, International Council for the Exploration of the Sea and Danida. Copenhagen.
11. XU, F.L. 1997. Exergy and structural exergy as ecological indicators for the development state of Lake Chaohu ecosystem. Ecol. Mod. **99**: 41–49.

Determination of the Italian Industrial Production Index from a Bayesian Point of View Using Electrical Consumption as Data

L. MAGNONI[a] AND M. MIROLLI

Department of Mathematics, University of Siena, via del Capitano, Siena, Italy 53100

The aim of this paper is to apply Bayesian methods to a field, economics, in which this approach is not very common. In particular, we examine the problem of determining the IIP (Italian industrial production) index by using the electrical consumption data. This relation between the IIP index and electrical consumption originates from a study by the Bank of Italy, and its aim is to determine in advance the above index with respect to ISTAT.[1] In this paper we follow this approach but use different statistical methods, that is, the Bayesian ones, and we present an algorithm that is an adaptation, from a Bayesian point of view, of the Kalman-Bucy filter.[2] The link between the IIP index and electrical consumption is a statistical model of stochastic filtering in which two unknown parameters are present. From a Bayesian point of view, we cannot estimate the "true" values of these parameters and then proceed with the determination of the IIP index, because, as they can assume a continuous infinity of values, a Bayesian statistician considers this estimation as a bet having null success probability. For this reason, we must consider the parameters as two random variables with given joint *prior* density and proceed, using electrical consumption data, to the determination of their joint *posterior* density throughout the application of the Bayes Theorem.[3]

The results reached are fully satisfactory even if we compare them with the values obtained by applying a frequentist statistical method often used for solving this kind of problems [4].

In this first approach we have considered a model which is an exemplification of the model of the Bank of Italy; in particular we consider:

$$x_t = \alpha \cdot x_{t-1} + w_t$$

$$y_t = \beta \cdot x_t + v_t$$

where x_t is the state variable, y_t is the observable variable, $\{v_t\}$ and $\{w_t\}$ are processes of standard white noise, which are independent, and are also supposed to be independent from x_0, which is considered a Gaussian distribution with a given mean and variance; α and β are unknown parameters with given *prior* density $f_0(\alpha, \beta)$. From a Bayesian point of view, this model cannot be considered a linear one because the parameters, being unknown, must be part of the state vector. Anyway we can consider it a conditional linear model; that is, if we knew the values of the random vari-

---

[a]Address for correspondence: magnoni@unisi.it (e-mail).

ables α and β, the model would be of a linear type. In order to take advantage of this conditional linearity, we propose a new adaptation of the Kalman-Bucy filter, that is:

$$E[x_t/y^t] = \iint_{\alpha\beta} m_t(\alpha, \beta, y^t) \cdot f(\alpha, \beta/y^t) d\alpha d\beta$$

where the function $m_t(\alpha, \beta, y^t)$ is given from the application of the Kalman-Bucy filter conditioned to the observations $y^t = (y_1, \ldots, y_t)$ and to α and β, and $f(\alpha, \beta/y^t)$ is the joint *posterior* density of the parameters that can be specified from the hypotheses on the model. In the same way we obtain the second moment of given the observations until time t[4]:

$$E[x_t^2/y^t] = \iint_{\alpha\beta} [P_t(\alpha, \beta) + m_t^2(\alpha, \beta, y^t)] \cdot f(\alpha, \beta/y^t) d\alpha d\beta$$

where $P_t(\alpha, \beta)$ is the function, representing the variance, given from the application of the Kalman-Bucy filter. Once we have solved the above integrals, we can evaluate the variance of the state process at time t, that is:

$$\text{Var}(x_t/y^t) = E[x_t^2/y^t] - E^2[x_t/y^t]$$

The application of the Kalman-Bucy filter to the model in exam, conditionally to the parameters α and β, gives:

$$m_t(\alpha, \beta, y^t) = E[x_t/\alpha, \beta, y^t] = \frac{\alpha \cdot m_{t-1}(\alpha, \beta, y^{t-1}) + \alpha^2 \cdot \beta \cdot y_t P_{t-1}(\alpha, \beta) + \beta \cdot y_t}{\alpha^2 \cdot \beta^2 \cdot P_{t-1}(\alpha, \beta) + \beta^2 + 1}$$

and

$$P_t(\alpha, \beta) = \text{Var}[x_t/\alpha, \beta, y^t] = \frac{\alpha^2 \cdot P_{t-1}(\alpha, \beta) + 1}{\alpha^2 \cdot \beta^2 \cdot P_{t-1}(\alpha, \beta) + \beta^2 + 1}$$

where the initial conditions $m_0(\alpha, \beta, y^0) = 0$ are and $P_0(\alpha, \beta, y^0) = 1$ (with y^0 we mean null observation). Finally we have to determine the *posterior* density of the parameters.

By applying the Bayes theorem and the independence hypotheses in the model we obtain:

$$f(\alpha, \beta/y^t) = f(\alpha, \beta/y_t, y^{t-1}) \propto f(y_t/\alpha, \beta, y^{t-1}) \cdot f(\alpha, \beta/y^{t-1}) \propto$$

$$\propto \prod_{i=1}^{t} f(y_i/\alpha, \beta, y^{i-1}) \cdot f_0(\alpha, \beta)$$

where $f_0(\alpha, \beta)$ is the parameters' *prior* density and the functions $f(y_i/\alpha, \beta, y^{i-1})$ are, for all $i \in \{1, 2, \ldots, t\}$, Gaussian having mean $\alpha \cdot \beta \cdot m_{i-1}(\alpha, \beta, y^{i-1})$ and variance $\alpha^2 \cdot \beta^2 \cdot P_{i-1}(\alpha, \beta) + \beta^2 + 1$.

We have given a recursive algorithm to determine the mean and the variance of the processes of interest. Results of analytical equivalence with another Bayesian recursive algorithm[5] for problems of combined filtering and parameter estimation, are

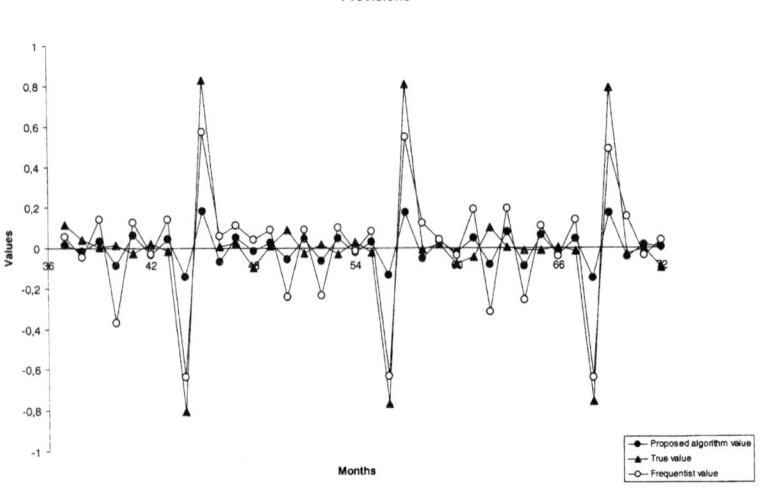

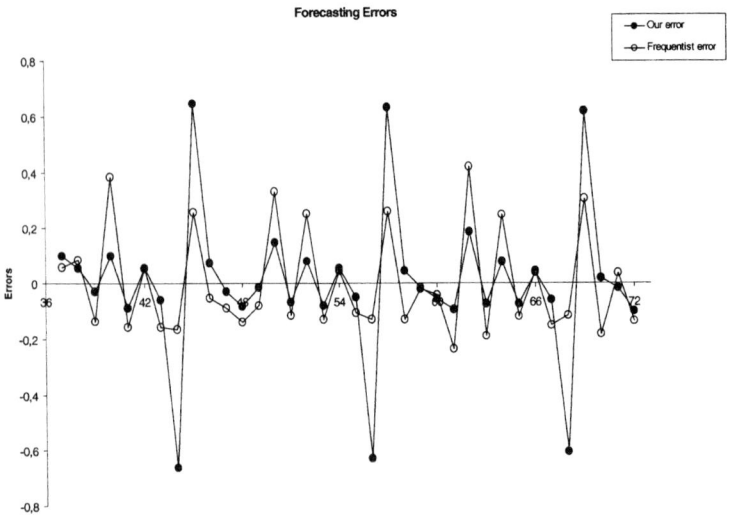

FIGURE 1. Comparison of our numerical results with those of frequentist statistics software.

proved.[4] In order to evaluate the integrals obtained, we have used a method of discretization of the prior density of the parameters because, for this method, robustness results are proved.[5]

We have applied the obtained theoretical results to the problem of determining the Italian industrial production index (IIPI) by using electrical consumption as data. This problem is one of stochastic filtering because ISTAT gives the IIPI's value with two months of delay. We have considered the noises present in the model to be equally distributed having Gaussian density with constant variance 0.01. The temporal measure unit is one month. The state variable x_t has been considered to be the natural logarithm of the quotient of the IIPI at time t with that of the month immediately before, $x_t = \ln(\text{IIPI}_t/\text{IIPI}_{t-1})$; in the same way, as observable variable y_t, we have considered the natural logarithm of the quotient of two consequential electrical consumption (EC) data, $y_t = \ln(\text{EC}_t/\text{EC}_{t-1})$. We have compared our numerical results with those obtained by using available frequentist statistics software (FIG. 1). We have used electrical data from January 1985 to January 1991; the first 36 for the first prevision, and then, adding one observation at time, we got the other previsions serially. We noted that the results obtained by the Bayesian method proposed are strongly penalized by the presence of data relative from August (a month in which factories are closed). In fact the mean square error of our method is 0.071, and with the frequentist approach we obtain 0.033; but if we consider the mean square error without the values for August we have 0.0059 while the applied frequentist method gives 0.031. This result is evident looking at the graph representing the forecasting errors. Finally we want to state our conviction that, with a slight change in the model, we will improve our results even for August data. We are now working in this direction.

REFERENCES

1. BODO, G. & L.F. SIGNORINI. 1987. Short-term forecasting of the industrial production index. Int. J. Forecast. **3**.
2. HARVEY, A.C. 1985. Forecasting, Structural Time Series Models and the Kalman Filter. Cambridge University Press. New York.
3. BILLINGSLEY, P. 1995. Probability and Measure. John Wiley and Sons. New York.
4. MAGNONI, L. 1998. Determinazione dell'Indice di Produzione Industriale mediante l'uso di metodi Bayesiani. Ph.D. thesis, Università di Siena, Siena, Italy.
5. RUNGGALDIER, W.J. & C. VISENTIN. 1990. Combined Filtering and Parameter Estimation: Approximations and Robustness. Automatica **26**: 401–404.

The Relational Project: A Goal for the Ecology of City Designing and Planning

VASCO MANCINI,[a,b] RITA MICARELLI,[c] AND GIORGIO PIZZIOLO[d]

[a]*Department of Cognitive Sciences, Florida State University, Tallahassee, Florida, USA*
[c]*Department of Architecture, Politecnico di Milan, Bovisa, Italy*
[d]*Department of Architecture, University of Florence, Florence, Italy*

RELATION, LANGUAGE, TIME

The concept of relation has, at all times, attracted the fascinated attention of scientists and philosophers from the time of ancient Greece in the West and even earlier in the East.

Relation has been described as bond, logos "of which men have no knowledge" (Eraclito), "hidden" bond, (*occult*, for Wittgenstein), "an unknown which produces unity" (Poincaré), "structure which connects" (mind according to Bateson), and essential element of becoming, present in all events. Of special interest has always been the relation of opposition. Parmenides saw reality as made up of oppositions of which one pole is positive and the other negative. One opposition in particular has attracted attention: lightness–heaviness, both in the real sense and the psychological sense. Far from establishing a void between two concepts, the opposition keeps them in mutual tension. It represents an epistemological continuum; and in the psyche of man, the poles switch places when either becomes unbearable. Oppositions do not convey a sense of structure but may be made to do so by relational mathematics and relational products.[1] The relation is the *quid* unique to the event taking place and may remain inside the facts following each event. In the exact, rigorous logico-mathematical language, "A binary relation R, between two sets x and y is a predicate with two empty slots. When an element x of X is put into the first slot and an element y of Y into the second, a grammatical statement follows which may be either true or false. If the statement is judged to be true, we write xRy.[2]" In this case the expression xRy and yRx tells us *a posteriori* what has happened, and whether the relation is from x to y or from y to x. From the direction of the relation, two results are produced (two facts), unequal and not interchangeable. A relational field constitutes the condition for the relation to manifest itself. From the event the possibility of the relation's manifestation, from the relation the fact, and so on, towards even more complex and tangled conditions of the taking place and evolution of our world; xRy and yRx represent the communication of the facts as they have occurred, and very economically and by convention according to Wittgenstein represent the two facts with appropriate propositional signs.

---

[b]Current address for correspondence: Piazza S. Stefano, Firenze 50122, Italy; 39-055-214428 (voice); mancini@dado.it or pizziolo@unifi.it (e-mail).

"One name stands for one thing and another for another thing, and they are connected together. And so the whole, like a living picture, presents the atomic fact." [4.0311]. "The totality of true propositions is the total of natural science (or the totality of natural sciences)."[3] [4.11]

Language follows fact. Its representation is by a process of progressive formation homologous to the real formation of the fact.

What takes place (facts) is projected into the proposition, which follows the events' becoming and the complex evolution of relations. As Langer reminds us, "words have a linear, discrete, successive order; they are strung one after another like beads on a rosary, and therefore we are required to string out our ideas even though their objects rest one within the other. This property of verbal symbolism is known as discursiveness. For this reason, only thoughts that can be arranged in this peculiar order can be spoken at all. Any idea that does not lend itself to this projection is 'ineffable,' incommunicable by means of words. That is why the laws of reasoning, our clearest formulation of exact expression, are sometimes known as the laws of discursive thought."[4]

Language is the outcome of a process triggered by the interaction between mind and experience during which the perceptions and conventional representations that make them communicable, starting from the connection name-thing more to propositional signs and to the articulation of more and more complex propositions. Whereas the processes of perception are inaccessible and not communicable directly, being primary relational processes,[5] the languages—being the representation of such processes—make possible the comunication of the facts that have occurred, which constitute our world. The relation is, together with time, inside both processes, the primary process of the evolutionary chain of new facts and the other, "derived" from the logical chains of their representation.

"By no means do all of our thinking compartments consist of propositions, and of those that do, or can by translation be said 'in effect' to do so, scarcely any operate with anything like formal logic. Even most of our conscious thought deals with images in fluid and intuitive ways. It is of course very difficult to apprehend the structure of our unconscious thought, but from what is known, it obeys a 'different logic'; its identification of opposites, ... is alone sufficient to cut it off sharply from any of the standard formal logical modes."[6] As Langer has shown, there are two main modes of handling symbols: the *discursive* and the *presentational*. And the function of symbols is to let us develop a characteristic attitude toward objects *in absenthia*, which is called "thinking of" or "referring to" what is not here. Symbols are the materials of thought, and their interweaving and other manipulations is human reason.

PLAY AND GAME

The elusive nature of the relation remains so in both cases, but in the primary creative process, the grand play which is the very evolution of the world, is beyond us, beyond our capabilities of control and prediction, whereas in the development of languages and the conventions for communication, the relation can be appropriated and made to enter other language games and modes of behavior (games, forever played by humans and many animals). Its nature does not change but its role does, since in this case, embodied, articulated, and ordered in the propositions of the language, it becomes the frame and ordering structure of the propositions. Therefore the relation plays two roles at the same time, the primary one of the play that creates the world, and the derived one of the language games. The purpose of language games is to discover the semantic relations between language and reality. From the first new facts always spring forth, from the second come rules and conventions (games with rules). One role of play is freely creative and unpredictable. The other provides internal structure and bonds from which come all operational rules for the articulation (operational) of language. The relation is "captured" in two ways in the unfolding of evolutive processes: in facts and in language. But its capture is never assured, never complete, and cannot force its nature nor limit its manifestations. The world, as Wittgenstein says, keeps unfolding and keeps being all that takes place, beyond our will. It is impossible to take the facts of the world's evolution apart and reassemble them in a mechanistic way; it is impossible to separate relation from time. All our dreams of control and domination of evolution are doomed to failure.

The distinction between play and game, which surfaced in the 1950s in the intellectual debates among many scholars of diverse training and disciplines, has brought about the separation between two contrasting evaluations regarding the two modalities of games. We tend, at great risk, to confuse the two, and to emphasize the game: the rules, the control, the predictability, man's domination of the world. This message, of course, comes from the more powerful and is addressed to the poorer, but perhaps freer, inhabitants of this planet.

Few are aware of this separation and distinction which is dramatically widening. Relations, once "captured" are progressively imprisoned in processes of "controlled" transformation of the world, as well as imprisoned in new "official languages" of scientific and financial power groups. The very development of such languages is controlled by such groups, wanting to divert the processes of the evolution of life by reducing them to pure operational matters, carried out with "no surprises," no creative jump in fields as diverse as living organisms, food, the environment, the landscape, esthetic taste whether individual or social.

This is an arrogant attitude which denies and turns logical thought into a banality, as Wittgenstein tells us: "How many kinds of sentence are there? ... There are *countless* kinds: countless different kinds of use to what we call 'symbols,' 'words,' 'sentences.' And this multiplicity is not something fixed, given once and for all: but new types of language, new language games, as we may say, come into existence and others become obsolete and get forgotten ... . Here the term "language-*game*" is meant to bring into prominence the fact that the speaking of language is part of an activity, or a form of life ... . How did we *learn* the meaning of this word ('good' for instance)? From what sort of examples? In what language games? Then it will be easier for us to see that the word must have a family of meanings."[7]

Because of all this we believe it to be urgent and important to keep open the horizons and continue to develop a relational epistemology (many times invoked in the course of this century but still immature). We can, in other words, rejoin our games, between play and game, mindful that the refusal to choose forever one or the other is the only way to live inside the temporality of the phenomena of becoming, the flux of the ever more complex intertwining of play and game. Relational epistemology is our common ground to dwell with both: with play, the "game" forever new, whose rules are never "frozen" and creativity abounds; and with game, "play" with fixed, secure rules from which stem many outcomes, always predictably finite, in spite of their illusory variety.

Such epistemology is itself a process going over the many itineraries of its formation (beginning with classical science, philosophy, and the experimental method) and at the same time leads us through novel itineraries accessible only on condition that we *throw away the ladder that brought us this far ...* as Wittensgtein suggests at the end of his *Tractatus*.[7]

THE LANGUAGE OF THE CITY

In the multi-cultultural, multi-ratial reality of the city today, there are multiple truths, which call into question the universality of true propositions. Instead of pursuing a global truth, it is more advantageous to look for "local truth." Bandler and Mancini proposed a definition of truth that reads: "truth is not an attribute of propositions, but rather a *relation* between a proposition and a database, in which 'database' is used here merely as a convenient abbreviation for an actual finite list of propositions." This family of "truths" suggests some kind of interrelatedness and invites us to make explicit what the relations between them might be. As for the "language games" in the architetctural language the relation between "language," activity, or "form of life" holds true, to which we need to add, for a compete definition of meaning, the relevant landscape.

From the "art of naming" to the unbreakable bond between name and thing, from language games from which derives the language and propositions of logic, to the theory of sets, of art itself, which renders the invisible visible of Paul Klee, to the "New Music" of Webern to the *Tractatus* of Wittgenstein, from the time of the last century's closing onward, the steps of relational epistemology are our steps. Sometimes the steps have been caught in the conflict between the human will to totally control the world's events (playing the game) and the creative force, free from control, driving "evolutive art," which can, today, consciously and totally appropriate art and science, and all those active in them.

No one can renounce this art. As designers and town planners we share with artists and scientists the wish to transform our activity, which statutory powers and other power structures in society would want risk-free, obedient and simply operational, into a creative process of "relational art," of "choral" experimentation, between people, social groups, and their living environment. The nature of this experimentation is the nature of play, free and unpredictable.

This task, intimately experimental in nature, addresses itself to individuals and social groups, and invites them to participate in this play of choral creativity through processes of perception and transformation whose epistemological premises are shared. It is worthwhile to introduce here some clarifying reflections upon this subject, which in the introduction has been lighlty touched upon.

That which in the introduction has been simplified, in presenting the elementary manifestation of relation between x and y, now shows itself in a new, vertiginous complexity of process, where the relation, originally a linear chain, intertwines itself in ever newer and more complex configurations, giving rise to *form* and *formation* at the same time. From the most elementary event conceivable (the collision of two particles) to the formation of such tangles, there occur transformations and events whose relational temporal *quid* of formation (gestalt) will never be captured by the instruments of traditional science (measurement, quantity) but as a becoming can be described, represented, and recognized scientificly, by means of the new relational epistemology.

The intriguing and apparently contradictory characteristics of relation (free and imprisoned, changing and held in stable form), while annoying those scientists wanting to get hold of the relational *quid* and "truth" with it, are the key elements for the construction of languages and communication making man's evolution possible. These languages transmit and communicate, establish ties and linkages in order to share the manner of transmission, the codes, the conventions. Nevertheless, the knowledge and use of these instruments, their continual evolution in the growing multiplicity of language formation does not open the door to the ultimate secret of the connection, a secret that cannot be cut open, nor disassembled, quantified and measured.

In order to make clear the necessity of looking in depth into the kind of relationality that leads beyond the consolidated rules of the world of forms ("I call series of forms the ordered series of relations," says Wittgenstein) and of games, and in order to come back to play, it needs be emphasized again the necessity in our projects of going beyond the deterministic model (the project as problem solving, starting from known data) and the systemic model (the project pursuing complexity by the progressive acquisition of order). Here we are not concerned with the kind of relationality deriving from deductive and known operations, but the one of exploration of new and unknown possibilities, of making up the rules as we go along, mindful that in being and participating in processes of transformation, we transform ourselves. It is true that we can verify in a given time interval the transformation that has occurred, but we shall never know what took place *while the process was unfolding*. The various states of the process (initial and successive), in a way represent the game (with rules), but the most complex and creative states of the transformation take place in the time of while, beyond the known rules and predictablity), and not in a mechanistic way but with a hidden dynamics of their own, relational not operational.

In this framework the concept of difference that Bateson put forth in the 1970s confirms what the thought of Poincaré and Wittgenstein had illuminated: quality and qualitative differences (beyond codification and measurement with known instruments) make up the *quid* connecting mind and nature and to the interactions of such differences Bateson ascribed the *novelty* of the new informatiom produced. It can only rise out of the qualitative and relational dynamics of *play*, it can only manifest itself in the time of *while*, in the secret locus of its creativity. Poincaré, Wittgenstein, and Bateson time and again illuminate this double face of the relation, pointing out that the co-presence expresses an unstable equilibrium, dynamic and fragile, that is the very equilibrium of evolution, always in danger of being jeopardized.

REFERENCES

1. MANCINI, V. 1998. Fuzzy methodologies in the social sciences. *In* New Trends in Fuzzy Sysyems. D. Mancini, M. Squillante & A.Ventre, Eds. World Scientific Publishing Co. Singapore.

2. BANDLER, W. & L.J. KOHOUT. 1986. A Survey of Fuzzy Relational Products in their Applicability to Medicine and Clinical Psychology. Knowledge Representation in Medicine and Clinical Behavioral Science. Abacus Press. Cambridge, MA.
3. WITTGENSTEIN, L. 1961. Tractatus Logicus Philosophicus. Bilingual edition. Routledege and Kegan Paul. London.
4. LANGER, S.K. 1942. Philosophy in a New Key: A Study in the Symbolism of Reason, Rite, and Art. Harvard University Press. Cambridge, MA.
5. FREUD, S. 1957. The Interpretation of Dreams. Standard Edition of the Complete Psychological Works of Sigmund Freud, Vol. IV. Hogarth Press. London.
6. MANCINI, V. & W. BANDLER. 1988. A Database Theory of Truth: Fuzzy Sets and Systems. p. 25.
7. WITTGENSTEIN, L. 1967. Philosophical Investigations. Oxford University Press. Oxford, England. As quoted in HARVEY, D. 1996. Justice, Nature and the Geography of Difference. Blackwell Publishers Ltd. Cambridge, MA.
8. BATESON, G. 1984. Mind and Nature. Italian edition. Adelphi. Milano, Italy.

Correlation between Greenhouse Effect and Exceptionally High Tides in Venice

N. MARCHETTINI, C. MOCENNI, V. NICCOLUCCI, AND E. TIEZZI

Department of Chemical and Biosystems Sciences, University of Siena, Pian dei Mantellini, 44-53100 Siena, Italy

INTRODUCTION

Global models are relevant tools for designing future scenarios for the greenhouse effect and global warming. But it is also important, over a shorter period, to underline the relationships between the greenhouse effect and local phenomena, as this relation is crucial to achieving sustainable development.

The natural equilibrium of the earth is only maintained if the energy balance is preserved. The atmosphere must radiate into space as much energy as the earth receives from the sun. This can only happen by means of infrared radiation. The band of wavelengths re-emitted by the earth is different from that of incoming sunlight. Atmospheric gases (CO_2, H_2O, O_2, CH_4, N_2O) absorb part of the radiation emitted by the earth, including the 4–100 mm band. Carbon dioxide does not adsorb incoming sunlight but absorbs infrared radiation emitted by the Earth in the 12–17 mm band. This is the essence of the greenhouse effect and explains why the carbon dioxide cycle is the cornerstone of an energy policy compatible with the environment and sustainable development.

The study of the carbon cycle of the biosphere is the study of the interactions between living organisms and their environment. Not only the atmosphere, but also the oceans, water and land organisms, organic material (i.e., the biomass, especially plants), and organic fossil material (natural gas, oil, coal) are involved in this cycle. Although carbon dioxide is only 0.03% of the atmosphere, it is nevertheless essential for life. Besides regulating the temperature of the earth's surface (and consequently climate), it is the basic material from which plants obtain carbon and convert it, by photosynthesis, into nutrients.

Two human activities, in particular, contribute dramatically to the increase in atmospheric carbon dioxide: the combustion of fossil fuels and deforestation, especially of tropical rainforest.[1] The former constitutes active production of greenhouse gases, formed *ex novo* from the oxidation of the carbon compounds of which fossil fuels consist. The latter is a passive action, which determines a drastic reduction in photosynthesis by reducing the plant biomass. It is well known that each of these factors has contributed to the climate variability of the past 400 years, with greenhouse gases emerging as the dominant force during the twentieth century.[2]

In this paper we examine the impact on the city of Venice of the high-water phenomenon, related to its increasing frequency and intensity observed in the present century. We analyze the future trends in the business-as-usual hypothesis, and we examine the influence of global climate change on this phenomenon.

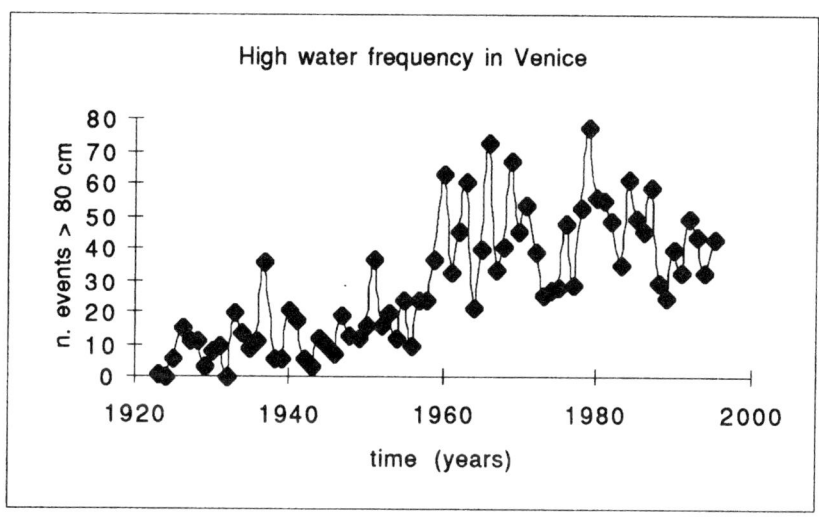

FIGURE 1. Venice city: high water frequency data greater than 80 cm from 1923 to 1995.

THE HIGH WATER EVENTS IN VENICE AND THEIR CORRELATION WITH CLIMATE CHANGE

The high-water phenomenon in Venice is due to the cumulative astronomical and meteorological factors. The astronomic tide, due to the gravitational force of the moon and the sun, is predictable. By contrast, the action of strong southeast winds which push the sea water to the deadend of the Adriatic Sea where Venice is located, is unpredictable; the difference between measured level and astronomic tide is called the *storm surge*.[3,4] Strong SE winds (Beaufort scale > 7, average sustained speed 30 knots) are characteristic—on average and in the considered geographical area—of the eastern sector (warm front) of the depression systems that transit over the Adriatic Sea.

FIGURE 1 shows that the yearly number of high water events is increased in this century.

This is obviously correlated with the increase in the sea level. The local increase in the sea level, is not significantly different from the Mediterranean average and from the observed global rise.[5] This phenomenon is due to cumulative effects of subsidence and eustatism in the Venice area[6]; the subsidence (2–4 cm per century) is the sum of (1) long-term geological subsidence, (2) subsidence due to the weight of the buildings, and (3) subsidence due to water extraction. Eustatism (11 cm per century) is the phenomenon of long-term sea-level change responding to geological changes: the local sea-level rise is thus estimated to be from 13 to 15 cm per century. Nevertheless, the measurements for Venice from 1900 to 1995 show an increment of 23 cm.

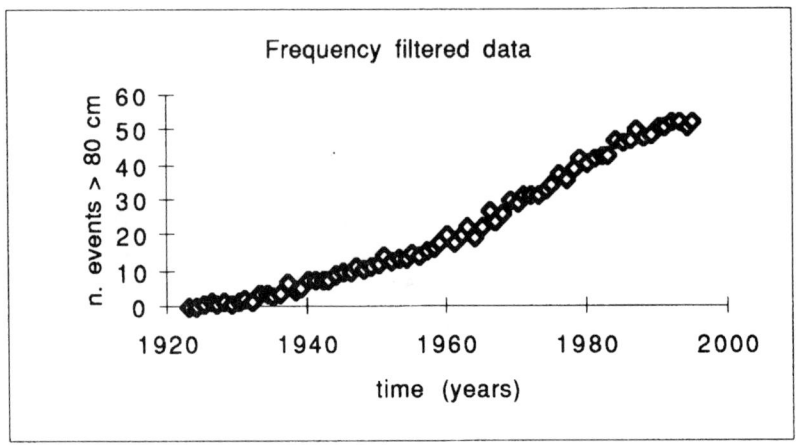

FIGURE 2. FFT-filtered high water data with respect to the sun and moon cycles and other periodic phenomena.

On the basis of the IPCC estimates, this is most probably an effect of global climate change, which is seen both in the rise in temperature and in the net increase of the yearly number and intensity of the SE storms in the Adriatic Sea. In fact, it is known that the yearly frequency of the intense meteorological perturbations that swept through Europe at the latitude of Venice has increased appreciably in the present century. For this reason we analyze the correlation between the annual frequency data of high water events in the city of Venice[7,8] and the global concentrations and emissions of CO_2.

CORRELATION ANALYSIS

Using a classical Fourier analysis, we filter out the oscillations due to regular cyclic events, like the known solar activity periods (7, 11, 22 years, etc.) and any other cyclic, nonchaotic components. FIGURE 2 shows the FFT-filtered time series of the observed high-water frequency which contains only the trend plus the chaotic component, the former being the high-water phenomenon itself.

RESULTS

There is a strong correlation between the increase of the FFT-filtered high water events in the Venice lagoon and the CO_2 concentrations and emissions. In order to perform this analysis, we use the classical definition of correlation coefficient $C_{x,y}$.[9]

The analysis shows a strong correlation between filtered data of high water events and both global concentrations and emissions of CO_2, giving correlation coefficient values of 0.98 for CO_2 concentrations data and of 0.986 for CO_2 emission data (see FIG. 3).

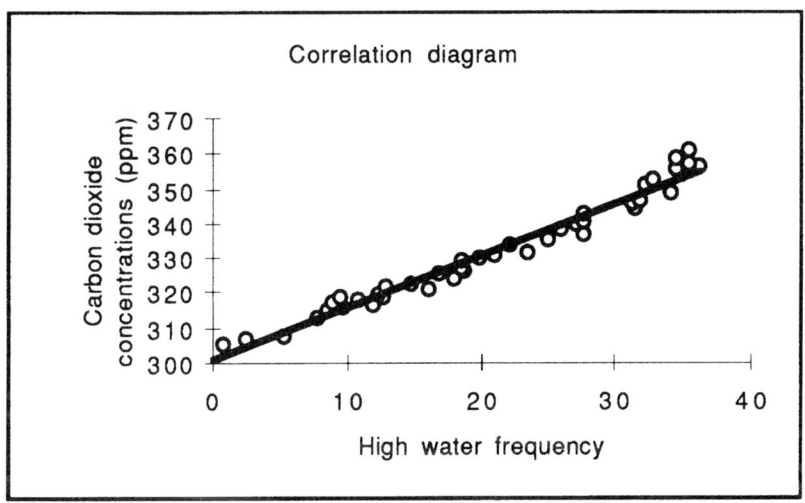

FIGURE 3. Correlation diagram of carbon dioxide global concentrations data versus exceptionally high tide (>80 cm) data in Venice in the present century. The line is the regression line of the two time series data.

CONCLUSION

The time series of exceptionally high tides in Venice apparently contain three components. The first is the well-known gravitational influence of the moon and the sun, while the second is the unpredictable action of the sirocco SE wind. The third cause of the increase of the high water frequency in Venice clearly emerges in observing the discrepancy between the sea level rise produced by the sum of the above phenomena (13–15 cm) and the effective measurements of this increase (23 cm) in the present century.[7] The real reason for this difference is not at all clear, but this paper provides proof of the great influence of global climate changes on this last component of the phenomenon. We have chosen to perform the analysis with global carbon dioxide time series because they provide the more representative evidence of the greenhouse effect. Moreover, the correlation between these data and the rise in temperature and extreme phenomena is well known.[10] The dramatic evidence of the above correlation has to be underlined, because of the outstanding artistic and natural importance of Venice.

ACKNOWLEDGMENT

This paper has been supported by Cofin. MURST 97 CFSIB grant.

REFERENCES

1. MYERS, N. 1989. The greenhouse effect: a tropical forestry response. Biomass **18:** 73–78.

2. MANN, M.E. *et al.* 1998. Global-scale temperature patterns and climate forcing over the past six centuries. Nature **392:** 779–787.
3. CAMUFFO, D. 1993. Analysis of the sea surges at Venice from A.D. 782 to 1990. Theoret. Appl. Climatol. **47**(1–2): 1–14.
4. CECCONI, G. *et al.* 1998. Climate record of storm surges in Venice. Proc. RIBAMOD. H.R. Wallingford, Ed. Wallingford, England. In press.
5. MARZOCCHI, W. & F. MULARGIA. 1996. Scale analysis to sort the different causes of mean sea-level changes: an application to the northern Adriatic Sea. Geophys. Res. Lett. **23**(10): 1119–1122.
6. RUSCONI, A. 1983. Il comune marino a Venezia. Ministero dei Lavori Pubblici, Ufficio Idrografico del Magistrato alle Acque, Rapporto no. 157.
7. COMUNE DI VENEZIA. 1996. Definizione dei rischi derivanti dalle acque alte per l'abitato di Venezia.Venezia, Italy.
8. SIA. 1997. Studio di impatto ambientale del progetto di massima per gli interventi alle bocche lagunari. Magistrato alle Acque e Consorzio Venezia Nuova.
9. SPIEGEL, M.R. 1973. Statistica, Etas libri. Milano, Italy.
10. IPCC. 1996. Intergovernmental Panel on Climate Change. Cambridge University Press. New York.

Ecodepuration Performances of a Small-Scale Experimental Constructed Wetland System Treating and Recycling Intensive Aquaculture Wastewater

S. PANELLA,[a,b] I. CIGNINI,[a] M. BATTILOTTI,[c] M. FALCUCCI,[a] V. HULL,[a] N. MILONE,[c] M. MONFRINOTTI,[c] G.A. MULAS,[a] G. PIPORNETTI,[c] L. TANCIONI,[c] AND S. CATAUDELLA[c]

[a]*Laboratorio Centrale di Idrobiologia di Roma, Ministero per le Politiche Agricole, Viale del Caravaggio, 107, 00147 Rome, Italy*
[c]*Dipartimento di Biologia, Università degli Studi di Roma "Tor Vergata," Via della Ricerca Scientifica, 00133 Rome, Italy*

INTRODUCTION

Italian aquaculture has recorded great development. Nevertheless, availability of new productive opportunities as well as suitable sites and adequate technologies for minimizing wastewater generation and maximizing water use are likely to be very important constraints for further development.

These aspects have been investigated in a project titled "Innovating models for valuable fish species rearing in rural contexts" (National Aquaculture Development Project 1994–1997), supported by a grant from the Italian Ministry for Agricultural Policies. In this paper some preliminary results of the project in progress are presented.

EXPERIMENTAL MODEL

An experimental aquaculture self-depurating system has been set up at the Laboratorio di Ecologia Sperimentale ed Acquacoltura di Roma (Università di Roma — Tor Vergata). The design of the system consists of[2,5]: (a) a hatchery; (b) a set of 10 tanks (64 m^3) for fish intensive rearing; (c) a ditch (100 m long and 1 m wide) with a gravel bed planted with *Phragmites australis* and *Typha latifolia*; (d) a 20-m^3 sedimentation tank connected with (e) a ditch (60 m long and 1.6 m wide) planted with *P. australis*, *Eichornia crassipes*, *and Scoenoplectus lacustris*; and (f) an artificial pond of 3000 m^3 (depth from 0.2 to 2 m) having a 200-m^2 platform lightly submerged inside. The pond has been colonized by aquatic macrophytes (*T. latifolia, P. australis, Scirpus lacustris, Myriophillum spicatum, Potamogeton crispus*) and microphytes (*Pediastrum tetras, Scenedesmus quadricauda, Gonphonema* sp.), by different zooplancton (*Brachionus calyciflorus, Keratella quadrata*) and zoobenthos

---

[b]Address for correspondence: 39-06-5140296 (voice and fax); panella@mailbis.inea.it (e-mail).

TABLE 1. Loading removal efficacy (%), reaction constant rates (K), and r-coefficients of predicted versus observed wetland outflowing concentrations (C_0)

	Loading Inflow[a]	Loading Outflow	Removal (%)	K (mean) (day^{-1})	K (20°C) (day^{-1}, a and b coeff.)	r coeff.
P-total	143	94	34	0.104		0.833[b]
P-inorg. diss.	39	16	58	0.330		0.709[b]
N-total	4325	2720	37		0.119 (2.7E+07; −5636)	0.907[b]
Σ N–inorg.	3522	2569	27		0.131 (5.4E+06; −6137)	0.899[b]
N-org.	767	280	63	0.490		0.604[c]
N-NO3	3424	2502	27		0.131 (3.5E+06; −5010)	0.900[b]
N-NH4	80	47	41		0.198 (3.4E+05; −4206)	0.686[b]
SS-org.	6909	5907	14	0.055		0.976[b]
C-org. total	12214	10541	14	0.019		0.994[b]
C-org. part.	2955	2246	24	0.139		0.941[b]
B.O.D. 5 (20°C)	4513	3002	33	0.140		0.920[b]

[a]Values in g/day.
[b]Level of significance > 99.9%.
[c]Level of significance > 95.0%.

(*Lymnaea auricolaria, Bitinia* sp., *Chironomus* sp.) species. Furthermore, two species of Mugilids (*Mugil cephalus, Liza ramada*) and a few Cyprinids (*Alburnus alburnus arborella, Scardinius erythrophtalmus*) were introduced.

The whole system is supplied at the hatchery inlet with 2 l/sec of fresh well water; The hatchery outlet water is conveyed to the ditch–pond depurating system; the outlet water of the pond is pumped (10 l/sec) into the rearing tanks, whose outlet is then collected and recycled in the wetland system previous mixing with the hatchery effluent.

For intensive rearing a 250-kg stock of sturgeon (*Acipenser naccarii*) and a 300-kg stock of rainbow trout (*Oncorhyncus mykiss*) were introduced in the rearing tanks in late 1995 and early 1998, respectively. Since 1995 the whole system has been periodically monitored for chemical, physical, trophic, and ecological parameters, in order to control the environmental evolution and the depurative performance, as well as fish growth efficiency. As a result of the pond management 1.5 and 2.5 tons of weeds were harvested and stocked as compost in 1996 and 1997, respectively. Fish biomass of 1.5 tons in the pond and 1.3 tons in the rearing tanks were estimated in early 1998.

RESULTS

Data analysis and remarks here developed are specifically turned to three questions regarding the behavior of the wetland system: (a) efficiency analysis of self-depurative performance, (b) kinetic rates and constants of waste removal processes, and (c) predictive estimate of the system performances.

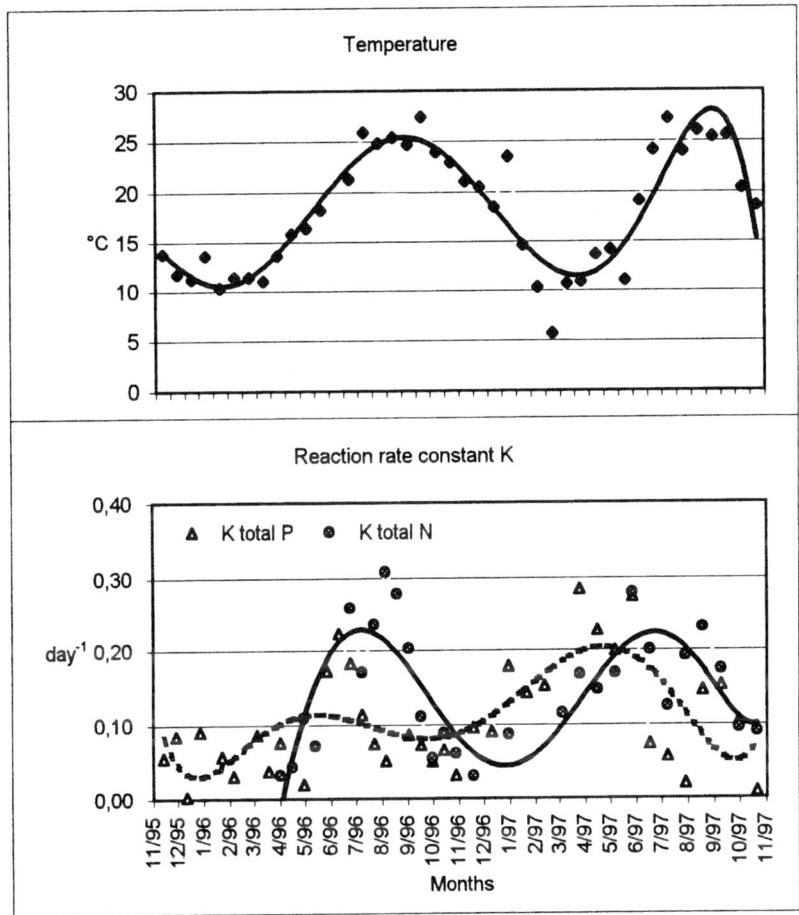

FIGURE 1. Seasonal trend of the reaction rate constant (K) of total P (*dashed line*) and total N (*continuous line*). Note the strict correlation of last K with temperature.

Removal Efficiency

Analytical characteristics of the influent and effluent water of the wetland system have been monitored every 15 or 21 days for over two years. Mass removal efficiency performed by the ditches and pond as a whole has been estimated on the basis of the concentrations removed daily (g/mc) and corresponding hydraulic loading on the system (900–1000 m^3/day).

Removal has been greatest (TABLE 1) for total organic N (63%), inorganic dissolved P (58%), and N-NH$_4$ (41%), while minor, but fair, performances resulted for total N (37%), total P (34%), BOD$_{-5}$ (33%), N-NO$_3$ (27%), and particulate organic C (24%). Negligible (14%) removal of suspended organic solids and total organic C was observed.

REACTION KINETICS

In this first step of data treatment, a first-order kinetic for no mixed fluxing reactors in steady-state conditions has been assumed.[3] Reaction constants (K) of removal processes of P, N, and organic components in the wetland system have been calculated by the basic relationship: $K = 1/t \times \ln C_i/C_0$, where K = reaction rate constant (days^{-1}); t = hydraulic retention time (days); and C_i and C_0 = concentration (g/l) of the specific compounds in the inlet and outlet water of the wetland system, respectively. On the basis of the wetland volumes and influent–effluent hydraulic flows data, the hydraulic retention times of the system on the whole were estimated to be in the range of 2.8–3.0 days.

Negligible or no temperature-dependent variations were observed in removal processes of organic components and phosphates. Conversely, removal rates of total N, total inorganic N, N-NO$_3$, and N-NH$_4$ appeared to increase significantly with temperature (FIG. 1). In these cases K values at different temperatures have been calculated using the van't Hoff-Arrhenius equation as the generic form: $K = a \times \exp(-b/T)$; where T = absolute temperature (°K). Coefficients a and b were esteemed[4] using an exponential regression of the already computed values of K (i.e., as $1/t \times \ln C_i/C_0$) versus $1/T$. K values at 20°C have also been determined (TABLE 1).

PREDICTIVE ESTIMATES

C_i, t, and K being known, predictive estimates of C_0 and removal efficacy of the wetland system can be performed by means of the basic equation of the form $C_0 = C_i/\exp(Kt)$. In the case of temperature-dependent processes, C_0 values have been computed[1] by replacing the K constant in the above equation with the corresponding expression of van't Hoff-Arrhenius—i.e., $a \times \exp(b/T)$—having estimated the a and b coefficients as previously described. In the other cases, the average of the K values calculated in relation to the observed experimental C_i, C_0, and t was replaced in the basic equation to obtain a predictive estimate of C_0. A close fit with high correlation coefficients between predicted and observed C_0 values have been obtained for the different water components investigated (TABLE 1).

CONCLUSIONS

Preliminary results obtained by the continuing project show good self-depurative efficiency on the part of the experimental, constructed wetland and demonstrate the suitability of water recycling and re-use for fish rearing. Further studies are in progress in order to assess the economic efficiency of the system as well as its positive externality in improving habitat restoration and eco-compatible aquaculture models in rural districts.

REFERENCES

1. BAVOUR, H.J. *et al.* 1989. Performance of solids-matrix wetland systems viewed as fixed-film bioreactors. *In* Constructed Wetlands for Wastewater Treatment, Municipal, Industrial and Agricultural. D.A. Hammer, Ed.: 645–656. Lewis Publishers. Chelsea, MI.
2. CICCOTTI, E. *et al.* 1996. Un modello di acquacoltura eco-compatibile: struttura e dinamica di un ecosistema acquatico artificiale per il lagunaggio ed il riuso di acque reflue. In Atti del VII Congresso Nazionale della Societá Italiana di Ecologia. September 11–14. Napoli, Italy. 831 – 834.
3. MOORE, W.J. 1967. Chimica Fisica. Piccin Publishers. Padova, Italy. pp. 270–272.
4. RESCIGNO, A. & G. SEGRE. 1961. La Cinetica dei Farmaci e dei Traccianti Radioattivi. Boringhieri Publishers. Torino, Italy. pp. 50–52.
5. TANCIONI, L. *et al.* 1998. Phytodepuration techniques in a small scale aquaculture system. *In* 2° International Conference, Advanced Wastewater Treatment, Recycling and Reuse. September 14–16. Milano, Italy. In press.

Fish Exclusion and PCB Bioaccumulation in Bear Creek, East Tennessee

M. PANZIERI[a,b] AND T. G. HALLAM[c]

[a]*Department of Chemical and Biosystems Sciences, University of Siena, Pian dei Mantellini 44, 53100 Siena, Italy*
[c]*The Institute for Environmental Modeling, Department of Ecology and Evolutionary Biology, University of Tennessee, Knoxville, Tennessee 37996, USA*

INTRODUCTION

Bear Creek, a small stream that flows approximately 12.9 km inside the Oak Ridge Reservation (ORR), Tennessee (USA) is contaminated by toxic chemicals released over the past 50 years. A temporal schedule of water quality, the main chemical toxicants, and fish bioaccumulation data has been determined for a suite of sites along the streams by Southorth *et al.*[1] Bioaccumulation for the fish species rock bass (*Amblopites rupestris*) and ambient chemical concentration measures were available only for the year 1993.[1] To explore effects of chemicals, a modeling approach is used to extend the investigation to other fish species and environmental conditions. The goal of the modeling is to examine polychlorobiphenil (PCBs) environmental concentration and bioaccumulation by fishes and relate it to observed fish presence along the stream. Because of availability of appropriate parameter values, bioaccumulation of PCBs in white sucker (*Catostomus commersoni*) and in bluegill (*Lepomis macrochirus*) were studied using an exposure simulation model developed by Barber *et al.*[2] the fish data for Bear Creek in 1993 are given in TABLE 1.

METHODOLOGY

We first assume that the toxicity of chemicals is independent, which implies that the total effects of the PCBs can be regarded as an additive function of the two concentrations. Then, we calculate bioaccumulation in fish of two PCB congeners, PCB 1254 and 1260, by an exposure model FGETS (food and gill exchange of toxic substances).

The PCB's *n*-octanol/water partition coefficient (Kow) is used to quantify chemical partitioning, while molecular volume is used to quantify aqueous diffusivity. Parameterization of the model for a particular fish species was made from the FGETS data base in the case of white sucker and by developing a set of parameters from the literature for bluegill. Because PCB 1254 and 1260 mixtures have different degrees of chlorination, parameterization of FGETS for these chemicals was obtained using Kow given by Hawker and Connell[3] for the polychlorobiphenyl present in the high-

---

[b]Address for correspondence: 39-0577-232088 (voice); 39-0577-232004 (fax); panzieri@envichem.chim.unisi.it (e-mail).

TABLE 1. Fish numbers by species along Bear Creek between 1988 and 1996[a]

Species Name	Stations (km)						
(common; *scientific*)	0.6	3.3	7.8	9.4	9.9	11	11.8	12.3
Bluegill; *Lepomis macrochirus*	15							
Rock bass; *Amblopites rupestris*	123	74						
White sucker; *Catostomus commersoni*	1	3	195	38	68			
White sucker 1993 distribution		2	69	12	33			
Total number of fish species[b]	20	16	8	7	9	6	4	4

[a]Data from Southworth *et al.*[1] Samples were made in spring and fall of each year. The second set of data is for the year 1993 only.
[b]Fish biodiversity in Bear Creek for the years 1988 to 1996.

est percentage in the mixture (Kow of pentachlorobyphenil for PCB 1254 and Kow of heptachlorobyphenil for PCB 1260). Calibration of the model was done with fish bioaccumulation measures made downstream, at km 0.6; and, starting from that site, the model was run to find concentrations of PCBs in white sucker and bluegill for all other sites where PCB concentration in water was measured and fish were present.

RESULTS AND DISCUSSION

The LC_{50} for sensitive freshwater and marine species is 0.01–10 mg/l for 7–38 days, whereas LC_{50} calculated using the model was 0.067 and 0.041 mg/l, for PCB 1254 and 1260, respectively.[3] Opperhuizen and Schramp[4] showed that PCBs kill guppies when the fish accumulate a whole-body concentration approximately equal to 0.7 mol/g (wt), corresponding to about 229 ppm. Concentrations in fish of both PCB congeners in Bear Creek do not exceed this value nor does their sum, so kills of mature fishes are not expected. Southworth *et al.*[1] exposed fathead minnows to water from Bear Creek at km 12.4 and found survival rates ranging from approximately 30% in March to 80% in August 1993. Tests of nine months' duration conducted with fathead minnows showed that reproduction was unsuccessful at water concentrations of PCB 1254 equal to or lower than 0.5 ppb, whereas for PCB 1260 a reduction of 20% in the second generation was found at concentrations in water as low as 0.4 ppb. Consequently, a considerable sublethal effect is probable for the observed levels of PCBs in Bear Creek (TABLE 2). Sublethal effects are important for population dynamics and can result in extinction of populations.[6]

The above-projected sublethal effects, if extrapolated along the gradient (TABLE 2), suggest the actual rate of presence and absence occurrence of white sucker along the stream. When additivity is assumed, the higher concentrations in fish upstream at km 11.8 and 12.3 suggest an absence of white sucker at these sites. A trivially smaller bioaccumulation of PCBs was found between km 7.8 and 9.9. (Indeed, in this site the largest white sucker population occurs).

To show that toxicity exclusion can occur, we applied the model to bluegill, which are present in Bear Creek only at km 0.6, and we found a fish concentration more than double of that found in white sucker at all the sites (TABLE 2). If additivity is assumed for the PCBs, accumulation for the upper reaches of Bear Creek range from 15 ppm to 17.25 ppm, which are close to the lethal concentrations.

TABLE 2. PCB 1254 and PCB 1260 water concentration (measured) and (computed) maximum bioaccumulation in white sucker and bluegill sunfish in Bear Creek in 1993

Station (km)	Water Conc.[a] PCB 1254	Water Conc.[a] PCB 1260	White Sucker[b] PCB 1254	White Sucker[b] PCB 1260	Bluegill[b] PCB 1254	Bluegill[b] PCB 1260
0.6	0.3	0.3	0.99	1.01	2.5	2.37
3.3	1	1	2.98	3.01	8.09	7.61
7.8	0.96	0.96	2.86	2.9	7.77	7.31
9.4	0.99	0.99	2.95	2.98	8.01	7.53
9.9	0.96	0.96	2.86	2.9	7.77	7.31
11	1	1	2.98	3.01	8.09	7.61
11.8	1	1	2.98	3.01	8.09	7.61
12.3	1.1	1.1	3.26	3.3	8.89	8.36

[a]Given in ppb.
[b]Given in ppm.

We have considered only one factor, PCB concentration, that could influence presence or absence of species in Bear Creek. We have demonstrated that there is a strong qualitative relationship between environmental concentrations of PCBs and the presence or absence of certain fish species in Bear Creek.

ACKNOWLEDGMENTS

We thank Drs. Mike Ryon, Craig Barber, and Melissa Weaver for assistance, constructive comments, and helpful discussions. The research was supported in part by a grant from the U.S. Environmental Protection Agency.

REFERENCES

1. SOUTHWORTH, G.R., et al. 1997. Monitoring ecological recovery in a stream impacted by contaminated ground water. Proc. War. Envir. Fed. **4:** 295–308.
2. BARBER, M.C., L.A. SUAREZ & R.R. LASSITER. 1991. Modeling bioaccumulation of organic pollutants in fish with an application to PCBs in lake Ontario Salmonids. Can. J. Fish. Aquat. Sci. **48/2:** 318–37.
3. HAWKER, D.W. & D.W. CONNELL. 1988. Octanol–water partition coefficients of polychlorinated biphenyl congeners. Environ. Sci. Technol. **22:** 382–387.
4. OPPERHUIZEN, A., & S.M. SCHRAMP. 1988. Uptake efficiencies of two polychlorobiphenyls in fish after dietary exposure to five different concentration. Chemosphere **17:** 253–262.
5. U.S. ENVIRONMENTAL PROTECTION AGENCY. 1976. Quality criteria for water. U.S. Dept. of Commerce, PB-263 943. Washington D.C. EPA Report N. 440/9-76-023.
6. HALLAM, T.G., G.A. CANZIANI & R.R. LASSITER. 1993. Sublethal narcosis and population persistence: a modeling study on growth effects. Environ. Toxic. Chem. **12:** 947–954.

Evolution as Computation

G. MICHELE PINNA[a] AND ELISA B. P. TIEZZI

Dipartimento di Matematica, Universitá di Siena, Via del Capitano, 53100 Siena, Italy

MOTIVATIONS

In order to study *dynamic* systems, one has to face an obstacle: how to *model* them, where by modeling we mean a representation of the system and its evolution. By a *dynamic* system, we mean an *open* system, hence a system in continuous interaction with its environment. Usually such a system is described by a number of formulae describing somehow the evolution of the system, and this mathematical model is built by means of *functions* (sometime quite complex ones). As mathematical objects, functions are deterministic: once the value of the variables is fixed, the result is unique, and this implies that the described system is also deterministic. If we describe nature by means of functions, we may conclude that our future is already determined. In order to overcome this difficulty, as we believe that nature cannot be represented by a function, we propose an approach driven by the following idea: the *evolution* of a phenomenon can be seen as a *computation* (in a precise mathematical sense) of a reactive system. A *reactive system* is a system that continuously interacts with the environment (what is "around") usually by sending and receiving stimuli (which can be anything—signals, data, events, or whatsoever). The shift in perspective can be better appreciated by stating the following: we try to model not only how a system reacts to external stimuli, but also the *interaction* between the system and its environment. A scientist performing an experiment also has to take into account the fact that he or she is *performing* an experiment! The idea is by no means new or original: for instance, in computer science an active research field called concurrency theory has as its subject the theory of interaction (there called communication) aiming at developing a paradigm analogous to the one developed in the 1930s for functions. Indeed, many of our ideas are borrowed from concurrency theory.

To come back to our problem, we can then imagine the whole object to be studied as being composed of two parts: one is the system (phenomenon) we want to study/model and the other is the environment (typically the surrounding of the phenomenon). We write $P \mid E$ for the object to be studied where P is the phenomenon, E is the environment and the symbol | can be understood as the communication medium (an interface) between the two parts. We will now make what we mean by computation a bit more precise. We say that P *evolves* under the stimulus a and becomes P' (we write $P \xrightarrow{a} P'$ to denote this evolution) if the environment offers the stimulus a. Of course, the act of offering a stimulus by the environment may lead to the evolution of the environment itself; that is, we stipulate that $E \xrightarrow{b} E'$ and a is some $f(b)$ for an appropriate function f. The symbols a and b denote the actions performed by the phenomenon and the environment, respectively. We call *computation step* the evolution

---

[a]Address for correspondence: +39 0577263711 (voice); + 39 0577 263701 (fax); pinna,tiezzi@mat.unisi.it (e-mail).

of both parts together and it will be denoted with $P \mid E \xrightarrow{obs(a)} P' \mid E'$, where *obs* is a function giving the observation. A computation is then simply a sequence (possibly infinite) of computational steps. It is worth noticing that the choice of what constitutes the phenomenon and its environment is arbitrary: the | can be "moved" (this goes back to the idea that the experiment is an experiment itself). We face the following problems, which we will discuss in turn (a) how to describe the object to be studied, (b) how to describe the computations, and (c) how to reason on the basis of the computations.

DESCRIBING THE OBJECT

We do not want to put forward any formalism to represent the phenomenon and the environment by stating that this is *the* appropriate one: one is free to choose any formalism she or he considers appropriate, but of course we have our own ideas and we clearly see that the formalism must match certain requirements we consider relevant. First of all, the formalism should describe *relations* instead of functions, which has as an outcome the capability of modeling nondeterminism. Nondeterminism means that our whole future cannot be determined today; at any point a decision can be made that precludes certain evolutions, and the decision is not necessarily determined by known parameters (i.e., parameters known by the observer). We do not want to discuss here the source of nondeterminism (the capability of making decisions independently), we observe that as a matter of fact this happens in reality. Of course one could argue that this is a deficiency of the "observation device," but we are quite convinced that it is not feasible to assume that knowledge is complete. Second, the formalism should be able to describe interactions, which we consider, as said before, the principal notion.

We consider an appropriate formalism to be the so-called *algebra of processes*, where a process is defined according to some grammar, and the evolution can be obtained using "rules." Not to be too abstract, we detail the ingredients of such an algebra. A process can be seen as an agent performing some basic actions or evolving according to some operations. Hence, a process is an object constructed according the following definition: (i) a basic action is a process; (ii) if $\{P_i\}_{i \in I}$ is a set of processes and $\mathcal{N}$ is an I-ary operation then $\mathcal{N}(P_{\min(I)}, ..., P_{\max(I)})$ is a process; and (iii) a process is defined only according these clauses. We illustrate with a simple example taken from the realm of concurrency theory: $a, \bar{a}$ are a pair of complementary actions, meaning that two agents want to communicate along a channel named a, the operation $\parallel$ takes two processes and defines in this way a new process, as well as the operation + (hence both have arity 2), · takes a basic action and a process and define a process, and finally **0** is a process with arity 0 which denotes inaction. A process is then the following object: $(a.a.\mathbf{0} + a.\mathbf{0}) \parallel \bar{a}.\bar{a}.\mathbf{0}$. Before formalizing we will provide an intuitive analysis of the meaning of these signs. **0** is clear: it does nothing, + behaves as the first or the second component and $\parallel$ allows communication. The meaning of a sign is given according to rules. A rule can be depicted as shown in FIGURE 1, where above the line are written the premises of the rule and below the line is the conclusion. Finally, the side conditions are conditions not expressible as possible for forbidden evolutions. Clearly $\cup_{i \in I}\{x_i \xrightarrow{a_{ij}} y_j\}$ are the positive premises; (what can

$$\frac{\bigcup_{i\in I}\{x_i \xrightarrow{a_{ij}} y_j\} \qquad \bigcup_{k\in K}\{x_k \xrightarrow{\rho_k}\}}{\mathcal{N}(x_1,\ldots,x_n) \xrightarrow{obs(a_{11},\ldots,a_{nj})} \mathcal{M}(x_1,\ldots,x_n,y_1,\ldots,y_n)} \text{ side conditions}$$

FIGURE 1. The general format of a rule.

$$\frac{P \xrightarrow{x} P'}{P+Q \xrightarrow{x} P'} \qquad \frac{P \xrightarrow{x} P'}{Q+P \xrightarrow{x} P'} \qquad \frac{P \xrightarrow{a} P' \quad Q \xrightarrow{\bar{a}} Q'}{P \parallel Q \xrightarrow{\tau} P' \parallel Q'}$$

FIGURE 2. The rules for the example.

be done) and $\bigcup_{k\in K}\{x_k \xrightarrow{/a_k}\}$ are the negative premises; finally $\mathcal{N}$ and $\mathcal{M}$ are constructor of the grammar (they group together parts). The interpretation is the following: if the premises are satisfied, as well as the side conditions, then the system can evolve according to the rule. A rule without premises simply stipulate an evolution that is always possible, provided that the side conditions are satisfied. Coming back to our small example, **0** has no rule, $\cdots$ has the following (without premises), $x. P \xrightarrow{a} P$, the rules for the + (two rules, according to the notion of left and right summand) and the one for the $\parallel$ are depicted in FIGURE 2 (there the τ is the result of the communication). Let us consider again the example $(a.a.\mathbf{0} + a.\mathbf{0}) \parallel \bar{a}.\bar{a}.\mathbf{0}$, here if the left summand is chosen we can observe two communications, and if the right one is chosen we observe only one communication.

We assume that for a particular evolution only one rule applies, but we don't require that at any time only one rule is applicable. We hope to have given the flavor of this formalism, and because of the shape of the general rule, to have suggested some evidence of its power. We conclude this section with an observation with a mathematical flavor: depending on the choice of operators of an algebraic process, it is possible to define calculi that are really powerful from the computational point of view, where powerfulness here means the capability of formalizing everything that can be formalized. The reader has certainly noticed that from the very beginning we have used notations in process algebra style. Indeed, one of the advantages of process algebra is that it is somehow tailored to describe objects together with their *context*.

DESCRIBING COMPUTATIONS

Before formally stating how to obtain a computation from a process, we will concentrate briefly on two problems (resuming the interpretation for P and E given at the very beginning): (1) who guarantees the evolution of $P \mid E$ and (2) what we observe (what is $obs(a)$). For the first problem, we assume that $P \mid E$ evolves *tout court*, whereas observations are given by a function of the stimulus that the phenomenon evolves from and the *result* can be very complex. Moreover, *time* (which we consider relevant) is a part of the observation. We further assume that time elapses and progresses. The assumption that time elapses is quite crucial: it implies that time cannot be internalized (i.e., cannot be enforced by the syntax of our formalism): it can be observed via computations.

As already stated, a computation is just a sequence of computation steps. Mathematics provide an adequate notion, namely the one of *transition system*. Formally a transition system is a triple $(St, \mathcal{R}, s)$ where St is a (possibly infinite) set of *states*, $\mathcal{R} \subseteq St \times \text{Observables} \times St$ is a relation stating how to reach a state from another (in the previous section we have written $a. P \xrightarrow{a} P$ for $(a.P, a, P)$) and s is the initial state. A computation is then a sequence $(s_1, a_1, s_2)(s_2, a_2, s_3)\cdots(s_n, a_n, s_{n+1})\cdots$.

We want to stress that this is a possible notion of computation, indeed we are assuming that the computation proceeds without jumps (the transition system is connected), but of course this is a choice. We think that here fantasy should not be limited, provided that the notion of computation preserves some reasonableness. The reader could have noticed that time doesn't appear explicitly here, but we can recover it either from the observables (timed algebra approach) or just by the position of a certain observation in the sequence.

REASONING ON COMPUTATIONS

We see two possible ways of reasoning on the basis of computations: one is a logic-based approach and another is an approach based on the notion of *simulation*. For what concerns the logic-based approach, we first recall that logic can be perceived as a formalism for writing and proving properties. Here, the idea would be to find the appropriate logical language and to consider the computations we are interested in as the model of this logical language, where "model" means that the theorems of the logic are properties of the computations. To be a bit more concrete, we have already noticed that a sequence (possibly infinite) of observations can be easily obtained from a computation (simply forgetting the objects $P_i \mid E_i$), and this sequence of observations can be equivalently sought as a function from time to observations (of course without the time). We still call this mathematical object computation. Then these computations can be studied with real-time logic; that is, systems of logic that are able to take account of the qualitative and quantitative aspects of time can be used to *prove* properties of the computation (by considering each observation as the set of propositions that hold at the corresponding point in time). In these systems of logic, it is possible to express, for instance, safety constructs (nothing bad happens) or liveness properties (something will happen). Propositions can be understood as mathematical constructs describing the state of the system observed. To *prove* means to be able to verify that a property is satisfied (holds) in the computation, proof that can be carried over using methods like model checking or by using proof systems. There are many possible systems of logic available here, depending on the kind of properties one is interested in. We believe that attention should be paid here to the complexity issue (a measure of the difficulty in proving properties), which can go far beyond what is considered computable.

The second approach one can consider is based on the idea of simulation and consists in reasoning on the transition system (the model of all computation) proving that it can be *simulated* by another model (maybe another transition system) of which properties are known. The simulation can be quite complex, and in fact the idea can go back to the one of definability in model theory, where certain theories are defined over known ones. In fact this notion is used in the process algebra setting to establish that two agents behave accordingly (in this case it is usually required that the first simulates the second and vice versa, henceforth it is called bisimulation). Again, the complexity of a simulation is a rather crucial parameter.

CONCLUSIONS

We are aware that reported here are just ideas taken from our own research experiences and proposed as a possible paradigm, mostly as result of discussions with Enzo Tiezzi and the members of his research group. So our main aim is to contribute to other research fields in a kind of cross-fertilization. We are also aware that we have not given credit to the researches who have contributed to developing the ideas motivating the paper. The list would be too long. (The bibliography would be much longer than the paper!)[b]

As final remark we want again to stress the fact that our ideas are tentative: at some point one should be pragmatic and should make choices, in order to obtain objects that can *feasibly* studied. The formalisms we have briefly illustrated are really powerful, sometime too much so; hence, the right level should be found. In the choice of the appropriate abstraction level, it should be clear that possible achievements are influenced by the interest of the scientist, again providing evidence that in theory (and not only in practice) the subject and object of an experiment have to be considered together. This is not at all in contrast with our discussion; on the contrary, it provides gives further evidence that these choices exist in reality, at least from perspective of the human being side.

---

[b]We refer the reader to a collection of volumes: one is the *Handbook of Logic in Computer Science*, Vols. 1–6 (S. Abramsky, D. Gabbay & T.S.E. Maibaum, Eds. Clarendon Press, Oxford, England, 1992–1997) and the other is the *Handbook of Philosophical Logics*, Vols. 1–4 (D. Gabbay & F. Guenther, Eds. Reidel Publishing Co., Dordrect/Boston/London, 1986–1989).

Mathematical Modeling and Numerical Simulation of Space-Dependent Multispecies Interactions

R.C. SOSSAE,[a] J.F.C.A. MEYER,[b] S. LOISELLE,[a] AND C. ROSSI[a,c]

[a]*Department of Chemical and Biosystem Sciences, University of Siena, Pian dei Mantellini, 44, 53100 Siena, Italy*
[b]*Department of Applied Mathematics, University of Campinas, C.P. 6065, Campinas CEP 13083-970, Brazil*

Classical Lotka-Volterra models continue to be present in applications and academic literature, mainly due to their versatility and their simplified description of several varied situations. This includes those variations of these models that present nonlinear characteristics describing several relationships both within and without each modeled species. On the other hand, however, there has been a steadily increasing use of those models for which, besides the evolutive aspects, spatial variations for population dispersal, migration, and chemotaxis are considered as well. In some cases, these diffusive-advective partial derivative equations or systems have taken the place of ordinary differential equations previously used as models for the study of population dynamics, both for one species as well as for multispecific cases.

On the one hand, mathematical tools for the consideration of these former systems have become more and more available, while, in the meantime, numerical tools made the simulations possible and reasonably reliable, increasing knowledge about the behavior of solutions of models for spatial and temporal variations including those nonlinearities present in the population dynamics terms. This did indeed direct research away from the relative comfort of analytic solutions, but brought and enhanced use of better modeling in the sense of more precise descriptions.

The purpose of this work is to define a diffusive-advective model that includes interactions between species in each population dynamic, simultaneously considering an environmental hazard that may vary in time or space. The motivation for this study is part of a joint project for the study of the Ibera region wetlands, in northeastern Argentina, in which a relatively untouched environment has recently been affected by the presence of several forms of human-induced impact.

These wetlands, located in the Corrientes province, are made up of a vast hydrographic basin with lagoons, floating islands of grouped vegetation and distinct biodiversity. The main threats for this region come, at present, from an increase in local agroindustrial activity, besides the need for pasture, the agrochemical treatment of rice fields, as well as the practice of controlled fires for creating new space for agriculture. To study this environmental preservation area, a project involving several universities in Italy (Siena), Argentina (Nacional del Centro, Del Salvador), Brazil

---

[c]Address for correspondence: 39-0577-232022 (voice); 39-0577-232004 (fax); rossi@spectrum.chim.unisi.it (e-mail).

SOSSAE et al.: SPACE-DEPENDENT MULTISPECIES INTERACTIONS

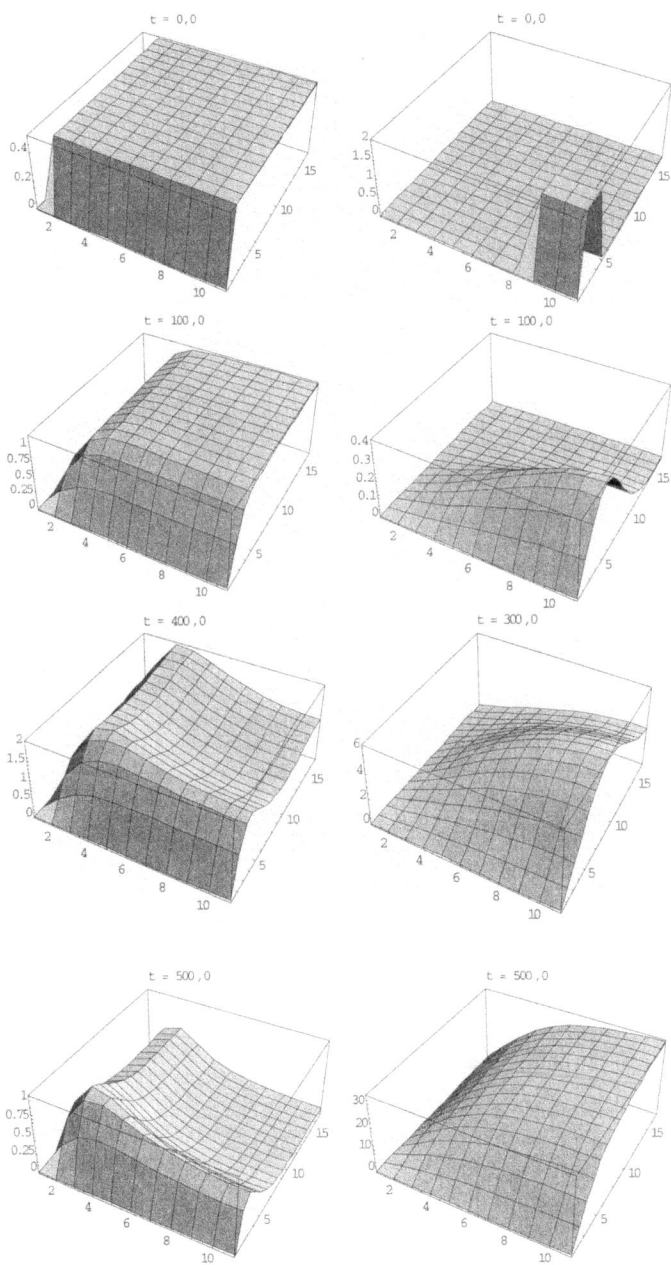

FIGURE 1. Model of space-dependent multispecies interactions.

(Federal in Rio de Janeiro, Federal in Rio Grande do Sul, State at Campinas), Spain (Cadiz), Portugal (Aveiro), and England (York), as well as joint researchers from other centers, has been submitted to the European Community as an International Cooperation Project (Project number ERB3514PL972697).

In order to undertake the studies of this model, the strategy decided upon has been to begin with a system that includes nonlinear Lotka-Volterra-type formulation of species interactions, generalized Verhulst population dynamics for the involved species, and for which population dispersal is modeled by the diffusion equation, with a term for decay that does not include a constant parameter, but that varies in space and time. Previous work has included the physical transportation and decay of diffusing pollutants in lakes of the Ibera region, so the idea is to include the temporary and decaying effects of these possible pollutants upon species that interact in the region. The models that this study intends to present, along with the contributions from members of the project as a whole and the consequent adaptations thereof, may be a very helpful tool in assessing medium and long-term effects of human activities in agroindustrial exploitation in those regions immediately close to the borders of the studied region, as well as establishing methods for similar studies in other wetlands.

The system that models this space as well as time-dependent species interaction is, for $P_1(x, y, t)$ as the prey and $P_2(x, y, t)$ as the predator:

$$\frac{\partial P_1}{\partial t} - \mathrm{div}(\alpha_1 \nabla P_1) + \mathrm{div}(V_1 P_1) + \sigma_1(x, y, t)P_1 = \lambda_1 P_1 \left(1 - \frac{P_1}{k_1}\right) - \eta_1 P_1 P_2$$

$$\frac{\partial P_2}{\partial t} - \mathrm{div}(\alpha_2 \nabla P_2) + \mathrm{div}(V_2 P_2) + \sigma_2(x, y, t)P_2 = \lambda_2 P_2 \left(1 - \frac{P_2}{k_2}\right) + \eta_2 P_1 P_2$$

In these equations, div $\alpha_i.\nabla P_i$ describes the population dispersal; div $(V_i.P_i)$ is the population displacement; $\sigma_i P_i$ stands for an induced mortality rate, a result of the above-mentioned effect of the pollutants in the region, while the right-hand side terms model the interspecific population dynamics in the classical way, that is to say that: λ_i is the growth rate for population P_i; k_i is the carrying capacity for population P_i; and η_i is the predation rate, affecting P_1 and P_2 with opposite effects.

This model is innovative in considering the so-called decay terms as space as well as potentially time-dependent.

The numerical scheme with which computer simulations are carried out uses second order finite elements in the spatial discretization and a Crank-Nicolson technique in time.

Preliminary numerical results were obtained using a rectangular domain—initially simulating a mathematical pond—with theoretical, albeit reasonable, values for dispersal, movement and species characteristics (FIG. 1).

The classical situation for the Lotka-Volterra system shows the predator cycle following that of the prey, but in this situation both the movement of the prey in the horizontal direction (along the x-axis) and the varying toxicity radically alter this picture, with prey surviving where toxicity is low and predators are not abundant. On the other hand, although the conducted numerical experiments did not provide for the movement of predators, these do in fact follow prey, reducing prey population levels. This is, of course, an expected result, but its importance stems from the fact that a map with varying toxicity can be obtained from other modeling and simulation software packages in which dispersing and transported pollutants may have their effect coupled to the work presented here, which describes the effects of human activities in wetlands upon interacting species.

This model can therefore be considered an auxiliary tool in assessing the effects of these activities in certain regions.

REFERENCES

1. KAREIVA, P.M. 1993. Local movement in herbivorous insects: applying a passive diffusion model to mark-recapture field experiments. Oecologia **57**: 327–332.
2. MEYER, J.F.C.A. & R.C. SOSSAE. 1999. Density-dependent population dynamics in dispersion and migration processes. Bol. Inst. de Pesca. In press.
3. MURRAY, J.D. 1989. Mathematical Biology. Springer-Verlag. Berlin.
4. OKUBO, A. 1980. Diffusion and Ecological Problems: Mathematical Models. Springer. Berlin/Heidelberg/New York.
5. SOSSAE, R.C. 1995. Population Dispersal and Migration Processes: Density-Dependent Population Dynamics. Dissertion (MSc). IMECC-UNICAMP. Campinas. 117pp.

Relations between Energy Consumption and Air Quality in an Urban Environment: Spatial Scenarios of Emission Reduction

LAURA VANNUCCINI, MIRIA BRACALI, AND RICCARDO BASOSI[a]

Departement of Chemistry, University of Siena, Pian dei Mantellini, 44, 53100 Siena, Italy

INTRODUCTION

The aim of this work is to set up a method to estimate air emissions from energy consuming activities in an urban environment and to explore the effects of measures to minimize impact on air quality.[4] The main energetic activities that produce emissions in urban environments are transport, industrial processes and heating of buildings. The amount of a particular pollutant arising from a combustion process is estimated by the method of "emission factors", that is, the emission is proportional to the "activity rate" of the process. The activity rate in transport is the number of kilometers covered by different kinds of motor vehicles and in the other combustion processes is the fuel (gas, gas oil, and oil) consumption. The emission factors used in this work are those of the EPA[1] and of CORINAIR[2] (transport). The input data needed for these evaluations are: fuel consumption and the number of kilometers covered by the vehicles in the area of interest. The method of "correlated variables" is used for spatial spreading of input data: it consists of attributing, to an aggregated value, the spatial distribution of the correlated variable. The distribution of occupied houses is used to spread fuel consumption for building heating in the household sector. All input data are distributed on a spatial grid to evaluate emissions for each pair of coordinates (i, j).

Several measures can be hypothesized to reduce air pollution: substitution of energy sources; introduction of better technologies; and, in the case of transport, reduction of road traffic and an increase of public transport.

These methods have been applied to the case of the city of Florence, Italy. It is a medium-large, historical city presenting acute episodes of air pollution mainly caused by CO, NO_x, and VOC (as a precursor of O_3). A number of measures have been hypothesized: substitution of fossil fuels with other less polluting fossil sources (liquid fuels with gas), substitution of residential gas furnaces with larger sized gas boilers, reduction in the number private cars and motorcycles and substitution by public transport, reduction of heavy transport and substitution of non renewable sources with renewable ones. The aim of these measures was a 10% reduction in pollutants.

---

[a]Address for correspondence: ++39 577 263545 (voice); ++39 577 263546 (fax); Errore. Il segnalibro non è definito (e-mail).

EVALUATING SPATIAL DISTRIBUTION OF EMISSIONS

Giving a region

$$Q = \bigcup_{k=1}^{K} Q_k,$$

the union of K distinct areas, and a square grid ($i = 1 \ldots I$ rows, $j = 1 \ldots J$ columns), giving

$$X = \sum_{k=1}^{K} X_k,$$

where X_k are K homogeneous values each belonging to Q_k, then, for each (i, j), the portion of X belonging to (i, j) is:

$$X^{i,j} = \sum_{k=1}^{K} X_k q_k^{i,j},$$

where $q_k^{i,j}$ is the portion of X_k belonging to (i, j).

The spatial distribution of X is a matrix D whose elements are:

$$D(i,j) = \frac{X^{i,j}}{X}, \quad \text{and} \quad \sum_{i=1}^{I}\sum_{j=1}^{J} D(i,j) = 1 \quad \sum_{i=1}^{I}\sum_{j=1}^{J} X^{i,j} = X$$

A variable Y, correlated to X, can be spread in Q using the expression: $Y^{i,j} = Y \cdot D(i, j)$, where $Y^{i,j}$ is the portion of Y belonging to (i, j). This method has been applied to the map of Florence to spread fuel consumption: the variable X is the number of occupied houses. FIGURE 1a shows the annual distribution of energy consumption for building heating. Figures 1b, 1c, and 1d show the corresponding emissions including the contribution of road traffic.

DISCUSSION

TABLE 1 shows the list of hypothesized measures for the city of Florence. The substitution of residential gas furnaces with larger size gas boilers has effects on CO emissions: the emission coefficient of larger sized furnaces is about half that of residential furnaces.

The substitution of nonrenewable sources with renewable ones decreases emissions and increases second-order thermodynamic efficiency.[3] In Florence, the efficiency of heating systems burning gas oil is about 3.8%, while that of solar thermal panels is about 0.32, hence 32%.

The transport sector gives the most important contribution to achieving a decrease of air pollution in a city such as Florence. Useful information for environmental policies could be deduced: for example, the reduction of CO emission could be obtained by limitation of private cars, whereas reduction of heavy transport and motorcycles diminished NO_x and VOC emission, respectively.

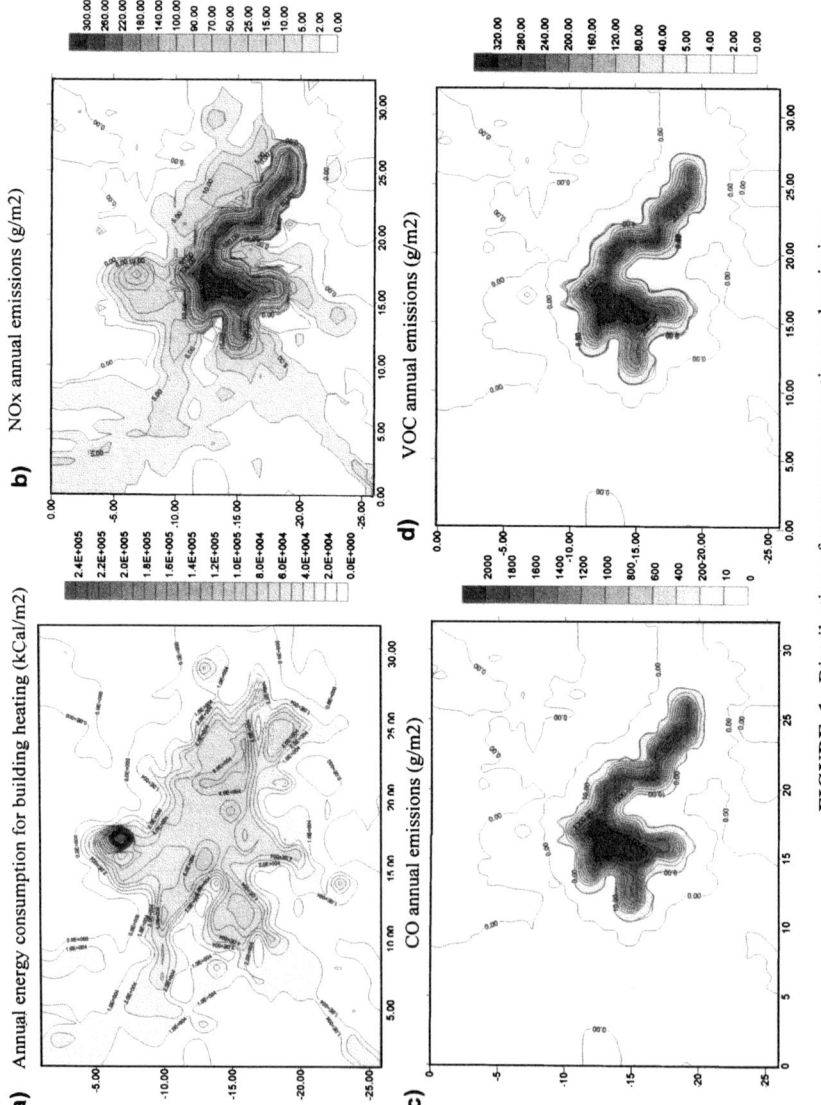

FIGURE 1. Distribution of energy consumption and emissions.

TABLE 1. Summary of measures and results: Target of a 10% reduction of CO, NO_x, and VOC emissions

Sector (Energy End Use)	CO	%	NO_x	%	SO_2	PST	VOC	%	CO_2
	Emissions (t)								
Transport (mechanical energy)	47683		6636		588	471	7631		
Industry (heat, industrial processes)	25		99		4	2	4		87652
Domestic, public services, commercial (heat)	140		525		162	41	33		509751
Total	47848		7260		754	514	7668		597403
	Effects on Emissions (t)								
Sector (Energy End Use)	CO	%	NO_x	%	SO_2	PST	VOC	%	CO_2
Total substitution of liquid fuel with gas	−15	0	−50	5	−159	−22	−2	0	−15508
Substitution of residential gas furnaces with larger gas boilers	−27	0	−49	5	−1	−12	−7	1	0
Substitution of 10% of private cars with public transport	−3486	62	−286	29		−12	−502	58	
Substitution of 10% of motorcycles with public transport	−295	5	3	0		0	−92	11	
Substitution of 10% of gas furnaces with solar thermal plants	−1	0	−3	0	0	0	0	0	−2737
Reduction of 20% of heavy transport (gasoline, gas-oil)	−1787	32	−586	60		−52	−264	31	
Total	−56112	100	−971	100	−160	−98	−865	100	−15508
Total emmissions	12		13		21	19	11		3

The measures regarding heating of buildings have a great effect on NO_x emissions, because its environmental effects are only concerned with the winter season.

The substitution of non renewable sources with renewable ones to produce sanitary hot water seems not to have a significative effect. This is possibly due to the negligible presence of these solutions in the urban environment.

REFERENCES

1. U.S. EPA. 1995. Compilation of Air Pollutant Emission Factors; Stationary Points and Area Sources. Fifth edit. US-EPA. Washington, D.C.
2. EGGLESTON, H.S., et al. 1993. CORINAIR working group on emission factors for calculating 1990 emissions from road traffic. Office for Official Publications of the EC. Luxembourg.
3. ZEMANSKY, M.W. 1991. Calore e Termodinamica. Zanichelli. Bologna, Italy.
4. VANNUCCINI, L. 1998. Valutazione di scenari di programmazione energetica in ambiti territoriali omogenei per il miglioramento della qualità dell'aria. Ph. D. thesis, University of Siena, Siena, Italy.

Index of Contributors

Agostiano, A., 215–219
Alia, L., 255–257
Antoniou, I., 8–28
Ardrizzo, G., 29–44
Arecchi, F.T., 45–62

Barone, G., 383–386
Barra, G., 220–223
Basosi, R., 276–279, 444–447
Bastianoni, S., 387–391
Battilotti, M., 427–431
Bellik, L., 235–240
Bellucci, A., 392–395
Bianciardi, G., 255–257
Bocchi, G., 63–74
Boero, P., 164–167
Bonechi, C., 235–240
Bracali, M., 444–447
Branca, C., 224–227
Branca, M., 228–235
Brunetti, A., 228–235
Busi, E., 276–279

Camastra, S., 249–254
Canziani, G.A., 292–301
Caravati, C., 228–235
Cataudella, S., 427–431
Ceruso, M.-A., 258–266
Ceruti, M., 63–74
Chandler, J.L.R., 75–86
Cignini, I., 427–431
Cimini, G., 154–157
Collina, E., 396–399
Crise, A., 392–395
Crispi, G., 392–395

D'Ambra, P., 383–386
D'Aniello, C., 220–223
De Santi, M.M., 255–257

Dejak, C., 400–405
Del Vecchio, M.T., 255–257
Della Lunga, G., 276–279
Della Monica, M., 215–219
di Serafino, D., 383–386
Donati, A., 235–240

Elia, V., 241–248
Emdin, M., 249–254

Falcucci, M., 312–319, 427–431
Faraone, A., 224–227
Ferrannini, E., 249–254
Ferri, G., 154–157

Ganadu, M.L., 267–275
Gastaldelli, A., 249–254
Gell-Mann, M., 1–7
Giampietro, M., 344–367
Giuliani, A., 258–266
Giunta, G., 383–386
Goldbeter, A., 180–193
Gonze, D., 180–193
Gould, S.J., 87–98
Gregori, G.P., 158–163
Gregori, L.G., 158–163
Guala, E., 164–167

Hallam, T.G., 302–311, 432–434
Hull, V., 312–319, 427–431

Jørgensen, S.E., 320–343

Landini, L., 249–254
Lasagni, M., 396–399
Loiselle, S., 440–443
Lubinu, G., 267–275
Ludovisi, A., 406–411

Luisi, P.L., 98–109
Luzi, P., 255–257

Magazù, S., 224–227
Magnoni, L., 412–415
Maida, V., 267–275
Maisano, G., 224–227
Mammoliti, R., 249–254
Mancini, V., 416–421
Mancusi, C., 280–283
Manetti, C., 258–266
Marchettini, N., 168–171, 422–426
Martini, S., 387–391
Mayumi, K., 344–367
Melis, A., 110–114
Meyer, J.F.C.A., 440–443
Micarelli, R., 416–421
Migliardo, P. 224–227
Milone, N., 427–431
Miracco, C., 255–257
Mirolli, M., 412–415
Mocenni, C., 422–426
Monfrinotti, M., 427–431
Morin, E., 115–121
Morowitz, H.J., 122–128
Mulas, G.A., 427–431
Mura, G.M., 267–275
Muscelli, E., 249–254

Niccoli, M., 241–248
Niccolucci, V., 422–426

Panella, S., 427–431
Panzieri, M., 432–434
Pastore, G., 344–367
Pastres, R., 400–405
Pecenik, G., 400–405
Picchi, M.P., 235–240
Pinna, M., 435–439
Pipornetti, G., 427–431

Pitea, D., 396–399
Pizziolo, G., 416–421
Pogni, R., 276–279
Pojman, J.A., 194–214
Polenghi, I., 400–405
Poletti, A., 406–411
Porcelli, M., 387–391
Prigogine. I., 8–28
Pucciariello, R., 280–283

Riccio, A., 383–386
Romond, P.-C., 180–193
Rossi, C., ix–x, 387–391, 440–443
Russo, R., 220–223
Rustici, M., 180–193, 228–235

Solidoro, C., 392–395
Sossae, R.C., 440–443
Svirezhev, Y.M., 368–369

Tamburro, A.M., 284–287
Tancioni, L., 427–431
Tiezzi, E., ix–x, 168–171, 422–426
Tiezzi, E.B.P., 435–439
Timashev, S.F., 129–142
Tosi, P., 255–257

Vannuccini, L., 444–447
Varela, F.J., 143–153
Villani, V., 224–227, 284–287
Vittoria, V., 220–223
Vrobel, S., 172–179

Webber, C.L., Jr., 258–266

Zbilut, J.P., 258–266
Zoroddu, M.A., 288–291